THE FASCINATED GOD

What Science Says to Faith And Faith to Scientists

By

Robert E. Zinser

This book is a work of non-fiction. Names and places have been changed to protect the privacy of all individuals. The events and situations are true.

ISBN: 1-4107-6846-5 (e-book)
ISBN: 1-4107-6845-7 (Paperback)

Library of Congress Control Number: 2003095761

This book is printed on acid free paper.

Printed in the United States of America
Bloomington, IN

1stBooks - rev. 09/15/03

Table of Contents

Preface vii
Introduction ix

Chapter 1. Assumptions 1
1st Everything happens for a reason 2
2nd Every cause is reasonable 2
3rd Something sustains all things 3
4th There is a creator 3
5th Everything has a purpose 3
6th There is a plan, a purpose, a teleology 4
7th There is someone in control. Miracles can occur 4
9th Prayer 4
10th The one in control can do good 5
The existence of evil 6

Chapter 2. Where Do You Begin 9

Chapter 3. The Assumptions Examined 13
1st.Assumption. Must there be a cause? 14
Relativity, Quantum Physics, Contingency 18
The Argument from Design 25
The Devolution of the Cosmos 30
The Evolution of Life 45
The Evolution of Human Life 60
Human Culture 84
So much for Intelligent Design 92
2nd Assumption. An understandable cause? 95
3rd Assumption. A sustaining cause? 96
4th Assumption. There is a creator? 98
5th, 6th & 7th. A purpose, a plan, someone in control? 99
8th Assumption. Miracles? 100
9th Assumption. Prayer? 103
10th Assumption. Do good through evolution? 104
Other Christian Assumptions. 105
Original Justice 105
Original Sin 106
Morality 108

Conscience 113
Existence after death 118
Transubstantiation? 121
Chaos, Good and Evil 121

Chapter 4. The Effect on Traditional Christian Faith 127
The Person of God 129
The Person of Jesus 132
Changes because of a new view of God 132
Summary on What is Left? 137
Evil, Free-Will Argument, Theodicy 138
The Search for Answers. Agnostic Scientists 141
Carl Sagan 141
Stephen Jay Gould 142
Edward Wilson 143
Stuart Kauffman 144
Lee Smolin 145
David Noble 146
Kafatos & Nadeau 146
Daniel Dennett 146
Paul Davies 147
Weinberg, Smoot, Dawkins & Crick 148
Richard Feynman 149
Ursula Goodenough 150
Chet Raymo 151
A.N. Wilson 152
Swimme, Searle, Wright 153
Summary of answers—Cold Comfort 153
The Search for Answers. Friendly Scientists 154
Kafatos, Nadeau, Pauli 156
Candice Pert 157
Frank Tipler 158
The Search for Answers. Philosophers and Theologians 160
Henri Bergson 160
Teilhard de Chardin 161
Popper, Eccles & Nogar 162
Ratzinger & Pius XII 163
M.B.Martin & Arthur Peacocke 165
Wolfhart Pannenberg 167
John Polkinghorne 168
Ian Barbour 170

Eugene Mallove 170
John Haught 171
Timothy McDermott 172
Angela Tilby 173
Diarmuid O'Murchu 173
Timothy Ferris 175
C.S.Lewis, Patrick Glynn, Walter Brueggemann, Karen Armstrong, Matthew Fox, Kenneth Miller, Denis Miller, Denis Edwards, Jerry Korsmeyer, Russell Stannard, & Steven Kuhl 175

Chapter 5. Toward a New Exposition of the Faith 179
God's Role Reexamined 181
Doubting the Old Theodicy 183
God Did Not. God Cannot 184
Then Who Is God? 186
The Role of Chance 187
Transcendent 188
Evolving and Growing 189
Interested 193
The Fascinated God 193
Revealing God 198
The Covenant in Community 204
Growth in God & Community 209
Faith, not Proof 210
God's Plan 213
Morality 214
Heaven & Hell 223
Resurrection 226
Judgement 231
Prayer 232
Healing Prayer 239
Real Prayer 240
Meeting God 246
In Jesus 247
In Sacraments 248
Incarnation 251
Who is Jesus? 253
The Trinity 260
Where Is Susan 262
The Church, Body of the Risen Christ 263

Revelation 266
In Teleology? 270
In Scripture 272
In Teaching Authority 276
Angels & Devils? 277
Summary 278
Conclusion 286

Preface

This work has been more than thirty years in production. It began with a vague dissatisfaction with the answers to the question of evil given in theology classes. It also began, though I did not know it at the time, with the few classes in the sciences then available in a pre-Vatican II seminary under Fr. Lawrence "Doc" Walsh, C.M. whose less-than-reverent reaction to the prevalent relation between science and theology influenced me more than I realized. This interest in science has continued, though I am by no means a scientist, and certainly not a philosopher or a theologian. I am a pastor and this has been the real stimulus to study and writing. Trying to pastor all the "Susans" (the questioner in the work to follow) in my life, in addition to my own questions and personal search for faith led to the approach contained here.

There are many endnotes, I have been told too many, in what follows since I am, as stated, not a professional scholar, scientist, philosopher or theologian and I had to prove to myself, and now show to you, that the positions I am reacting to are actually held by others. On the advice of others I have moved much material which was in the text to the endnotes. This gives the reader the choice of plowing through more material or keeping it simpler.

Gathering these opinions has been a long process because of the demands of an active ministry. It is only in the past few years that I have had the luxury of time to collect these thoughts and try to put them in context, though often interrupted by funerals, weddings, pastoral visits and doorbells. If the work seems disjointed at times, I apologize, but my purpose has been to get these ideas into a form that can be shared with others and so seek their reaction and critique. If the editing process has been too short (it seems long to me) and the effort required of the reader is therefore greater, it is because I hope this is a "work in progress" and that a dialogue can be established leading to greater clarification of the issues involved. It is also because reconsidering the faith in such a radical manner is so complex a topic, much of which could not even be dealt with or went unrecognized in the prevalent culture in the Church, that one just has to stop somewhere and let the dialogue begin.

This "being a part of a process" has been true in Catholic circles since the beginning. Edward Schillebeeckx, no stranger to controversy, says that,

> "the Catholic theologian is always only one small voice within a great movement which began with Christ and the apostolic Church with its Scripture and has continued throughout the ages…[and] knows from the very beginning that he is subject to criticism—not, in the first place, to possible criticism from the teaching authority of the Church, but to the criticism of history. He is, in other words, conscious that he is simply *taking part in* a great social undertaking which is theology and that he can only play a very subordinate part in this undertaking."[1]

My thanks to those who have read this material and given me the benefit of their corrections and suggestions: my questioning Father and my faithful Mother; Thomas & Peg Zinser, Alan & Kathy Zinser; Sr. Patricia Bober C.S.J.; Kathy & Cal Prewitt, Jan Parker and Linda & Carson Coil from Dardenne; Joan Douglass, Judy Mantych and Carol Eck from Bellefontaine; Helen Reiss from Baden; Srs. Mary McGrone from Carondolet, Margaret O'Gorman of the Franciscans and Mary Dolan from the Dominicans; Fr. Robert Brungs and Donald Merrifield, of the S.J.'s; Robert Macke; Professors Robert Richards from the University of Chicago and Charles Granger from The University of Missouri St. Louis; Bill Massa from out in left field; Msgr. James Telthorst, Kathleen Zinser, Amy and Tom Zinser; and Barbara Bowman, Zelma Glass &Vernita Nash from St. Teresa of Avila.

Introduction

Every pastor knows the question will come up eventually. It is never the first question asked, of course. The first is very specific, practical and certainly not theological. Take the example of Susan who wanted to know how to deal with an abusive husband. It had taken years for her to recognize that there was a problem and that the problem was not her, except that she had tolerated the situation for so long. It had taken another long period before coming to talk to someone about her questions. Of course the conversation did not start there. It began with parish news, went to a description of how the children were doing, lurched forward with disclaimers that the pastor should not think that she had a bad marriage nor that her husband was a bad person. However, he did slap her around from time to time, not that she did not bring it on, reverting to her old habits of blaming herself. Finally there was the question of what to do about it. Patiently the pastor brings Susan to reveal where she thinks the problem is, with herself or her husband? Over that hurdle, he may have to talk about the church's position on separation, divorce, receiving Communion when one does not feel "worthy," or any other moral or church discipline problems that may be on her mind. Then he can refer her to agencies or trained counselors that may be able to help her in the situation. However, none of this is the real question that she needs and wants to discuss with her pastor who presumably knows something of the answer, but the question is still not asked outright. Maybe at the end of the first meeting, almost as if it is an afterthought, or perhaps not until several meetings have occurred, will Susan state her question. It will indeed come as a question but it is a theological position statement. "WHY IS GOD DOING THIS TO ME?" she may ask, thereby stating that god is in fact responsible for her problem. Or the question may be, "For what is God punishing me?", either feeling that she is indeed guilty - Lord knows her husband has told her often enough how bad she is and why she needs to be beaten - or subtly stating that she is not guilty and therefore god has no right to punish her, does god? Or there may be an excusing of god by posing the question as, "Is God trying to teach me a lesson?", or "Maybe this is really God loving me and disciplining me?", just as the Book of Job counseled, "Happy indeed the one whom God corrects! Then do not refuse this lesson from Shaddai. For he who wounds is he who soothes the sore, and the hand that hurts is the hand that heals."[2] This answer in the form of a question is often coupled with "I know god will never give me more than I can bear," which has even been put to music.

There may even be a genuine search for the good that is supposedly to come from accepting the situation. A typical reaction is, "Maybe God is doing this so that I will learn never to abuse my children?"

A man injured in an accident who manages to gain a new perspective on life and new values may credit god with causing the injury so that he might learn.[3] Sometimes god is just tacitly present in the question. When Barbara Walters on TV's "20/20"[4] interviewed a deaf woman who had an accident that blinded her as well, she asked at one point, "Do you ever question `why me?' or do you think it happened for a reason?" Walters seemingly had no other options to offer. Though "who" might have that reason was not specified, it was presumed that god is the obvious choice. Shakespeare's Macduff, on learning of the murder of his entire family first questions how a good god could allow this, asking, "Did heaven look on and would not take their part?" The presumed explanation comes immediately to hand, which is,

> "Sinful Macduff, they were all struck for thee!
> Naught that I am,
> Not for their own demerits, but for mine,
> Fell slaughter on their souls."[5]

Few have questioned his reasoning over the centuries.[6]

Whether posed as a question or a statement, underlying is always the theological position that god is somehow involved, causing the events for god's own reasons, and the pastor is being asked to explain what those reasons might be.

Perhaps it is not Susan, but her son who comes to the pastor with his own problem. He is not doing well in school. What can he do? The pastor will question him about his own efforts. Obviously if he is not trying there is a ready solution, but, he answers, he is trying to do his best, it is just that his best is not good enough. In fact, he has been tested and found to have a learning disability. Certainly the pastor can and should try to help the boy to be reconciled with the situation, make the best of the remedial helps offered, and still love himself in spite of the problem. But he had better find out if the mother's question is also the son's. If god is understood to have a role in causing his disability, if the boy is thinking, "god made me this way", then the pastor's merely repeating the cliché "and god does not make junk" is hardly going to resolve the question. The youth has only one opportunity to go through life and he may have to do it with less chance of ever making the big money, having a high position in business or a leadership role in society.

It can happen, of course, in the movies, if your name is Forrest Gump, but in real life, the boy has a substantial beef with god. For what hidden purpose has god decided that he will not be as smart or quick as other men? Where is the justice? Where is the much-vaunted love of god for him?

Susan's sister and brother-in-law have another question to ask of god. Why would he make it impossible for them to have children when they both want them so much? For that matter, Susan's husband also has a complaint. As hard as he tries and against every resolution he ever makes to himself, he still finds himself hitting the woman he loves. Why did god make him like this? Why does god not change him?

All this suffering, and all this blaming of god for it. All the questioning, but is it truly the right question? All the explanation, but only barely satisfactory. And the pastor *will* try to explain. It is so clear to him, especially, and maybe only, if the problem is not his. For has he not seen many examples of god bringing about exactly what these questioners presume god is trying to accomplish. How many sinners have changed their ways because they found themselves being punished for their sins? Too late sometimes, but not always. Bigots have sometimes come to value the very cultures they hated, abusive husbands have been known to change when they felt the loneliness caused by the wife's departure. Perhaps god is inefficient since more people seem to never learn their lessons from the consequences than actually reform, but maybe that is their own fault. God's purpose is sometimes accomplished, so the means must be valid.

And how much good has indeed come from facing evils. How much love has been learned and shared because of the birth of a mentally retarded child? How many have been inspired to join a Peace Corps or volunteer their time in a helping agency because of their viewing of the misery of others on television? How many family reconciliations have been accomplished because of the lingering sickness and immanent death of a relative? All right, maybe the answer is, "not much and not many", but it does happen! Sometimes! So god is justified in bringing these evils about because of the potential good that can come from them. We are told that it is the humans' fault if the good does not in fact occur. So, maybe it is not god's inefficiency, just the poor material god has to work with.

However neither the questioners nor the questioned, really reach the right question, although they may spend considerable energy on the question as it was phrased. The pastor will refer to a catechism, or to popular books that seek to explain why there is evil in the world and god's role in it. Some books will just state boldly, "God permits some evil like illness, poverty, only to bring good out of evil"[7] and let it go at that. Others will go a little further, saying,

> "God allows suffering to come upon us so that by this means he may lead us to salvation. God always has a holy purpose in everything he does, even though we do not understand it."[8]

The catechism just cited then expands on this explanation by saying,

> "God leads through suffering to the lesson that we must turn aside from evil. Suffering can be a wholesome punishment through which we come to see our own fault. God also sends us suffering to purify us...Through sickness, poverty, and other troubles he wishes to cleanse us from faults and imperfections...Suffering is at the same time a testing: we are to show that even under suffering we fulfill the holy will of God. When we bear suffering as sent to us by God, we become like Christ."[9]

The latest Roman Catholic catechism is more subdued and more sophisticated, though hardly more helpful when it deals with the question of evil by saying,

> "Only Christian faith as a whole constitutes the answer to this question (why does evil exist): the goodness of creation, the drama of sin, and the patient love of God..."[10]

In this catechism, god is not so immediately involved in causing the evils that befall people, but is still ultimately involved as creator. Could not god have created a better, or even perfect, world? Yes, the catechism answers,

> "...but with infinite wisdom and goodness God freely willed to create a world `in a state of journeying' towards its ultimate perfection...With physical good there exists also physical evil as long as creation has not reached perfection."[11]

This distance then allows the catechist to say, "In time we can discover that God in his almighty providence can bring a good from the consequences of an evil...caused by his creatures.",[12] begging the question of those "evils" not caused by creatures. It also allows the catechism to quote saints who presumed that god was the immediate author of their woes as if they, too, were taking this longer view.[13] The ultimate answer given is the same that the questioners are presuming,

> "We firmly believe that God is master of the world and of its history. But the ways of providence are often unknown to us. Only at the end...will we fully know the ways by which—even through the dramas of evil and sin—God has guided his creation to that definitive sabbath rest."[14]

The pastor will then, following the guidance of the masters, help their people to make some sense as to why god would allow, or even cause, such an evil situation to befall Susan, or her son, or the childless couple, or Susan's husband. And when all else fails, their faith may be questioned, for if it were deep enough and informed enough, they are told, they would just trust god and not question god.[15]

For some, however, this is not sufficient. They will not accept the standard answer that, yes indeed god does these things, but god means well and sometimes it even works out. No, they feel, a god who acts this way is either not a good god or does not exist. Every pastor who knows a Susan who does try to grow through her experience with the presence of evil in her life; who does believe that god means only the best for her; and perhaps is even able to look back from a vantage point in later life and exclaim that she is indeed better off because she faced that problem; every pastor also knows many Susans who left the Church and abandoned their belief in god, either asserting that there is no god who would do this to a beloved creature or that that god is irrelevant to their life.

WHAT ***IS*** THE QUESTION?

All of them, pastor included, catechisms included, have not asked the real question, "IS god involved in these events?" They have all presumed that god is responsible, on some level, for everything that happens. In response to this presumption, they may blame god or even abandon faith in god's existence, but more likely they attempt to absolve god of any wrongdoing. Ultimately this absolution is indeed going to be required, for a god who is all-powerful and all-wise IS responsible for whatever happens in god's handiwork. A human author of a work may execute something that contains unintended elements. Composers have sometimes come to grief when their songs are played at a speed unintended and something "demonic" is exposed. Unless a subliminal message was truly intended, the human author is hardly responsible. God cannot make this claim. Can god really claim that the presence of evil in the world was wholly unintended or not foreseen, and still be all-wise and all-knowing? Can god claim that free will requires that there be evil as well as good when this same god is given credit for promising a kingdom, a heaven, where there will be no evil, either physical or moral, and yet the inhabitants are presumably perfectly human with a functioning free will? No, ultimately if you create something flawed and you could do better, you are responsible for the flaw. A truck manufacturer who mounts a gasoline tank in a dangerous position is responsible when it explodes on impact. A tobacco company that knows its product is dangerous cannot claim that it is absolved of all guilt because the customers brought their problems on themselves. And a god who would be called "good" cannot be one who has anything to do with hurting the creatures, however remotely.

Why, then, is the presumption that god is somehow involved not questioned? The answer is historical. That god must be involved on some level has always been the presumption and it has the sanction of longevity. A big lie will fool many people not because it is so obviously wrong, but because they never even ask if it might be wrong. The same process is true of anything that people believe or have believed for an extended period in religion, science, politics, conspiracy theories, urban myths, or any other area of life, and/or have held for personally important reasons. This process of presuming and then building on the presumptions will be the subject for chapter one.

Most Christians are familiar with the expected answers to these questions. They have probably been taught them by family and religion teachers, have heard them in sermons and homilies, they hear them in the songs in church, read them in religious magazines and they hear them as a part of the secular culture around them. I hardly need to describe them further.

These answers are not, however, the way that I approach the questions posed by the Susans who have come to me. Why I do not use them is a long story which will be argued in the rest of this book. What I do use I will attempt to summarize now. If you agree with me, as many laity, priests and ministers do, although sometimes reaching the same conclusions as I from another direction, then "go in peace," for you do not need this book unless you want another reason for saying what you believe. If you do not agree with me, or at least want to know why I say what I do, then you have a bit of a trek ahead of you as I lay the groundwork and then draw conclusions.

If Susan were to place her question to me, I would, gently, take the position that god cannot be held accountable for the negative things in her life because god has no control over them. This is because god cannot impact physical things and bend them to god's will, either for positive or negative results. God is not the creator of the universe and is not, therefore, liable for questioning. The universe has no cause, including the philosophers' "First Cause," about which more in the text. Humans were not specially created by god, are not the end toward which the universe has been evolving, and are not innately immortal. There is no such thing as a soul, the brain being responsible for all our cognitive and volitional capabilities. Humans are accidents, just as are all living things, and life itself is an accident of the universe, not something brought about by god. There is no one in control of physical forces, except those that humans have learned to master which, by and large, are rather puny. There is no one in control over the direction of evolution, there is no purpose behind it, and in fact there is no direction at all. Miracles therefore do not occur in the physical world, and any prayer made to god to work one, or to control sickness, the weather, the lottery, or any other physical event is misplaced since god has no such power.

Well then, you may say, we know who you are and what you believe. Specifically you must not believe in god, you are not a Christian and you find no place for faith in god. But if you say that you are mistaken. I have faith that god is a personal being who cares about us and works with us, and who holds out to us values to be lived by which surpass those our own history provides. Moreover, I believe in the Church, I believe in prayer and I believe in god's promises, especially of eternal life. Why I say these things is what follows in the rest of the book.

So you decide, gentle reader, whether this is something you want to pursue. It has taken me a long time to reach these conclusions and to back them up by research. It has been worth the trip for me, and I hope it will be for you, should you decide to come along.

By now the reader is asking why god is not spelled in the conventional manner. In the following chapters, I will use the word "God" with a small "g", except in citations. This is not to show any disrespect for god but rather to remind ourselves that we are investigating who this god is and this may, or rather will, require that we be open to a change of mind. This process of changing our minds about the nature of god has been recognized throughout Christian history. Augustine and Aquinas and all the other great theologians have always warned that everything they wrote or ever could write about god would be inadequate and, in fact, "wrong", not because they had made some error but because god always exceeds what the human mind can conceive. In our own day we have become more aware of the bias toward thinking of god as a male and jolting our sensibilities each time we see the word without the capital "G" may serve to remind us to avoid this fallacy. However, there are other ingrained ideas about god that we may also have to set aside

1

There are certain assumptions that religions make which are basic to all their doctrines. They may be stated as articles of faith, whose answers are found in revelation, or they may just be some basic human presumptions, but they are made by many religions in somewhat the same way that science takes some things as "givens." Science assumes, for example, that the same physical laws that apply on earth will also apply elsewhere, in fact everywhere.[16] There may be different phenomena in the center of a star than are observable on earth, but the hypothesis is that if the conditions could be duplicated in the laboratory, or the particle accelerator, the same physical laws would be observed. We assume that if conditions in some alien world are the same as those on earth, then the same reactions will occur.[17]

Science assumes or presumes that things are the same everywhere, *and* that they can be understood. The physicist Timothy Ferris points out that,

> "...it is the *faith* of science that nature is rationally intelligible...each act of observation and each scientific model based on observation puts a frame around a piece of nature...but our belief in the model remains forever tentative...The model is not reality; it is but a painting, and it has a frame."[18]

The size of the frame will influence our understanding of the reality inside and outside this frame.

The picture is similar in religion as in science Things are self-evident until proven mistaken.[19] Faith is something experienced, or even intuited, not primarily taught as a discipline. In most societies, everybody "knows" something is true and it is very hard to change that "knowledge". For example, centuries ago in a mixture of science and religion, the Catholics of Europe did not acknowledge a supernova clearly visible to all since it became a million times brighter than usual. It was recorded in China, the Middle East and America, but not recorded by those who just "knew" that stars did not change and who never questioned that assumption.[20] However, we live in a post-Enlightenment time and we ask questions of everything. We are not alone in this, even from the earliest times. St. Anselm spoke of "faith seeking understanding." I will consider some of these assumptions that "everybody knows" and ask whether in the light of our knowledge today they can still be held.

ASSUMPTIONS. THINGS EVERYBODY KNOWS

It would seem that the first assumption would be "that there is a god".[21] However, I submit that the first assumption made was not that there are gods, but that all things are explainable. Either something occurs because of human exertion and influence or through some other cause. Since the earliest humans may have attributed a life force to every agent, the cause would either be human or something other than human but personal and motivated, though not necessarily a god in our sense. It might be an ogre, spirit, ghost, or demon, but it was not just a physical law. No one could imagine any other kind of agent than one something like us. The causes just "were", as humans just "were" and everyone just knew it to be true. Later, at the dawn of philosophy at least, the cause was sometimes divorced from a being like ourselves and even made an impersonal philosophical concept as it still is today. So the first religious assumption that we will investigate is that, **"Everything happens for a reason or cause"**.

The second assumption flows from this philosophical concept but personalized once again, which is, **"Every cause is rational"** meaning that something has a reason to cause this effect and the reason can be known. This may again lead to the assumption that there is a god or gods, or in any case, something not human operating, or at least not like a normal human.[22] Whatever is not being done by a human person is being done by some other kind of being for its own purposes, be it blessing or curse, and whatever is beyond the capability of a human is surely of this order. The more extraordinary the effect, the higher the divinity must be who can do it. A human being can drown a neighbor's cow, so another kind of being would not have to be too dissimilar to do the same, a troll under a bridge perhaps, or a minor demon. The latter might even work through a human being. It takes a much higher being to cause dew on the ground and a mighty high god to cause a thunderstorm, a Zeus perhaps or a Thor complete with thunderbolts. Through myth and tribal experience and legend, the group identifies itself with a particular god and explains why things are as they are. A seafaring people will have a Neptune, an agricultural people will have an Astarte and Baal. When one tribe or nation conquers another, it is seen as the victors' god overpowering the god of the defeated. It is not only the Greek warriors who are victorious over the Trojans, the Greek gods are victors, too. In fact, the humans are merely the pawns in the wars of the gods.[23] The victors often assumed some of the culture of the vanquished, including their pantheon of gods, and the number and qualities of the gods is refined, perhaps being eventually subsumed into just a few, or only one. However, the function of the gods is always the same, to explain why something happens that is obviously beyond human capability.

The actions that are most important to humans are not the chaotic but the orderly. It is vital that the migratory animals follow the same route so they can be ambushed. If seeds sprout only occasionally after being buried, there will be no harvest. Since these phenomena and many more, such as the arrival of the rain at the proper time and the fertility of the females, are beyond human control, but human existence, as well as the existence of all life on earth, depends on them, the rational beings controlling them are thus responsible for the necessary rhythms. This leads to the third assumption, that **"there is something (or someone) who sustains all things in harmony"** or conversely, withholds sustenance.

It is a short step, though not a necessary step, to the most important assumption of all, namely that this sustaining force is also a creative force, bringing things into existence that were not. This is most evident in the birth and generation of living things, but can be extrapolated back to the origin of all things, the world itself. The fourth assumption is then, **"there is a creator."** Not every religion has reached this conclusion, of course. Some, perhaps most, treat the world as eternal and the gods just manipulate it. Others just never ask the question.

If you accept a creative role for the gods, something else becomes obvious. Since there are few truly unique things in the world, but usually more than one of each thing and often-uncountable numbers of objects of the same kind, there must be some purpose in the mind of this creator. Things are being made with some utilitarian purpose and not usually merely capriciously. There is little room for experimentation or chance. The gods may toy with us. They may make an occasional monster to harass us but there is always a reason known to them. The fifth assumption is "**everything has a purpose**". Even today we often hear people ask a question about "what is a mosquito good for?", showing our preference for answers with a purpose rather than "it just happened that way". Graham Greene's bishop in the so-called Catholic novel, *Monsignor Quixote*, opines that, "a mosquito may be likened to a scourge in the hands of God. It teaches us to endure pain for love of Him. That painful buzz in the ear—perhaps it is God buzzing."[24] True, a human being can certainly do things for a purpose that is either unworthy or limited, or both. I may cut your fence and release your cow so that I can get even with you, or to eat the cow, or just because I think cows should be free to wander where they will. These may be good or bad purposes, and they are perhaps not part of a consistent plan. They may be just impulsive actions. The gods, too, can be fickle. Certainly the lesser deities can act merely for their own amusement, and even the highest can have a bad day and take it out on their human subjects, but usually there is some reason and purpose.

Conspiracy theories are not new to the world. Eventually someone will start to suspect that there is a master plan being carried out. Perhaps things are not merely random, or rational but capricious, many of them may be coherent with a future goal. Humans can act this way, why not the gods? We make the assumption that there is more than whimsical purpose at work; **there is consistent purpose, a plan**. It may be a plan to benefit the planner, in which case the gods are using us and we are only pawns, or it may be a plan to build up the god's tribe or people or nation. Either way, the more that the divine agent carrying out his or her purposes is consistent, the more the pattern becomes not just a purpose or plan but **a teleology**.[25] Things are caused in order to serve some end purpose, which is more than a vague or partial plan but a well-worked-out script. History is going somewhere and the direction is determined by the purpose of the acting being.

The more powerful this being, the more control it will have over events occurring in the universe. The assumption grows that "**there is someone in control of everything**." Most of this being's ways of acting are regular and orderly and even predictable by humans. It is just that the humans are not privy to the plan and to the purposes behind why things happen. We do not know for what purpose the tree fell on the fence and the cows got out, but we at least understand the means by which it was accomplished. We know about wind or wood rot. It seems that while the gods may have other means at their disposal to carry out their designs than humans do, we have seen just about all of their bag of tricks, though a good Midwestern tornado might be a surprise to some tribes in other lands. However, the gods can suspend the operation of events or initiate them at their whim. A rain cloud can be sent to break a drought. A woman can get pregnant past the normal age that the gods usually grant it. A normally fatal illness can be relieved if the gods will but change their minds and modify their purposes. In other words, it is assumed that **"Miracles can occur."**

People assumed that, just as with humans, the gods can change because of something internal or something external. They can have a special fondness for someone and make them rich, or a hatred or indifference and make them poor. They can care about a whole nation and make them the victors. They can even become angry or forgiving, or just show off a little and change the course of events. All these are the same internal emotions that govern humans, and just like humans, they can be bribed, flattered, coaxed, reasoned with, shown the error of their ways or how their plan might be better advanced by some other course of action. Providing the gods with these external reasons to change constitutes ***prayer***.

At its best, prayer is conversation with god as with a friend. It makes no demands on the friend and expects nothing from the friend. It shares the events of the life of the person praying, reveals the self behind the mask kept up for others, and invites reaction from the other in a spirit of trust and openness. Since it is conversation, it also allows the other to reveal themselves. It celebrates the relationship and the goodness of the other person. Unfortunately, most prayer is not on such an exalted level. Much of human communication with the deity is designed to advise god on our needs and how those needs can be met. This presupposes that the god cares and might well respond, or at least one can sweeten the pot by adding a little bribery to the request. If the god will grant this request, whether for some specific favor or just for the god's good will, the petitioner will live in a certain way, erect a monument to the god, dedicate or sacrifice one's first born, buy a chalice for the missions, add a wing to a hospital.

As the religion becomes more sophisticated, such overt attempts to buy the god's favor may lessen and more emphasis be placed on flattery. By prayer one pays homage and attention to the god, recognizes god's power and majesty and figuratively or literally grovels at the feet of the god, just as I might do before a mighty king. This too may not be consciously recognized as our motive, but there is little doubt in our minds that you would be really pushing the limits to expect that god will do something for you unless you pay some attention to the god whose attention you seek. True, god may spontaneously do something for you, rescue you from the bonds of alcoholism perhaps, without this being preceded by prayer, but most often it is the person who asks who receives. The squeaky wheel gets the grease. And even if this time the god did act first, some response from the human is expected. You do not just take the goodies and presume on god's good graces. The nine lepers who are cured of their affliction are evidently cured permanently.[26] There is no indication that their miraculous recovery is revoked because they did not return, but the one who does give thanks is obviously in a better position with god. God offers the Covenant with

Abraham and the patriarchs spontaneously with no strings attached, but the strings are put backwards into the story at a later date, perhaps centuries later, since everyone "knows" that this is not how god normally works. Indeed, the Covenant of Sinai is explicitly connected to the Ten Commandments, just as we would expect.

Underlying all these positions on god or the gods is not so much the assumption that the god wishes to do good for the petitioner, at least if properly motivated, but that **"the one in charge CAN do good"** After all, there is no sense in asking someone for something that it is beyond his or her ability to provide. If the god is in charge and makes things happen to promote the god's own plan, then the god can make them happen differently, more to the liking and benefit of the one beseeching the god's benefices.

Although the gods can be the source of good for their adherents, they are also responsible for the evil things that happen, since either they or some human agent are responsible for everything.[27] This evil is subjective, of course. What is evil for one person may be a benefit to someone else. One horse owner's loss is a bettor's gain. One country's loss in war is another nation's victory. These relative evils must be explained, of course, and they are explained in the various religions. The real problems are those events that everyone would find "evil" in the sense of never or rarely bringing good to the humans; a devastating flood, for example, an earthquake, sickness, poverty and ultimately death. How are these to be dealt with, or at least explained? Of course they, too, are somewhat subjective since a country ruined by a natural disaster is a blessing to an invading neighbor; the sickness of the king weakens his grip; and heirs have been known to pray devoutly for the death of the estate holder. Of course the bacteria and maggots consider a good human plague a real godsend, but the viewpoint of critters is usually not considered in judging the goodness of an event. Beyond these considerations, however, there is a general understanding that these happenings are "evil." And this evil is as real as the good, the curse as important for human life as the blessing. The "**existence of evil**" is also assumed by the religious outlook and it, too, is variously explained by the different theologies. However, it must fit into the assumption that the one in control, the one with the plan, is in control of the evil as well, at the very least in its final defeat. There is also the assumption, or at least the suspicion, that the one in control is also the master of evil and uses it as a tool to achieve their purpose.

CONCLUSION OF CHAPTER 1

These assumptions underlie all of Western religious conviction and faith and most often are taken as a package. If the first assumption, that there is a cause for everything that happens, is accepted, this postulated cause takes on, sooner or later, personal dimensions, the cause becomes a divine being. While Oriental religions may leave plenty of room for blind chance and fate, the Western religions usually do not. Even in the East their own assumption is often not taken seriously but efforts are made to influence "someone". In both traditions, it is true, things can "just happen" without rhyme or reason and therefore cannot be understood or controlled but merely endured or adjusted to. Nevertheless, the assumption that there is a reason behind every event and phenomenon usually prevails and leads to a questioning of what that reason is, and who it is that holds it. In the West, even those who deny the existence of a personal god still put the "what" of evolution in the role of the ultimate cause in place of the religious "who". I, too, accept this first assumption that there is a cause, for it is the basis for all of modern science that presumes that causes, at least in the form of explanations, can be found for everything that exists. What we must do is ask how the investigation of the "what" influences our understanding of the "who" if indeed one exists.

2

The following are the position statements that I have labeled as "assumptions" to be investigated:

1. Everything happens for a reason or cause.
2. Every cause is reasonable or understandable.
3. There is someone (or someones) who sustains all things in harmony or conversely, withholds sustenance (contingency)
4. There is a creator.
5. Everything has a purpose.
6. There is consistent purpose, a plan, and not just a plan but a teleology
7. There is someone in control who can carry out the plan.
8. Miracles can occur.
9. Prayer effects physical changes.
10. The one in charge can do good.
11. God is related to the existence of evil.

It is my contention that the assumptions described in the previous chapter are generally held by many if not most religious worldviews. I do not mean this to be a comprehensive list, there may be more, and a given religious or philosophical system may hold all or only some of them. In fact, there are so many variations that I invite the reader to examine their own approach to these assumption in their own view of life, whether they would describe their view as religious or not. This may be helpful as we proceed. Try to choose just one of the various options listed but read through each item since some may build on the prior positions.

1a. There is a cause for everything that happens.
1b. Nothing can come from nothing so there must be a cause.
1c. There is a cause for most things that happen.
1d. There is a cause for almost everything that happens.

2a. There is a logical explanation for everything.
2b. There must be a logical explanation for everything.
2c. There is a logical explanation for most things, but some just happen.
2d. There is not a logical explanation for everything.

3a. There is a first cause which itself has no cause.
3b. There is a first cause, which must keep on working.
3c. There is a first cause, which need not continue to operate.
3d. There is no first cause.

4a. If God does not keep things in existence they don't exist.
4b. If God does not move things, they do not change.
4c. Things exist until something destroys them.
4d. Things move without direction from God.

5a. Something which "must exist" is required so other things can exist.
5b. Nothing would exist unless this "something" makes it exist.
5c. Things can just "happen", there is no need for a Creator.
5d. Things did just happen, there is no Creator.

6a. There is a reason why things exist.
6b. There is a reason why things exist, but it may be known only to God.
6c. There is no reason why things exist.
6d. There need not be a reason why things exist.

7a. There is a plan for human beings, including one for me.
7b. There is a plan for the universe, but not necessarily one for me.
7c. There is no grand plan for humanity.
7c. There is no plan for the universe.

8a. There is someone in control, working out this plan.
8b. There is someone in control, but it is a loose control
8c. There is no one in control, things just happen.
8d. There is nothing in control except physical or mathematical laws.

9a. The one in control often works miracles.
9b. The one in control sometimes works miracles.
9c. There is no one in control and there are no miracles, but things occur which cannot be explained.
9d. There is no one in control and everything can be explained.

10a. Prayer works.
10b. Prayer works, but God expects us to do our part.
10c. Prayer has no effect
10d. Prayer has no effect and it is harmful to believe it does.

11a. The one in control always does good.
11b. The one in control can do good but often stays inactive.
11c. There is no one in control, so they cannot change things.
11d. There is no one in control except physical laws.

12a. God has told us what is good and evil.
12b. God has told us what is helpful and harmful.
12c. There is good and evil but based on physical reality.
12d. There is no good or evil.

WHERE ARE YOU?

The reader has probably noticed by now that those with a religious faith, or adherents of many philosophies, will have chosen one of the first two possibilities in many or all of the sets. Those who are agnostics or atheists will have chosen one of the last two. Scientists who are agnostics or atheists will always have chosen one of the last two, but scientists with a traditional religious faith may have chosen some from the first two and some from the last. On the other hand, of course, you may have been all over the map.

If you consistently chose one of the last two of the sets, you may well be conversant with modern science, but let us note that none of these positions can be proven through science. I realize that each statement may not express your position, but consider for a moment that "there is no first cause", "there is no plan for the universe" and "there is no good and evil" are not within the purview of science. Scientific findings can and do cause doubt of these things, but there is no priority for these positions. That is not the purpose of this work. My purpose is to examine whether or not the first choices in each set are valid. By what right do religions make statements such as "nothing can come from nothing", or "there is a creator".

Those who agree with the "scientific" statements must be honest enough to at least be agnostic in distinction to atheistic and admit that the best that can be held here is "I doubt it" instead of "I know that it is not true." Those who chose the "religious" statements must be just as honest in facing challenges to what has been believed.

We are going to take the assumptions listed and examine them in the light of modern knowledge.

3

I have proposed that there are presumptions made in any area of human knowledge, but science since the 16th century has challenged every field to submit its assumptions to the test of comparison to empirical data. No longer is it permissible to appeal to a Galen as the court of last resort in anatomy, for example. It is perfectly all right to appeal to the authority of an expert, but the expert's opinions must be in agreement with external data, or so one assumes. If the results of observation of the object under study do not agree with the theory, then it is the theory that must change.[28]

For centuries it was assumed that science would be in accord with religion since it reflected the mind of god, the same god who was revealed in religion. They were "Two Books" with the same author. Faith would be bolstered and possibly explained by the findings of science, but certainly not contradicted. Seeming contradictions could be explained by false findings or by a lack of information on the part of science. It could be conceivable that faith would have to change in some aspect, but only in the details and not in the major tenets of the faith. God's direction of the workings of the universe may include things that were not previously known, such as electricity or magnetism, but this only deepened one's awe of the power and extent of god's marvels rather than questioned god's direction of them.

Today most agree, conversely, that faith itself must be in agreement with the results of science, or at least not contradictory to them. In response to Darwin, there was denial, of course, and an effort to cast out the new heresy, or there were claims that once science had more information the faith would be vindicated, but there was also an effort to use the new findings to enrich the theological positions. These responses, says Arthur Peacocke, a noted commentator on the relation of faith and science,

> "...were not based on any mood of defeatism or any sense of accommodation of Christian truth to a new and overwhelming force. Rather, they were based on the conviction that has always motivated the best and, in the long run, the most influential theology - namely that, to be intelligible and plausible to any generation, the Christian faith must express itself in ways that

> are consistent with such understanding of the nature of the world as is contemporarily available."[29]

Science and faith, it is claimed, can approach the same material from different viewpoints, but there can be no contradiction.[30] This idea of change as merely the accumulation of new facts that can be fit into the old faith pattern or paradigm is popular in the Church since it seems to pretty much leave things as they are. No basic doctrines need to be changed, just explained in a new way or from another slant. Historically this is not what occurs in a so-called paradigm shift. Usually, though not always, according to Hans Kung,

> "...new models of theological interpretation...come into existence...because the traditional interpretive model has failed, because the `problem solvers' of normal theology, in the face of a changed historical horizon, can find no satisfying answers for new major questions, and `paradigm testers' set in motion a `extra-ordinary theology' alongside the normal variety."[31]

So it is my intent to examine the assumptions of religion and ask the question whether there is a disparity with the findings of science; how the model of faith currently held, at least in the Catholic tradition, responds; and if it does not respond well, to hazard a guess at a direction in which theology and faith need to go in a new paradigm.

THE FIRST ASSUMPTION, THERE IS A CAUSE FOR EVERYTHING.

The first assumption, as stated, is not that there is a god but that everything happens for a cause or reason.[32] It presumes that there is a history of cause and effect behind every phenomenon. For the most part these causes are considered open to scientific inquiry and it is the glory of modern science that many of the common occurrences in the universe have been explained or at least there is a direction in which to look for the cause. "Why does it rain" is no longer answered with a primitive "science" speaking of the floodgates of heaven opening. It is now answered with an explanation of evaporation and condensation. It is significant that at no point is it considered necessary to invoke the activity of a god in the train of causation leading to a downpour. There may be gaps in our knowledge of natural phenomena, just watch any weather report and compare it with what actually occurs. The findings of Quantum Physics present other gaps in our knowledge, and as Ken Wilbur in a book titled, *Quantum Questions* says,

> "No one doubts that we are confronted with a causal connection whose causal components are in the main unknown to us. Occurrences in this domain are beyond the reach of exact

> prediction because of the variety of factors in operation, not because of any lack of order in nature."[33]

And not because, in the view of most, a hidden god has a hand in the picture.

Of course many Christians do still insert god in the equation. Let any flood occur, or even a long heat wave, much less a very destructive hurricane or tornado, and the comment, "maybe somebody is trying to tell us something" will be heard. There is no doubt who this someone is, nor the tenor of what is being said. God is ticked off about something, just as at the time of Noah. Some, Susan among them, still feel that god is somehow the beginning of the physical processes leading to illness or misfortune and some will carry this to a great extreme. A woman who questioned why god would cause her to have a flat tire on a dark and dangerous highway found an answer to her question when two rather scruffy and intimidating men stopped behind her, only to reveal themselves as perfect gentlemen who changed her tire and sent her safely on her way. Why did god cause the flat tire? To give her an opportunity to preach to them about the Lord during the whole time they worked on the vehicle. One wonders whether the gentlemen will ever stop again given the aggravation they had to endure, and most would wonder about the preacher's theology, putting god into the chain of events leading to her running over a roofing nail. The author of a book titled, *Realities* and subtitled "The Miracles of God Experienced Today" would not shrink from putting god in this position, however. She tells the story of the unending rain hindering the construction of their convent. Her Sisters "explained the rain by natural causes which satisfied our reason - even though the Scripture so often states that the weather...and the storms are governed by God..."[34] It was not until the Sisters recognized that god was in control and admitted their guilt in not doing so that the rain stopped.

An evangelical minister tells of meeting a woman much like Susan who had been abused and is on a plane bound for a doctor and a divorce. He theorizes that, "perhaps He worked it out for me to miss an earlier flight this morning just so I could tell you of Jesus' love for you and how much He wants to help."[35] This is akin to the opinion of some abolitionists' that Africans seized after a mutiny on the *Amistad* around 1840 were learning Christianity because, "God had brought the *Amistad* blacks to the United States for Christians to make them His `most honored ambassadors to the dark continent of Africa'."[36]

Even though some rather sophisticated Christians may find themselves saying something of this sort, when pressed most will agree that god probably has better things to do than mess with the weather. Rearranging airplane departures and having people captured, transported to Cuba, rescued from a drifting ship, winning a Supreme Court decision and returning to Africa all seems a little complicated, too. Certainly any god worth their salt should be able to do such things, but that these events must happen by divine decree is certainly an assumption in need of investigation.

A FREE LUNCH?

In the Church this position on the need for an ultimate cause continued, but by the 1950's and 60's, the "Steady State" Theory was very popular among physicist-cosmologists and by then it was evident that the old philosophical and theological argument that "*ex nihilo, nihil fit*", or "nothing can come from nothing," and that therefore a divine actor was needed, was being discredited. This theory called for a continuous "creation" of matter from nothing. Fred Hoyle "suggested that matter might be continuously created by a hypothetical `creation field' akin to a gravitational or electromagnetic field."[37] This obviated the necessity of envisioning the universe coming into existence at some beginning point in time or space. The theological or even magical overtones of a one-time beginning offended many and the Steady State theory did not need these explanations. Creation from nothing was still required and the theory was vague on the mechanism responsible, but it was obvious to the theorists that this would be a run-of-the-mill occurrence that needed no divine intervention. It has since become apparent that the theory's predictions do not fit observations, leaving the Big Bang theory that we will soon examine in possession of the field.

IS IT TRUE THAT THERE MUST BE A CAUSE?

For centuries, the scientists of the world would have no problem with the concept that the world was made by god out of nothing. For one thing, there was no idea of an expanding universe, no reason to think that it was any different now than it had ever been, at least on the larger scales. For another, many of these investigators were Christian believers who wanted to leave a place for god in the scheme of things, indeed to use the universe as a textbook in which they could learn of god through god's actions. The Steady State Theory at least did that, although the Roman Church had a problem with it as we will see.

As more and more "laws" were discovered, supernatural agency was less and less invoked to explain why things behaved as they did and god was pushed into the background as a proximate cause of physical phenomena, but was always given the place of honor at the beginning of everything. Deism set up a clockwork universe that needed no ongoing intervention from god, but god was certainly the one who wound up the mechanism at the onset. "A Deist in general believes in God, but he may be either an absentee God, or he may be one who cares directly for men by natural laws. The Theist is sure that God exists for the same reason that the Deist is sure - because one cannot explain this universe if there is no adequate first cause, an infinite designer capable of producing such a marvelous world."[38] The Founding Fathers of the United States were often formally Deists, like Franklin, Jefferson, and John Adams. "There is a God, they said, but he is to be found through reason rather than through revelation. God created this world but he did not interfere with its workings; a man's heaven and hell were of his own making."[39] Yet they often also believed in a god who actively intervened in human affairs, not only by giving the natural law and letting it run its course, but also by "actively helping good men attain happiness."[40] This view was due to a growing knowledge of science but a lack of information about the beginnings of the universe, and because many were believers who wanted to spare god a place, but also because the concept of creation from nothing and the need for a first cause was so ingrained in their understanding.

As data accumulated that indicated the expansion of the universe, "...by 1929 [Edwin Hubble] had information enough to allow him to announce a very simple law. It seemed that the Universe of galaxies is expanding, and that the radial velocities of the galaxies are simply proportional to their distances from the sun."[41] It had only been four years earlier that Hubble had shown that there were indeed other galaxies than our own and now he was speaking of galaxies moving relative to one another. Objects in the universe were not just moving, but moving at a certain rate, and it was soon found that aside from local motion influenced by neighboring galaxies or clusters of galaxies, they were all receding from one another at that rate.[42] It was possible to "play the tape backwards" and arrive at the conclusion that there had been a beginning of this expansion. With the growing scientific mindset that since everything we observe can be explained without recourse to divine action, the explanation for the inception of this totality of the universe was also sought without any supernatural component. It was not long in coming.

QUANTUM: NO FEAR

Quantum physics is a term that strikes fear into the hearts of non-physicists (and, I suspect, into the hearts of many a physicist as well). One physicist notes that, "much of the way that quantum theory describes the world may seem at first sight to be nonsense - and possibly it may seem so at the second, third, and twenty-fifth sight as well."[43] Einstein himself deplored it as counter-intuitive and impossible, but it could not be denied since it was, in spite of its seeming contradiction of common sense, confirmed repeatedly through experimental affirmation of its predictions. It just works too well to discard simply because it contradicts some of our most cherished beliefs. It turns out that the universe of matter with which we are all familiar is not built of little hard blocks of material something like billiard balls, it is composed of energy/matter which means, as Einstein pointed out, that matter and energy are simply two manifestations of the same reality, $E=mc^2$. [44] That reality seems to be little "packages" or "quanta". An object that is warmed up emits energy, like a stove or radiator and what it emits is called "black body radiation."[45] This radiation is not emitted in a seamless stream but in little "packages" or "quanta" and so for example a wave of light can be described as a series of little particles of light called "photons,". At the same time, a single quantum could behave as a wave under some circumstances. There is a "wave-particle duality."[46] Now that is certainly absurd. How could the same thing be a wave and a particle?[47] The notorious paradox of Schrodinger's Cat brings out the counterintuitive and seemingly absurd understanding of reality inherent in the quanta and quantum mechanics.[48] Not only that, the darned things are notoriously tricky to pin down. Where they are and where they are going and how fast, and what color they are, and which direction they are spinning and even "*if* they are" are all properties that classical physics from Newton and Maxwell and others can determine, but not absolutely, it turns out. You can figure out these things about a collection of $500x10^{10}$ billion of the pesky quanta, making up some large object, but you can never know precisely about just one, not only where it is but even if it is. It is the ultimate "now you see it, now you don't." Yet, in experiment after experiment the findings of quantum physics are born out while the classical physics reaches a barrier it cannot surmount.[49] I will have to say more about this phenomenon later, but bear with me for a while.

Cosmologists today are no longer confined to the philosophy department and to Thomistic theologians. They are found in the laboratories and research departments and they have no problem making the following statement of Paul Davies,

> "How can something come into existence uncaused? The key to achieving this seeming miracle is *quantum physics*. Quantum processes are *inherently* unpredictable and indeterministic; it is generally impossible to predict from one moment to the next how a quantum system will behave. The law of cause and effect, so solidly rooted in the ground of daily experience, fails here. In the world of the quantum, spontaneous change is not only permitted, it is unavoidable...It is no longer entirely absurd to imagine that the universe came into existence spontaneously from nothing as a result of a quantum process."[50]

How is this possible? How do you get something from nothing, from a vacuum? You redefine the vacuum to be something a little different than Webster says it is, namely "a space absolutely devoid of matter." In quantum physics there is indeed no matter present "*on the average*" but "the vacuum is actually full of activity. According to the laws of quantum mechanics, it is possible for particles and their opposite numbers, the antiparticles, to appear briefly out of nothing, then interact quickly so as to wipe each other out and disappear...a `quantum fluctuation'."[51] As K.C. Cole says, citing another physicist, "The answer to the ancient question, `Why is there something rather than nothing' would then be that `nothing' is unstable."[52]

This "something from nothing" evidently happens all the time. "At the subnuclear level, the quarks and gluons which make up the neutrons and protons of the atoms in our bodies are being annihilated and recreated on a timescale of less than 10^{-23} second"[53] As far as we know, normally nothing comes of these fluctuations, the entities destroy each other and the balance of zero continues, (though some are looking into extracting usable energy from the vacuum, i.e. from nothing).[54] In fact there never was any other balance than zero, either when nothing was present or when two opposing objects were present. They always cancelled each other. Universes (probably?) do not spring into existence every 10^{-23} seconds, however. The fluctuation that began our universe was not the normal pattern, though how often the pattern deviates from the normal and spawns a universe we may never know. In the case of our universe, however, something occurred which allowed the duration of existence of the particles, and far more than one pair, to be greatly expanded. "Quite by accident enough fluctuations occurred close enough together to trigger the process..."[55] Normally the balance is preserved by having opposing particles, matter and anti-matter, created whose charges balance each other. The same may be possible with an innumerable quantity of pairs, but the situation would be short-lived since they would immediately destroy each other on contact. Various explanations have been devised to explain why our universe seems to be composed only of matter instead of antimatter, mostly revolving around a process that does call for mutual annihilation of the matter-antimatter pairs, but resulting in one extra matter particle left over per 30 billion destroyed.[56] The balance is preserved for these survivors by pairing the positive energy of the matter against a negative energy of gravity, a debt that is yet to be paid by the universe, but which maintains the delicate balance. The net sum of energy in the universe is therefore zero, as it is when matter and antimatter are in balance.[57] All this is explained to some degree of satisfaction by quantum physics and the theory of an inflationary universe that we will consider later. All the details are subject to verification and to change, but scientists are confident that even if this explanation turns out to be faulty in the details, or even in great part, there is a perfectly understandable explanation of the appearance of the universe "out of nothing" and that we have found or will find that explanation.[58] Nowhere in it will be a divine "*fiat*" or "let it be".

To quote another cosmologist, Trinh,

> "Quantum uncertainty allows time and space, and then the whole universe, to arise spontaneously from a vacuum. At the Planck time (10^{-43} second), the universe was only 10^{-33} centimeter across, 10 million billion billion times smaller than an atom, and quantum mechanics, which governs the subatomic world, could do its work. The universe does not require a first cause. It appears thanks to a quantum fluctuation...This description of the creation of the universe strangely resembles the creation ex nihilo evoked by many religions. The major difference is that the appearance of the universe, thanks to the magic of quantum mechanics, no longer appears to require a first cause or the existence of God. Its emergence may be explained by purely physical processes."[59]

It would seem, then, that the concept that there is a cause underlying everything is not going to be upheld by appeal to the "common sense" notion that nothing can come from nothing. The general feeling is that we have found or we will find an explanation. There is no way to preach the gospel to a scientifically literate person by saying that we can prove the necessity of the existence of a creating god by appealing to the doctrine of *ex nihilo*.

TIME, FOR A CHANGE

Is there a need for a "prime mover" as the Thomistic philosophers insist?[60] This is a different question, but it relates to the assumption that there is some cause of everything. Some say yes because the chain of causality needs to lead to some ultimate explanatory cause. Others say yes because of the problem of time. If you have an infinite progression of time, does anything really happen? If so, "when" did it happen, and relative to what? If we need not have some pre-existing and non-contingent agent to explain how the universe came into being from nothing, do we need some such being to explain how it moves in time?

Going back in the chain of causality required ancient philosophers to come to a first in line, not necessarily temporally but one which is not contingent.[61] Still, underlying these concerns was the question of time and space and infinity. Thomas Aquinas, following Aristotle, recognized that time is a measurement of change in beings, and a measurement is done by an intellect. "If there were no mind there would be no time."[62] Since god has an intellect, there is something outside the time we know in the universe, but since god does not change, it is "an endless duration without change, an "all-at-oneness" (*tota simul*).[63] Thus, Aquinas distinguishes the time we experience as not that which god as an eternal being "experiences," but he is sufficiently vague as to leave plenty of room for confusion. Time is certainly not something extrinsic to god for Aquinas, but an attribute of god that is then applied to god's creation. It exists as long as there is an intellect to perceive it, but the something that *we* call "time" exists only when there are changing entities, and the change can be measured. God does not change, so the "time" which exists with god is not the same time as we experience. Aquinas teaches that both space and time are constructs of the created universe, perhaps of the mind, but with some external reality. This universe may have existed for an infinite time or it may have had a beginning. Without the revelation of the creation event it would be impossible to know which. This comes close to the modern understanding of space and time, but Aquinas does not pursue it to its conclusion.

Space, too, is something real for the philosophers. Aristotle, believing as a good Greek in the existence of concentric spheres comprising the universe, held that the outmost sphere was not *in* anything, but contained everything inside itself. Aquinas followed suit. For Euclid and his followers, space was also flat, meaning that parallel lines never meet and you could send them out from yourself, two light beams for example, into the surrounding space which was waiting to receive them. To him space never gets bigger or smaller and is probably infinite. If it were not, you could throw a ball or a lance through the edge of space into, what? Few could conceive of there being anything on the boundary of space but more space. True, god was said to exist outside of space and time, but how this could be was mystifying. Perhaps god was "outside" them in the sense that they did not affect or limit god.

With Copernicus, and then Galileo, the idea of a limited spherical universe unraveled, and with Newton, space was no longer the empty place where something happened, it now had mechanical properties as well.[64] As long as an object is just coasting along, you cannot tell anything about it; not its speed or whether it is moving at all. This is because some measurement must first be made relative to something else. To say that it is moving at 100 mph means it is moving at this speed relative to the earth, perhaps, which is considered stationary. When the object accelerates, you can then tell something is going on without reference to another object. How can this be? Newton assumed that there was something absolute against which the motion of an object could be measured. It was true that if you measured the speed of a ball against the earth, this was not its true speed since the earth itself was moving around the sun. If you were observing from the sun, the speed of the ball would be its speed as measured on earth plus the speed of the moving earth. However, there was also an "absolute space" against which all motion could be measured. The space inhabited by the matter of the universe was not moving, and was in fact a real thing that Newton seems to have equated with god.[65] This absolute space was separate from absolute time that also existed. His concept of space was questioned by Gottfried Liebniz and by the philosopher Bishop George Berkeley, but most people's view, if they had one, was that space was a "something" in which things happened. "Until the early years of the twentieth century, neither philosophers nor astronomers had questioned the notion that space was absolutely fixed - an arena in which the stars, the planets, and all the other heavenly bodies played out their motions."[66]

In the late 1800's, Ernest Mach held that inertia was the result of the matter of the universe, the stars and other matter, pulling on each other and on every object within the universe, and not the effect of an "absolute space". James Maxwell found that the number measuring the speed of light was not dependent on reference to another object, but came up a constant number,

300,000 km/sec, in his mathematical equations. Well then, said Newtonian physics, there must be something against which you are measuring, and so the ancient idea of the ether was resurrected. This would be a substance filling the entire universe, though it had never been detected, through which light moved at a constant speed relative to the unmoving ether. As a consequence, if you are on the leading edge of the earth moving into an unmoving ether in which light was moving toward you, you should get a higher number for the speed of light, just as the oncoming police car with the radar gun going 30 mph measures you as doing 90 when in fact you are doing 60. The officer has to subtract the difference because you are both moving toward each other. If he were moving perpendicular to you, or not moving at all, he would not have to subtract, he could tell your true speed relative to the stationary earth, or the stationary ether. When Albert Michelson and Edward Morley did the experiments, there was no change in the speed of light.

It was Albert Einstein who formulated the explanation of this strange state of affairs, namely relativity.[67] The famous illustration of the moving train shows what happens. Hang a light bulb from the center of a train car and turn it on. Film the event and run it back in such slow motion that you can see the light wave move through the car (make it a dusty car so you can see it). To the person inside the car who turned on the light, the wave will reach both ends at the same time since the speed of light is independent of the speed of the train and the distance to each end is the same. However, to a person standing outside the car, the distance to each end is not the same since the car is moving. So they will see the light hit the back of the car before it hits the front. To the person in the train, the light hits the ends at the same time, but to the person outside it does not. The time it takes is relative. The distance, or the space, between the two events is also relative to where you happen to be standing. "Just as it is no longer possible to talk of *the* time interval between two spatially separated events, such as the arrival of two pulses of light at opposite ends of a railway carriage, so it is no longer possible to talk of *the* distance between two spatially separated objects."[68]

In fact, space and time are not only no longer the discrete entities they were once thought to be, they are united as one in a single concept called space-time.[69] Stephen Hawking points out,

> "It used to be considered obvious that time flowed on forever regardless of what was happening; but the theory of relativity combined time with space and said that both could be warped, or

> distorted, by the matter and energy in the universe. So our perception of the nature of time changed from being independent of the universe to being shaped by it. It then became conceivable that time might simply not be defined before a certain point...If that were the case, it wouldn't make sense to ask who, or what, caused or created the big bang. To talk about causation or creation implicitly assumes there was a time before the big bang singularity." [70]

If both time and space existed prior to the universe then you can if you wish talk about god causing the universe to come into existence in time. If there were no time or space prior to the beginning of the universe, then there could be no change since time exists if something is changing. There is a "before" and an "after". If there is no "before" the Big Bang, since time itself is a construct of the universe, there need not be causality in time. All the arguments that there must be some external agent to make something begin to be in time are suspect. Time itself, and space as well, or rather space-time, are attributes of the universe. "Before" there was a universe there was no space and no time. Indeed, time itself may be nothing more than a construct that our minds use to make sense of the universe. This means that time that Thomists called for to be outside of created beings was itself present in the universe, not in something or someone outside of the universe.[71]

NO TIME? THEN CONTINGENCY?

Does there have to be a cause in the sense of a prime mover, or because nothing can come from nothing, or because both space and time require a grounding in some being who is in some sense outside of themselves? In the light of modern science, it does not seem to be the case that such a causative agency is necessary. There is a long history of trying to prove the necessity of some kind of "god" from the contingency of the universe. "Contingency," again, merely means that the being is capable of either existing or not existing, unlike a non-contingent being which must exist. The Ontological Argument put forward most famously by St. Anselm, holds that there must be some being which is not contingent and which, in fact, has "existence" as a necessary aspect of it's nature. This argument has been rather thoroughly rejected as a proof for the existence of god and I will not try to disprove it here.[72] The whole idea of contingency is that something that could or could not be either can never come into existence or, if time is infinite, would have already gone out of existence, or someday will do so. From what we have seen and what follows it becomes obvious that things can indeed come into existence out of nothing and the universe may well wink back into nothingness. There is no need for an external causative agent to enable the universe to do either. "Why does anything at all exist?" is an intriguing question but one that perhaps has no answer and certainly does not need a "necessary being" to provide the starting place for an answer. Of course, the same question can be asked about god's existence. If there is no need for a "necessary being", is necessity true of god or "why does god exist?"

What about on the other end? If there is no need for a causative agent in time, no need for a Prime Mover, and no need for a non-contingent being to explain the existence of the universe; if it can just pop into existence out of the vacuum; does there have to be a purpose to it all, and by implication a rational creative being who has something in mind to accomplish? Must there be a "Final Cause and hence a designer?" Richard Dawkins asks,

> "What final cause can be cited to bring this hierarchy of reasons to a close? Aristotle had an answer: God...the *for-which* to end all *for-whiches*. The idea, which is taken up by the Christian, Jewish, and Islamic traditions, is that all *our* purposes are ultimately God's purposes."[73]

THE ARGUMENT FROM DESIGN.

We have seen that the old philosophical conclusions calling for a prime mover and insisting that nothing can come from nothing are no longer compelling arguments, but then they were never anything more for the Christian than backup arguments. The Hebrew scriptures certainly believed in a creative god who set everything else in motion long before they heard of the Greek reasoning about prime movers and such.[74] It is also true that while the evidence from science arguing against the *necessity* of a prime mover is compelling, it does not preclude the *possibility* that there is such a first cause. If a first cause is not necessary, what of a final cause or Designer, one who has a purpose in mind and orders things so that this purpose will be attained? The Judeo-Christian traditions, and most of the mythologies of human history, have accepted, perhaps from prehistoric times, a rational cause of all that exists based on experiential evidence. Looking around the world it seems clear that there is evidence of a designer. Some rational being, it seems obvious, intends for whatever purpose that things should be as they are.

It was this god Aquinas hoped to make known, even though he used the language of the Greeks to do so. Aquinas' Fifth Way of demonstrating the existence of god is the teleological argument in which,

> "...he argues that many material and therefore nonintelligent things of different kinds co-operate in producing a stable world order and stable subsystems. They thus achieve an end or purpose. Being material, they cannot consciously bring about this end. So the situation implies the existence of an intelligent Author of nature who guides things in such a way that they achieve this end."[75]

This way of believing would go on for a long time before a more Platonic view would begin to emerge once again.

> "Although Kepler, Galileo, Descartes, and Newton assumed that the metaphysical or ontological* foundation for these laws was the perfect mind of God, that assumption would be increasingly regarded, even in the eighteenth century, as ad hoc and unnecessary...What would endure in an increasingly disguised form was the assumption of ontological dualism which allowed the truths of mathematical physics to be regarded as having a separate and immutable existence outside the world of change."[76]

David Hume in the eighteenth century asked some questions of Aquinas' argument, as did Immanuel Kant and others. As Barrow and Tipler point out, despite these occasional questionings, the majority of natural scientists prior to the publication of Darwin's "Origin of the Species", especially those who worked with living organisms,

> "...regarded the living world as organized into a Great Chain Of Being with single-celled organisms at the bottom of the Chain, mankind somewhere in the middle, the Angels above him, and God at the top...this picture of living creatures was static; the species were created to fit into this ordering at the beginning of time and were ordained to remain so ordered for all time. God's purpose never changed since He was unchanging. A species could never become extinct."[77]

With the publication of the *Origin*, and the whole atmosphere that had developed of counting god out of the equation, there was a reaction disputing the evolutionists' position that there was no necessity for a creative and sustaining agent, in fact a designer god.

---

* Ontology is a branch of philosophy dealing with beings and existence. An ontological change, for example, is a change from one being or state of being into another. The ontological argument is about the existence of God.

It is William Paley whose name is most associated with the design argument of these more modern times. Prior to Paley, most of the arguments in favor of the existence of god from design had spoken of the orderliness of the cosmos. "...Aquinas stresses cosmic orderliness whereas Paley ...emphasized biological adaptation."[78] Scientists of the time were brought up with these assumptions and found evidence of god's role of designer in their discoveries. Richard Owens was a British anatomist who ascribed the name "dinosaur" to a whole new group of fossil animals. He and his supporter, William Buckland, felt that "the superb design of ancient organisms proved the constant superintendence of a loving deity rather than a natural process of increasing excellence, from initial crudity to current complexity." According to Buckland, the design of dinosaurs,

> "...shows that even in those distant eras, the same care of the common Creator which we witness in the mechanism of our own bodies...was extended to the structure of creatures that at first sight seem made up only of monstrosities." [We therefore cannot view their excellence in design,] "without feeling a profound conviction that all this adjustment has resulted from design and high intelligence."[79]

Of course Darwin had recently made biological adaptation the centerpiece of an argument that implicitly denied the necessity for a divine Designer and some of his supporters made this view more explicit than Darwin had dared. In 1902, in response, Paley wrote,

> "In crossing a heath, suppose I pitched my foot against a stone, and were asked how the stone came to be there, I might possibly answer, that...it had lain there forever...But suppose I had found a watch upon the ground...[I am forced to conclude] that the watch must have had a maker."[80]
>
> "Paley rammed home his argument by multiplying up his examples. He went right through the body, from head to toe, showing how every part, every last detail, was like the interior of a beautifully fashioned watch."[81]

Not only did he show the wondrous coordination of the body of animals, he also put into words the reaction to the theory of evolution that most hoped would eventually prevail. What else but an intelligent designer could possibly fashion so intricate a mechanism? If you find such perfection and craftsmanship in an obviously man-made object, you know that a person made it, not by accident, but by design. Surely, the same argument holds weight when finding such a piece of work that could not be of human artfulness. Most tellingly, he and many others *could not imagine* any other agent who could have brought this about.

Though the Catechisms of the past certainly use the argument from design,[82] there is no overt mention of the argument from design in the new official *Catechism of the Catholic Church.* It says things like "God is the sovereign master of his plan. To carry it out he also makes use of his creatures cooperation",[83] and, "The witness of scripture is unanimous that the solicitude of divine providence is concrete and immediate; God cares for all, from the least things to the great events of the world and its history"[84] but the modern catechism is not about to repeat older works and their unquestioning assertion that god handles all the day-to-day details of this caring, carefully saying that,

> "We firmly believe that God is master of the world and of its history. But the ways of his providence are often unknown to us. Only at the end, when our partial knowledge ceases, when we see God `face to face' will we fully know the ways by which...God has guided his creation to that definitive sabbath rest for which he created heaven and earth."[85]

The presumption that there is a god who is the creator of all else that is, and who created it and sustains it for some purpose of god's own has been challenged by philosophers and defended by philosopher/theologians for centuries. It was felt that this must be done because of the revelation of the Scriptures, but also because the make-up of the universe demanded it.

The whole question wraps into one the first six assumptions that we considered earlier, but it all hinges on whether or not there must be a cause for every effect; whether something can come from nothing. If there must be such a cause, then we can argue over whether or not that cause is rational, whether it is a personal god, whether it is a personal god who gets involved in the workings of the creation or just leaves it to go its merry way. This then is the first step in our investigation of the assumptions on which Christian faith is based. The way we answer the question of "cause" will affect our answers to the other presumptions as well.

HOW DOES THE DESIGN ARGUMENT FARE AGAINST MODERN SCIENCE?

Until recently there was no reason to doubt the evidence of our senses and our `common sense'. Things in our experience do not arrange themselves into useful objects; there is no spontaneous generation. If something changes there is always a cause, and in the realm of human endeavors it is always a rational cause. However, according to modern findings, things do change in ways that seem to defy our experience, or even our credulity. Richard Dawkins complains,

> "Nowadays theologians aren't quite so straightforward as Paley. They don't point to complex living mechanisms and say that they are self-evidently designed by a creator, just like a watch. But there is a tendency to point to them and say `It is impossible to believe' that such complexity, or such perfection, could have evolved by natural selection. Whenever I read such a remark, I always feel like writing `Speak for yourself' in the margin.[86]

The first chink in the armor of the design argument was the discovery by Galileo that there were other worlds, moons around Jupiter, which did not fit easily into the neatly ordered pattern of Ptolemy.[87] The discovery that the planet Venus went through phases like the moon showed that it revolved around the sun and not the earth. This was troubling, but not conclusive and since the then-science of astrology held that the heavenly bodies had various subtle influences on human life and affairs, it would have been presumed that there was some purpose for these objects and that that purpose was aimed at humanity.

It was the design in living things that was the real basis for the design argument, as we saw in Paley, and the new sciences of geology and paleontology that brought about some new doubts. If the world was not always as it is now, what was the point of the previous, now-supplanted designs? The Biblical flood could not explain all the changes in the morphology of the earth, and they seemed to have no bearing on human life. The discovery of ancient animals that no longer existed and had lived before human beings posed another question. Why had these species existed when there seemed to be no purpose to them?

None of these arguments and explanations would have been devised in the first place if there were not an overwhelming conviction that creation had been done by a rational designer, but there *was* such a conviction and, more importantly, no other process known which could explain how these beings arose, until Darwin and his likeminded contemporaries. Richard Dawkins writes,

> "Living bodies...seem to have purpose written all over them...The apparent purposefulness of living bodies has dominated the classic Argument from Design invoked by theologians from Aquinas to William Paley to modern `scientific' creationists. The true process that has endowed

> wings and eyes, beaks, nesting instincts and everything else about life with the strong illusion of purposeful design is now well understood. It is Darwinian natural selection."[88]

Dawkins also points out that, "[Paley's] hypothesis was that living watches were literally designed and built by a master watchmaker. Our modern hypothesis is that the job is done in gradual evolutionary stages by natural selection."[89] Clearly, modern scientists feel that the question has been answered about the need for a creator, and in fact the need for a cause of any kind.[90] What are the findings of science that deny the necessity of a designing creator? We have already seen that there is no demand for a *creator ex nihilo*, but just because quarks, or whatever is the basis of matter in the universe, come into existence; there is no demand that they form anything. Nor has science reached the conclusion that in every universe that could come into existence there will be any organization or complexity. Indeed, the theological argument was that there was not even a real possibility of anything but chaos unless there was the imposition of order from outside. True, theologians would have to admit that chance could have produced the watch on the heath, but the probability is vanishingly small. However, given the initial conditions of our universe, a subject to which we will return, we now know that organization can arise from "utter chaos", and in fact has.

THE EVOLUTION OF THE NON-LIVING UNIVERSE

According to the bible as most people have understood it, the universe came into existence through the word of god. This was good enough when there was no other information and one could believe the evidence of one's senses that the earth was the center of all. Then with Galileo Galilei's 1609 telescope, the victory of Copernican or sun-centered astronomy was all but assured.[91] The now visible landscape of the moon and spots on the sun brought into question the vaunted perfection of the heavenly arena. Not only that, the discovery of moons around Jupiter made it even more credible that the earth and its moon were indeed revolving around the sun just as Jupiter's moons obviously revolved around that planet. So forcing Galileo to recant was a losing battle and by the end of the 17th century there was not an important astronomer who would uphold the Ptolemaic system, or the Tychonic (from Tycho Brahe) theory of "ether," though it would take another century before the word filtered down to seminary professors and many academics that the world had changed. The fears for the faith that brought opposition to Galileo were justified fully by the time that Isaac Newton's "*Principia*" showed what moved the planets, and it turned out not to be the motion of celestial spheres moved by the angels, but the very same physical laws that govern movement on earth. The heavens were a physical system which could be explained mathematically and without the constant intervention or oversight of god, even though Newton felt that god had to step in now and then to correct for the observed motion of the planets that did not quite fit his calculations.

The Copernican/Galilean system was the understanding of the universe through the 18th and 19th centuries. We have already encountered the Steady State theory in the discussion of creation from nothing. In this scenario, "the idea was that as the galaxies moved away from each other, new galaxies were continually forming in the gaps in between, from new matter that was being continually created."[92] You would think that the Church would like the Steady State theory with its call for a continuous creation. Most theologians, including Aquinas, and many philosophers had no problem with a universe that was infinite in time. The theory allowed humans to be at center stage and the rest of creation to be our servant, as we had always thought. As one physicist remarked about his understanding of the world in his youth, "I took for granted that the landscape that we passed through on our pilgrimage was static and even rather indifferent. God had created this universe as a stage for the adventure of life to be played on."[93] However, the continuous creation of matter out of nothing was not what the Church had in mind when it talked about god sustaining everything in being. In accordance with the biblical picture, the Church was interested in one moment of creation, if it could not have the biblical six days. The Big Bang fit this picture better than the Steady State hypothesis. Stephen Hawking rather flippantly mentions a Vatican conference at which Pope Pius XII, the head of the church which had attacked Galileo, and which Hawking facetiously feared would attack him if it knew what he was up to, instructed the physicists not to bother inquiring too deeply into the Big Bang, "...because that was the moment of Creation and therefore the work of God."[94]

Fortunately for the Church, and also ironically for those who disliked the Steady State idea because of its quasi-theistic sound, the Steady State explanation was running into trouble. Edwin Hubble had put forth the idea that the galaxies were all receding from one another and therefore the entire universe was expanding, but not in the same way as in the Steady State theory. If it was expanding as Hubble described, then you could trace the expansion back to a beginning. The "...profoundly important consequence of the fact that the recession velocity of a galaxy varies in proportion to its distance...is that the universe had a beginning. Because of the proportionality between distance and velocity, each galaxy has taken exactly the same time (equal to distance divided by velocity) to reach its current position from its point of origin. If the sequence of events is reversed, the galaxies would all meet at a single point at a specific time in the past.'[95] This led to the idea of an initial, enormous explosion, the Big Bang, which gave birth to the whole expanding universe.[96] In the Steady State theory of continuous creation, though not necessarily divine creation, it was barely possible that everything arose from the same point, but certainly not from the same time. With a Big Bang universe, you could even look back in time, by looking farther into space, and see the evolution of galaxies. The discovery in the 60's of quasars, far different from anything in the older universe of today, showed a primeval universe just beginning its evolution. Not what one would expect of a steady universe. There should be old galaxies in the past and young ones now, but there were not. On top of that, in 1964 a microwave radiation with a temperature of 2.7°K was found coming from all points in the sky. The Big Bang theorists had predicted this leftover radiation at exactly that temperature.

So today we have the Big Bang Theory to explain how the universe came into being, and before we can continue with our investigation of how the religious assumptions hold up against the evidence of science, we must understand what it is that this theory holds.[97] Some readers may already agree that the universe could have come into existence out of nothing and needs no designer, creator, or external agent of any kind. If this is your position, you may already be familiar with the following material on the origin of the universe according to the Big Bang theory. If you do agree with this conclusion and are familiar with the science, you may as well skip to page 44.

THE BIG BANG

A picture was recently released[98] taken by the Hubble telescope in space. It is an exquisitely detailed view of one part of the universe, and not a particularly special part. In fact, it was carefully chosen *not* to be special and covers only one 15 millionth of the view from earth, yet it shows thousands of galaxies of every description. Not thousands of stars, but galaxies, each one composed of millions and millions of stars. In fact, there are clusters of galaxies embedded in tremendous balls of gas.[99]

Some of these galaxies are spirals, like our Milky Way. Look at a picture of one of the spirals. We would be just a point in one arm of a spiral in our galaxy, and by "we" is meant not earth alone but our whole solar system.

So we, the earth, would be just a point on a point in one of these arms. The Milky Way is itself just an average galaxy with 200 billion stars of which the Sun is sort of small. The large galaxies contain more than a trillion stars. Where did these spiral galaxies come from? When and how were they formed?[100] We can begin with our own, the Milky Way. The *Companion to the Cosmos explains,*

> "Our home in the cosmos, an island of stars (hundreds of billions of stars broadly similar to our sun), gas and dust held together by gravity to form a disc galaxy some 30 kilo parsecs* across surrounded by a halo of visible globular clusters, and embedded in a much more extensive halo of dark matter, detected only by its gravitational influence."[101]

It is thought that the structures we see in the universe today formed from the amalgamation of smaller packages of gas clouds. Many of the stars that form the galaxy are second generation stars, or even third generation. We know this because they contain components that were not available to the first generation stars, the iron that forms much of our planet for example, and the carbon that forms much of us. These elements were forged in the first generation stars that, on their death, released them to space where they were gathered up again into our own galaxy and its stars. We even see the beginning of the galaxies in quasars. Looking at a quasar we are seeing the universe as it was at the young age of one billion years or so, since quasars are thought to be the cores of galaxies so young they have not formed yet. It was quasars that helped drive the nails in the coffin of the Steady State. These were perhaps the first structures in the universe, the result of gravity gradually pulling matter together to form lumps. Since gravity is the result of a warping or distortion in space, the matter falls into what today are known as wrinkles in the fabric of space-time.[102]

Prior to the formation of quasars, a milestone in the history of the universe, atoms of hydrogen and helium, plus a few other odd atoms had been coming together at a rather leisurely pace for 999,500,000 years, give or take a few, and had been slowly cooling. As they came closer to each other and the density increased, there was in the young galaxies a warming until quasars "lit up" in nuclear fusion, the same process that fuels our sun. This occurred about two or three billion years after the beginning of the universe.[103] Other atoms that were not gathered into the quasars and the stars continued their cooling. This cooling had been going on for billions of years, from shortly

---

* A kiloparsec is a measure of distance equal to 3.2616 light years (a parsec) times 1000.

after the beginning of the universe. Working our way backwards from this time, we arrive at a period when the universe was only 500,000 years old. The materials composing it were not cool enough to allow any gathering of the elementary particles into atoms. "Before then the universe was still too hot for an electron to fall into a quantum orbit around a nucleus, and the cosmos was a boiling sea of loose electrons and nuclei."[104] However, the nuclei and electrons were certainly cooler than they had been several hundred thousand years before the half-billion mark. If today we take the temperature of the space around us, independent of galaxies and their stars, we find that there is a background, called "Cosmic Microwave Background Radiation" (which is a "black body radiation" mentioned previously). This radiation is awfully chilly by our standards, but is far from absolute cold. This is the microwave radiation which fills the universe at a temperature of 2.7° Kelvin. This "heat" has been likened to, "...coals found in a circle of rocks in the woods, a strong clue that something hot was there before..."[105] "The radiation itself dates from...when the universe had cooled to a temperature of about 4000°C..."[106] Today the radiation is separate from the rest of the materials that make up our universe and provides a background which reacts hardly at all with matter. Before the period of 100,000 to 300,000 years after the beginning, matter was so energized and so dense that the radiation could hardly move any distance at all without running into protons, neutrons or electrons. It takes many interactions before a photon of radiation hits just the right circumstances so that it interacts with one of these elementary particles, but whereas in our own time things may be just right every few centuries or millennia, at this time in the history of the universe they might be just right every fraction of a second. It might take billions of near misses before the direct "hit" occurs, but there are billions of near misses every second. These conditions can be duplicated in the lab today and the mixture of particles and radiation is called a plasma. "With the decline in temperature the plasma turned into normal (un-ionized) hydrogen gas, which is transparent, allowing the radiation to propagate through it freely." [107] After this time, the radiation met fewer and fewer particles per volume of space. Interactions still occurred, as they do today, but with much less frequency. This allowed the particles to begin their cooling to a temperature where, in several hundred thousand years they would be cool enough to form atoms, complete with a nucleus and electrons. This plasma state had endured since approximately the three-minute mark in the age of the universe.

Prior to that three-minute mark (give or take some seconds), we find a period of assembly of protons and neutrons into the nuclei of helium.

"Calculations indicate that during the first few minutes about 25 percent of the nuclear material would have formed into nuclei of the element helium, with a little deuterium and lithium, and negligible quantities of anything else. The remaining 75 percent would have been left unprocessed in the form of individual protons, destined to become hydrogen atoms."[108]
In fact, we find today that the universe is made up of approximately 25% helium and 75% hydrogen, which is one indication that the theory is correct. That leaves 0% for iron, silicon, carbon and oxygen, the stuff that composes our earth and us, but then we *are* merely a trace element in the universe. Before one second, there are no nuclei of anything. It is just too hot and energetic for the protons and neutrons to form into nuclei.

LET'S TAKE A MOMENT
A word about the notation that can be used to refer to events prior to this time. We could say "one half of a second" and "one one hundredth of a second," but very soon this will be cumbersome. If we divide a second by 10, then we will have .1 or one tenth, and so on. If we divide the tenth of a second by 10, we will have .01 or one one hundredth of a second. Or we could write this last figure as 1×10^{-2} meaning that we write 1.00 and then move the decimal point two places to the left. Why bother? Because .000,001 - one millionth - can then be written as 1×10^{-6} or even more easily, just 10^{-6}. We will use this shorthand in the following material.[109]

So where are we when the universe is one one hundredth of a second old, or 10^{-2}? Conditions are rather hot, in fact it is 2×10^{11} °C or 200 billion degrees Centigrade [notice that a positive exponent means you add a zero in back of the number instead of in the front]. This is approximately 360,000,000,000° Fahrenheit, or 36×10^{10} °F. All kinds of particles are forming from quarks, the basic components of all matter, and being torn apart again, but as we get closer to 1 second there will be an overwhelming preponderance of protons and neutrons which will then start to form into nuclei, although without the electrons which will make them atoms. This has been going on since about 10^{-6} seconds. Between 10^{-4} and 10^{-6} seconds, the neutrons and protons, the, "...basic constituents of matter in the universe were created as quarks [combined] in groups of three, forming neutrons and protons. Before this, the universe was a seething soup of quarks, too energized and too densely packed to form into nucleons."[110] By 10^{-10} second, the universe was about the size our solar system is today, its temperature was 10^{15} Kelvin, and all the matter and energy now in the universe was there. In fact, that was the only "there" there was since space-time is all the "there" there is and space-time is a quality of the universe. The soup contained quarks, muons, W and

Z particles, neutrinos, photons and whatnot. These identities can be produced today in the particle accelerators that can reach such high-energy states.[111] When we go back farther to 10^{-32} seconds, the universe is the size of a softball and its temperature is 10^{27} K. There is speculation that quantum fluctuations that took place at this time provided the disturbances or "wrinkles" in space-time that would later accumulate matter into galaxies and even larger structures.[112]

All this seems similar to working backwards in a discharge of TNT or some other violent chemical reaction, or maybe to a nuclear explosion, but there are significant differences. An explosion in our universe today is just that, an explosion *in* space. It draws its energy from the degradation of a previously existing entity. The energy that binds atoms to molecules, or in the case of nuclear, the energy that binds the atoms together, is unleashed. The Big Bang is not like this. First, it is an expansion *of* space and time, or rather space-time. It is not an explosion into a previously existing space surrounding it, but an expansion into truly nothingness. Not the nothingness of a place that happens to contain no "thing" but into a nothing of no "thing," no "place", no "time", no "space", no "nothing", "...a void, so absolute as to mock any human concept of emptiness..."[113] This is why the background radiation is all around us. The Big Bang did not happen "over there somewhere," it, "...happened everywhere because it was everything...hence the radiation is everywhere and goes in all directions, and will continue to do so as long as the universe exists."[114] This is also why the entire universe is rushing away from our local neighborhood of galaxies, and of course we from them. It is not that we are all flying apart from a central point, and so the lines from each to the center are diverging, like streams of fire from a skyrocket. We *are* the central point, *and so are they.* We were all there at the Big Bang and each point of the universe acts as though everything else is receding from it. Second, the Big Bang is not the destruction of anything into its constituent energy, quite the opposite it is the *con*struction of matter and radiation from what preceded it.

There is another way to trace out the evolution of the early universe, and that is according to the separation of the four fundamental forces; gravity, the weak force, the strong force and the electromagnetic force, all of which were presumably originally united in one force, but separated as the universe cooled. This whole scene is going to reach the same conclusion as the argument I am making in another manner and so it is not treated here.[115]

At 10^{-35} of a second, the Standard Theory calls for something called "inflation".[116] The idea of inflation arose as an answer to some questions about the way things are today. One question is why the extremely small, and still "virtual" universe did not just collapse under its own gravity back into nothingness. Another is what is called "isotropy" which means that the universe is the same everywhere, give or take a little isolated clumpiness like a group of galaxies. However, the clumps are themselves a problem. The universe, "...is isotropic, it appears the same to an observer in any direction...one would have expected zones of very disparate density coming from an initially chaotic `soup' of particles...Yet the distribution of matter in the universe is not nearly as uniform as the background radiation. How did galaxies...emerge from primordial homogeneity?"[117] In other words, scientists evidently would expect the effects of "chaos" to result in a universe that does not look the same in every direction. Since it does, they worked out the mathematics that would explain why this is so. Then, having worked through that problem, they have difficulty explaining why it is not smoother than it is. The suggestion of "inflation" is designed to explain both of these qualities of the cosmos.[118]

At 10^{-35} second, the universe was about 10^{-50} centimeter, which of course is exceedingly small. In one 10^{-30} of a second, it expanded by a factor of 10^{50} or more, to about one centimeter, which doesn't sound very impressive, but if it had been a centimeter at the beginning and expanded by the same factor, it would be about 620 trillion, trillion, trillion (10 more times) miles wide when it slowed down. This solves the isotropy problem because the universe *begins* small enough to be crossed by a photon, so all parts are in balance with each other, and everything is smooth. However, the random fluctuations that occur *during* the expansion period, short as it is, are inflated as well and become the "wrinkles" that will develop into the structures of the universe we see today. The inflation theory, according to Hawking,

> "...explains the paradox of an extraordinarily uniform early universe...that somehow evolved into the evident lumpiness of the present universe...[and] it explains why the universe is expanding...The universe is, according to inflation theory, much, much larger than anyone ever suspected; hence our tiny corner of the cosmos is even tinier than we thought."[119]

As we saw when considering whether something can come from nothing, it is possible for virtual particles to spring into (virtual) existence out of the quantum fluctuation of the false vacuum. "...The inflationary hypothesis, and the Big Bang theory, presumes that the universe comes into existence out of the vacuum state, or out of a completely unvisualizable `point nothing' capable of generating enormous energy."[120] All the forces of electromagnetism, the strong and weak forces that today bind the atoms, and the force of gravity are one force, somehow. In fact, there is nothing *but* energy. The radius of the universe today is on the order of 3×10^{27} centimeters. At the point we are considering, all of what is now both matter and energy, the entire composition of the universe, was contained in a ball with the radius of 1 centimeter. How did it all get there?

NOT YOUR 8-POUND ORECK

A false vacuum, which we have seen before, is an excited true vacuum. Paul Davies explains,

> "...The energy of the false-vacuum state is stupendous: typically a cubic centimeter of space would contain 10^{87} joules. Compare this with the paltry 10^{-18} joules or so that an excited atom possesses...given the extreme conditions of the Big Bang, these figures are plausible. The huge energy...has a powerful gravitational effect...The energy of a cubic centimeter of false vacuum would weigh 10^{67} tons...the huge false-vacuum energy is associated with an equally huge false-vacuum pressure...The crucial property...is that the pressure is *negative*. The false vacuum doesn't push; it sucks. A negative pressure produces a negative gravitational effect - which is to say, it *antigravitates*...The net effect is to create a repulsive force so large that it can blow the universe apart in a split second. It is this gargantuan inflationary push that causes the universe to double in size as rapidly as every 10^{-34} second."[121]

Of course, a false vacuum normally gives rise only to virtual energy or particles. It is theoretically possible in laboratory conditions to give a virtual electron, for example, enough energy from outside the system that it becomes a normal electron with a normal life. Without this, it "...typically survives for only about 10^{-21} seconds...[and] may travel a distance of 10^{-11} centimeters."[122] However, there is plenty of energy around *inside* the system, since the false vacuum, "...is alleged to have contained within itself tremendous energies...every cubic centimeter of volume in this false vacuum would have to contain 10^{95} ergs of energy..."[123] This is enough energy to supply the immense number of virtual particles.

Then according to the most widely held theory, inflation occurs, which has been likened to the common process of making an ice cube when, "...something analogous to a sudden crystallization of ultra-cool water to ice occurred. Heat was liberated...producing a momentary temperature rise."[124] The heat is the source of energy that the virtual particles need to become real particles. "A remnant of these primordial particles went on to provide the 10^{50} tons of matter that makes up...the observable universe."[125] There is no imbalance, no violation of the second law of thermodynamics such as the creationists charge, since the energy that went into producing the universe from a virtual universe is exactly balanced by the gravitational energy still contained within it, but it is a debt waiting to be repaid. "The gravitational energy of the universe exactly cancels its nongravitational energy - the two energies having opposite signs. The net energy content of this kind of universe could be exactly zero."[126] "The positive energy locked up in mass is canceled by the negative energy of the gravitational field."[127] If it was repaid, by all the matter collapsing back on itself, for example, the net energy would again be obviously zero. The false vacuum "wanted" to decay into a true vacuum, as most false vacuums actually do, but it took on a debt that manifests itself as our universe, and as ourselves in fact. It is thought that the concept of quantum gravity will be the key to understanding this mechanism.

SUPERSTRINGS

Stephen Hawking points out that, "we do not yet know the exact form the correct theory of quantum gravity will take. The best candidate we have at the moment is the theory of superstrings."[128] The cosmological theory called "superstrings" even tries to go back beyond the traditional Big Bang.

> "According to the superstring theory, the universe began in ten dimensions. However, perhaps this ten-dimensional universe was in a false vacuum and therefore unstable. If the ten-dimensional universe did not have the lowest energy, then it would be only a mater of time before it made the 'quantum leap' to a lower-energy state. We now believe that the original expansion of the universe had its origin in a much greater, much more explosive process: the breakdown of the ten-dimensional fabric of space-time. Like a dam bursting, the ten-dimensional fabric of space-time ruptured violently and rapidly re-formed into two separate universes of lower energy: a four-dimensional universe (our own) and a six-dimensional one. The violence of this explosion could easily generate enough energy to drive the inflation process forward. The standard Big Bang expansion

> would emerge only later, as the inflation process slowed down and made the transition to a traditional expanding universe."[129]

The theory of superstrings was developed to try to resolve the conflict between the general theory of relativity which works well for large objects, and the quantum theory which works well for small. However, they are not easily reconciled and in fact yield conflicting results when applied to the same object, the first moments of the universe being the most extreme example. Although this superstring theory has its problems, it does show the continuing effort to understand and the possibility of doing so.[130]

SAY WHAT?

The reader may be thoroughly confused, as is the author, and overwhelmed by all those exponential numbers and the concepts discussed. What does it all mean? For our purposes, it is enough to say that it is rather obvious that nowhere along the line from the present back in time to inflation and beyond that to the Big Bang itself is there a point where the physicists say, "then a miracle occurred". Hawking claims that when quantum gravity, or some other explanation, is developed, "the laws of science would determine the universe completely; one would not have to appeal to some agency external to the universe to determine how it began."[131] There are many examples of what are frequently called *self-organized emergent systems*,

> "...in which structures arise without the imposition of special requirements from the outside. In an astonishing variety of contexts, apparently complex structures or behaviors emerge from systems characterized by very simple rules. These systems are said to be self-organized and their properties are said to be emergent. The grandest example is the universe itself, the full complexity of which emerges from simple rules plus the operation of chance."[132]

The inflationary hypothesis is by no means proven[133] and may be replaced in the future by a better hypothesis, but it does show that there is confidence that the emergence of the universe and its subsequent evolution into what we see today can be explained with no recourse to anything outside itself.

One can read all kinds of objections in the Creationists' literature claiming that all of this Big Bang and quantum vacuum business is not proved and is in fact just a made-up explanation of the origin of the universe designed to deny god's rightful place as creator.[134] Their argument strikes a chord in the hearts of all of us who are not so familiar with modern physics, especially since so much of cosmology today and of quantum physics in particular is counter-intuitive. Can you really have something come into existence out of a vacuum and claim that it came from nothingness? Surely there is some causality involved in this episode or at least some pre-existing "something" from which it comes! Suffice it to say that this is not the case as physicists understand it.[135] This present work is not a science book and I will leave the proof up to the experts in the field. I am not trying to explain why the scientific community takes the position it does. I am trying to take what they say at face value and examine what happens to the faith if they are correct, and scientists are indeed saying that these things occur even though they may make no sense in our mind-set developed from exposure to the world as we experience it. Quantum effects are evident in the laboratory, in the operation and design of computers, and in the everyday life of many researchers. Scientists would be the first to say that they may not have the final answers, but if they cannot explain everything with the metaphysical sureness we would like to have, they can explain things well enough to say that what they are describing are real phenomena which put the lie to our more familiar (Newtonian) ideas of causality and reality. Our age may be analogous to the time of Galileo and the period following. He did not have all the answers and some he thought he had were false, but he was close enough that his understanding of reality presented a challenge to the old assumptions on which the faith of the time had been built. It was futile then to ignore the general direction of his findings and it is futile now to do so with the findings of our age.

NOW, WHERE WERE WE?

Let us finish our picture of the events after the Big Bang. If the inflation occurred at 10^{-35} second, what happened before that? At 10^{-43} second we reach what is known as the Planck moment. The universe's temperature is 10^{32}°K and its density is 10^{96} times that of water. "All matter is dissociated into its most primitive constituents, which may be simply a sea of identical strings existing in a ten-dimensional space-time...even the distinction between spacetime and matter becomes nebulous."[136] It is presumed that gravity has just broken its bond with the single unified force that existed at the instant of the Big Bang. Nobody knows for sure because there is no quantum theory of gravity. "The actual instant of the beginning envisions...a moment when an infinitely small point of space was packed with matter squeezed to an infinitely high density. This condition of infinities is referred to as a singularity."[137] This is where cosmologists like Hawking wax poetic, saying for example,

> "Into a void, so absolute as to mock any human concept of emptiness, appeared a single point of raw potential. And at the very instant of its creation, this point, bearing all matter, all dimension, all energy, and all time, burst out, spewing forth its contents..."[138]

Where did it all come from in the first place? From the hand of god or from a hiccup in the quantum vacuum? As we saw in the treatment of "ex nihilo", there is no reason to doubt that it could have come from nothingness without any help from "outside". Once it appeared, it needed no help in evolving into what we see today. How often does this kind of thing go on? Was this a one-time occurrence or does it go on all the time? Of course we can't know since if it did occur again, the universe spawned would be entirely outside our own and we would never know of its existence. Unless, of course, there could be a "wormhole" to another universe, akin to the wormhole of "Deepspace Nine", although that one only goes to another part of our own universe. It seems that this singularity did happen at least once, for we are here to wonder at it.

Once the universe had reached an inflated state and had cooled enough for the formation of atoms and molecules, another era in the evolution into the cosmos with which we are familiar had begun, the evolution of stars and galaxies of stars with all their attendant clouds of matter and "dark matter", whatever that is. Whatever it is, dark matter makes up more than ninety percent of the universe so that what we can see and measure is but a small part of reality. How this development of galaxies occurred need not concern us here. It is enough to point out that no one can realistically deny that the process can be explained to a great deal of precision by the classical physics of gravity, centripetal force, orbits and the like. Even Ptolemy, Copernicus, and Galileo had reasonable explanations that did not require intervention from some outside, divine force. Indeed, many of the ancients presumed that some divinity had brought it all into existence and then pretty much left it alone to follow the laws that had been laid down. Today we would call the primordial state the "initial conditions" but would consider them the results of random quantum effects and, ultimately, chance.

The evolution of the universe from a singularity to the present flies in the face of the assumption that everything has a cause, prime or final. It seems that not only something, but everything that exists in our universe, came out of nothing and for "no reason". No prime mover had to be a "ground of being", and there did not have to be a designer of any kind. The findings of science today give a plausible explanation for the appearance of the universe from nothingness.[139] Once such a universe appears, it need not evolve in the manner in which ours has gone. The parameters that allow for the development of matter, and then the appearance of quasars and the subsequent galaxies are very narrow. The density of the universe is very close to the critical density that allows the universe to last long enough for anything to develop. The homogeneity allows for a universe that is not so chaotic that nothing can evolve, but not so smooth that everything is exactly the same. There are seeds of clumpiness that allow for the formation of local clusters of matter into galaxies. It is amazing to some that the universe is here at all, and now that it is, that it developed into anything. The force that holds the nucleus of the atom together need not be the value that it is. If it were either stronger or weaker, atoms either would not form at all, or not for a long enough period for the atom to interact with other atoms to form structures, or it could be so tightly bound that it was inert. There are a number of other critical factors, all of which must be just right or our universe would not function as it does. For some, this means that there must be someone, or at least something, which sets the initial conditions which guarantee that something does happen. For theists this is god of some sort and has managed to make believers out of skeptics and agnostics.[140]

THE ANTHROPIC PRINCIPLE

Those who support what is known as the "anthropic principle" maintain that somehow "...the universe contained the seeds of the conditions required for the appearance of an observer. It tended towards self-consciousness through the creation of intelligence."[141] This need not be imposed on the universe through some outside creative agent. Some theorize there is an almost "backwards causality" or "final cause" brought about by the seeming necessity of an observer for quantum phenomena to be actualized. Indeed, there is a "strong" anthropic principle theory that maintains that the existence of intelligent life is a condition for the actual existence of the universe and that therefore intelligence must develop if the universe is to exist. What "actually existed" before (human?) consciousness is undetermined. The "weak" anthropic theory merely states that our universe must tend towards intelligence since here we are, intelligent life, or in another version, the universe must at least be open to the possibility of intelligent life since it has developed it. It claims that we can make inferences about the nature of the universe from the fact that it has produced intelligent life. There is even a "moderate anthropic principle" espoused by John Polkinghorne that calls for a "fine tuning" of the universe to allow for the development of life. The judgment of science at this time seems to be contained in this analysis of the theory by Daniel Dennett,

> "Believers in any of the proposed strong versions of the anthropic principle think they can deduce something wonderful and surprising from the fact that we conscious observers are here. ...That we exist *so that* the universe as a whole can exist, or even that God created the universe the way he did so that we would be possible. Construed in this way, these proposals are attempts to restore Paley's argument from design..."[142]

an argument that is rejected by all but fundamentalists and creationists. The theory of Inflation has removed the "necessity" for this argument by showing the possibility that any number of initial conditions may be "smoothed out" during an inflationary process and would result in the conditions we see now.

On the other hand, the "many worlds" hypothesis maintains that universes are popping into virtual existence all the time, and some into actual existence.[143] Some have the conditions necessary to sustain themselves but most do not. Those that do may continue to exist as ours does, but perhaps do not have the critical conditions to form any structures such as stars and therefore never amount to anything. Or they may form stars or something like them but never develop beyond this stage and therefore nothing intelligent ever develops to wonder about its existence. Even a universe just like ours may well not develop intelligence. Our universe did just fine without us. This would be another form of the weak anthropic principle that would merely point out that our universe must have certain characteristics or else we would not be around to notice them. This, of course, tells us nothing we did not already know and so the theory is worthless. The "many worlds" situation points out again that there is no need for a cause or a design. If some chance quantum process makes it possible for an almost infinite number of universes to pop into existence, one at least will be ours. If an infinite number of universes exist, then an infinite number of universes will have stars and other characteristics of our universe, including consciousness. Big deal.

Back to our expanding universe. Of course, the formation of galaxies is not the end of the question, for complex as the universe is, there are beings in it that are much more intricate and hard to explain than stars, namely every living thing. That atoms and molecules could sort themselves out into stars within galaxies is amazing, but rather straightforward once you know the forces at work. Can the same be said for living beings? Or must there be an intervention, ála Genesis, to explain the presence of life? Or of intelligent life? Must there be some extraneous cause at work?

THE EVOLUTION OF LIFE

If you skipped the material on the evolution of the universe and landed here, there is another decision to make. If you also agree that life, including intelligent human life, can be explained without benefit of a creator/designer, then you may as well go on to page 92. There are many, however, who can agree that the physical universe can be explained without reference to god, but feel that the emergence of life demands an intelligent or at least living begetter. Others will admit that life needs no such explanation but balk at including human intelligence (or any other intelligence, for that matter) in the range of things that can be scientifically explained without the intervention of an intelligent agent. This section of the present chapter will explore these questions.

We have been considering whether the assumption that everything has a cause is valid. The first question was whether the appearance of the universe was grounds for declaring the assumption valid and the conclusion is that it is not since the universe *can* and in the opinion of most people of science today, *did* come about "*ex nihilo.*" No cause, at least in the sense of an external agent or a cause in the normal sense of the term, is needed. It is true that coming into existence and then remaining in existence can be two different things. The "many worlds" hypothesis posits many different universes coming into existence with vastly different or slightly different initial conditions than exist in the universe we know. Some will wink out of existence as fast as they enter, some will persist for longer periods, and some will last a long time but are not productive of anything like we experience. With the theory of inflation, we saw how the initial unstable conditions can turn into very long-term (unstable) conditions, the situation in which we find the universe, and ourselves, today. On the other hand, Thomistic theologians have long separated the creative action of god from the sustaining action but insisted that both must be continuously applied from outside the universe since the universe has no "power to exist" of its own.[144] So, once in existence on its own terms, not the philosophers' terms, was any guidance necessary for the universe to take on shapes and structures such as we see now rather than remain in a state of essentially formless chaos? Evidently not. Explanations based on physical principles alone are available and the reader can refer to the story of the development of stars and galaxies, clusters and superclusters.[145] However, we are living proof that not only galaxies and the stars that compose them developed, but from them also arose structures far more complex, the living beings that inhabit our planet and may inhabit other planets as well.[146] Does the refutation of Paley's arguments about the need for a watchmaker apply to the complexity of a living being? I have stated that it does not, but then what *does* explain the evolution of living beings and the complexity that comprises them? We need to understand the theory of evolution before we can make any judgments.

WHENCE LIFE?

Before Charles Darwin published his *Origin of Species* in 1859, "...organisms were deemed to be as timeless as the perfect triangles and circles of Euclidean geometry. Their individual members came and went, but the species itself remained unchanged and unchangeable." [147] The whole idea of a "substance" held by Aristotle and Aquinas meant that there was a set classification of a kind of being and that this never changed, and certainly would never change into some other substance. An individual object might be changed, such as wood changing from a tree into a table, but a certain kind of tree never changed into another kind of tree, the substance "oak tree" to the substance "palm tree" for example, or one animal species into another. Much work had been done on classification of all the different kinds of animals and plants, but there was no thought of change of one into another, although Carl von Linnaeus hinted that species may change in his *Systema naturae* of 1737. At the beginning of the nineteenth century, Jean Baptiste de Lamarck "...proposed that species may take on new forms in response to their needs"[148] but this was done through acquired characteristics, an animal, for example the giraffe, developing a new mode of action such as stretching to reach the higher leaves, and then passing that characteristic on to its progeny. The whole species changed in how its essence was expressed in the individual, not into another species. In Thomistic terms it was "accidental", not "essential." Lamarck's theory was however a monumental break in the dominant ideas, and came as a result of the classification efforts that preceded it. There were so many species that were very similar to one another that there had to be a connection. Not only that, fossils had begun to reveal the same relationships. How had they come about? Lamarck gave an explanation in his theory of the inheritance of acquired characteristics that was very powerful. Darwin explained the same facts and observations, but did so through another mechanism entirely, that of natural selection.

> "All the theory of natural selection says is...if within a species there is variation among individuals in their hereditary traits, and some traits are more conducive to survival and/or reproduction

> than others, then those traits will (obviously) become more widespread within the population. The result (obviously) is that the species' aggregate pool of hereditary traits changes. And there you have it."[149]

Darwin did not know the correct mechanism that passes on those variations, but it was not long before Sir Ronald Fisher's work, *The Genetical Theory of Natural Selection* united the genetics of Gregor Mendel and Charles Darwin's natural selection.[150] Parents pass on to their offspring various genetic characteristics that determine what the individual will be like. The environment then works on the variations between individual descendants. If one is better suited to survive and to pass on its characteristics to its progeny, then that gene combination will often be preferred over the other and will come to be dominant. In addition to the variety brought about by the combinations from the contributions of two parents, the parent gene may be changed from the one operating in the parent itself.[151] Background cosmic radiation, for example, "...induces mutations in biological organisms and that helps drive evolution."[152] Something called "DNA Microsatellites" are, "repetitive sequences made up of various combinations of the four DNA bases...repeated over and over, like a genetic stutter." Considered "junk" at one time, it now appears that they may be, "prone to grow or shrink in length and that these changes can have both good and bad consequences for the organisms that possess them."[153] In our own day, chemicals in the environment often cause genetic mutations. Most often this will be a disadvantage to the descendant, since of course the original gene was at least good enough to allow the parents to pass on their genes and now the offspring does not have that gene but another. And statistically, and certainly according to Murphy's Law, if something can go wrong and be a disaster, it probably will. However, sometimes the change is for the better and the new genetic inheritance is an advantage to its recipient. Or the change is at least not fatal to its bearer but is just meaningless and gives no advantage or disadvantage. However, if the environment changes, the mutation may turn out to be either advantageous or disadvantageous. If the giraffe is a leaf eater and a parent passes on to its offspring a gene for a longer neck, but forage is abundant, the changed animal is no more or less likely to survive and pass on its longer neck to its young. If there is a drought, however, and leaves are more scarce, the one able to browse higher up the trees will be more likely to survive long enough to reproduce. (That this process explains the long neck of the giraffe is disputed, but the principle is sound.) The animals become over time more suited to their environment. It is equally likely, of course, that the change could have been for a shorter neck. If the environment remained the same, no problem, but if a scarcity arose and browsing higher would have saved them, we would find only fossils, not descendants. You throw the dice and you take your chances. "Darwin was offering a skeptical world what we might call a get-rich-slow scheme, a scheme for creating Design out of Chaos without the aid of Mind."[154] If the environment remained static through enough generations, or if the changes were helpful if the environment changed, the creatures might become a completely new species with the accumulation of enough other changes. The originals may, of course, continue to exist in some other area where conditions were favorable to them, or they might become extinct, replaced by their own descendants. It is also possible that a genetic mutation may occur allowing the eventual new species to inhabit the same macroenvironment as the original species but having different behavior that effectively separates them. For example, a species of frogs may eat only mosquito larvae but through a mutation leading to individuals which eat only nematodes they give rise to another species if the prey live in different microenvironments of the same stream and the frogs rarely meet to mate. This is called "sympatic" speciation in contrast to "allopatic", the first meaning they live in the same place and the latter meaning there has been a geographic separation.

An investigator in the mold of Aquinas might note the differences and conclude that they were merely "accidental" in the sense of being only on the surface. If they were extensive enough, the underlying reality would be deemed to be another (Thomistic) substance altogether and the animal or plant would be worthy of a new name. However, this philosophical system is no longer very helpful.

All that is needed to explain the vast range of beings on our world is some mechanism of change, which Mendelian genetics provided, plus a vast amount of time, which geology provided, plus a means for preferring some changes over others, which natural selection provided.

> "Some people find it hard to believe that a heartless, brainless, spineless bacterium floating around in the primordial ooze could have evolved into a multibillion-celled animal that can agonize over lost loves, debate the nature-nurture question, and exceed a score of 10,000 in Pac-Man. This is a classic case of misplaced incredulity;...our evolution (is) not really that amazing...great evolutionary strides are all but inevitable. That is what makes natural selection one of the most appealing theories ever: great complexity follows from a few simple assumptions."[155]

Darwin himself speculated that in "some warm little pond, with all sorts of ammonia and phosphoric salts, light, heat, electricity, etc." things happened over immense time periods to produce life. J.B.S. Haldane speculated that the oceans were a primordial soup which gave the right mix.[156] Darwin's "warm pond" may actually have been a volcanic vent at the bottom of the sea or the interstices of granite rock under tremendous pressure and high temperature. We may not be that far away from generating life in the laboratory, though without new findings in paleontology we may never know if this or that possible scenario is in fact the one that occurred, or indeed if only one possible avenue was followed. No matter how life began, it was certainly very primitive.

After the emergence of life, or concurrent with it, there is the living cell. The earliest cells were perhaps "...a kind of double-layered soap bubble made of fatty lipid molecules enclosing a network of molecules that had the power to sustain itself and reproduce."[157] Beyond cells is the congregation of cells into organisms and organs of more complex creatures. How this or something like it happened is also open to question, though many explanations have been put forward.[158] There are explanations of evolution readily available that can be consulted, but they all come down to a few basic principles. First, highly complex arrangements of cells into organs occur through very small steps over immense periods of time.[159] Indeed, the very primitiveness of the first life speaks to this. Certainly life as we think of it now could not have arisen in such complexity in one moment from a primordial soup or any other vector of genesis. However, something much more primitive and hence much more robust could, and evidently did. Second, the less efficient are removed from the scene. We will consider each of these in turn.

THE UTILITY OF THE EYE

There has always been an objection to evolution that the sheer complexity of an organ such as the eye removes any possibility that it just spontaneously arose. The same can be argued for the DNA sequence itself; for the emergence of plants on land; and for the subsequent invasion of the land by air-breathing animals. We will consider only one, the eye, since it is popular in the literature and gives us the flavor of the evolutionists' argument.[160] The reasoning that forms the explanation is based on the seemingly obvious truth that any little improvement is better than nothing at all. Certainly it would be advantageous for an organism to develop a fully functional, focusing, color sensitive eye such as we have. The chances of this happening in one fortuitous accident are well nigh zero. Yet, it is not so difficult to see that just a little ability to sense light would be good, too. If you are a being made up of only one cell, like a paramecium, you can travel but to where are you going to travel? Certainly you would want to go where there is a higher concentration of nutrients and/or a lower probability of being the nutrient for something else. Once you can distinguish light, if you live in the sea, you can go toward that light or away from it, depending on which is better for you. And so the creature that benefits from a lucky mutation that has just a little sensitivity to light has an edge over the brothers and sisters who do not. Of course pure chance might bring about your sudden demise before you can clone yourself or mate, whichever is your preference, and then we would have to wait for another fortunate mutation in someone else, but the chances are at least better that you will survive to pass on your advantage to your descendants. Over time those with this new sensitivity to light, however small, are going to be more apt to survive than those which lack it. So we have a new kind of critter happily swimming and procreating in the sea. Now if a little sensitivity to light is good, more is better. More may give you more time to escape or to be the first to get to the dinner table. So if you are a multicellular animal with light sensitivity in one cell, it is even better if you have it in two cells. Moreover, if sensitivity is good, the ability to distinguish light from the sky and light reflected from another object, especially a hungry object, is even better than that. Think what 100 cells could do, especially if they all face in a slightly different direction, and so an object moving across your new field of vision can be sensed as in motion. And what if you could form an image of that other object so that you could avoid it or mate with it as the situation arose. Now maybe it would take a billion years or so to develop these capabilities, but so what? There is plenty of time.[161] No step is more than just a tiny little adjustment to what already exists. No eye springs full-blown into existence, but the result is the same. And the, "...geological time is awfully long...the number of generations that separate us from our earliest ancestors is certainly measured in the thousands of millions. Given, say, a hundred million [generations with a beneficial mutation] we should be able to construct a plausible series of tiny gradations linking a human eye to just about anything!"[162] Indeed, the eye has evolved not once but more than forty times. The eye of the octopus is radically different from that of other creatures. The eyes of insects and the eyes of mammals perform similar functions, but evolved separately.[163] "There has been enough time for [the eye] to evolve from scratch fifteen hundred times in succession within any lineage...The time needed for the evolution of the eye...turns out to be too short for geologists to measure!"[164] Some object that there are many gaps in the evolutionary history of most species, enough to make a whole series of special creations a possibility. There does not seem to have been enough time for a species living today, they say, to have evolved from a fossil ancestor which is quite different in many aspects. Yet time and again an intermediate fossil is discovered showing that the change did indeed occur and often in a surprisingly short time, geologically speaking. It often does not take much mutation to make profound changes.[165] Some *lobelia* flowers have a second set of petals that are sepals in other *lobelias*. It turns out that this, "...sepals-to-petals was precipitated by the mutation of a single gene...What this says from a biological and evolutionary standpoint...is that major changes in plant shape can happen quickly...The evolution of the very first flowers, too, may have been a relatively simple event, genetically speaking."[166]

The same pattern of evolution is true of all the other questions. If it is helpful for a plant to grow on the beach away from the competition of other plants in the sea, even if you need a more rigid framework to support your leaves, then it is also helpful to stiffen it just a little more and live a little higher up the shore. You can even lose something you had already developed. Jared Diamond explains,

> "The long list of ancestral traits that were lost or reduced in the course of human evolution includes tails, body hair, wisdom teeth, the ability to synthesize vitamin C, the size of our teeth and appendix, the thickness of our skulls, and the bony browridges over our eyes. Likewise, snakes lost their legs, whales lost most of their sense of smell, and the dodo and many other birds on remote, predator-free islands lost their power of flight."[167]

When our diet no longer required the advantage of the appendix, it wouldn't hurt if you were the victim of a mutation that reduced it in size (though there is some evidence that it still fulfills a useful role in the immune system). Such a mutation may well have been fatal in a remote ancestor who needed this organ, but it made little difference to you.

ONWARD AND UPWARD?

This brings in the role of progress in evolution. We like to think that we are the high point of evolution and that there has been a steady progress towards us, almost as though it were planned that way. In our more modest moments we may admit that certainly beings who descend from us may be quite different, even a different species, and also much more advanced. Then they will feel like we were just signposts on the way to them and they were the ones intended all along. Actually, no one is intended. There is no plan to evolution and no "progress". "Darwinian evolution is not a theory of progress, but only a mechanism for building better adaptation to changing local environments..."[168] Every organism that manages to survive its environment and pass on its genes is a success. As Annie Dillard puts it, more poetically and theologically,

> "Look...at practically anything - the coot's feet, the mantis' face, a banana, the human ear - and see that not only did the creator create everything, but that he is apt to create anything. He'll stop at nothing. There is no one standing over evolution with a blue pencil to say, `now that one, there, is absolutely ridiculous, and I won't have it.' If the creature makes it, it gets a `stet'...Utility to the creature is evolution's only aesthetic consideration."[169]

WE MUTATED. NOW WHAT?

There is another leg that Darwinism stands on in addition to the transmission of mutations. If change in genes were the only mechanism of evolution, there would be no species at all since every slight difference would be preserved except those that were lethal, though most mutations are lethal or meaningless. When a beneficial change does occur, what makes the difference in who ultimately survives is natural selection. A current and good illustration of the process of natural selection is somewhat "unnatural". Our drugs are producing resistant strains of bacteria at an unprecedented rate, selecting out those mutants that can survive the drug.[170] Those that cannot are eliminated, but often some survive to carry on the struggle against their human host.

Natural selection has two sides. The first is that the organism with the beneficial mutation has a better chance of both surviving and of reproducing. The modification must yield both advantages or it is not beneficial. If the individual lives to a ripe old age because it is better in dealing with its environment, but cannot reproduce, the mutation will not be transmitted. If the individual could reproduce in large numbers but does not survive long enough to do so, again the variation will not affect any descendants. The giraffe whose neck is a little longer than the "normal" giraffe will be able to forage higher. The frog with a little extra skin between its toes will jump from a tree for safety just like all the others, but a miss will mean a leisurely journey to the ground rather than a long fall and flat frog. Both the giraffe and the frog had hit upon a better way of doing business and unless the females found their appearance offensive, they had a better chance to live long enough to pass on that advantage. That is the positive side of natural selection. There is also a negative side if the environment is not rich enough to support a large number of organisms in the animal's particular niche. If the amount of leaves in reach of all giraffes is sufficient, both the long necked and the shorties can make a living, but if not, something has to give. In the long run it will be those who cannot reach food in hard times. If there are no predators who like the taste of frog, then you can have both the gliders and the hoppers. If there are frog eaters about, then it will be those who can escape more often that will live to woo the ladies or wink at the guys. And this is the dark side of natural selection for ours is always a limited world and there are often not enough resources to go around but there are plenty of predators. Something not only has to give, something often has to disappear. As Stephen Jay Gould notably phrases it, "Selection carves adaptation by eliminating masses of the less fit - imposing hecatombs of death as preconditions for limited increments of change."[171] Not only individuals but entire species will often disappear from history unless special circumstances apply, which they seldom do. If there is a mutation that spreads in a species in that some animals have it and others do not, and if a local population is somehow isolated in a different environment, on an island for example, it is possible that both the original species and the daughter species will exist contemporaneously, but not usually in the same niche in the same locality. Still, other than these special circumstances the normal course of events is that the earlier species will be replaced by the later. And replaced is just a polite word for out-competed, vanquished and eventually extinct, which is itself a polite word for dead. As Annie Dillard says, "Evolution loves death more than it loves you or me...We value the individual supremely, and nature values him not a whit."[172] In fact, many if not most species have disappeared leaving no descendants at all, and are brought to extinction *not* by competition with close relatives but through some natural disaster that just wipes out whole masses of species.[173] All this from a process that some would hold has a noble Designer.

Darwinian natural selection then presents some real problems to those who might want to find some director of the process. One can talk about direction in the process of evolution, but in the most minimalist of terms. Daniel Dennett writes the following sentence, "So Paley was right in saying not just that Design was a wonderful thing to explain, but also the Design took Intelligence." It sounds as though he has gone over to the camp of the creationists, but Dennett follows this statement with another,

> "...this Intelligence could be broken into bits so tiny and stupid that they didn't count as intelligence at all, and then distributed

> through space and time in a gigantic, connected network of algorithmic process."[174]

Not the kind of Designer that the Creationists have in mind.

There can indeed be a "direction" to evolution forced upon it by its own history. The eyes of insects have the nerve behind the light receptor, a nice sensible place. The mammalian eye has the nerve in front of the receptor, which is certainly poor design but given the history of the evolution of that arrangement, that was where it ended up. Primates might be able to do well with three arms instead of two, or even with two arms but four feet, but since our ancestors had only four limbs, the fork in the road had been taken and we had little choice but to follow it. There can also be direction to evolution because of "forced moves" imposed by the laws of physics. Certainly, a 500-pound flying predator would be the ruler of all it surveys, but in the real world, it could never get off the ground. The bones of birds are hollow and the animals relatively lightweight because they have to be. So there is room to speak of direction in evolution; even the word "purpose" has its place and is especially helpful when used in a question such as "what is the purpose of this structure in this animal." However, nowhere do we reach a point where we must call on the intervention of an intelligent designer or the mediation of an agent with a vision or purpose. There is no grand scheme of things.

ALGORITHM?

A word about an "algorithm" since it came up in the quotation from Dennett. Webster defines it as a "step-by-step procedure for solving a problem or accomplishing some end..." Evolution is certainly a step-by-step process, the steps *as if* written out for any idiot to follow with *some* product guaranteed. The italics are necessary in speaking of evolution because "algorithm" can carry with it the notion of intelligent design leading to some preconceived result, as in "but who wrote the algorithm?" This is certainly not the process involved. In any algorithm "...the underlying process always consists of nothing but a set of individually mindless steps succeeding each other without the help of any intelligent supervision; they are 'automatic' by definition: the workings of an automaton. They feed on each other, or on blind chance - coin flips, if you like - and on nothing else."[175] Of course some algorithms *are* intelligently designed to work on blind chance to produce some desired and predictable result such as the winner of a coin-toss contest[176] while other algorithms are designed to do away with blind chance; long division for example. There is no indication that the process of mutation and natural selection is written down by anyone or that anything other than *some* product will be produced, but what that product is is beyond prediction of any kind.[177]

HALF WAY MISSING LINKS?

There is also the objection similar to that about the eye, raised by those of Darwin's time and by creationists in our own, that we never see the intermediate steps or the "missing links" between the purported ancestors and modern versions with a developed organ.[178] Of course we never saw these things actually working out in Lamarck's theory, either, but then Lamarck was not the theological problem Darwin was. The skeptics ask, for example, where is there a bat-like creature with only half a wing? Surely these species must have come into existence fully developed, and at the direction of an intelligent Designer and not by evolution from a previous species. For many, these objections did not come from a concern for lack of scientific evidence but from a preconceived philosophical notion. Certainly in the Catholic Church there was a strong Essentialism in which the underlying identity of a thing was an "essence" of the category of things to which the individual belonged. Very reminiscent of Plato's "ideal forms", passing through Aristotle and Aquinas,[179] the idea held that, "...Essences were unchanging, and a thing couldn't change its essence, and new essences couldn't be born - except of course by God's command in episodes of Special Creation."[180] In 1950, Pope Pius XII warned against some views of reality which, influenced by evolution, viewed as open to change things that were held to be absolute and unchangeable, and warned of Existentialism which did not take into account the unchangeable essences of things.[181] This same form of thinking is found in the apologetics of the famous Archbishop Fulton Sheen. He writes,

> "In a perfect manner, everything in this world has been made according to certain ideas existing in the mind of God from all eternity...These ideas, which from our point of view are multiple but really are one in the divine mind and identical with His Being, are called Archetypal Ideas..."[182]

So, many who objected to Darwinism did not do so merely because of a lack of evidence of intermediate forms, but because they did not *want* species to evolve from others. Nevertheless, there *are* many animals that have, not half a bat wing, but adaptations that are on a continuum between being earth-bound and air-worthy. Flying squirrels can drive a dog absolutely crazy if he is familiar with squirrels that can only leap a short distance from tree to tree and encounters one that can glide across a clearing. There are even frogs and snakes that can glide or parachute through the air, and reptiles have been doing this for a long time. More than 250 million years ago, *Coelurosauranus* even developed new bones, though not attached to the skeleton, to support a glider's wing.[183] They are not half of a bat, but they do show the way that each slight improvement in equipment can be advantageous to its bearer.[184] As Richard Dawkins puts it,

> "Five per cent vision is better than no vision at all. Five per cent hearing is better than no hearing at all. Five per cent flight efficiency is better than no flight at all. It is thoroughly believable that every organ or apparatus that we actually see is the product of a smooth trajectory through animal space, a trajectory in which every intermediate stage assisted survival and reproduction." [185]

Birds obviously benefit from feathers as instruments of flight, but the first creatures sporting them may well have been dinosaurs[186] using them as body ornamentation. A feather also makes a great insulator (try a goose down comforter sometime), and who knew it could also be a superb flight instrument!

"Whoa," opponents counter, where are the *fossils* of those animals we see today that are midway between the very primitive and the modern?[187] Since there are gaps in the fossil record, they say, these in-between forms must not have existed! This is certainly a specious logical argument since one can never prove a proposition by the lack of evidence for its competitor. Until recently there was much evidence describing how the kings of Assyria lived in their palaces, but almost none about how the peasants in the countryside lived because excavations had not been done in these areas for lack of interest in the common people (also because their farmsteads held few art treasures to exhibit in a museum). Because there were "missing links" did that mean that there were no farmers or common people in those days? Certainly not. On the other hand, there *are* many examples of the finding of fossils that do fill in the gaps along the continuum from one species to another descended from it.

A WHALE OF AN EXAMPLE

Stephen Jay Gould cites the history of the whale since it has been a favorite example of the lack of intermediate steps between a land animal and this totally aquatic mammal. The case of the whale has become "...a classic case of a supposed cardinal problem in evolutionary theory - the failure to find intermediary fossils for major anatomical transitions or even to imagine how such a bridging form might look or work."[188] Links have been recently found which present, "...the sweetest series of transitional fossils an evolutionist could ever hope to find." The oldest whale found, *Pakicetus*, from 50 million years ago, lacked the hearing mechanism of modern whales, had the teeth of a terrestrial relative and was found in what had been river mouths and shallow areas. Unfortunately, the fossil's limbs were not found so it was not the proof sought after. Then *Basilosaurus isis* was found, named as a lizard but actually an early whale, which had complete hind limbs but so reduced in size as to be non-functional. It was already committed to life in the sea and so still was not a good proof to the skeptics. In between the two was *Indocetus ramani*, certainly a whale but with hind limbs most likely able to support the weight of the animal on land. And then the clincher was found in Pakistan, where *Pakicetus* had been found, *Ambulocetus natans*, a whale which powered its swimming by the same up and down motion as modern whales, but through large hind feet rather than a horizontal tail fluke.[189] In fact, the animal is not only the "missing link" in how there could be a transition from a land animal to a fully aquatic animal, it also shows the results of the constraints of history. The land-based ancestor, like modern mammals, had much more flexibility in the vertical motion of its spine than in the horizontal motion. You are a land-based mammal. Pick something up in front of you, and then try to bend sideways and pick it up. You can propel yourself through the water like a dolphin, another mammal, if you bend at the waist and undulate the body. Try to move from side to side like a fish and you may as well be a broomstick. Mammals just don't bend that way. So why did the whale evolve a horizontal fluke driven by vertical motion of the spinal column when fish all propel themselves by a horizontal motion of the spine? Because that was the equipment they had to work with, inherited from their terrestrial ancestors.[*] [190]

Many other examples can be given[191] but this one will suffice to deny validity to this objection to evolution as the explanation for the development of species from other species. There are indeed formerly "missing links" showing the transitions from ancient forms to modern forms. It should also be remembered that today's versions of animals, including ourselves, are transitional forms to something else. We do not see ourselves or our

---

[*] Contrary to Ned Land in *20,000 Leagues Under the Sea* who explains that, "...at the beginning of time, these whales could swim faster than they can now...Because in the old days they had vertical tails like fish, and they moved them from side to side in the water. But the Creator saw that they swam too fast and he twisted their tails, and from then on they have had to move their tails up and down in the water and they haven't been able to swim as fast

contemporaries in other species as somehow incomplete and on the way to something different, but in fact, that is what we are, if we and the other living beings inhabiting our world today survive ourselves. Nor were the species that preceded us incomplete in themselves. They were what they were and that was good. There was no design of which they were only the rough draft.

THERE WAS ENOUGH TIME, ALREADY.

Another objection often heard is that there would not be enough time for these changes to occur.[192] We have already seen in the example of the eye that there has been more than enough time to evolve visual acuity many times over, and the same is true for every other structure or sense organ. The objection that there was not enough time is most frequently heard from those who secretly believe that the earth is only about 4000 years old or so, or who have little concept of just how long a million years really is and how much can go on in such a period. Archeology can give us an example and a hint. Ancient sites of human occupation are often found on "tells" or mounds, because our ancestors chose easily defended raised sites, or else they built the hill themselves. However, their remains are not found on top of the site but within it, covered sometimes from debris of another city built atop theirs at a later time, and often just from the natural accumulation of dust on top. Any house cleaner knows how quickly dust can accumulate in just a few weeks. If 1/100th of an inch was laid down on an abandoned city in a year, in a century there would be an inch of dirt over the site. In 10,000 years, 100 inches of covering, and in a million years you will have to dig through more than 830 feet of soil. (You probably won't have to dig that much because dust blows off as well as on.) A lot can happen in a million years, even in a passive accumulation of sediment. In the day-to-day competition for survival and millions of generations (for most animals and plants) things can occur in a relative "hurry."

LIFE IS DIFFERENT

Let's consider more directly the objection that evolutionary theory can indeed explain the development from the simplest living things to the most complex, but that it cannot explain the transition from non-life to life. Ironically, in the Middle Ages it was presumed that life sprang spontaneously from other living things, from dead matter and even from the mud. To the Thomists this was not a problem, for god infused a "form" into each living thing, and a living "soul" into each human being, just as the Bible and Aristotle said. For the bible, "...Life is not considered as a function of the organism...The spirit is, rather, a mysterious reality, comparable to the wind. When God breathes it into the creature...it comes alive."[193] Later philosophers were more sophisticated and less willing to grant god such a direct role. Henri Bergson proposed an *élan vital*, a vital impulse infiltrating the universe, "...an insubstantial essence that permeated and animated the inorganic molecules of cells and brought them to life."[194] Along Bergson's lines, other philosophers who felt that, "Contrary to [mechanism] the development of living things depends primarily upon the internal impetus, rather than upon an external shock...and contrary to [finalism], development leads to original, free forms of life, and not to results read off from an ideal blueprint."[195] These and other explanations of the difference between living and non-living make sense to us because there is such a seeming chasm between ourselves and the other forms of life that we see around us, and the objects that are not alive. Michner's "The Covenant" tells of the Zulu ruler Shaka killing his advisors in an effort to understand the difference between this living man and, on the other side, this dead man. Then killing pregnant women in an effort to understand how life arose.[196] Surely there is something "vital" that escapes at death or conversely is mysteriously emplaced at conception. Of course we are a little confused when we get down to the level of viruses. Are they living or not? It all depends on your definition, and that is the key to the answer to the objection. If we could go back in time to when "life began", what would we see? A clear-cut division? Presumably there was some kind of material, a molecule perhaps, which could reproduce itself on the same order as a crystal that can add new units to itself and grow, though no one would say that a salt crystal is alive. Perhaps there was a mutation, an impurity, which entered in and made this molecule more efficient in gathering the materials from the environment around it that it needed to grow. It passed on this mutation to the new parts of itself and these carried on much more expeditiously. This continues for a million generations, which may not have been much calendar time at all, and other mutations occurred every once in a while, say on the order of once every 1000 replications. By the millionth generation there have been 100,000 changes. Let's presume that "generation 1 million" is alive by our definitions of life. At what point did we pass from non-living to living? At 1 million less 1? Less 200? Obviously with so many small steps it would be impossible to say. The problem, however, is not with the process of evolving from non-living to living, but with our definition of life. Once again, the objection that Darwinism cannot explain the transition from non-living to living is refuted by calling, not for the sudden appearance of something entirely new, but for the accretion of small changes which by the end of the process have become something new.[197]

Sometimes the changes may not have been so small. There is much evidence that there were occasions when whole entities such as viruses or bacteria were incorporated into other organisms and led to great change in the functioning of the resulting entity. Mitochondria in vertebrate cells may once have been free-living bacteria and the same is true of the chloroplasts that convert sunlight into nutrients in plants. These may have originally been cyanobacteria. The mixtures of their DNA functions resulted in entirely new lines.[198] Once again, we have to conclude that there need not be an intervention of a causative agent, much less a designing agent, to explain the origin of life or its development into species.

YEAH, BUT. HUMAN EVOLUTION

Another response to evolution takes it seriously for everything but humankind. The moderate fundamentalists can allow for the evolution of the physical bodies of animals and plants, since they can interpret the Bible to allow for this, though they see the hand of the Designer at every step. Still, they say, the special creation of "man" as described in Genesis must mean that our bodies are the exception. Almost no one takes this extreme position anymore and most recognize that our bodies are most certainly located on the evolutionary line we can trace from reptile, through mammal to placental mammal, to primates like ourselves. I will not bother to argue with those who deny any such descent since those who do so are probably not open to any other viewpoint but their own and trying to have a conversation with them is decidedly unproductive. However, others will accept that our bodies, similar as they are to other animals, are indeed the product of an evolutionary history, but surely there is something about us that differentiates us from the rest. Darwin, probably out of consideration for the feelings of his contemporaries, left room for a special nature in humans, though his heart was not in it. Alfred Russell Wallace, a co-discoverer of natural selection with Darwin, felt, as Richard Leakey explains him, that supernatural intervention was called for because humans were just,

> "...too intelligent, too refined, too sophisticated to have been the product of mere natural selection. Primitive hunter-gatherers would have had no biological need for these qualities...and so they could not have arisen by natural selection...Scientists such as Wallace...were struggling with conflicting forces, one intellectual and the other emotional. They accepted the fact that *Homo sapiens* derived ultimately from nature through the process of evolution, but their belief in the essential spirituality, or transcendent essence, of humanity led them to construct

> explanations for evolution which maintained human distinctiveness."[199]

At least Wallace thought he had reached his conclusion on scientific grounds. Others had a preconceived notion stemming from philosophical concepts as well as biblical, that there was something special in humans that gave them a life different from other animals. Pius XII in 1950 wrote that it was fine to make inquiries into the role of evolution in the development of the human body, but, "...the Catholic faith obliges us to hold that souls are immediately created by God."[200] Paul VI reaffirmed this position later [201], and the latest catechism states that "The human body shares in the dignity of `the image of god': it is a human body precisely because it is animated by a spiritual soul."[202] Writers from other Christian denominations seek to make the same point when, for example, they assert things like an additional human spirit, a *neshamah* in Hebrew, in addition to the source of life, the *nefesh*, which gave life to all the animals.[203] Both of these life-infusing entities follow from a supposed biblical notion that is actually at variance with the Greek, even though the Greek philosophers powerfully influenced the Church. To the Greeks, at least with Plato, the soul really did not belong in the body but was imprisoned in it.[204] For the Old Testament, "...the Hebrew idea of personality is an animated body, and not an incarnated soul." [205] Thomism may have wished to preserve the Greek idea of the person being basically the soul which could do very well without the body, thank you, but Thomists had to be true to the understanding of scripture which held what the Catechism still maintains, "the unity of soul and body is so profound that one has to consider the soul to be the `form' of the body: i.e., it is because of its spiritual soul that the body made of matter becomes a living, human body; spirit and matter, in man, are not two natures united but rather their union forms a single nature."[206] In this philosophical view a human being is a unity and cannot be separated into body and soul and still be a complete human being. The Old Testament would have made no distinction at all. If there was separation from the breath of god there was death and non-existence, or at most a shadowy kind of existence in Sheol that was almost worse than not being. Because of Christianity's insistence on resurrection of the dead, a distinction was made between body and soul. It was the soul that really was made in the "image of God", having been made directly by god, and it was the soul that kept one in existence after death, at least until it could be reunited with the body. The latest Catechism says, "The church teaches that every spiritual soul is created immediately by god - it is not `produced' by the parents - and also that it is immortal: it does not perish when it separates from the body at death, and it will be reunited with the body at the final resurrection."[207] This is another reason to posit a soul, something that could remain in existence, since the body clearly died and disintegrated. And we desperately want to remain alive after our death.[208] Archeologists found a child buried perhaps 80,000 years ago who appears to have been carefully positioned in the grave.[209] Neanderthal remains of 58,000 years ago have been found in Iran, which were possibly ritual burials. A thirteen-year-old boy was buried with a deer antler in his upturned hands and in Iraq a man was found with flower petals scattered around him.[210] This may, of course, have been done for the same reason we might put a favorite article in the casket with our loved one, not because we feel they will need it, but as a gesture of respect. Still, the prevailing opinion is that, "It is not likely that people would have lowered fine tools and cooked food into a pit unless they believed that these things could somehow be used by the dead, that consciousness, or a spirit, or *something* persisted in the graves."[211] It would not be long before the Egyptians and others were burying their dead, at least their important dead, with provisions obviously meant for the afterlife. The soul is not only the element that separates us from the animals and forms us in the image of god, it is also the principle of our continuing existence after physical death. As such, some surmise, it surely cannot be the result of something so earth-bound as evolution and natural selection.

There was and is also the objection to Darwinism in explaining our appearance on the scene for the much less worthy but very powerful motive of wanting to be on top. If we are just the product of the same forces that formed the beasts, where is our vaunted superiority? "Darwin's idea thus also threatened to spread *all the way up*, dissolving the illusion of our own authorship, our own spark of divine creativity and understanding."[212] Copernicanism had removed us from the center of the universe, Newton had threatened our free will, and now Darwin made us into monkeys. Too much.

BUT, WE ARE CONSCIOUS!

A soul is also needed, it has been claimed, to explain those powers that we do not seem to share with other animals. Obviously, these dissenters may admit, the structure of our bodies, our circulatory, nervous and sensory, reproductive, digestive and respiratory systems have been formed by evolutionary processes. Equally obviously, the role of the brain in coordinating all these systems, as well as the hormonal system and its effects on our emotional life, can be allowed to follow the dictates of evolution. Where some draw the line is in our cognitive and volitional abilities. Here, they say, there is something distinctive that can only be the result of a non-physical agency.[213] Aquinas had certainly furthered this thinking, and according to him,

> "..the angel and the separated soul are at once perfectly intelligible and perfectly intelligent...They are, in fact, self-conscious, but when in a body, the soul is not aware of itself...Death frees the soul from the shackles of the body; it renders actual the knowledge of its own substance that the soul used to possess only virtually."[214]

One would think that this contradicted the insistence that neither body nor soul was an entity unto itself, only forming a "nature" which could be expressed in a unique individual when they were united. Thomas really wants to be a Greek, and any good Thomist can certainly give some convoluted explanation for the seeming inconsistency, but for Thomas, all the "higher powers" of humans were to be found in the soul.[215] It was Rene Descartes who brought "dualism" to the fore, claiming that the mind and the body were two separate entities. "Descartes founded the image of the human mind as a sort of nebulous substance that exists independent of the body."[216] For him the brain collected sense data and presented it to the mind in some mystical manner and then took the orders from the mind and put them into action in the motion of the body. The brain was certainly a fantastic piece of machinery, and coordinated all the input into a coherent picture, but the mind was certainly not the brain. It has been likened to a spectator in a movie theater and "the ghost in the machine" and it resided somewhere in the brain, probably the pineal gland, Descartes thought, but at the same time it was "ontologically disconnected from space and matter; it is conceived to hover, somehow disembodied, outside the realm of space and matter."[217] This mind need not be equated with a god-created soul, but whatever its origin it is not the result of the forces that operate on matter. For Benedict de Spinoza, too, though he does not follow Descartes in all things, there is something of the mind that is immortal and his "Ethics" contains the proposition that, "The human mind cannot be absolutely destroyed with the body, but there remains of it something which is eternal."[218] The mind as immaterial has long been accepted and assumed by writers in and out of the Church. In 1966, the writer F.J. Sheed did not feel it necessary to prove his description of the mind. He uses Einstein's mind as an example, which was, "like some monster machine, producing ideas which have revolutionized the world. Yet it is not itself in space. Its ideas, which have brought the conquest of outer space within our grasp...are themselves not in space."[219]

I hardly do justice to Descartes' and Spinoza's thought here: it is quite philosophical and complicated, but it is also dead wrong as we shall see, so we need do no more than state Descartes' basic position. There was opposition to him, mostly on philosophical grounds[220] and Bergson and others contradicted him by placing the intellectual power inherent in the universe itself and not completely detached from matter, akin to Spinoza except that for Bergson the intellectual power of the human mind resided somehow in god. "To Bergson, the world is an instrument in the hand of its Master, who would have men His image and likeness, His partners - `gods'."[221]

Today there is confusion among the majority of people. Among Americans, 45% say that they believe in evolution, though only 9% accept that humans and other species have evolved with no divine intervention,[222] and 47% said in 1993 that they believe that humans were created by god less than ten thousand years ago.[223] The latest Catholic Catechism makes no pronouncement on the issue of the seat of consciousness and intelligence, but what else is the soul for? A number of Catholics I interviewed expressed the understanding that the brain is what thinks and is conscious, not a soul, but when asked what that means about the soul's ability to be conscious after death they were confused and frequently changed their minds. It is clear that the work in brain research and the obvious changes that occur when the brain is damaged by stroke or accident has influenced peoples' opinion about the function of the brain, but it is fair to say that there is much confusion.

I THINK, THEREFORE I AM A...WHAT?

One area where the confusion surfaces is in the matter of Artificial Intelligence and whether computers will ever think. When *Time* magazine reported on a chess match with the world champion pitted against the computer, Deep Blue, it brought up the point that, "The better these seemingly soulless machines get at doing things people do, the more plausible it seems that we could be soulless machines too." [224] *Time* also turns the question around and asks whether as computers mimic increasingly the functions of the human brain, will they really have mind and consciousness? "The lights are on, but is anyone home?" The article brings out the questions being raised about whether or not humans have something "extra" that machines will never have, and if so why we have it.[225] "God" and "soul" are two parts of a possible answer, of course, but most investigators, while not necessarily ruling out that scenario, continue to look for other explanations. The confusion over whether or not a machine can ever be gifted with true Artificial Intelligence, and if they are, whether the intelligence will be any more artificial than our own will probably be with us for a while. Margaret Wertheim speculates in her book about cyberspace and the internet[226] that science has made reality restricted to the physical, having done away with heaven as "up there" among the celestial spheres. On the internet the human "spirit" can fly beyond boundaries, perhaps the closest we can get to the non-physical, because we desperately want to believe in something of ourselves that is not tied to the physical.

IF YOU SKIPPED, WELCOME BACK

At this point we have reviewed several positions on evolution that have in the past just been, somewhat uncritically, accepted by theists and by most people in the Western world. Certainly almost every adult Catholic and many other Christians have somewhere run into the concepts that we have been considering, and perhaps have incorporated them into their world-view, consciously or unconsciously. If so, we certainly do not expect to see a whole species of animals or plants evolve into another, or several other, species. Most people are vaguely aware that some people believe this has happened in the past, but it does not affect us at all and the theist can happily believe that things are as they are because god made the various species that way in the "beginning". Certainly most believe that of us humans. The theist again may well believe that god created us separately from the rest of nature, or at least creates a soul or some principle of human life that differs from whatever animates the other beings with which we share the earth. And most people evidently think that there is something special about our consciousness that is not shared with other animals, and certainly not with plants. This may or may not, in people's minds, be due to an inanimate or spiritual soul, but that is worlds apart from the next best thing. As has been seen, these assumptions do not hold up to scrutiny and we have to continue to ask what the findings of science have to say about these questions. It will take a little more background but we will see about this "soul".

WHAT ABOUT ANYTHING "SPECIAL"

The assumption of the special creation of species that remain ever the same cannot be maintained as is obvious to anyone exposed to Darwinism. Despite the creationists, there is just too much evidence already present in Darwin's "Origin of Species" to deny that, "...that is how species originate: by `descent with modification' from earlier species-not by Special Creation."[227] Since Darwin's time so much supporting evidence has been gathered that even the Popes have called it a "Theory" and not just a "hypothesis." Of course, the idea of a species has had to be refined from

Lamarck and the philosophers. The bottom line is still a category of animals or plants that cannot interbreed with another, at least not without producing a hybrid. For example, with the process of the rise of the isthmus of Panama, individuals of one shrimp species were separated from others, first by changing conditions and finally by a physical barrier. Over time the genes changed in different ways. Now, "no shrimp across the isthmus mate readily - they are all separate species."[228] There is still considerable continuity between a parent species and its great granddaughter species. There is no "essence" of Atlantic shrimp, now changed to the essence of Pacific shrimp. No essence of a fish species that is suddenly transformed into the "essence" of an amphibian species. There is a succession of small modifications, the possessors of any one of which can usually breed with the preceding form. Indeed, these are its potential mates since the mutant alone possesses the modification. The sport of the litter whose eyes bulge upward a little and whose descendants will look something like mud skippers today with their fleshy appendages and froggy eyes on top of their heads, may not be the most beautiful fish in the school, but he will probably mate and pass on his genes, if only by ambushing some sweet young thing. There is always the chance that his eyes bulge because this tendency has been building up in the population because of another mutation in the female genes that cause them to think he is the most beautiful hunk they have ever seen. And those at the end of the process (which of course really never ends) are significantly different from those at the beginning. Indeed the new species often replaces the old because of the effects of natural selection, unless the two are separated by distance or ecological niche, a process that we can see operating today, though on "varieties" instead of species. The varieties, of course, may well become new species in time. Golden Retrievers do not have an especially good nose, but then they do not need one. A dead duck leaves very little scent on the water. If you want to hunt rabbits, go for the beagle. Both are just varieties of the species "*Canis*" and can mate with each other. Let the ice caps melt and the world turn into a whole lot more swamp with a greater need for eyeballers than smellers, throw in half a million years or so, and *voila*, perhaps a new species which has been so removed from the rest of the canine population by human intervention in selecting those depending on sight more than scent that it can no longer reproduce with a beagle. Make "human intervention" read "natural selection" and you have the same result. The conclusion, of course, is that we must set aside the ancient belief that each species was created of itself and remains itself throughout the ages.

GRADUAL? OR NOT.

A little aside on the pacing of evolution. There is an ongoing discussion among evolutionists on whether there is a very gradual and constant tempo of mutation and selection in most species at all times or whether there is what has been called "punctuated equilibrium" meaning that the normal condition of a species is *stasis* with rather large leaps occurring infrequently.[229] The truth is probably in the middle, as usual, with some species, especially those not very mobile but living in rather stable environments, changing slowly in response to conditions or the pressures of predation, or even to the changing preferences of the opposite sex. Other species may change very little over a long period of time because they are already flexible enough to tolerate changing external conditions or because they can move to suit their needs like the dinosaur *Jobaria tiguidensis*, a 40,000 pound, 70 feet long plant eater that remained the same for millions of years. The coyote today need not change much if it can be equally at home in a prairie, a desert, or the common area of a subdivision. A mountain goat does not have to change to accommodate even an ice age if it can just walk down or up the mountain to find suitable conditions and climate. On the other hand, some species may indeed change radically over a short period, and very many species may develop in a relatively short time as has occurred, but many will simply die out to be replaced by another organism, another species, which can make its way in the new situation.

There are "Neo-Darwinians" whose position is that random mutation of genes followed by selection of those fit is the only mechanism of evolution. They seem to be the mainstream teachers. Some of these, I presume, agree with punctuated equilibrium and some not. There are others who take what they call a "system view of evolution" in which random mutation and natural selection are only one avenue of evolution, and not the most important at that. Scientists such as Stuart Kauffman and Fritjof Capra insist that evolution is driven by, "life's inherent tendency to create novelty, in the spontaneous emergence of increasing complexity and order."[230] Other means of achieving the results we see are genetic recombination in which genes are shared with other organisms and symbiosis, or the inclusion of one organism by another and their mutual dependence. Human mitochondria would be an example. This system and the meaning of spontaneous emergence of complexity are complicated, not yet proven, and not universally accepted, though they are gaining in adherents. The understanding of the processes and pacing of evolution is still being refined, but the conclusions that concern us remain the same.

ARE HUMANS "SPECIAL"?

The presumption of special handling for humans if not for other species must be denied since there is so much evidence that we are the result of a long history of mutation and selection, no different from any other animal species. The evidence is all around us, indeed inside of us. "Mutations are random changes in the way the nucleotides are put together resulting primarily from the chance bombardment of the master molecule [DNA] by ultraviolet light from the sun, cosmic rays, nearby chemical reactions, and random quantum processes during reproduction."[231] Evidence is found in our DNA today in the form of parts of the chain that have no purpose. Do we explain them by saying that god just had a bad day, or put them there to preserve the mystery? Or do we see them as the result of mutations that did no good, but did no harm either or that once had a purpose but have since been replaced by further mutations? Obviously the explanation without the supernatural is preferred. There are even "parasite" genes that we could totally do without but which hitch a ride on our chromosomes.[232] Are they there because of a genetic accident or placed there at our special Creation? The evolution of the mammalian ear shows an odd and random process in its development. When we think of "development" we often think of progress because that is usually what we see when things change through human instrumentation. Nobody makes a *worse* mousetrap on purpose. Evolution, of course, is not a conscious designer. Most of the changes or mutations will be detrimental but we will not see them because the bearer will not survive. Most mutants die in the womb or the egg, or in B movies, and those that survive to birth cannot compete with their own better-adapted species-mates. What will remain to be noticed will be the improvements, and so we think that evolution is heading in some certain direction "on purpose". Nothing could be further from the truth, as seen in the equal possibility of losing some ability, eyesight in the cave salamander for example, if the mutation happens to occur in new environmental conditions where it makes no difference. And so, we can trace the movement of an embryonic jawbone in mammals as it splits from the jaw and moves backwards to form the middle ear of the adult animal.[233] The inner ear is a strange affair and although it obviously works, could have been engineered in a much more straightforward manner. Why do the former jawbones move? They do not move in the reptile forbearers of mammals, but the developing brain of the mammal pushes the bones backwards. Since the jaw worked without them, some of the mutants survived and the bones became available for incorporation into another organ entirely, the perception of sound. Our inner ear is formed in the same way as the earliest mammals of 160 million years ago. We are in the same continuum as all the other non-human and even non-primate mammals whose inner ear develops in this manner and to call for a special creation of human beings means one must explain why such a pattern was perpetuated. One biologist states it this way: "The fact that species share homologies* is an argument for evolution, for if they had been created separately there would be no reason why they should show homologous similarities."[234] If god was starting over, why not fix the mistakes? Far more likely that our ear is as it is because we inherited it from our ancestors, small, furry and rodent-like as they may be. Any book on human evolution will give more examples which all show our relationship with other primates and with progenitors much farther back in time and development than they. No, we are nothing special in this way. No wonder that Darwin was opposed by those who insisted on humanity's uniqueness. As Trinh Thuan explains it,

> "Dethroned from his central position by Copernicus, reduced to insignificance in an infinite universe, and remote to God in Newton's mechanistic universe, the western man of the nineteenth century could console himself with thoughts of his divine lineage. Was he not, after all, the descendant of Adam and Eve, themselves created by God? Even though no longer occupying the center of the universe, he remained the chosen child of God. Even that consolation was denied him after the publication in 1859 of Charles Darwin's *Origin of the Species.*"[235]

We may not like it, but this seeming demotion was and is the judgment of science. Humans of the *Homo sapiens* persuasion descended between 500,000 and 34,000 years ago from *Homo erectus* in Africa, who had descended from *Homo habilis*, which arose sometime prior to 2.5 million years ago. *Homo habilis* was something new in the primate world, with a larger brain and a changed tooth structure to accommodate both plant eating and meat eating. It was "on the cusp between the ape...and the human being"[236] pardon the pun. *Homo erectus* was firmly in the direction of modern humans in its anatomy and in the length of time needed for the maturing of its infants. It was this species that first extended its range beyond Africa to regions in Asia and Europe.[237] Between half a million and thirty four thousand years ago there is a mixture of various forms, including the Neanderthals, and "despite many differences among the individual

---

* A homology is a likeness of parts of two different species due to common origin from a remote ancestor. Why do bats have five fingers in their wing when this is not necessary? Because they evolved from an ancestor common to all mammals which had five.

specimens, they all have two things in common: they are more advanced than *Homo erectus*-having larger brains, for instance-and more primitive than *Homo sapiens*."[238] Interpretation of how these forms of *Homo* were related to each other and to moderns is still in flux, but the "out of Africa" school seems to be predominant.[239] About 50,000 years ago, the Neanderthal populations of Europe began to be replaced by modern *sapiens* in what is called the "Upper Paleolithic Revolution", and evidence points to Africa as the starting point for the replacements, *sapiens*, though this is far from certain. The hypothesis of "mitochondrial Eve" suggests a common origin of modern *sapiens* from the ancestral population, sometime around 260,000 years ago.[240] The point, however, is that there is no abrupt demarcation of who is human and who is not. The most primitive in the line are somewhat apelike and somewhat humanlike. There is no point at which we could say, this one is a human being but its "parents" were not. We see the same progression as we see in the other animals' evolution and the Special Creation of the human body is not called for as an explanation, indeed it is ruled out by the evidence. It is equally obvious that there is no need for the special creation of a human soul as the principle of life, any more than we have to explain other forms of life by recourse to some spiritual or at least immaterial element such as a "form" or "essence". One naturalist observed that,

> "...links between man and his subhuman ancestors were described nearly a hundred years ago by Charles Darwin in his book *The Expression of the Emotions in Man and Animals*. We can guess that it offended more people than even his *Origin of Species* because it is so specific. A preacher thundering from his pulpit about the uniqueness of human beings with their God-given souls would not like to realize that his very gestures, the hairs that rose on his neck, the deepened tones of his outraged voice, and the perspiration that probably ran down his skin under clerical vestments are all manifestations of anger in mammals. If he was sneering at Darwin a bit...did he remember uncomfortably that a sneer is derived from an animal's lifting its lip to remind an enemy of its fangs? Even while he was denying the principle of evolution, how could a vehement man doubt such intimate evidence?"[241]

We may wish to be different from the rest of the primates and propose a soul either out of this wish or because of the wording of the Bible in speaking of the breath of God being breathed into the first human, combining this with Greek philosophy of matter and form, but there is no empirical evidence from our history that requires such an intervention. Nor is there any great leap evident in the history of our immediate ancestors that requires some supernatural agency. Except, perhaps, one? At least that is what the critics of evolution hope.

BUT, "I THINK"

If we accept that no external causality is necessary to produce life; that species are not static but dynamic and evolve from other species through the process of mutation and natural selection; and if we accept that we as a species are no different from any other in the evolution of our bodies; can we exempt our brains from this process? Or must we claim some privileged position for our consciousness If Plato, Aristotle, Aquinas and all their ilk were wrong, in inventing forms and essences, was Descartes at least, right in making the mind distinct from the body? Scientists do not see any reason to appeal to a soul or some other explanation. James Trefil points out that,

> "...the traditional response to the question of the difference between humans and animals...was the assertion that only humans possess a soul. In essence, this has the effect of removing the question of human-animal differences from the realm of scientific inquiry, a step I would be extremely reluctant to take."[242]

To understand why, we must continue our investigation.

The human brain is a wonderful organ with a hundred trillion neural connections doing all kinds of things we are not even aware of. Certainly there are living things that have no brain, plants, for example, relying exclusively on chemical signals to regulate themselves. One-celled creatures have no need to send signals to distant places in their bodies and can get along quite well with enzymes, which humans also use. Still, chemicals take some time to do their work, to travel by diffusion to all parts of the organism. Certainly the first animal to develop a neural pathway for signals was at a great advantage in moving much faster toward or away from a stimulus. And if one nerve is good, two are even better. And if one way to "read" your environment is good, two are better. If I can sense through nervous tissue the pressure transmitted through the water when a predator is nearby, as fish do with a "lateral line", it would help immensely if I could also see its precise location through visible light, and at the same time see where my escape route lies. Yet, receiving two sensations will do me no good unless I know that they both come from the same source. Somehow I must integrate the feeling of pressure on my side with the object I see with my eye. Not only that, I must also throw into the "picture" what my appropriate response should be. If it is a danger and I try to mate with it my line comes to an end. If it is a mate and I run away, my line also fails to continue, plus I miss out on some of the best fun to be had. Evolving a mechanism to join my sensations into an image of my environment and combining this with instinctive behavioral reaction is certainly in my interest. Of course I do not have to be aware of what I am doing, nor understand the process that is taking place within me. A thermostat does not have to know a Kelvin from a BTU to register heat, nor does it have to decide what to do about it. In fact, once it makes a connection between two terminals, its job is done and it makes no difference if what occurs then is the ignition of a furnace or the ringing of a bell to tell Newt to throw some more wood on the fire. It can all be automatic, and a neural mechanism that accomplishes a similar connection can be nothing but a living machine, call it a brain. Of course the more sensations I can put together, the better my internal representation of the external environment will be. The more information I have, the better the chance that the appropriate response can be chosen and the more choices I have for a response the better, but then this will require some means of choosing among alternatives. If I am a dog with three masters (or three servants from the dog's view), which one seems more likely to pity my sad face and share their meal? No dog owner will doubt that their dog makes these judgments. In the absence of the first choice, it will move on to the next best sucker. Few dog owners have ever felt the necessity of inventing a soul or mind to explain the dog's action.

The actual development of such brains is much more complex than the scenario above and took millions of years to accomplish, but the basic pattern is correct. Natural selection favored those who could better adapt to the conditions around them, and brains, being excellent adaptors, became increasingly capable.[243] Sometimes an entire species disappeared when the conditions changed because the brains were "hard-wired" and too inflexible and there was not enough time to adapt before the adverse circumstances killed off the whole lot. People wonder why deer will freeze in the road in the headlights of their car (or worse, so will the skunk), but they did not evolve with headlights in their world and have had maybe 100 years of experience of them. Somewhere out there may be a deer with a mutation that equates "headlight" with "wolf" and causes it to run. Give that deer a fair amount of luck and a sexy personality and throw in some time and it may be the great-great-great-etc. granddaddy of all the deer in the county, all of whom run from headlights.

It would be better if you could cut the time needed to respond to changing conditions. This was done in a primitive way just to take into account the movements of the predator and one's own evasive movements. There has to be some continuity to my perceptions so that I can know that what I sense is the same enemy, but it is moving, and that it is now behind me instead of to my side because I have moved. It helps even more if I can predict where it will be in another moment. The better I can "predict" what it will do in response to my movements the better I can plan a countermove. In fact, the more I know that there *is* an "I" the better off I am. Once more, this began in a primitive way so that I don't swallow my own tail thinking it is another's. Knowing the "I" does not have to be conscious, nor do I have to recognize "myself" in a mirror to avoid biting my own tail. The ability has wondrously evolved, especially in the highest primates.

To repeat, how all this is actually accomplished is the story of the evolution of life and of all living things on our planet. It is an extremely complex story and has taken an immense amount of time, accompanied by countless branchings, setbacks and false starts, but the process has brought about everything that exists. In this scenario we have reached a point where we have a brain in an animal that is capable of sensation and reaction; capable of controlling all the other parts of the body; that manufactures a picture of the world around it to some degree of accuracy; that can choose among various options of response; that can predict the future; and that has some sense of "self",[244] at least enough to know what is self and what is enemy, mate or food. And we have not progressed even to the level of a sardine. Any dog or cat owner will testify that their pride and joy can do all these things and do them surprisingly well. A bonobo or chimpanzee can do them immensely better. No one mentions the need for a non-material explanation of how they do it. And no material explanation requires an intervention by a creative designer.

EMERGING TALENTS

It is indeed the judgment of science that there is no such required intervention. Richard Dawkins, in explaining why a baboon does not walk out over a cliff on a plank which his weight will upset throwing him into the abyss, postulates that in addition to a possible innate fear and in addition to any prior experiences of the individual monkey, it well might, as we would, picture the future if it walked on the plank, and avoid it. Dawkins explains,

> "The imagination, the capacity to simulate things that are not (yet) in the world, is a natural progression from the capacity to simulate things that are in the world...Whether or not it was originally designed to run on into the future, its ability to do so, to simulate things not only as they are but as they may turn out to

> be, is a natural, almost inevitable consequence of its being a model at all...Natural selection built in the capacity to simulate the world as it is because this was necessary in order to perceive the world...Having built in the capacity to simulate models of things as they are, natural selection found that it was but a short step to simulate things as they are not quite yet - to simulate the future."[245]

The imagination is not something mystical, nor is it reserved to our species.[246] Nor is the ability to choose among various options or to learn from experience restricted to us. We may use these abilities differently, at times, than other species, (though most often our own behavior is just as "instinctive" as theirs) but it is the same basic mechanism. To say "mechanism" should not be interpreted as "mechanical" in the sense of "determined". In our brains, and in the brains of at least some other animals, "...What matters is the *pattern* of neural activity, not the detailed functioning of individual neurons. It is at this collective level that new qualities of self-organization appear, which seem to have their own rules of behaviour that cannot be derived from the laws of physics governing the neural function."[247] This may or may not be true of bees and wasps, but we often think of "higher" animals as nothing but machines, at least the animals we do not contact. When we become more familiar with their behavior and "personality" we know that some have moods, they are not consistent, they can learn from and anticipate events. Or at least there are many indications that this is true. There is a degree of flexibility, if not to say "freedom" in their behavior that is hard to explain if they are mere automatons. This is explained, not by granting them some non-physical governing agent but by recognizing that complexity brings about in structures facets that were not predictable from the features of the components. "...It is a general property of complex systems that above a certain threshold of complexity, new qualities emerge that are not only absent, but simply meaningless at a lower conceptual level. At each transition to a higher level of organization and complexity new laws and principles must be invoked..."[248] The human brain is certainly a higher level of organization. "Most scientists feel there is no need to introduce new and mysterious factors into the problem; by and large, they assume that mental phenomena emerge from the awesome complexity of the brain's neural interconnections. With some 100 billion cells and 100,000 billion possible connections, the brain is the most complex structure known to science."[249] Complex, but not magical. The human brain is more complex than other animals, but not by all that much in some cases, and it maintains the pattern that can be traced directly back to whatever is in second place and then down the line. Eugene Mallove was speaking of the human mind when he said,

> "...It seems that the evolution of neurological abilities that let animals cope with their environment has conspired to give rise in humanity to a surfeit of mental abilities."[250]

The complex brain of the chimpanzee cannot be explained by the working of its parts either. There are emergent qualities to both animal and human brains that could never be intended by evolution (especially since nothing is ever "intended") but gave immense advantages to their recipients and therefore would be selected for in the competition of life.[251]

Consciousness itself is one of these emergent qualities of the brain. A "Time" magazine article explains,

> "Descartes was right in one sense: the mind is not a physical object, and while it exists within the brain, it has no particular location...However, Descartes was profoundly wrong, it appears, in his assertion that mind and body are wholly independent...Consciousness may be nothing more than an evanescent by-product of such mundane, wholly physical processes - much as a rainbow is the result of the interplay of light and raindrops."[252]

We may or may not want to deny consciousness to the orang-utans and chimpanzees, but let's say for argument that they are not conscious as humans are. The orang-utan, on finding a lost screwdriver, has been known to deliberately not call attention to it and, when the keeper is out of sight, take the cage apart with it. They "...use sticks as probes, rakes and levers; they sometimes bend, break or chew on them to manufacture more efficient or specialized utensils."[253] Of course, whether this is thinking, consciousness and planning depends on your definitions, but it is at least on the way.[254] Some primates have demonstrated the reasoning ability of a human two-year-old.[255] How much tinkering with the genetic apparatus would it take before they could match a twelve year old? Or a forty year old? A considerable amount, to be sure, but the evidence seems to say that it could be done. In fact it has been done. The brain capacity of an ape is about 400 cubic centimeters, that of *Homo habilis* about 800 cc's, and *Homo erectus* was about 900 cc's. Our own brain capacity is about 1350 cc's. No one has reason to doubt that it was brain development and not a supernatural intervention that gave rise to the primitive humans. To claim that such mediation is necessary to go from 900 to 1300 is unreasonable. Our own IQ's (not cc's) have increased, on average, by three percent per decade for the past few periods, especially in spatial recognition. Is this because the soul has gotten better at its work or because the brain has been impacted by the environment?[256]

So the emergence of life, the development into species, their evolution into other species, including the human species, like the appearance of the universe itself, does not need a cause outside itself. Indeed, scientists say that, "given a planet of earthly size, distance from a central star, and composition, life of simplest grade may originate with virtual certainty as a consequence of principles of organic chemistry and the physics of self-organizing systems."[257] Francis Crick, one of the discoverers of the DNA helix, feels that, "DNA almost certainly originated fairly close to the origin of life when things were necessarily simple or they could not have got going."[258] Get going they did, and Crick goes to some length to show that a designer need not be a part of the process, pointing out that, "biologists must constantly keep in mind that what they see was not designed, but rather evolved." Nor does the appearance of a brain and its integrating functions, nor its ability to present an image of the real world inside itself for manipulation and reaction, call for something non-physical.

WOE IS ME! WHO IS "ME"

One more objection can be made to this conclusion. To "whom" is this image presented? It certainly feels like there is a "homunculus" or a little person inside us, maybe akin to a computer operator on the Internet. All the information is presented to him or her. One may admit that there might even be a "computer program" in the brain that filters out irrelevant material and highlights the important items. Then "he" or "she" decides what to do. So if we grant to the brain all the "hardware" responsibilities of linking my arm muscles to the pain receptors in my finger so that I pull away from a fire without having to think about it; and if we even grant to it the "software" duties of filtering so that we pay little attention to the bus going by, but we do hear the tornado siren; and if we concede that the brain may have a role in constructing the picture of the world that we "see", what does the brain do with this picture? Is there a "self" as we fervently hope there is and that we experience? "...The message from the front line of brain research could hardly be more bleak"[259] for those who want an immaterial "self". If something happens to the brain, "I" am altered.[260] The famous story of Phineas P. Gage who in 1848 had a 13-pound iron rod blasted through his brain has often been told. He survived the accident seemingly without permanent damage. All his sense organs still functioned. Whatever the brain does to integrate it all seemed still to operate normally for he knew where he was and who he was, but he was not the same. "So radical was the change in him that friends and acquaintances could hardly recognize the man. They noted sadly that 'Gage was no longer Gage'."[261] What had changed were those things that we would claim are the "self". If the soul is like a little person sitting in the brain and conducting the orchestra, Gage's tamping rod must have gone clean through the little rascal. If you find the story of Gage interesting, you ought to read about the "man who mistook his wife for a hat."[262]

Then what gives us this feeling of being an individual "self"? There is no one area of the brain that has been identified as the source of this sensation, certainly not Descartes' pineal gland. Probing the brain can locate areas that bring the taste of salt, the memory of a face, even the experience of the transcendent, the so-called God-spot, but not any one place where the subject says "that's me". In fact, the evidence is that the sense of self is diffuse over the brain. Some researchers believe that,

> "...both 'you' and your awareness of a stream of events in a certain time sequence, are *created* by the confluence of these many parallel streams of data processing...Conscious awareness...is not the presentation of (possibly edited) data to a mythical subject (mind), but *is* the sum total of the data streams taken together."[263]

Our sense of self is a result of the operation of the brain and "merely" adds a new dimension.

> "Any objective awareness of self is better understood in terms of a priority decision-making and action system which re-represents information from representational systems elsewhere in the brain. Our objective awareness of self appears, in other words, to be a selected subset of the full content of consciousness - it is this subset that we read out as the content of self in any given moment."[264]

How this can happen cannot, it is true, be fully explained at this time, but consider for example this speculation from one in the field,

> "...Many cognitive scientists and neuroscientists are convinced that quantum theory could provide a large part of the explanation. If there is a quantum mechanical wave-phase aspect on the level of synaptic connections that proves to be fundamental to a mathema-tical model of overall human brain function, this could provide the now missing substantive validity to the claim that complementarity is a fundamental aspect of human consciousness."[265]

To explain how such a quantum mechanical effect known as "complementarity"[266] could produce the feeling of consciousness and self is beyond our scope here. Suffice it to say that scientists do not believe they will need to ring in any non-physical agent such as a soul to explain it, nor any causality outside the universe. This reduces the self to an illusion, although a pleasant one. As one researcher puts it, "The more we learn about the brain, the more it seems 'we' are cast in the role of passengers rather than pilots."[267] Antonio Damasio, cited above for his book, *Descartes' Error*, has a more recent work, *The Feeling of What Happens,* in which he speaks of a "core self" which we may share with many other animals but which is largely transient and barely "conscious", along with an "autobiographical self-consciousness" which includes our memories and weaves a picture of an enduring self.[268] This kind of conscious awareness of a self may be unique to humans, though it may also be shared with some primate cousins, or dolphins, or whales, or whatever, but in any case it is a development and refinement of a physical, though "emergent" quality of the workings of the brain. In fairness to those who still question whether the working of the brain can explain all our mental capacity, scientists cannot yet explain, "What Goes On In Your Head When You Have a Thought?...We really do not know" as the authors of a recent book on consciousness put it.[269] However they do present a good case that, "it is the amazingly complex material structures of the nervous system and body that give rise to dynamic mental processes and to meaning. Nothing else need be assumed - neither other worlds, or spirits, or remarkable forces as yet unplumbed..."[270]

DO "I" KNOW "ME"

A further illustration of our unity with the rest of the animal kingdom in the matter of brain function is the role played in directing our actions and motivation. We are sometimes under the illusion that all our actions are perfectly free and "rational", though when presented with extreme cases and even when confronted with evidence from our own "ordinary" actions, we can admit that there is often unconscious motivation going on. While it is true that humans can often overcome their unconscious drives and emotions, there is no denying that they are there. Sigmund Freud began the trend that has led us from considering ourselves as rational souls to emotional beings whose feelings often precede our reasons. Researchers into the brain often concentrated on the electrical activity since that is what could be measured. We are becoming more aware of the role of chemicals in emotions and therefore in our behavior. Candace Pert, in *Molecules of Emotion* reaches the conclusion that,

> "If we accept the idea that peptides and other informational substances are the biochemicals of emotion, their distribution in the body's nerves has all kinds of significance, which Sigmund Freud, were he alive today, would gleefully point out as the molecular confirmation of his theories. The body is the unconscious mind!"[271]

Dr. Pert continues,

> "In the popular lexicon, [the] connections between body and brain have long been referred to as 'the power of mind over the body.' But in light of my research, that phrase does not describe accurately what is happening. Mind doesn't dominate body, it *becomes* body - body and mind are one.[272]

Why does a mother love the child she bears and love the father, (whom she may have cursed mightily during delivery)? Researchers Hrdy and Carter report that,

> "During the last third of pregnancy, a cascade of endocrinological events readies and motivates mothers. [Oxytocin which produces uterine contractions,]...released into the brain is known to promote calming and positive social behaviors, such as pair bonding."[273]

Hence the urge to cuddle the baby. This is not deterministic and certainly mothers are known who reject their children, but it does affect mothers' behavior. The effect of hormones on human love and bonding has been found to be quite extensive. True, human cultural pressures guide the expression of love to a great extent, but it is the rush of phenylethylamine (PEA) that makes it feel so desirable, though the effects do not last for more than a few years. Then endorphins keep the couple together when the PEA is no longer effective, along with the ubiquitous oxytocin that makes it feel so good to make love or to just cuddle. These same chemicals cause other animal species to make their little animals. Other chemicals called pheromones are used for communication by many animals, from stimulating the opposite sex to warning away rivals. Since these are so ubiquitous among mammals, it would be strange if humans were not also affected, and there is evidence that indeed we are.[274]

Robert Wright in *The Moral Animal* gives many illustrations from evolutionary psychology of human behavior in mating, in family life and friendships, in business and in many other everyday situations of our innate inclinations and strategies which are obviously derived from evolutionarily directed tendencies of our brain. We all want to express our feelings of closeness to our friends, and we find ways to do this. Why is there a difference in the means used to express these feelings to friends of one sex, usually the opposite from ourselves, from the means used to express many of the same feelings to those of our same sex? If holding hands and petting is a good communication technique toward a person of the opposite sex, why is it not for another of the same sex? This is not something about which we often make a conscious decision, nor do the people attracted sexually to their own gender. One just feels right and the other does not. True, we *can* think about it and change our behavior, which most animals don't seem to do, but the point is not our differences from animals but our similarities. We have these tendencies because they *tend* to be helpful in the game of life, usually the reproductive game. As Wright says,

> "It all feels so rational, and in some ways it is. But that doesn't mean it isn't in the service of Darwinian ends...To an evolutionary biologist, what seems natural is...that human brains evolved not to insulate us from the mandate to survive and reproduce, but to follow it more effectively."[275]

It is also true that we can choose not to do these actions when it is not appropriate, for example when a married woman's karate champion husband is around, but a baboon can do the same in the presence of a larger rival. We can also bring higher inhibitions into play such as the unwillingness to hurt the other by a sexual signal that might give a false declaration of commitment, and it is unlikely that the baboon ever has these thoughts. Marilyn vos Savant makes a good point in a popular magazine when she suggests that it is the psychological which produces the chemical reactions.[276] It sometimes comes down to which came first, the endorphin or the feeling of love. All the while, we are having these psychological processes, and acting on them, we still feel the same biological urges as the baboon and possibly feel foolish if we do not follow up on them. If our decision-making processes are of such a higher realm than the animals, why should this be so? It serves no useful human purpose to not be able to put away the feelings once we have made a decision. Once I have decided that it is better to end a relationship, why should I feel a loss, and maybe keep feeling it for a long time or even not be able to do what I know is best? There is no logical or utilitarian reason why this should be so, but many can testify that it is so. It would be much more logical and comfortable to put away our feelings once our "higher intellect" has decided, but we cannot deny our heritage even if we can direct it.

I have been reporting on the effect of chemicals on our moods and actions, but the structure and function of the brain itself seem to also have their effect. There is evidence that the prefrontal cortex of the brain,

> "...determines both our general outlook on life and whether we respond positively or negatively to events and experiences...The right side governs a physiological loop that produces negative, inhibiting feelings, while the left commands a loop for positive, outward-reaching emotions. Research now suggests that a person's natural temperament - optimistic, pessimistic, extroverted, or introverted - may depend on which side of the prefrontal cortex is more active."[277]

These researchers do not conclude that the action of the brain is somehow having an effect on a "soul." It is the brain alone that does these things.

MORE DIFFERENCES NOT SO DIFFERENT

In our conduct of business and commerce we pride ourselves in being worlds apart from our primate cousins. None of them has ever won a seat on the stock exchange. However, Ken Boulding points out that according to Robert Trivers'

> "...theory of reciprocal altruism, some of our very distant, very hairy ancestors exchanged goods and services, like food and grooming, before they were acutely aware of *anything*...If Trivers is right, and the impulses underlying barter are rooted in our genes, it means that genetic evolution built division of labor not only into our bodies, but also, more grossly, into the structure of our societies. Cultural evolution, impelled by the pursuit of individual self-interest, then carried the social division of labor to greater definition."[278]

Humans have refined the practice of barter and economics, but they did not invent it.

Even the "pursuit of truth" so quintessentially human, is not without its evolutionary roots.

> "Scientific thinking has almost certainly been with us from the beginning. You can even see it in chimpanzees when tracking on patrol of the frontiers of their territory, or when preparing a reed to insert into the termite mound to extract a modest but much-needed source of protein. The development of tracking skills delivers a powerful evolutionary selective advantage. Those groups unable to figure it out get less protein and leave fewer offspring."[279]

How far is it from sitting quietly and contemplating whether a stick can be used as a termite-extracting tool or whether another must be found, to sitting quietly and planning whether a dovetail joint would be preferable to a butt joint? And how much farther is it to contemplating the mysteries of the universe? Some distance, surely. No chimp has won a degree in systematic theology. However, there is a continuum from one end of the thought process to another, not a gap that must be filled by another unrelated (spiritual) mechanism.

I SPEAK, ERGO I AM.

Humans' ability to use language is often seen as a distinguishing feature of the species from other animals, including the primates. It is certainly true that no other species speaks as we do, they do not even have the physical capacity for forming words, though they are capable of making sounds. In spite of all the research done over the years to teach animals, if not to speak at least to communicate in human language, it has not been proved that they can do so. It is interesting that if the teaching is done to a young animal it can progress much faster than one whose learning begins as an adult. Humans are not much different in this and our vocabulary is fairly well set before we get out of our teens unless a special effort is made - a selling point for programs designed to increase your child's vocabulary. However, to say that the primates do not have language as we do is certainly not to say that animals cannot communicate with one another. Nor is it to say that primates cannot develop linguistic skills other than a spoken language. They certainly use body language, and can be taught to use symbols to express rather complex meanings. "...The arresting fact here is not that so many language skills are uniquely human; it is that speechless animals have linguistic abilities at all. The fact that they do - which is surely no longer in doubt - strengthens the case for thinking that our minds, like our bodies, are largely something we have inherited from our animal relatives."[280]

There are also no non-humans with a degree in mathematics, but rhesus monkeys can tell which group of objects is larger than other groups, chimps can perhaps top that. Anecdotally, I am convinced that my beagle, who gets two halves of a pig ear in the evening, knows when he has one coming and also knows when he has had two. Although he may ask for a third he takes "no" for an answer, which he will not do after one.

It is not necessary to say that our consciousness and our motivation are just like other animals. They do not have to be to make my point. All that needs be said, and at this point perhaps all that can be said, is that our consciousness and the basis for much of our motivation is not entirely different than other animals, and that it is obvious that ours evolved from theirs.

The objectors are not finished just yet. Much of what we do is done by all animals with a brain. Much of the remainder is done by other primates, though perhaps not quite as well. And all of this may be done without consciousness and a sense of self. There doesn't seem to be any reason for us to be conscious either since we could function almost as well without it. If this is the case, how can natural selection prefer the conscious? It is not so mysterious as it sounds. Consciousness, which Ken Boulding calls "know-what" (in distinction to DNA's "know-how" to build a body or a brain), knows how to build an image and do something with it.

> "...it entails conscious understanding, like the understanding - attained, presumably, somewhere between a chimpanzee's degree of complexity and a human's degree - that prey will drop dead if clobbered. Once you get know-what...this produces a profound change in the evolutionary process. With the development of the human race, evolution goes into what I would call a gear change."[281]

There is not such a large distance between the two in image making and processing. The consciousness makes for better imaging and planning of responses and it brings along with it a capacity for feelings and emotions, which itself is not unknown in the animal world.[282]

Emotions opened up a whole new arena for evolution. The organism can be motivated in many new and subtle ways and can relate to others of its kind in a new manner, often very helpful in survival. We have already seen the role of emotions in reproduction. The emotion of fear triggers in us, and evidently in many other animals, either the reaction of flight or that of defense, whichever seems most productive. A lack of fear in the face of a life-threatening situation can be fatal. This is not to say that an appropriate non-emotional response is not possible. That a plant has emotions is much harder to believe than that an animal might, but the plant can react defensively to stimuli. There is even evidence that a tree can warn the next tree to take precautions, too. So could an animal, including us, without the addition of emotions. However, the emotion adds a certain intensity that may enhance the response. The appearance of emotions in brains may well have been serendipitous, another emergent quality, but once it was present it could be used by natural selection. The benefit in pleasure or enhanced response time may be accidental, but it is certainly also helpful. In humans, it is what makes life "worth living". It certainly seems possible to have emotions without having a human-style consciousness. It is also possible to have our consciousness without emotions. Phineas Gage of iron-rod-through-the-head fame was a good example of this. Persons suffering from brain damage can be found today with this condition from other causes, and there certainly seems to be an evolutionary advantage to having both. Phineas did not do near as well for himself after his injury as he had before. We are more successful, at least so far, than animals with less consciousness, and humans without emotion often cannot function in society very well. So the advantage of both has been selected for by the same Darwinian processes that give the cheetah speed and the beagle a great nose. "All the mammals - and many other animals as well - experience emotions: fear, lust, hope, pain, love, hate, the need to be held. Humans may brood about the future more, but there is nothing in our emotions totally unique to us."[283]

There is another advantage to language and consciousness that also has great survival value. Having them allows for checking signals with others about what is real in the world. This is the function of culture. Certainly animals learn from one another, and learn from experience, but much less successfully than humans do. Once again, we can see a development of this ability through the ages that reaches a new plateau in us.[284] Researchers have noted that our ability to learn from each other may have developed even before we split from other primates. Once language was possible, the ability to pass on knowledge was greatly enhanced. All culture is designed to pass on ideas of how the world operates and what is real and what not. Even abstract art bears a message to the viewer that the "real world" may not be what it seems and an invitation to open up to other possibilities. The same is true of economic systems which all have a foundation in values based on an interpretation of how the world operates.[285] Religion, too, proposes that there is more to life than meets the eye and that there can be more meaning to human actions than is apparent. All these views of reality are transmitted by language and symbolism. A far cry from the chimpanzee's training of its young to fish for termites, but on the same continuum.

GENES & MEMES

The entity that regulates the development, formation and evolution of our bodies is called the "gene". Richard Dawkins has pointed out that while the genes are still very much with us and operative, another element has entered human evolution, the influence of culture. "Dawkins coined the term *meme* to refer to cultural replicators that are `capable of being transmitted from one brain to another.' Memes are ideas, fashions, art, etc. - entities that evolve extragenetically and extrasomatically. Memes suppose brains...to manipulate them in the intricate process of memetic evolution."[286] A meme may be a belief, a belief in god as the sender of rain for example. If this belief contributes to the welfare of the society and its culture then it will survive. If not it will be replaced by some other belief, just as a physical attribute that contributes to the survival of the creature's body may be passed on but one that is deleterious will usually not be. "A faith, like a species, must evolve or go extinct when the environment changes"[287] Ken Boulding says that the founders of religious movements are mutations in the memes of their culture.[288] Clifford Geertz,

> "...argues that a need for meaning systems is built into the human condition. As the only major animal lacking an elaborate system of instincts man depends upon his culture for survival. Geertz contends the evolution of human culture and the evolution of the human organism have gone hand in hand...In Geertz's theoretical framework there are a number of different meaning systems with which man copes with the complexity of reality: science, common sense, ideology, history, art. But the ultimate in meaning systems is the religious one, precisely because religion addresses itself to the most basic questions that man must ask - that is to say, questions about the ultimate nature and structure of reality."[289]

Susan Blackmore in *The Meme Machine*[290] treats of religions as "memeplexes" and gives a rather unflattering analysis of the development and effect of the religion meme(s). Though she admits that science is also, "a mass of memeplexes", she concludes, "that science, at its best, is more truthful than religion" since it allows for testing of the truth of its memes, while religion does not. Whether one agrees with her conclusions or not, the idea of memes as a cultural factor is helpful if not accepted by everyone.

Capitalism and Communism are also memes that have had tremendous influence on human history. It is obvious that Communism as practiced in the Soviet Union and elsewhere did not have survival value. It remains to be seen if a mutation of it will arise. This idea of memes seems helpful but still a fiction, especially when memes are spoken of as having a life of their own as Dennett does when he says, "the haven all memes depend on reaching is the human mind, but a human mind is itself an artifact created when memes restructure a human brain in order to make it a better habitat for memes."[291] Still, there is a reality here on the order of Dawkins' speaking of "selfish genes" as if they and not the individual that they direct are the important entities. This way of speaking about these realities can be pushed too far or it can be dismissed, but either would be a mistake.

However, for our purposes the question is not whether memes or genes or the organism that contains them is the ultimate and controlling reality. Our question is, could culture evolve or did human culture need some kind of (divine) intervention to arise? Once again there seems to be no reason to doubt that the capacity of the human brain to go beyond perception of sense data and the application of innate programs evolved from the earlier primate brain. Hints of art, language, science, imagination and creativity are found in our primate cousins. Not culture as we know it, or language as we employ it, but the rudiments;[292] enough to start from and arrive at us.

Once again we see distinctly human characteristics that have been proposed as reasons for insisting that there must be some extraneous cause since they could not have arisen on their own and certainly not "from nothing". However, we have traced the development of complex entities from the Big Bang through the formation of the physical structures of the universe, through the appearance of life, its evolution and exploration of innumerable niches in the environment, including that of a conscious human person and found no place where we must insist on an external agent or a cause outside of itself. "A modern neurobiologist sees no need for the religious concept of a soul to explain the behavior of humans and other animals."[293] Nor is there a need to bring in any process that produces consciousness other than the process of evolution. Frank Tipler puts the conclusion very bluntly when he says,

> "I therefore regard a human being as nothing but a particular type of machine, the human brain as nothing but an information processing device, the human soul as nothing but a program being run on a computer called the brain...Many people find this extreme reductionist approach to life not only wrong but repulsive"[294]

Daniel Dennett brings out the implications,

> "...All the achievements of human culture - language, art, religion, ethics, science itself - are themselves artifacts...of the same fundamental process that developed the bacteria, the mammals, and *Homo sapiens*. There is no special creation of language, and neither art nor religion has a literally divine inspiration. If there are no skyhooks needed to make a skylark, there are also no skyhooks needed to make an ode to a nightingale..."

Dennett, like Tipler, recognizes that this is not a popular position to maintain adding,

> "...Some people find this vision hateful, barren, odious. They want to cry out against it, and above all, they want to be magnificent exceptions to it. They, if not the rest, are made in God's image by God, or, if they are not religious, they want to be skyhooks themselves. They want somehow to be *intrinsic sources* of intelligence or design, not 'mere' artifacts of the same processes that mindlessly produced the rest of the biosphere."[295]

As much as those who believe in god want to have god playing a role in our coming into existence or at least in becoming what we are, it looks like it need not be.

CHOICES

However, it is not just consciousness that has led us to believe that we are vastly different from the rest of the living organisms on our planet. Even if we allow the possibility that other organisms are "conscious" of their environment in some manner analogous to our own awareness, and that our own perception of the external world and of our place in it is different in degree but not in kind to theirs, there is yet another facet of what it means to be human that must be explored. In some animals the perception of "other" and the appropriate response of the "self" (I am not speaking here of an animal knowing that it has a "self") is "hard wired" into their brains and though there may be a great deal of complexity in their response, at times resembling the deliberate choosing of options, this is not the case. When a crawfish is threatened it backs away with a quick flexing of the tail. At what precise point does it "decide" to do this? And then when does it decide to move straight back or veer to the right or to the left? Not at the first presentation of the threat, or it would be jumping at every motion outside itself, even though the menace may be nothing more than a leaf floating through. Well, maybe when the menacing presence is within one foot? One inch? When? Does it choose before it makes the first move which direction it will take, or does it decide enroute? Investigation shows that there is really no conscious choice involved but rather a neurochemical trigger that reaches a certain threshold and causes the avoidance behavior. No different than ourselves when we suddenly "decide" to take a breath, or that we have had enough to eat. Why not one forkful *before* we "decided"? Or one after? These things are not normally a matter for conscious decision, though we feel that if we want to, we can make them such. Whether or not we really can may be open to debate. If I decide that I will decide when to stop eating, have "I" really made that decision when I think I have or has my unconscious (brain) made it for me and then convinced me that I made it? Experiments have been done in which the subjects were asked to note at what second they decided to flex their finger. Instruments determined that the brain sent a signal to the finger *before* the mind decided to act. The brain also delays the feeling of the movement giving the conscious mind the illusion that it decided and then the action took place. "Actually, by the time the mind orders the finger to flex, the impulse has already been dispatched."[296]

FREE WILL

Sociobiology is a branch of science which holds, in the words of Edward Wilson, that,

> "Human beings inherit a propensity to acquire behavior and social structures, a propensity that is shared by enough people to be called human nature. The defining traits include division of labor between the sexes, bonding between parents and children, heightened altruism toward closest kin, incest avoidance, other forms of ethical behavior, suspicion of strangers, tribalism, dominance orders within groups, male dominance overall, and territorial aggression over limiting resources. Although people have free will and the choice to turn in many directions, the channels of their psychological development are nevertheless - however much we might wish otherwise - cut more deeply by the genes in certain directions than in others...The brain, sensory organs, and endocrine systems are prescribed in a way that predisposes individuals to acquire the favored general traits of social behavior."[297]

Be that as it may, we do feel that in some measure we have "free will" and can make choices and decisions that cannot be predicted, and we also feel that animals cannot do this. If it is true that we have free will, does this require the addition of something non-physical within us, a soul perhaps? Or do we really not have a free will at all and fool ourselves into believing it? Or is there some middle ground? Certainly, for the determinists of an earlier time, especially after Newton showed us the reductionist universe of blind adherence to rigid physical laws, free will is not possible. Eugene Mallove observes,

> "Since the hollowness of atoms was far from apparent, we can perhaps excuse the naive exuberance of eighteenth-century scientists who imagined a billiard-ball universe...In principle, an omniscient being knowing the exact position, mass, and velocity of every atom at one instant could, by using the newly developed theories of dynamics and gravitation, predict the future course of

> all matter. Free will was merely an illusion, and determinism ruled the cosmos."[298]

The crayfish was merely following a predetermined course and if you knew all the parameters you could predict when it would flex its tail and which direction it would go. You could do the same for a human being. Yet, it certainly does not feel right to deny that we have free will. Most people would argue that the best way to exercise this free will in making choices is the use of pure reason and logic. One should sit down and list all the options available and the possible outcomes of each course of action. Then one adds up the pluses, takes the negatives into account, and makes the best decision. Of course it rarely if ever happens just this way, we merely think that it does. In our better moments most of us would admit that this is not our experience of decision-making. There is not some dispassionate little person sitting atop or outside the brain making use of the computer-like abilities of the brain to do the calculating and then making a decision. How many times we hear that someone did not know who they were going to vote for until they were in the booth. How often we make up a picture in our mind of the best possible mate for us and end up falling in love with the opposite of our dream partner. There is definitely something unconscious going on in our decision making process and, in fact, it often has the deciding voice.[299] Good thing, too. Every action has so many possible consequences and ramifications that we could spend a lifetime making the decision to floss or not to floss if it were not for something about us that short cuts the process and brings us to a decision. We sometimes act on "hunches" or "intuition" and recognize that we have done so. This seems to be an awareness in these instances of the process that goes on in every decision-making situation. Antonio Damasio in *Descartes' Error*, talks about "somatic markers" that our brain puts on the various options we consider, emotional responses that propel us, more or less consciously, towards or away from a particular choice.[300] Some of these may be innate, such as a fear of snakes may well be, though most are learned from prior experience, especially of negative consequences of our actions. The crayfish learns by experience, perhaps, that avoiding a floating leaf is a waste of energy, while avoidance of a crayfish-eater is most likely innate. The crayfish makes a decision without being aware of the somatic process by which it does so and it has no mechanism to challenge the decision of its brain. The chimpanzee can decide more consciously to use an inferior stick because a better is not available. A human can consider a far wider array of choices and overrule the unconscious or conscious emotional tags that the process has placed on the options under consideration, as we do when we put aside immediate gratification, which we understand would feel good, for something better in the future. We have much better equipment to use than the crayfish and our decisions can be much better informed than theirs because of this, but once again there is not a completely different process involved but a refinement, brought about through the process of evolution, of something common to all animals with a nervous system.[301] There is a continuum, not a quantum leap, (in the incorrectly used phrase implying a totally unpredictable change.)

More accurately, there *may be* a quantum leap. With the advent of quantum mechanics and the introduction of a fundamental uncertainty into reality, some introduce the hypothesis of a "randomizer" bringing chance, or a quantum effect, into the decision-making process. Tipler theorizes that,

> "...there is a growing consensus among cognitive scientists that all human originality is due to an essentially random mixing of ideas in the human creator's mind, with subconscious elimination (natural selection) of the bad ideas...Some chance must exist at the physics level, if the consciousness level is to be ontologically not determined. At the consciousness level, we are completely unaware of the operation of the randomizer, just as we are unaware of neurons firing. We think *we* have made an undetermined decision whenever the randomizer operates, as indeed we have. It is the agent, and not the randomizer that makes the decision...The various levels of implementation interact with each other in a human being...[the] randomizer...is constantly being ordered to change the probability weights in the decision matrix."[302]

Others do not speak of chance but give an explanation of an inwardly determined but not externally predetermined process, based on quantum laws which brings about the selection of a "top level code" in the brain, compiled from external stimuli and memories which then composes both a picture of external reality and a suitable response.[303] The introduction of chance or randomness, however defined, into the process of decision making shows why it is so much preferable to choose as we do rather than the "pure reason" advocated by the philosophers. If there were no mechanism for taking a "leap" toward one option over others, we would be as bogged down as the pure logician, a Vulcan like Spock maybe, in pursuing every possible avenue and never making a choice. There has to come a time when we "just do it", and how could this be possible if the brain were as mechanistic as some have thought it to be? Quantum uncertainty, the quantum leap, provides a possible explanation of how we can suddenly land on one option over others. The theories involving quantum theory's effect on choosing among options are not proven by any means, but most scientists presume that they are at least on the way to providing an explanation of the workings of the human brain that will explain our consciousness and experience of free will.[304] Since natural physical processes can in all likelihood explain these things, these processes arose as a result of the action of evolution, mutation and natural selection. I said earlier that we do not feel that animals make conscious decisions and have "free will" such as we do, and perhaps most do not, maybe none do. Still, they certainly do make decisions that cannot be predicted by a Newtonian mechanistic explanation of their behavior, and a quantum phenomenon cannot be totally predicted. There is always a probability, however small, of another result, another choice. A researcher into animal consciousness cites the actions of the common sparrow confronted by food-for-the-taking, but located in a garden patrolled by a cat.[305] The bird evidently "decides" whether and when to call together comrades who will provide more eyes to watch for the danger, but who will also eat some of the food. The bird "decides" whether to call only a few or many other birds; or to take the risk and dine alone. Granted, the bird may not be conscious of weighing all the factors and coming to a decision, but the decision is reached, and if we changed "sparrow" to "hungry human", the sequence of "thoughts" may well be the same. The only difference, presumably, would be that the human *might* be aware of making a decision and the bird *might* not be, and the human may take into account a wider range of possible scenarios than does the bird. We do not know about the bird's consciousness, but we all know from experience that the human might well make a decision to face some danger alone or to call for backup, and later, after the fact, be unaware of having consciously thought through all the particulars. Or the human may swear later that he or she coolly reached a conclusion based on logic, and really believe that he or she did, when witnesses know that they acted rather spontaneously. It is not such a big step after all from decision-making among our animal relatives and freely choosing a course of action ourselves. The conclusion remains the same. At no point in the explanation of consciousness and free will must we appeal to a non-physical source of empowerment such as a soul to accomplish these deeds, nor to a cause or designer beyond evolution and natural selection.[306]

James Trefil in *Are We Unique* asks, "...when we have understood the limits both of other animals and machines, will there be anything left that is uniquely human?"[307] His answer for the comparison with animals is that there is something unique to humans and it can be physically seen in the structure of the brain that has evolved like any other living organ.[308] As for computers, though the brain is much more than a digital computer, and Trefil feels that it will possibly never be duplicated by a machine, this is a very open question, because the brain itself is still a physical system through and through. Robert Jastrow in *The Enchanted Loom* is a little less sanguine about the competition of brain and computer and suggests we had better find some way to coexist unless we want to be replaced.[309]

NO CAUSE FOR CAUSE

I have mixed a few topics in the investigation so far, considered a few twice, and have wandered rather far afield. The first question is really quite simple: "is the assumption true that there must be a cause to explain the existence of the universe and its contents?" Since there was no other explanation available to the ancients, or even to people just a few centuries ago, and since in our experience of the way things operate we feel that there is always a cause, in fact a whole chain of causality, for every event, it was natural to presume that the universe must have a cause as well. The obvious question was what caused the cause, and the best that could be arrived at was that existence was part of some "uncaused cause's" being. This entity need not be a god or even a conscious being, but it had to have had no cause itself, or else that would be the first cause. You had to stop the parade somewhere. And certainly nothing could come from nothing. It either had to come from something pre-existing itself or it had to be eternal, time being considered pretty much a pre-existing condition itself though not a being. Now one could call the quantum vacuum the "uncaused cause", but this is hardly what the philosophers had in mind. There is no way one can call a vacuum a being, even a false vacuum. After the considerations above, however, we are left with the conclusion that there need not be a cause for the universe or for anything contained within it, ourselves included. Nor for any of our qualities. All can evolve without guidance and planning or anything resembling "cause". And if there need not be a cause, *a priori* there need not be a designer. Though evolutionists like to say that there is a designer in some sense, namely the process of evolution, this again is hardly the designer that Paley had in mind as his watchmaker.

NOR PURPOSE, RHYME OR REASON

Most scientists would be the first to admit that their findings do not *deny* that there is a designer whose purposes are being worked out in the universe, it is just that there is no scientific reason to propose one. This was not true in the past when there was no known mechanism that could have brought the evident complexity of the universe, and especially of living things, into existence. No one could fault Newton and others for presuming the Designer's existence. Howard van Till, a proponent of the necessity for a designer, preferably god, says,

> "Evolutionary processes, claim the proponents of naturalism, are composed of a string of accidents; surely a string of accidental events could not function to achieve a specified purpose...(But) questions of purpose must of necessity entertain the possibility that the universe and its temporal development achieve their significance in their relationship to a nonphysical realm...Questions of purpose are fundamentally religious questions, they are not questions that natural science alone is capable of addressing."[310]

This appeals is a favorite tactic of proponents of the existence of a Designer. None of us wants to be irrelevant. The loss of something that defines us and our role, the death of a parent, loss of a spouse, a limb, a job, all these things threaten us because they force us to redefine who we are. We sometimes act as if our sole purpose for working is to make our fortune, earn our daily bread, or some other limited purpose, but if you deprive us of the feeling that our work is valuable and goes beyond these limited purposes, thus giving our lives value, we rebel. One can make a good living putting the proverbial bumpers on the proverbial assembly line cars, but workers have often insisted on the rotation of duties, following one vehicle through the whole line, for example, as a way to experience the joy of accomplishment. We need purpose and meaning in our lives, and in order to have it there must ultimately be purpose and meaning in the universe. If not, then I will certainly die before my accomplishments do, but no matter what my successes may be they, too, will one day fall apart. Even if they last until the end of the universe, they will eventually end and come to nothing, and with them I come to nothing as well. So we not only need to live meaningful lives, we need the universe itself to have some purpose and meaning. One way to attack the possibility that the universe arose from nothing for no reason at all is to appeal to our need for purpose. Van Till claims that a string of accidents cannot reach a specified purpose, and certainly they cannot. But who is to say that there *is* any purpose in the universe? Or to the universe? We like to think that there is, but there is no necessity that it be so. We like to judge things by, "what is it good for?", though in our better moments we recognize that we should not apply this standard to other human beings, but give us a gnat buzzing around our head or a virus taking over our cells and we wonder what god could have been thinking to make such a creature. What is it good for? Not only must it have a purpose, that purpose must be for our welfare, since we are the darlings of the universe and the apple of god's eye, and when something does not promote our welfare we are perplexed. We can usually find some purpose for things, even if the gnat's purpose is to be food for the indigo bunting that so delights us from its perch on a dead tree limb. Nevertheless the gnat does not have to have a purpose, it just is. Just as we can, sometimes with difficulty, understand that the value of the mentally handicapped human being is not dependent on his or her utility to society, the handicapped, we realize, are valuable because they exist, so we can sometimes extend the same courtesy to other species. Good thing, too, since there does not seem to be a design leading to the existence of anything at all, it just is. The universe itself need not have a purpose, and if it does have one, that purpose totally eludes us, though we often try to make its purpose to be the arena for human activity. Proponents of a designing god claim that the universe exists to make possible the fulfillment of some purpose in the mind of god, but they can merely propose it, not prove it. Indeed, if god's purpose is a good one, god certainly takes a roundabout way of getting to the goal. On the other hand, the non-believing scientist cannot prove that there is no purpose, but can surely show that the universe need not have a purpose in order to exist. There need not be a cause, or a design, or a purpose to any of it, ourselves included. The bottom line is that there need not be an ultimate explanation for the existence of the universe. The assumption that was made early on that there is a cause for everything may hold for the day-to-day operation of the sensible world, but it does not hold in the larger picture. We have seen in the understanding of quantum uncertainty that our grip on those things that we experience in our sensible world is somewhat tenuous as well.

CAUSALITY AND CONTINGENCY—AGAIN

Contingency is another favorite argument of the philosophers against the apparent lack of need for an initial cause or a prime mover. In response to Julian Huxley's agnostic or even atheistic pronouncements, the philosopher Raymond Nogar dismissed him by saying,

> "Huxley misses the necessity for transcendent being, i.e., existence not fully contained by the materials of the universe, because he mistakes the apparent dynamic order of evolutionary unfolding for existential self-sufficiency. He is so enamored with the marvelous order of evolution in the universe...that he misses the absolute requirement of its contingency, its fundamental creatureliness. That requirement is God."[311]

Contingency means that the thing under examination does not have the reason of its own existence in itself. It requires some prior agent to give it a "ground of being." The argument is a variant of the ontological argument for the existence of god that has been roundly rejected among philosophers and theologians and is very close to the appeal to personal witness that says basically, "I do not see how something can exist without a cause, so therefore it does not exist without a cause." That is the same as saying, "I do not see how an eye could have evolved, so therefore there must be a designer for the eye." What is so "absolute" about the requirement of contingency? To say that a creature cannot exist of itself because it is a creature which demands a maker is to answer the question before it is asked. The question is, is the universe or anything in it a *creat*ure-which-is-by-definition-*creat*ed? The argument from contingency always did beg the question of how the prime mover, or the uncaused cause, or god, got *its* existence. One always had to say, "it just is", which is precisely what we can say about the universe. Van Till claims that "...a `vacuum state,' for example,...is not the true absence of anything; it is merely a certain nonmaterial state of this cosmos. Even self-creating universes must begin with some form of `self,' thereby leaving the ultimate question of origin unanswered."[312] On the contrary, the quantum vacuum, though it is a factor in our universe, "exists" (though that is not the right word for something [also the wrong word] which is an absolute nothingness), "before" (also inaccurate) there is time and space, which only come into existence (finally the right word - almost) with the Big Bang. The quantum vacuum is not a non-material state of this cosmos or of any other cosmos. Van Till wants a personal, non-material, transcendent god but he cannot have one by using the quantum vacuum as an instrument of his god. Science cannot prove that there is not a nonmaterial being (Van Till's God) apart from the quantum vacuum that is self-creating, but then in terms of contingency, what is the difference between the two?

THE SECOND ASSUMPTION, THERE IS A *RATIONAL* CAUSE.

The second assumption, that every cause that was presumed to exist in the first assumption, is a rational cause, is very similar, of course, but it does add another dimension. The prime mover of the Greeks did not have to be a rational being, and the prejudice of some of the philosophers against the myriad gods and goddesses of their civilization's myths led them to deny that it was. Certainly most people identified the most fundamental forces they knew of, whether a prime mover or a thunderstorm, with a rational, personal being. However, we can make short work of this assumption. If it need not be true that there be a cause of any kind, then there need not be a rational cause, either. I repeat that this is *not* to say that there is not a rational being who has caused everything that exists, but it is to say that in the light of scientific cosmology today there *need* not be such a being.

THE THIRD ASSUMPTION, SOMEONE SUSTAINS CREATION

As long as no one knew the cause of many physical phenomena, it was natural to assume that the gods were the explanation of all existence and the cause of all change. If it snowed at an odd time or in a strange place where snow was not normal, it must be that the gods caused it. Aside from the extraordinary, they were also involved in the ordinary. Besides the strange events, they were also the background for the mundane. If you had a great crop, the gods had smiled on you, and if you had just an ordinary crop, they were satisfied with you and just keeping things going normally. In the ancient myths from which the myths of Genesis took their inspiration,[313] "the gods wanted somebody to take care of their garden and cultivate the food that they needed, so they created man."[314] In the Genesis story, at least as later interpreted, it is not god who needs the food, but the humans. Either way there must be continuity. Neither humans nor crops do well with constant upheaval. This understanding of the workings of god in the natural world persisted for at least millennia and if it was questioned we have no record until the time of Socrates and Plato. Surely it was questioned, since even when people claim today to believe that god is "in charge" and pulling all the strings, they often do not act like it. Still, it was the Greeks who removed god from the day-to-day operation of the universe yet gave god the role of sustainer, either as a personal being or a prime mover who or which provided the "ground of being", not just in the beginning but on an on-going basis. In the Iliad and the Odyssey, everything happens at the whim of the gods. By the time of Thucydides, almost nothing did. "In a few hundred years, history passed from god-driven to human-driven"[315] This particular antitheist philosophical position was never popular, however, and the gods remained in their everyday role of making things happen. Augustine declared that,

> "God is the unchanging conductor as well as the unchanged creator of all things that change...He is ordering all events according to his providence. This will hold good until the beauty of the completed course of time...Shall have played itself out, like the great melody of some ineffable composer."[316]

The most recent *Catechism of the Catholic Church* continues this understanding when it says that god, "...at every moment upholds and sustains them in being, enables them to act and brings them to their final end."[317] This builds on centuries of understanding god as being a sustaining and guiding force in the universe. On the Protestant side, a creed called *The*

Belgic Confession of Faith of 1561 believes that god "...rules and governs...according to his holy will, so that nothing happens in this world without his appointment..."[318] All of these confessions of faith not only insist that god sustains and guides all events according to god's own purposes, but also hold that if god were not to do so nothing could happen, nothing would occur. Indeed, the universe would slip back into the nothingness from which it came. It was not until the beginning of the scientific revolution that this premise was really challenged. And since the revolution mainly occurred in the Establishment of the Church, even then the new findings of how and why events came about tended to be seen as evidence of god's subtlety as a cause rather than a denial of that role. Still, the more that was known, the more god was removed from the day to day workings of the cosmos and relieved of the role of puppet master to the elements. "Before Newton, a scholastic philosopher, certain that an arrow arced toward its target because, as Aristotle taught, it was constantly acted on by a mysterious force, or impetus, could easily believe in a God who also moved things along by according them his sustained attention. Such a God might look after one if properly addressed...But after Newton the laws alone sufficed."[319] Gradually god was pushed farther and farther back in the chain of causality, and sometimes in history. When evolution was first proposed many who did not react adversely against it attempted to at least keep god somewhere in the causality loop. God as sustainer was insisted on, not as one who was continuously active, for it was obvious that things happened for physical, chemical and mechanical reasons and did not require a fresh intervention from god to occur. But the ancient Greek notion that there must be a sustaining presence or ground of being was brought into play again. With modern findings about the timetable of the universe's coming to be, god's activity was pretty well limited to the extreme beginning of the universe. Since physicists are fairly comfortable that they can find answers to questions of "why is it like it is" back to the Planck time of 10^{-43} second after the Big Bang, this can result in god having an extremely tiny window of opportunity for involvement. This is obviously not the god of biblical faith. So the argument is put forth that if evolution can explain how the feather of the bird and the eye of the newt developed, what can explain how evolution itself came into existence? Surely, they say, it is an instrument of god who intended to work through the proximate causality of evolution. This is similar to the argument of some strong anthropic principle adherents that god must have set the initial conditions since they seem so finely tuned to produce the universe and life within it. Evolution could have taken so many lines that lead nowhere that the fact that it leads somewhere must mean that god is directing it. This proposition has a long history and is so logically self-evident that it is very popular for those who are, or wish to appear, sophisticated in scientific thought but still maintain the theist position. Arthur Peacocke, who is certainly knowledgeable in science, attempts to reconcile the difficulties presented by the removal of a need for divine meddling (causation) but to save the role of sustainer when he asserts that,

> "The stuff of the world has a continuous, inbuilt creativity...So, if we identify the creativity of the world with that of its Creator, we must emphasize that God is *semper creator*...But to speak thus is to recognize also that God is creating now and continuously in and through the inherent, inbuilt creativity of the natural order, both physical and biological - a creativity that is itself God in the process of creating. So we have to identify God's action with the processes themselves...And this identification means we must stress more than ever before God's immanence in the world."[320]

What he says is true, of course, only if we must identify the creativity of the universe with the creativity of god. If we need not make this identification, then there need be no continuous or sustaining activity on god's part. There seems to be no reason to deny creativity to evolution, if only because of the weak anthropic principle.* If it were not able to bring about life and then intelligent life, we would not be here. In our universe the random action of evolution is capable of bringing into existence any number of forms. In some other universe perhaps it cannot, but then there is no one around to ask why not. In our universe, evolution is capable of bringing into being forms that are self-sustaining and capable of mutating into a whole succession of forms. If this were not possible, the universe itself would possibly not be self-sustaining either and would wink out of existence, either immediately or after some time. The philosophers may require some "ground of being" for anything else to continue in existence, but evolution does not seem to know of this requirement.

THE FOURTH ASSUMPTION, THERE IS A CREATOR

What, then, of the creator, the fourth assumption?[321] By this point in our consideration it is obvious that there is no need for a creator of any kind. Again, this is not to say that there is *not* a creator, and perhaps a personal deity who created from nothing with a rational plan in mind. Such a being

---

* Remember that we defined this above as the claim that we can make inferences about the nature of the universe from the fact that it has produced intelligent life.

could create by setting the initial conditions of the universe with full knowledge of where the process would go - a very reductionist theory. Or such a creator could be the playful sort who knows the general tendency that creation will go but not the particulars and is content to "let it rip". Maybe a nudge in the right direction would be required from time to time, but such a fun-loving creator would get its kicks from seeing what developed. This has been the approach of many who try to reconcile the old philosophical demands for a prime mover with the findings of the role of chance and chaos in the universe. Or a creator could have made the whole thing in six days in exactly the manner recorded in the Bible. None of these positions can be ruled out, but the modern Catholic, or any Christian, is going to have to face the fact that it might be that *none* of these possibilities is true. The universe can explain itself without reference to a creator. All the Paley's in the world cannot show the necessity for such a being and none of Aquinas' proofs add up to conclusive evidence.

THE FIFTH, SIXTH AND SEVENTH ASSUMPTIONS, THERE IS A PURPOSE TO CREATION, INDEED A PLAN, AND SOMEONE IS IN CONTROL.

The fourth assumption of the theists, a creator, is shown to be just that, an assumption that has been made. We can explain why the assumption was made from historical developments in humanity's efforts to understand the universe and especially our own existence and place in it, but we cannot show that the assumption is a necessary conclusion. Quite the opposite! There is no reason to believe that the assumption is true except those religious assertions themselves. And without a creator, without a prime mover, and without any necessary direction to the development of the universe, it is rather obvious that there need be no one who has some purpose in mind for every action (assumption #5) and certainly not a coherent plan or teleology (assumption #6), at least in the manner in which this has been postulated previously. [322] Lacking a goal, there is no need to invent a being who is pursuing said goal or ultimate purpose throughout history, and therefore no one need be in control (assumption #7) and capable of making the corrections in the direction of evolution to reach this end. Not in the daily working out of the details; not in the progression throughout eras and eons; not in decisions prior to a beginning setting physical (or moral) laws or initial conditions. Assumption #7, that there is someone in control [323] would then be as meaningless as the other assumptions.

THE EIGHTH ASSUMPTION, MIRACLES CAN OCCUR.

Then is it true that miracles happen?[324] Certainly there are many Christians who find miracles every day, in fact are trained and encouraged to "expect a miracle" and taught that you will see it because god is constantly manipulating the universe for the benefit of god's loved ones. Angela Tilbey writes that for much of her early life,

> "…I tended to have the typically evangelical view that the things that happen in my life are directly caused by God, that by being religious one enters into a secret world determined by laws of spiritual causality. Parking spaces appear in crowded high streets for certain kinds of evangelical Christians; cancers are cured by prayer; cheques arrive through the post just when the mission hall is about to close down."[325]

After much consideration of the content of her faith, Tilby ends with the view that,

> "The kind of religious reductionism which dissects it into narrow bands of that caused-by-God and deprived-of-God is a denial of the interrelated givenness of things. God's sun shines on both the just and the unjust. When Hawking and others sneer at the kind of religion that can only see god as cause and explanation, they are right to do so."[326]

Eugene Mallove reaches the same conclusion, that finding god continually intervening in creation is a devaluation of the universe.

> "A purely theistic viewpoint treats nature as a footnote to a transcendent divinity, one clearly not to be approached intimately through science…The theistic need for supernatural events - miracles - may simply represent a fundamental misunderstanding of and dissatisfaction with nature as it is, or should we say, as it is incorrectly imagined to be. Nature, viewed intimately with a deep sense of awe, not devalued as base, naive materialism, is seen to express "miracles" continuously."[327]

Mallove, Tilbey and Hawking are hardly the first to object. David Hume questioned the wisdom of believing in miracles in his *Essay On Miracles*, preferring to believe in what is experienced as a guide to what is possible. What is experienced is always to be preferred over what is reported by others when what is reported is at variance with the common experience. Our experience that the dead do not rise is to be preferred over the report that at some time in the past or over in some other village the dead do rise. This is firmly in the camp of the scientific mindset which is fundamentally opposed to ceding ground to the unknowable and which, moreover, has never seen or heard any credible evidence that a miracle has taken place. Dawkins explains that

> "…my thesis [is] that events that we commonly call miracles are not supernatural, but are part of a spectrum of more-or-less improbable natural events. A miracle, in other words, if it occurs at all, is a tremendous stroke of luck. Events don't fall neatly into natural events *versus* miracles."[328]

One explanation for the "miraculous" is, once again, mere chance. Anything is possible, given enough time. With a lot of time, but not an infinite amount, given the constant motion of the molecules making up one of the Rocky Mountains, the entire mass of the mountain could move twenty miles in one leap. Maybe one already did, since the Bible says that the "mountains skipped like rams",[329] but what are the chances of this happening in our lifetime? Or in the lifetime of the universe? Very slim, but not zero. The improbable is not the impossible.

How many battles have been fought in all of human history? The figure must be quite large, given our proclivity for war. And how many clouds could look like a cross? Not as many as could look like a cotton ball, but still a significant number. So what are the chances that the commander, Constantine for example, of one of these armies could look to the sky and see a cross, win the battle (50/50 that he will win), interpret it as an omen, and leave for the history books his conviction that a miracle occurred and in the sky was a message, "*in hoc signo, vincit*" or "in this sign, conquer". Now the chances that the general's name would be Constantine are mighty slim, indeed, but the chance that *someone* could have such an experience are quite large. It would not be too surprising if somewhere in Turkey another general saw the same cloud formation and saw nothing unusual in it, but then he was about to take a nap, not go into battle.

Many "miraculous" events can be explained in this manner. Even for one individual, the chances of something unusual happening in the course of their lifetime are good, and if they are "expecting a miracle" they are very likely to find one. There is a range of probability for anything happening. Dawkins explains that,

> "Somewhere in this range, too, are those coincidences that give us an eerie spine-tingling feeling, like dreaming of a particular person for the first time in decades, then waking up to find that they have died in the night. These eerie coincidences are very impressive when they happen to us or to one of our friends, but their improbability is measur[able]."[330]

The same is true of coincidences of healing, of course. A ruptured aorta near the heart is often fatal, the victim bleeding to death internally in a very short time. How many people die of this disorder every year? So the chances that one of them will be fortunate enough to survive because the aorta somehow seals again after the initial release of pressure may be small, but they are far from zero. It must happen to at least some every year, year after year, but when it happens to your relative, maybe the only person you personally know who has suffered an aortic aneurysm and survived, it is remarkable. If anyone in the family happened to be praying for their relative, it is a "miracle". Just as miraculous, maybe even more so, is the cure of the incurable following an avalanche of prayer directed at that individual by their family, or perhaps by their entire church. Never mind that every Pope prior to the present one has had every Catholic pray for him at every Mass they attended, yet they died anyway. And never mind that "...some illnesses are psychogenic...Conceivably, endorphins - the small brain proteins with morphinelike effects - can be elicited by belief. A placebo works only if the patient believes in it as an effective medicine. Within strict limits, hope, it seems, can be transformed into biochemistry."[331] No miracle there, just good old mind over matter, or mind over mind since if we can make ourselves ill, we can heal, too, and if the brain is a part of the healing process, what surprise is that?

THE NINTH ASSUMPTION, PRAYER.
Prayer for divine intervention itself would be meaningless then, except on the human level of exciting the production of healing chemicals through the physical process of hope, if all the purported effects of prayer for an intervention by a creative-sustaining-intervening god can be explained by other means. As long as there is belief in prayer's efficacy it may have some value because of the change in the pray-er that it can bring about. However, it may be there is no supernatural agent able to change the physical world whose deliberations can be affected by the prayer, no one whose mind can be changed, certainly no incomplete being who can be bribed by something it needs. Assumptions #8, that miracles can occur, and assumption #9, that prayer is effective, go the same way as the others. No one can say that they are incorrect, but no one can say with any assurance that they are legitimate statements. The jury is still out, and indeed can never render a verdict since no one can prove that something is not possible. Science can show only that something need not be so, and religion can only say that it *believes* that it is.

How does the theist respond to these findings? One can deny them, of course, and claim that indeed there is need of a prime mover and of more than that, a conscious, purposive being who has the power to bring the universe into existence and the will to have it progress towards its goals. However, this believer can speak only to himself or herself and to others of like mind. Hardly anyone knowledgeable of science will feel that they must listen, at least not from the perspective of science, and this person is one of those the believer wants to converse with. After all, the scientist is as much a potential brother or sister in faith as anyone else, but contact is precluded if the believer just dismisses what the scientist knows to be true. The theist can seek to find a niche for god in the whole process, either inserting god into the gaps of scientific knowledge or claiming that there is something outside the domain of science of which they alone have knowledge. However the "god of the gaps" approach does no service to the cause of conversation since the hearer will likely dismiss it for the futile and self-serving attempt it is, nor does it do a service to the god the theists seek to honor, but succeed only in diminishing. While scientists may well be backing off from a purely reductionist view of the constituents of life and conscious life, leaving room for the possibility of realities and laws arising at levels beyond the molecular, these too are seen as consequences of physical processes including evolution and present no need for the intervention of a divine being. In the tradition of "if you can't lick `em, join `em", as well as in the interest of honesty, it seems that the only course for the theist is to accept the findings of science at face value and then ask ourselves what our faith looks like after these untenable assumptions have been set aside. This I will attempt to do in the last chapter of this work.

THE TENTH ASSUMPTION, THE CREATOR CAN-DO.

Before we proceed, however, there are some other topics that must be investigated, other assumptions that flow from those already treated. The first of these is one already identified and labeled as assumption #10, that the one in control can do good and that (it, he, she) intends to do so. There is not much to be said about this assumption from the perspective of science other than there is no compelling evidence that there is any intelligent agent in control. If there is such a one, it is a matter of faith that this being wills to do good for its creatures. That faith is well established, though it has always faced the problem of a lack of evidence. There is no lack of anecdotal testimony for god doing good, but there is also dearth of complaints that someone prayed for good and received evil. Susan is one such person whose prayers have not been answered and whose questions inspired this investigation. Such stories are found in the Bible itself, particularly complaints in the Psalms and the prophets. We have already seen how merciful the process of evolution is to individuals. They are sacrificed wholesale. And not only individuals. Whole species, whole families of species have vanished from the scene, some slowly, some in the blink of a geological eye, leaving no descendants. For unknown numbers of soft-bodied creatures especially, there is not even a trace through fossil evidence of their existence. A controlling, merciful being should presumably be able to do better in designing a system, though those of faith can always counter that the ways of god are inscrutable and we need only wait until after death to understand how all of this was really for the good. Who can dispute that? On the other hand, who can believe it?

THE ASSUMPTIONS IN PERSPECTIVE

We have covered the main assumptions that underlie the faith and seen that science, though it cannot refute these assumptions, makes them tenable only with some faith statement. That is, the faith can make some statements that science cannot absolutely refute, but there is no evidence requiring belief in them. For someone to propose these items as true, one can only say that they believe them to be true. For example, the statement that god is the creator in the same sense as the faith has always believed cannot be disproved, nor can it be proved. One can only say that they "believe" that god created the universe. Of course "how" god created would be drastically different if one is going to be true to the findings of science, but "that" god created can be insisted on as a legitimate faith statement. Legitimate, but increasingly suspect. Of course the same is true for science, or any other human endeavor. The cosmologists can say that they "believe" that their explanation of the Big Bang is the manner in which the universe came into existence; that all the evidence points in this direction; that other explanations such as the Steady State and the biblical creation in six days are no longer tenable. Still, they would be the first to say that new evidence may dictate adjustments or even radical revision of the scenario. This may seem to play into the theists program, especially that of the creationists, but in fact it does not. The honest person of faith wants to take into account all the evidence available, and from the evidence available it is just as wrong for the believer to conclude that there is some definitive proof of their position in favor of god as creator, for example, as it is for a non-believer to conclude that the evidence disproves the possibility of a god. What I am trying to do here is to understand science and the picture that it presents, give it its full credence, and then ask how that picture impacts on the world-view built up on faith in god.

AND THAT'S NOT ALL JUST YET.

We have considered the basic assumptions upon which all religious faith has been built, often without scrutiny or unconsciously. We have seen, or at least I have seen, that these assumptions do not fare very well in many important aspects. There are other corollaries and consequent assumptions and positions that need to be investigated in the light of modern knowledge which follow from the major positions. I propose to do so now, recognizing that I can only choose the more general for investigation. The following will not be an exhaustive look at all the tenets of the Christian and Catholic faith, just those whose consideration will best effect the conclusions I hope to show in the remainder of the work.

ORIGINAL JUSTICE

The notion called "Original Justice" follows logically from a literal acceptance of the biblical story of the creation of humans. It holds that we were not created in the situation in which we now find ourselves, surrounded with and infected by "evil". God, it is presumed, did not create us in this state; in fact, the Genesis stories are told precisely to explain how we ended up in such a condition and to absolve god from blame. Therefore, it stands to reason that the first humans must have been in a much better condition, physically, morally, and relationally. This is what we find in the biblical stories of the Garden of Eden. The inhabitants live in a utopia at peace with all the animals, plants and the whole of creation, nothing like the way the world is today.

Is any of this possible? Well, possible, but not likely. Maybe god picked out one instant in the continuum of human evolution and placed one couple, or a whole tribe, in some special circumstances from which they, regrettably, fell. Few seriously believe that a historical recollection of such an event is behind the biblical story. Since there is no reason to believe that this did, in fact, happen, and every reason to presume that it did not, we should dismiss this position from our faith-picture, at least as anything more than symbolism.[332] F. R. Tennant recognized this in the `20's and pointed out,

> "Evolutionary theory and the researches of archaeologists and anthropologists have made it hard indeed to accept the traditional and biblical account of the Garden of Eden. It is true that it will probably always remain possible to insert Adam into the gaps in our knowledge of prehistory. But the story has lost its plausibility." [333]

Humans arose in a continuum of species, some of which evolved the attribute of consciousness and thought to a greater degree than others, but all of which were in the same relationship with the other beings around them, which usually means predator and prey. The idea that suddenly a group of primates-turned-human lived in some idyllic vegetarian state and perhaps in some garden of delights, with no need to work for their living, no pain in childbirth, no reason to fear being eaten by some predator, just does not fit the evidence of our origins. Tennant again, though the scenario has not changed since his day,

> "...The picture is more like this: of men struggling upwards out of barbarism, ignorance, and animality. Men have not so suddenly blossomed on the evolutionary tree...Taking the evolutionary picture, it seems unlikely that our remote ancestors, whose life no doubt tended to be pretty nasty, brutish, and pretty short, lived in harmony with God until a rupture came."[334]

ORIGINAL SIN

Original sin, many feel, must have occurred, since at one time, even by the scenario presented by evolution, as well as the image of the Bible, our ancestors were animals innocent of moral evil,[335] but we certainly are capable of moral choice now. At some point we crossed the Rubicon in our ability to make these choices, chose wrongly, and that was the first or original sin. Good point! But,

> "The traditional theological doctrine of `the fall' as a disobedient act by the original man and woman (Adam and Eve) in which they fell from grace and which so altered their state that they transmitted the state of `original sin' pseudo-genetically to all succeeding mankind is clearly at odds with the scientific account - which sees in man only an emergence from the consciousness of the higher mammals to self-consciousness, language, and deliberate choice. The meaning of what is now regarded as the genesis myth of `the fall' has been widely and profoundly interpreted existentially in our times by both Biblical and systematic theologians as a myth of man's present *state* of alienation from god, and of disharmony both between men and between man and nature."[336]

In fact, though, it is not the concept of original sin that science denies but the explanations that have been given for its origins, effects and for its transmission. Who can deny that humans are affected by something that might as well be called sin, and we seem incapable of rooting it out of our nature. Charles Colson, a good Calvinist, points out that,

> "History continues to validate the biblical account that man is by his own nature sinful - indeed, imprisoned by his sin. And we are not reluctant prisoners. Like Augustine, we actually delight in sin and evil."[337]

Where then does it come from? The Catechism agrees about the omnipresence of sin and offers its own analysis of the cause.

> "Sin is present in human history; any attempt to ignore it or to give this dark reality other names would be futile. To try to understand what sin is, one must first recognize the profound relation of man to God, for only in this relationship is the evil of sin unmasked in its true identity as humanity's rejection of God and opposition to him, even as it continues to weigh heavy on human life and history."[338]

The Catechism stops short of insisting that the Fall happened as described in Genesis, but it does insist that,

> "The account of the fall in Genesis 3 uses figurative language, but affirms a primeval event, a deed that took place at the beginning of the history of man."[339]

Science cannot quibble with this, either, since it is obvious that other animals do not "sin" as humans do. To claim that a bird taking a fish from another bird in mid flight is the same moral act as mugging an old man in the park is certainly anthropomorphism. However, we once again see a continuum of activity from other animals to ourselves and this must certainly be factored into the understanding of Original Sin. Robert Wright sees this going both ways. One must take into account that humans are but animals when imputing sinfulness to them, but also,

> "...for all the emphasis in popular treatments of sociobiology on the `biological basis of altruism,' and for all its genuine importance, the idea that John Stuart Mill ridiculed - of a corrupt human nature, of `original sin' - doesn't deserve such summary dismissal."[340]

Another researcher, Lyall Watson, reaches the same conclusion that,

> "It begins to look as though there is something in `original sin.' There is an inherited, genetically related system that is unrelentingly selfish, ruthless and cruel. And Saint Augustine is right, we are never going to be without it."[341]

Still, science's explanation for the presence of "evil" in humanity is a far cry from the explanation given in most religious education classes, and the remedy for it is going to be different from those often proposed. Suffice it to say that the idea of original sin would never have arisen if the facts of our origin through the evolutionary process had been known. The idea that there was anything but a continuity from our ancestors into ourselves just would not occur to anyone.

The corollary to humanity starting out good is that all of creation began in the same condition, that all things were originally good. Since it is seemingly out of balance now, something must have happened. The "something" was, in this biblical view, the same culprit as before, humanity. Prior to the original sin, creation was in balance. Was there ever a time when all of nature was a paradise? On the one hand, yes, since there can be no morality when there is no free will, so if paradise is defined as a world in which there is no moral evil, then it was the Elysian Fields. However, a universe that changes "for the better" by killing off the losers wholesale, and sometimes hundred of species at a time, is hardly a universe that we would consider "good". And that has always been the way the universe operates.

MORALITY

Someone might object, maybe there was no garden or a special time from which we fell, but if there is goodness in the universe now, or at least in humanity, how could it get there without some creative activity of god in bringing it about? If the world and its processes are essentially amoral, how does morality develop, if not from god? First let us look at the phenomenon we call altruism,* the willingness to sacrifice oneself for the sake of another. The highest example for Christians has always been god's willingness to share godself with humanity, reaching a pinnacle in the giving of self on the cross in the death of Jesus. All human altruism is seen as a reflection of god's unselfishness. We have to be taught to love and care and share. Just ask any parent about the spontaneity of their three year old's sharing. We are amazed by an animal's seeming altruistic behavior because we feel that this is a human action, arising not from the animal side of our nature but from the "spiritual." Doing good is of god and god has instructed us to imitate this "way." Is god's example and invitation the explanation for altruism in humans? Or is there once again a scientific explanation of the development of this behavior that does not require a divine intervention? Many think so. Robert Wright says, "Altruism, compassion, empathy, love, conscience, the sense of justice - all of these things, the things that hold society together, the things that allow our species to think so highly of itself, can now confidently be said to have a firm genetic basis." [342] Why and under what circumstances a man or woman will give up their own good for another can be understood through the same mechanism that causes the worker bee to give its life for the hive, or the prairie dog to call out a danger signal even though by doing so it places itself in the center of the bulls-eye for the approaching hawk. And that mechanism is what has been dubbed the "selfish gene". For it is not only the organism that seeks to reproduce itself, it is the genes that compose that organism. This is not a conscious "seeking", of course, it is just that those genes that can constitute an organism that is reproductively efficient will automatically be reproducing themselves as well. On this level, it sometimes makes sense for the organism to surrender its own well-being, or even its life, if that means that the gene will survive in another organism possessing it. The bee is genetically identical to all its hive mates and so the loss of one individual does not mean the end of the gene's existence. Quite the contrary, stinging the bear coming for the honey may well mean the very existence of the gene, while not stinging it and losing the winter's supply of food will preserve the one bee until the end of its life, but may also assure the demise of the gene in a very short time. Since bees have been around for millions of years while the life of one individual is less than one season, the strategy seems to work well from the gene's point of view. And since the bee itself has no point of view about the future or its own death, all is well. Ants are

---

* Webster's definition 1: unselfish regard for or devotion to the welfare of others. 2: behavior by an animal that is not beneficial to or may be harmful to itself but that benefits others of its species.

similar. "Because of a quirk in the way the queen ant reproduces, worker ants are closer to their siblings, who share on average three-quarters of their genes, than they would be to their own offspring, who would have only half their genes. So, becoming sterile workers and laying down their lives for their sisters may in fact be the best way to transmit their genetic heritage.[343] Male spiders are known which are not only willing to put themselves in a position where the female may eat them but actually encourage being eaten.[344] Similar reproductive "reasoning" is true of the prairie dog, of course, and the same is true of the human as well, except for the complication that the human often knows what sacrifice he or she is making. Our first point, however, is why the person should make the sacrifice in the first place, and at least part of the answer is that it is built into us. Unfortunately, so are the limitations. The prairie dog will not call out to save a mouse since if they share any genes at all, and certainly they do, their history is too different for the genes to recognize the connection. In fact, the prairie dog will not sacrifice itself for other prairie dogs that are not in his or her rather immediate family. And a human father or mother may work himself or herself literally to death in support of their own offspring, but are less likely to do so for the offspring of someone else. They may work a second job to help out a brother or sister who is in need, and maybe a first cousin, but a third cousin once removed is on their own. Altruism evolved as a mechanism for the survival of the genetic information that forms the organisms, but not for the good of the entire species. It is possible, of course, that a gene might evolve which caused every organism possessing it to recognize and to care for every other organism owning that gene. It is even possible that the world could produce an organism with a gene that made it solicitous of every other living organism in the world, but so far none has arrived, tree-hugging environmentalists being about as close as we humans come. It is still dog-eat-dog except for the closest of relatives, and even then probably only for a brief period in their lifetime. Mother animals are known to kill and eat their own offspring in order to have the nourishment to feed the others, or to deliberately withhold food from the runt if not enough is available for all. Some even reabsorb the fetus if conditions favoring its survival are not good, in order to maintain the strength of the mother to try again later when things improve. Most fathers are even less attached to their own offspring than the mothers are. Some animal fathers never know which are their own anyway, and some that could know don't bother. Certainly there are fathers that are models of parental concern, seahorses being a notable example, and penguin fathers stand around in the cold for weeks until mama returns. Then, sometimes, watch out for the mother! Surely there are "loving" couples who both sacrifice themselves for their young. But last year's crop had better leave the nest, or else. The genes can make us very solicitous at times, but their memory is often not long. We humans are inheritors of all this, but at the same time we often transcend it. The sacrifice of Jesus for all of humanity is again the model, but so is the soldier in war who is willing to lay down his or her life for the fatherland (or motherland) and all its citizens, to most of whom the soldier is not related. The whole point of boot camp is to convince the, usually young, soldiers that they are related to every citizen of the country. They are all brothers and sisters. This is done not only to lie to them that it is to their (genetic) advantage to give of themselves, but also to convince them that they are not related to the enemy and therefore incur no (genetic) disadvantage by killing them. The fact that the soldier could be more closely related genetically to the enemy than to the people being defended is conveniently not brought up. Certainly there were many Semitic American soldiers involved in the Gulf War who did great damage to the survival of genes shared with the enemy while advancing the welfare of genes found in African Americans, or Anglo Americans. They did not see it that way, and this is because at this point in human evolution it is no longer only the reproduction of the physical genes that determine our behavior, though this certainly influences it strongly still. Dennett explains,

> "Our reproductive ends may have been the ends that kept us in the running till we could develop culture, and they may still play a powerful - sometimes overpowering - role in our thinking, but that does not license any conclusion at all about our current values. It does not follow from the fact that our reproductive ends were the ultimate historical *source* of our present values, that they are the ultimate (and still principal) *beneficiary* of our ethical actions."[345]

On the larger scale than preserving our family genes, we are also interested, and sometimes even more strongly, in preserving our family *values*.[346] These have been dubbed "memes" as we saw above and our evolutionarily built tendency to altruism can now be placed at the service of memes as well as genes.[347] My point here is the question just how "good" creation is because it is capable of producing self-sacrificing behavior when such actions are so circumscribed by conditions and selfish "motives"? To say that goodness is something inherent in the universe is a hard position to maintain. So what of the assumption that god has laid down for us a code of morality and, conversely, that we are moral because god has commanded us to be, or taught us, or inspired us, and not because we evolved to be good? The basis for a moral code is certainly enshrined in religious documents with a divine mandate. Other powerful figures have also seen it to their advantage to "lay down the law". Hammurabi is famous for his code of law, and Napoleon's still forms the basis for the law of the Franks. Why do they promulgate such legal codes? For the regulation of their society, certainly, but this is not so altruistic as it would seem. It is much harder to raise an army if everyone in the country is a law unto themselves. It is difficult to conduct business in a lawless society. Indeed, it is even better for the reproductive advantage of the monarch if all others must follow laws of marriage and property rights of the husband over his wife, but of course the king is exempt. God, too, is exempt from the moral code god lays down. While it is not right for one human to harm another in order to test their loyalty, it is perfectly alright for god to kill all of Job's children, and his sheep to boot, in order to make a point. Aside from this, is it true that the basis for our morality is something extrinsic to us, possibly god? Or is it built in through the same mechanism that we have found throughout, the influence of evolution?

We do not speak of morality among animals, since they act only on "instinct", which is not something rational. And we do not look for a divine origin of their behavior, especially when that behavior is so obviously selfish, the belief in the innocence of animals belonging to a long-past age. Our morality, we feel, is of another order. However, is there such a dichotomy between the actions of the other animals and of ourselves? Morality is a guide for actions. What forms the basis for the actions of animals? Certainly tried and true instinctive behavior that leads to the acquisition of food and mates and the avoidance of enemies. What is the basis for our own actions?[348] Usually the same as other animals, for we surely realize that most of the time we do not stop to consider the morality of an act, we do it first and perhaps think about it later. Indeed, we are experts on rationalizing our actions, and especially *ex post facto*, so that what we have done we now justify. Few would first justify, for unselfish reasons, enslaving another person for their own good and then go ahead and do so. However, many, once the institution was established, justified their actions by appealing to the supposed good done to the enslaved. In order to do that, one had to postulate the inability of the enslaved to govern their own affairs. Never mind that African kingdoms and culture flourished when Europe was a backwater of the world, beginning perhaps with Nubia as a cultural rival of ancient Egypt. Like the animals, we do what we must do to further our own ends. Unlike the animals, we then feel a need to explain. Why this need? What can explain the development of a conscience?

CONSCIENCE

The Catechism explains that the conscience is the voice of god that is implanted in the reasoning process of humans, presumably in the soul. "By his reason, man recognizes the voice of God which urges him 'to do what is good and avoid what is evil'."[349] The Catechism adds, "Deep within his conscience man discovers a law which he has not laid upon himself but which he must obey. Its voice...Sounds in his heart at the right moment...For man has in his heart a law inscribed by God..."[350] Inscribed, yes, but, it seems, by the forces of human evolution, and subject to the same limitations as the other animals. We must also remind ourselves that humans do not evolve only through their genes any longer. There is also an evolution of culture and "memes" and these certainly impact the conscience of those living in the culture. The whole culture of the German society during the Holocaust was such that many of its citizens, readily justified and participated in the persecution and execution of the Jews.[351] A conscience based on god's wisdom would certainly never counsel such actions, so either what we call "conscience" is not based on god's placing it within us or it is easily overridden, and so, useless. Mark Twain said much the same thing when the queen in *A Connecticut Yankee in King Arthur's Court* wants to have a person killed who offended her. The Yankee muses, "We speak of nature; it is folly; there is no such thing as nature; what we call by that misleading name is merely heredity and training."[352] If conscience is so malleable, what is its purpose and where does it come from? From the same mechanism that has always guided us, the forces of evolution. Conscience, much as we hate to admit it, is nothing more than a survival tool for culture, and often for the individual, and will serve that purpose whether or not it serves Judeo-Christian "morality". Why should we be imprinted with the dictate of conscience (read *feeling*) that one should not kick a person when they are down? Because we may be down sometime. Take away the feeling of vulnerability, that it-could-be-me, and we will kick the person at our feet. When the Germans felt that they could never be in the position of the Jew in front of them, often a Christian conscience did not stop them from victimizing the other.[353] We have agonized in the 20th Century over how these things could happen in "civilized" countries but the evolutionist knows that the veneer of culture is thin and underneath are the same processes that guide the behavior of any animal. If society has become more "moral" it is not because the process by which we make moral decisions has changed, it is because we have become smarter about what makes for a workable society. Before the advent of mass communication I could ignore the plight of peoples in distant parts of the world and still feel that society works.

Now it is much more obvious that unrest in other countries does impact my own. I may tell myself that I am concerned because I am a good person and my spirituality is highly developed, and these may well be true, but I am also concerned because it seems to be happening closer to me - too close for comfort. As long as my village was my world I couldn't care less about those in the next valley, and may well despise them. When my village is the world, my perspective changes. When the village develops a meme to further its own values and ends, it will enforce it through inculcating the proper feelings (read conscience) in its inhabitants. If watching a news report on starving children in some other country can stimulate a feeling of concern and pity and a desire to help, which it does because watching brings the child into our circle of "family", you can be sure that any organization which is seeking funds to help that child will develop an appeal designed to intensify those feelings. And if a pastor with a burst boiler wants to take up a special collection for replacement, it helps if the parishioners feel that they are a family and that it is the family home that is threatened. Even better if they feel the cold themselves. None of the studies done on why we behave as we do conclude that in any area of human psychology we must claim a radical difference between the human species and other animal species. Differences, yes, even from our closest primate cousins, because our culture is so very much more complex even than primates, and certainly from tadpoles who have no culture at all, but not *radically* different. The same engine that directs their behavior directs ours, it is only the gears that differ. And none of the scientific studies of human behavior feel the need to say at any point, "there is no explanation for this behavior other than the influence of some outside agent". The basis for our morality is ultimately not a list of commandments written on stone, but a string of genes written on DNA. If we are less barbarous than our ancestors, it is because our "family" is larger and our genes dictate that we should care for family. There are plenty of places around the world where humans are at least as barbarous as Attila the Hun. And why? Because one group at least has defined another group as "not family".

Let us not fool ourselves that these alleged changes in the level of morality are the result of purely logical processes, for our "consciences" do not direct us to behave in certain ways through inspiring thought, nor by urging us to follow the dictates of the lawgiver (except through fear of consequences) but by inspiring feeling. We do not justify war as an institution because we have thought it through and reached the conclusion that it is a good thing. Quite the contrary. It is usually only when we feel fear and dread of the enemy, or when we feel the rush of adrenaline in the heat of battle, or perhaps when we experience the feeling of victory and domination, that we begin to glorify the role of the soldier and speak of "just wars".

In the past, thinking that we alone have a moral sense and that animals do not, and believing that this moral sense was not grounded in our physical selves (save for the "natural law") but was a gift from god to guide us (or perhaps test us) on our pilgrim way to holiness, we felt that the origin of any true moral code was in god. Those who accepted it would find happiness because god could only reveal truth, so following the truth could not but bring happiness, while those who transgressed it were rejecting god and god's wisdom and would inevitably be unhappy with the results. Now we see the evidence that our moral feelings, including our conscience, are the result of the process of evolution and firmly grounded in the behavior of the other animals as well. Aside from our ability to reject the dictates of our genes, there is little or no difference in the processes that lead us to certain actions. Most of what I do, I do for the same reason as the brutes, i.e. something in my makeup is urging me to act in this manner. I may later rationalize some action that I now somewhat regret, or I may explain my actions by making up a reason, but at the time I did it I was mostly reacting to a stimulus. I may, for example, trip a fleeing purse-snatcher so the police close behind can nab him. At the time I probably just reacted, especially if something inside me discerned him as an enemy. Upon retelling the events later on, I might explain to a reporter that I wanted to save potential future victims from being robbed. I had no such motivation at the time, but later I myself really believe it. This is even more likely if my action brought some serious harm to the person I tripped. Now, feeling somewhat guilty at the disproportionate consequences, or feeling attacked by a questioner wanting to know if I intended those consequences, or in any case *feeling something*, it is more important than ever that I claim loftier motives than, "my genes made me do it." Amid all the evidence that indeed our genes, or at least our cultural memes make us do much of what we do, the origin of a moral code through revelation and not through the results of evolution working on our feelings, all of which have a physical origin, is very suspect. The editor of a recent theologian's effort to reconcile some of the differences of science and religion points out that,

> "Up until now, theologians have made themselves feel reasonably secure by hiding behind the Kantian split between theoretical and moral knowledge and by consigning science to the former while reserving privileged access to the latter. Even if natural scientists exclude God from their worldview, theologians have thought they could earn an honest living by

> appealing to the realm of value and morality. No longer, says [Wolfhart] Pannenberg...Pannenberg makes it clear that the challenge goes both ways: `If the God of the Bible is the creator of the universe, then it is not possible to understand fully or even appropriately the processes of nature without any reference to that God. If, on the contrary, nature can be appropriately understood without reference to the God of the Bible, then that God cannot be the creator of the universe, and consequently he cannot be truly God and be trusted as a source of moral teaching either.' If we fail to make the case that the God in whom Christians place their faith is also the God of created nature, then we can claim no privileged access to God through ethics."[354]

Obviously Pannenberg and the editor want to make that case, but "Nature," in this case our moral conscience, can indeed be appropriately understood without reference to god. Robert Wright challenges Christians and others who would make a case for a god-given conscience when he points out that even our feeling that moral behavior carries the weight of divine authority may be the result of evolutionarily produced memes and that,

> "...the claim that these condemnations have moral force may be just a bit of genetically orchestrated sophistry. Extricating the wisdom from the sophistry will be the great and hard task of moral philosophers in the decades to come, assuming that more than a few of them ever get around to appreciating the new paradigm."[355]

SO WHERE ARE WE?

To sum up the investigation so far, we have seen that though we cannot say absolutely that there is *not* a cause for the existence of the universe, there is no reason to say that there is one, either. The universe is perfectly capable of popping into existence out of nothingness, containing all time and space within it. It's existence is "contingent", not necessarily meaning coming from the mediation of another's existence or relying on god to give it existence, but in the sense of not containing existence as a necessary quality of itself. It need not be and it need not continue to be, but it *can* be even if no other being exists. And just as there need be no cause at its beginning, there need not be any goal to reach either, no final cause. Though thousands and thousands of books have been written about the need for a Designer, neither the present state of the universe nor the development of life on earth, or even consciousness, demands such an agent. These findings are the keys to the whole thesis of this work. If there need not be a Creator-Designer, then there need not be a Sustainer. Obviously if there does not have to be a creator who brings the universe into existence, there does not have to be any on-going creative activity to keep it thus. And if there is no Designer, then there is no design or purpose to the universe, nowhere it is directed to go, no goal it is created to meet. It may be that there is a purpose, and therefore some being with a goal who had something in mind when creating, or at least shaping, the universe. Certainly now that conscious agents have arrived on the scene from within the universe, namely ourselves and any others there may be, we will attempt to impose our own purposes on the universe. The universe is amenable to being directed, but it can also get along quite well without, thank you. In fact it very much looks like it has been without purpose or direction for some 15 billion years and if you count ourselves as a prize it has not done so badly for itself. Much as we might wish to be the goal of the universe, however, it looks like we are serendipitous rather than intended and any idea of a teleology may just be wishful thinking.

We have also noted that if there is anything or anyone in control, it is a very loose sort of management. Some theists deny this, of course, and find god in control of every event, but most explain the state of things with reference to our free will and contend that if we are to be truly free to make our own decisions, god must necessarily exercise a loose hand. And again I must insist that either of these camps may be right, but the third alternative is that there is no control, no hidden agenda, no "making all things right in the end." Given the sheer volume of random events, fortuitous or calamitous circumstances, and, yes, mayhem and violence in the universe, even the theists may be well advised to go for this third option.

Hearkening back to Susan and her problem with evil, to which we must eventually return, for every miracle and saving grace there seem to be a myriad of accidents, coincidences and happenstances, most of them bad. Given all this seeming lack of rigid control and lack of evidence of a sovereignty over the works of nature, the potency of prayer directed to changing our environment and the circumstances of our lives is certainly questionable. Far from revealing a god working through nature and the physical laws to bring about god's own purposes and plans, to reward and punish, to teach and direct, and ultimately to bring about a new creation, the universe seems to reveal a monumental witlessness, bringing about through chance and accident an array of forms who survive or perish in accord with mere chance, following no direction and going nowhere. Those survive who can, and no mercy, compassion or benevolence is recognizable in a universe that can create or destroy stars or species and neither revel in their beauty or bemoan their loss.

Along the way I have examined the ideas of original justice and original sin, the contention that all things are created good and all that is is valuable; the persuasion that it all means something and is going somewhere and that likely we are the explanation and the goal; and finally that there is the same meaning to terms like "good" and "evil" as they apply to our behavior as is given in Christian Moral Theology. None of these hallowed assumptions of the Christian faith come out as necessary conclusions, and in fact are more than suspect as valid positions.

I could go on, but have been urged not to do so on many more topics. If you would like to consider the question of angels and devils, look up this endnote.[356] The idea of a "final judgment" at the end of life or the end of the world is treated in this endnote.[357] The "end of the world" is similar and is discussed in endnote number[358].

WILL "I" BE AT THE JUDGMENT?

I do want to treat the widespread belief in some manner of existence after death, with or without a judgment. For Christians, as well as many other religious traditions, this means the continued existence of the individual personality in some form continuous with the manner of being that the person enjoyed on earth. Other traditions may not insist on an abiding presence of the personality but the animating force now enlivens a new individual, and sometimes not necessarily a human individual. Francis Crick writes,

> "The interest of human beings in the nature of the world, and about their own natures in particular, is found in one form or another in all peoples and tribes, however primitive. It goes back to the earliest times from which we have written records and almost certainly from before that, to judge from the widespread occurrence of careful human burial. Most religions hold that some kind of spirit exists that persists after one's bodily death and, to some degree, embodies the essence of that human being. Without its spirit a body cannot function normally, if at all. When a person dies his soul leaves the body, although what happens after that - whether the soul goes to heaven, hell, or purgatory or alternatively is reincarnated in a donkey or a mosquito - depends upon the particular religion."[359]

For Christians it is the same individual who remains in existence and lives a life similar, more or less, to the one we know. Indeed, this life on earth is only a prelude to the real life that is to come, just as the earth we now live on is only a prelude to the heaven that will replace it. Our true life and true home is yet to come. Augustine says, "We are Christians and our homeland is not here. Like good children, let us turn our steps homeward, that our course may be approved and guided to its conclusion."[360] A preacher was heard to say, very memorably, "we are not human beings having a spiritual experience (of God), we are spiritual beings having a human experience. [361] This presupposes that someone intends that we should live forever, for it is rather obvious that death and the cessation of existence would otherwise be our normal lot. True, Thomistic theology goes along with the Greeks in positing an immortal soul which has immortality as part of its nature,[362] but it is taken for granted that this is not an open-ended immortality. It is open on the back end, but not on the front. We may be destined to live forever, starting from the time we come into existence, but we have not existed eternally prior to this event. The one who created the immortal soul intends that we should live forever. Since we have seen no evidence of the existence of such a thing as a soul, however, and science certainly does not see a reason for suggesting one, nor does it find a reason for proposing a creator of a soul or of anything else, what reason do we have for this hope of enduring beyond death? George Ellis was speaking for religious doctrines in general, and not just about belief in resurrection and eternal life when he wrote that,

> "...in my view it would be a mistake to dismiss [religious convictions] entirely because of [the possibility of hallucinations or self deception]: we have to take seriously claims of this kind not merely because they are made persistently in a great variety of circumstances and cultures, but because (whatever their cause) they demonstrably have the ability to alter people's behaviour decisively and permanently."[363]

Certainly belief in resurrection has such power. What else but some (perhaps divinely revealed) information would even make us dream about such a thing. Well, wishful thinking, of course, but could there be more to it? Evolution theory says there is certainly more to it, but it needs no divine revelation. Once we had become conscious *and* aware of self we became aware of the inevitability of our own death. It must have dawned on our ancestors at some point that this was both something to be feared and to be fervently avoided. Even before consciousness there is an innate avoidance of things that can bring death, for any animal that lacked such a tendency would not survive long enough to reproduce. On the other hand, fear of it could easily reach such proportions as to also render the organism incapable of surviving. There had to be a balance of a healthy fear and a watchful daring. Consciousness threatened this balance. Something had to evolve in the conscious animal that could restore the equilibrium, and it did. There seems to be something built into our psyches that conditions us to the belief that our death is somehow an illusion. Carl Sagan reports that,

> "Sometimes I dream that I'm talking to my parents…When I wake up I go through an abbreviated process of mourning all over again. Plainly, there's something within me that's ready to believe in life after death. And it's not the least bit interested in whether there's any sober evidence for it.

Many report going farther than dreaming and Sagan points out that, "More than a third of American adults believe that on some level they've made contact with the dead."[364] As a scientist, he sees no reason to think that his dream has a basis in reality, though as a human he would certainly like it to have. "If some good evidence for life after death were announced, I'd be eager to examine it; but it would have to be real scientific data, not mere anecdote."[365] Genetic evolution would do well to place such a thought in us, and *meme*tic evolution would have reason to do so as well. The old Marxian saw about religion being an opiate for the people can certainly ring true. "Societies that teach contentment with our present station in life, in expectation of post-mortem reward, tend to inoculate themselves against revolution." In addition, "…fear of death, which in some respects is adaptive in the evolutionary struggle for existence, is maladaptive in warfare. Those cultures that teach an afterlife of bliss for heroes - or even for those who just did what those in authority told them - might gain a competitive advantage."[366] "[Expensive royal] funerals are a striking example of a practice that is downright wasteful in squandering hard-earned community resources, but that may be adapted at the social level as evidence of the reality of an afterlife. Thus they might motivate individuals to forego selfishness in this life to obtain the rewards of their societally adaptive cooperation in the next."[367] Edward Wilson adds another advantage conferred by such belief,

> "…The idea that there is some purpose to life, or that another life awaits us after the earthly one is over, may prevent an existential crisis that would induce prereproductive suicide, and may keep postreproductive parents happily providing for their children and thus for their genes."[368]

All of these explanations are only guesses, of course, since no one knows the paths taken by evolution to instill this belief in us, if any. Still, they show once again that its presence need not be the result of divine activity or revelation. The idea seemingly imbedded in us can be explained through our old standby, evolution. There is evidence that religious feelings, like other emotions, are generated by the limbic system in the brain.[369] To the extent that such belief advances reproductive success for individuals or societies, it will be fostered vigorously by both genetics and culture, neither of which, as Sagan pointed out, cares a hoot if it is true or not.

> "It has often been pointed out that Plato's curious theory of reincarnation and reminiscence, which he offers as an explanation of the source of our *a priori* knowledge, bears a striking resemblance to Darwin's theory...Darwin himself famously noted the resemblance in a remark in one of his notebooks. Commenting on the claim that Plato thought our `necessary ideas' arise from the pre-existence of the soul, Darwin wrote: `Read monkeys for preexistence'."[370]

There may be reason to hope in our continued existence after death, but those reasons will be found through faith, not through empirical evidence or through a lack of another explanation.

Belief in an existence after death naturally brings up the question of Eschatology and of heaven. Once again, I chose to treat this at some length only in an endnote since it is but one more item about which science can say little but about which the faith can prove nothing.[371] It is something that believing scientists have written of, however, though I believe with the same assumptions.

I have also included in a note an extended treatment of the Catholic idea of the "transubstantiation" of bread and wine into the body and blood of Jesus since it is an important doctrine of the faith. However, to understand the challenge of science to the explanation of how this is possible is a very long process involving the underlying identity of objects in the universe. Suffice it to say that the Greek-Thomist explanation of "forms" and "essences" has no place in science. If you would like to get into it, look up the endnote, but I warn you it is practically an appendix.[372]

CHAOS AND ITS COUSINS

We do need to take a look at the concept of "chaos"* and its role in bringing about seemingly unpredictable results since there is some confusion among theists about whether this is an indication of the presence of a god or not. "A chaotic system is one which, although in a strict mathematical sense

---

* For these purposes, chaos is defined as "vastly complex effects arising from simple causes." "Simple laws can lead to behavior so irregular that it is to all intents and purposes random."

deterministic, is nevertheless so highly sensitized to minute disturbances that meaningful prediction over the long term is precluded. The tiniest disturbances amplify and amplify until they wreck the predictability of the system; its behavior is essentially random."[373] "Because of chaos, it is impossible to predict how some systems will develop without knowing the *exact* values of all the properties of the system at the start - for example, the precise position and velocity of every object in the system. In practice this is impossible..."[374] This is not to say that a universe which contains chaos is not deterministic and reductionist, for scientists are busily deciphering the laws of chaos itself.[375] Even with the presence of chaos in the system, sufficient knowledge of the initial conditions, coupled with the knowledge of the laws of chaos as well as Newton's laws of physical causality would not enable one to know the expected outcome. As Kauffman says, "Only God has the wisdom to understand the final law, the throws of the quantum dice. Only God can foretell the future. We, myopic after 3.45 billion years of design, cannot," [376] though the god he refers to seems to be more a literary device than a real being. There is, as we have seen many times before, room for god, but certainly not the necessity for god. If the universe is completely predictable and following laws that are the result of the initial conditions of the universe, certainly, the theist can ask, "and who set those initial conditions?" The answer that no one need set them can certainly be greeted with skepticism and the skeptic may point out that if the initial conditions were only slightly different there would have been no development in the universe. All might remain as chaos or fall back into nothingness. This objection is made with a dash of derision since this is obviously not what had occurred, but the answer received from science would be that the skeptic is precisely right. If the conditions were not right ours would be a barren or a stillborn universe and we would not be around to ask embarrassing questions of it. In another universe the conditions may be more favorable and here we are in that other universe asking those questions. Nevertheless, the skeptic continues, there is order in our universe and one who knew the precise initial conditions can know exactly how it will develop. "Perhaps", comes the answer. But there is also chaos, which may well be a requirement for development. "The existence within the equations of classical physics of a chaos so complete that any hope of returning to unrestricted determinism seems utterly dashed makes perfect sense within the logical framework of complementarity. Any definition of order must also feature disorder as its complementary construct."[377] Too much order and you have an unchanging, static universe that never gives rise to new forms. Too much chaos and you have such disorder that nothing ever comes into form, or if it does it does not last long enough to evolve.

Just the right balance is necessary. "Networks in the regime near the edge of chaos - this compromise between order and surprise - appear best able to coordinate complex activities and best able to evolve as well. It is a very attractive hypothesis that natural selection achieves genetic regulatory networks that lie near the edge of chaos."[378] Who put this balance in? Certainly a god may have, but it may also be that any universe that lacks it never goes anywhere. There is still the question of whether it is possible for anyone to know all the initial conditions of the universe, including Kaufman's "god". If to forecast the weather on earth one would have to know where each molecule of the atmosphere is and how fast it is going, as well in which direction it is going, to know the fate of the universe one would have to know the whereabouts and the pedigree of every quanta of energy at the moment of the Big Bang. Is this possible? Certainly not for us, but how about for god? By many definitions god is omniscient and possesses such knowledge, but it is not quite so simple. According to Heisenberg's Uncertainty Principle and the findings of Quantum Mechanics, one cannot know the precise location and velocity of any quantum, much less every one. This is not just because of a lack of the proper instrumentation, but also because the quantum does not *possess* a position until it is measured, or a velocity until it is gauged, but only a range of possibilities. The properties of place and velocity are complementary and exist independently of each other, though they become contradictory if you try to join them. The same is true of chaos and order, perhaps. The universe is full of both and they both have an effect, but you cannot treat them both together without doing damage to one or the other concept. Does god have a knowledge that allows knowing not just an indefinably large amount of information, but information that does not truly "exist"? Must god be credited with a knowledge of where a thing is and what it is doing when it does not "know" these things itself? The theists may make such a claim without fear of contradiction, for who can prove "no", but must be aware that they are claiming far more knowledge than a Newtonian universe needs.

There are more assumptions of the faith that could be considered in the light of modern knowledge, but these will have to suffice for now, save one.

GOOD AND EVIL

The last time we saw Susan, she was worried because her god, who was supposed to be in control of everything and able to help her in her problems was losing some control. It seemed that a "myriad of accidents, coincidences and happenstances, most of them bad," had control of the universe rather than her god. We have examined a number of other suppositions that she entertained and that make up her faith and, while no finding of science today can categorically deny any possibility of the kind of god in whom she would like to believe, or that this god has some kind of control over the universe, the amount of control that this god might have, or at least the amount that god chooses to exercise, has become less and less than she previously thought. We are faced with the possibility that the universe is nothing more than a zero-energy-sum nothingness without meaning or purpose, and that we are too. Not "knowing" this, the universe goes blithely along following its own lights to nowhere. And although we would like to find meaning and purpose, and even design, in the universe and certainly in our own existence, there does not seem to be any compelling reason to believe that there is any. About the best that one can do is marvel at the capability of the universe to develop such complexity and stand in awe of the raw power involved which from our point of view is stupendous, even though from an "outsider's" view it all comes to naught. In addition, with all the violence, unknowing and unfeeling mayhem and struggle, with all the chance and random activity, with all the good reason for hopelessness and pessimism that does and always has affected the universe, feelings of awe for its order and law are easily swamped. Any control that god may have seems so remote and restricted, except to those who insist on finding god's hand in what is really the mindless working of chance and evolution, that there is little reason to turn to god at all. And what of evil? There is none. If there is not goodness, purpose and design, then there is no evil, either. What befalls any particular grouping of quarks and molecules (Susan for example) is of no significance to the universe and in fact the grouping itself is not a being but a fiction to which we put a name for our own convenience, being mere fictions ourselves. Science does not conclude this, and "god knows" most scientists rebel against it, looking themselves for some meaning in the teachings of religion or making a religion out of mathematical laws, the concepts of physics and cosmology, or awe of the universe or their own existence. Carl Sagan expresses his own answer to this pessimism and says that,

> "Whenever I think about any of these discoveries [of modern science], I feel a tingle of exhilaration. My heart races. I can't help it. Science is an astonishment and a delight...Nearly every scientist has experienced, in a moment of discovery or sudden understanding, a reverential astonishment. Science - pure science, science not for any practical application but for its own sake - is a deeply emotional matter for those who practice it, as well as for those nonscientists who every now and then dip in to see what's been discovered lately. And, as in a detective story,

> it's a joy to frame key questions, to work through alternative explanations, and maybe even to advance the process of scientific discovery."[379]

Here is both the awe experienced by Catholics before the "tabernacle" as well as the commitment to discipleship. In their less poetic moments, many scientists have to admit that this falls short of the hope that religion has offered. They would have to agree on the judgment made by Fr. Raymond Nogar on Julian Huxley's search for morality and meaning,

> "For Huxley, evil is simply man's choice not to fulfill his maximum evolutionary capacity. Like the primitive `noble savage', the Greek `virtuous man' and the mid-Victorian `gentleman', Huxley's evolutionary man has no meaning beyond aspiring to an arbitrary, sophisticated, aristocratic ideal, an ideal which the deep absurdities of existence reveal to be illusory."[380]

They would have to say that this absurdity that Nogar finds in Huxley is in fact the reality, there is no meaning, no good and no evil. However, we are not satisfied, or at least we have not been satisfied, with finding meaning within the universe that will ultimately "pass away". We want to find meaning placed within it from someone whose own existence gives meaning to the existence of others, but there is none evident from science. Therefore, there is no good and there is no evil. From some relative position there is good and evil. If we set up a goal or purpose arbitrarily, the good of the human race for example, then those things that contribute to this end are good and those that do not promote it or work against it are evil. But the good of the human race is not a purpose of the universe. It cares not if we exist, makes no special provision for our continuation, and does not need us. And certainly this is true of any one individual. We ourselves are totally unfeeling about the fate of the vast majority of caterpillars. What happens to one of them is neither good nor evil, it just happens. It may take on more significance for us if it is the last individual of its species. We mourn the loss of the last dodo bird and the last passenger pigeon. We put up memorials in their honor, but the death of the second last, or the thousandth last does not particularly move us. This is because we (mistakenly) do not equate its welfare with our own. We must be honest and apply the same logic to our own fate as judged from a non-human viewpoint. What difference does it make to the caterpillars if there are humans on earth or not? In fact, they would be much better off without us. They would not describe our demise as "evil". And if it is all so relative, then it is all so meaningless.

Susan has no more right to question her fate and raise her fist to the universe than do the extinct dinosaurs. In fact, the universe could answer by pointing out that it "gave us existence" for at least a little while. Be grateful for small favors. If Susan wants to credit a god with ultimate control and then blame the god because this control is not exercised in her favor, she is free to do so, but she is not required by the evidence to do so and is in fact discouraged by the facts of how the universe works.

4

We have now examined the received faith tradition's assumptions in the light of our knowledge today. Given these new insights into the nature of reality, what are the changes in the faith or in its explanation that must be made to adjust? Following on the last chapter, this may be nerve-wracking for the traditional believer like Susan, questioning as it does what she has been taught and taken for granted, but it is necessary to get the full picture of what needs to be reexamined before we can answer her question about "why is god doing this to me." From what has preceded this section, I presume that the reader will already know many or all of the propositions of the Faith that will have to be rethought. In the interest of saving you time I will state these items briefly and refer you to endnotes if you care to have more information. However, I encourage the reader to refer to these notes.

THE CHALLENGE OF SCIENCE

We are faced with questioning some very basic tenets of the faith and our understanding of god. We can make all the claims we want to make but there is no way to prove that god had a plan in mind, nor that god created the universe at all, much less in conformity with some preconceived purpose or teleology. Worse, far from being able to *prove* any of these traditional contentions, it looks like there is simply no reason for *making* these claims. And worse yet, making such claims, especially in their crasser forms, not only exposes Christianity to derision, it makes it incomprehensible to many scientifically literate people whose number will continue to grow as modern science is disseminated more widely. It is true that there are many scientifically literate people of faith, and there are many without faith who realize that science's ability to make undemonstrated, apodictic statements about god's existence or lack of it, or god's working in the universe is limited. A recent commentary points out that, "science's necessary silence on these questions doesn't prove that there isn't any infinite cause - or that right and wrong are arbitrary conventions, or that there is no plan or purpose behind the world," and takes Richard Dawkins and Stephen Jay Gould to task for acting as if their guesses that there is no cause or purpose are really facts. The commentator admits, "...our ignorance doesn't prove anything one way or the other about divine plans or purposes behind the flow of history. Anybody who says it does is pushing a religious doctrine."[381] The fact remains that there is no way to prove the theistic position either, and as we have seen it is perfectly possible to explain all that exists without reference to god as god has been explained in past apologetics. Unless the faith takes science into account and still has something to say, it is going to be irrelevant to most people's lives. Even those formerly called "biblical archeologists" no longer look to their science as a proof of faith. William Dever says that, although he studied theology throughout graduate school, he has reached the conclusion that theology and belief cannot find confirmation in archaeological discoveries. The Pope's pronouncement that Catholics can and should subscribe to the theory of evolution is a good stop-gap effort to put off the inevitable, but the fact remains that the faith is being examined in the light of new knowledge and is often found lacking.[382] The Christian way of life and view of morality is going to make, seemingly, less and less sense. The system of morality long espoused by the Church based on the two wheels of natural law and divine revelation has had both wheels severely damaged. God's ability to make any difference in the events of world history or in our daily lives has always required rather nuanced explanations to deal with what often seemed, even in the past, as ineffectiveness. Now, with the king-pin of god as creator and sustainer of the universe as well as the ultimate arbiter of meaning and purpose (not to mention reward and punishment) being removed, for many the axle has fallen off the wagon of faith. There is no room for miracles and little room (none actually) for any intervention by god in the aspects of our lives that require manipulation of the physical universe. Witch doctors and faith healers are out in many cultures in favor of chemical treatments even of the mind; alternative medicines by and large do not refer to spirits and supernatural forces nearly as much as formerly, although some New Age theories do. While it is recognized that these traditional practices may well contain truth which modern science may or may not have discovered as yet, the presumed source of their curative powers is no longer divine intervention. Even the Sacrament of the Anointing of the Sick is celebrated with much more expectation of spiritual healing than physical.[383] People still pray to win the lottery, but most really heart-felt prayer for god's intervention is *in extremis*, when all medical efforts have failed, no rain has fallen for a very long time, one is trapped in some inextricable situation, or one is otherwise past hope. Even then, fewer and fewer people really expect god to answer. Even god's ability to affect the realities of the universe in the Eucharistic sacrament is challenged and though it is true that there is a resurgence in some right wing Catholic circles of devotion to the reality of Christ's physical presence in the bread (the wine is rarely mentioned), polls indicate that fewer and fewer Catholics actually believe in this and the theologians have admitted that other explanations than transubstantiation are needed. God is farther and farther removed from any responsibility for the workings of the physical universe and more and more things are recognized as being included in the merely physical.

Much of the faith has presumed that god is intimately involved in the physical, being responsible for its existence and determining, directly or indirectly, its workings. If this is questioned, and it certainly is, then more and more aspects of the faith are in question. We have already examined the main presumptions of the faith as listed in the first chapter. The third chapter examined whether these presumptions can continue to stand and in the main found more reason to reject them than to retain them. Now we are admitting that the faith must be re-examined and very often changed in the light of these findings. The remainder of this chapter will list more of the former articles of faith that must undergo this scrutiny.

THE PERSON OF GOD

Obviously, faith in a god who is the cause of all that exists and who is active in sustaining and directing the workings of god's created universe will not be tenable.[384] Faith in an all-controlling god who is ultimately the cause of all that is and all that happens must give way to a faith in some other kind of god (or no god at all). Theologians have tried to describe the so called "attributes of God" and we must ask if they can still be ascribed to god.

God as a personal being, who always existed and yet exists outside of time, who can only be known by analogy and is the only god there is, are all attributes which have to be addressed. Studies in science, philosophy and biblical history and interpretation do not always support the established explanation.[385]

In the normal Catholic's idea of who god is, god is a personal being who has always existed. God is transcendent to the universe but by choice, because of our contingency, is also immanent. God is the one and only god and yet is somehow three persons forming one being. God created everything for god's own purposes and so can be understood, if only haltingly, in the image of god found in creation. God intended that the result of the creative activity would be a conscious being who could respond to god's love and receive god's self-communication; and to this end implants by some special act of creation a rational soul in each human person. In the light of the investigation we have undertaken, which of these must be seen from a new direction?

The idea of god as a personal being was pretty much presumed from pre-historic times. In fact there were considered to be many personal beings who were gods and goddesses. It was not until the (probably Greek) understanding of a prime mover or a ground-of-being that the possibility that the ultimate explanation for everything did not have to be personal was even considered. Plato, Lucretius, Aristotle and others could imagine a being who would in no way resemble the Hebrew's or Christian's god, or especially the gods of their fellow Greeks, but their ideas were never the common understanding. In our understanding today of the manner of the coming-to-be of the universe we need not posit the existence of either a personal or an impersonal god. If the faith testifies to the existence of a personal god, that is a matter of faith and not of science or philosophy and will depend on the experience of this god by those who encounter god's presence.

It is notable that the idea of "person" developed from the early Church's reflection on the being of god, especially in its doctrine of the Trinity. Then this concept of the identity and worth of the individual person was extended to human beings. The quality of humans was judged by their correspondence to god. Now it seems that we may have to reverse this process and understand god by analogy to human beings, but there is nothing new in this.[386]

The quality of god as being "eternal," may need some retooling given the understanding of time that is involved in this basically philosophical concept.[387] The qualities of "transcendence" and "immanence" will stand, but with some adjustment on the implied relationship to the universe.[388]

The concepts of monotheism and trinitarianism are beyond the comment of science but historically much of what has been written on these subjects, notably by Augustine and Aquinas, tried to make some sense of them by reference to god's creation somehow containing images or vestiges of the inner life of god, which as we saw above must be rejected, but both thinkers admitted that were it not for revelation we would know nothing of these things, and this is certainly the case for us today. If we remove god as a cause of the existence of the universe, then there is no "vestige of the Trinity" to be found in it, nor an image of god to be found in humanity, nor any of the other attributes of god to be inferred from the attributes of the universe or in human history.[389] If we believe that god is "beautiful", for example, it is felt that we do so because we have somehow experienced this quality in god, not because we have first experienced it in the universe and then claimed that it must have its source in the divine being since nothing comes from nothing. Conversely, we have such a concept as "beauty" from the experience of the universe around us and we can then apply it to something we experience in god, recognizing that it is an analogy. The beauty of the universe, which of course is something very subjective in us, is not a "pale imitation of the divine" quality. Pale it may be in contrast to what we are calling beauty in god, but it is not in imitation. It evolved on its own with no reliance or patterning on the beauty of god or of one of Plato's eternal ideas in the mind of god.[390] It is likewise true, it would seem, that not only can we not infer something of god by extrapolation from the universe that reflects its creator-god, but also god cannot reveal godself through expression in physical creation. There is no "mind of god" that can be known by the way it speaks in the make-up of creatures.

God as the creator of the universe is a bedrock foundation of the Faith, but in the light of the previous chapter it is a question whether god should be so described. Some try to preserve some role for a creator, but whether this is legitimate must be examined. For an extended discussion of god as creator, and Scripture's intent when putting god in this role, see this endnote.[391]

To be theological for a moment, we may be able to express something of the relationships of the persons of the Trinity as Augustine and Aquinas and others do by saying that the Second Person is something like an idea in a human mind that seems to take on an existence of its own, even though it remains in, or inheres in, the human mind. So the Logos is the Word spoken by the Father in an analogous manner to our idea being spoken by our mind to our mind. We can make god understandable to our limited human minds by the use of analogies, but we cannot explain the universe as resulting from a creative idea in the mind of god in a manner analogous to the Word being an idea in the mind of god.[392]

THE PERSON OF JESUS

Since the questioning of the assumption that god is a creator of the universe has led us to at least question whether god can impact physical matter at all, certainly the idea of god becoming physical matter must also be examined. Many Christians call this the "Incarnation" of god in Jesus and it is a bedrock of Christianity. Still, it will have to be revisited. If it must be changed, then the meaning and message of Jesus must be looked at differently. Included in this description is a hint at where my argument is going, which is to say that the reason god does not intervene in the physical universe is not that god chooses not to in order to allow us free will, or for some other reason, but that god *cannot* affect physical matter, so you may wish to be sure to read this endnote.[393]

CONSEQUENCES OF A CHANGED VIEW OF GOD

If, as I believe, the role of god as the creator and master of the universe must necessarily be called into question, there are profound consequences for a whole range of beliefs and behaviors which are based on god as a creator. Foremost is the identify of human kind. We are so accustomed to defining ourselves by our relationship (or lack of one) with god. If we are creatures, then god must have some attitude toward us. If creation is done for the benefit of the creatures, then god must be benevolent. If humans are created for the service of god, that is another kind of correlation. The Judeo-Christian-Muslim traditions have certainly assumed that we humans are the pinnacle of god's creation. If our investigation calls into question god's role as creator, our role is changed as well.[394] We are obviously not the goal of evolution and the universe was not made for our delectation.[395] We find the universe interesting and beautiful because we are designed by evolution to do so. We are on the top of the food chain and able to carry out the use (and the destruction) of other species because that is where our evolutionary history has led us. There is nothing that sets us apart from the life chain we inhabit in any other way than a squirrel is different from a tadpole. Therefore, finding god's love for us in the way the universe is configured for our welfare is a bogus conclusion. The universe supports us "because" if it did not we would not be here, and of course it could withdraw that support without a qualm. One good-sized asteroid to the mid-section would show just how privileged a species we are. So if god has taken some special interest in us we must find other ways to understand it and see the results.[396]

If god is not the creator of the universe, we are going to have to find some other way to relate to the rest of creation.[397] The Second Vatican Council called for justice in sharing the goods of this world by saying, "God destined the earth and all it contains for...all people so that all created things would be shared fairly by all."[398] If god did not create, then god made no conditions. If we are of the universe, we have a greater claim than god to say how it will be used. I certainly believe in sharing the wealth of the world with all its inhabitants, but do not believe we can make this a moral directive by reference to god as the creator. Another approach is needed. Reverence for other life forms has been based on god's claim as creator to demand such. Claiming that the spotted owl should be saved since it is one of god's creatures can be a fairly effective tool in the environmentalists' arsenal.[399] The argument was always open to the objection that the logger was also one of god's creatures, but given the negative feelings many have for their fellow humans and given that most loggers are not very cute, this was not nearly as effective. Now if we say that god had nothing to do with making these creatures, nor ourselves, but, to put it anthropomorphically, god just stumbles upon them, then why should god have a say in how we treat other inhabitants of the universe. Francis of Assisi may be a quaint character who considers the birds his brothers but the fact remains that he as a human is an omnivorous predator who under other conditions would consider the little brother to be a suitable candidate for lunch *and would be entirely true to his nature to do so.* Walden may be a picturesque specimen out on his pond, but Peabody and his strip miners are behaving just as much according to "natural law" when they haul West Virginia away to the blast furnaces. There is no divine design being violated, no gift of god being misused. If the faith leads us to believe that there are moral mandates to be observed and that there is a right and a wrong way to relate to the rest of the universe, and it does, it is going to have to find some other basis for its position than that of god as creator. Again, there are other aspects that will need to be changed, but you get the idea, so I have put more in an endnote.[400]

The efficacy and legitimacy of prayer, especially the prayer of petition for god to change something in the universe, is likewise called into question. Prayer as a request presumes the divine person addressed has the power to grant the request. In what sense does god have such power?[401]

It is questionable, also, that there is some plan operating in the universe, leading it in some predetermined direction, no matter how general that direction is proposed to be.[402]

If the understanding of prayer will need to be modified, certainly the possibility of miracles, especially when they are seen as an exercise of the divine power over god's creatures, will have to be investigated.[403]

Especially in the Roman Catholic tradition, morality has been based not only on Revelation but also on the existence of a "Natural Law", placed by the creator in the hearts and consciences of god's creatures. Whether there is a Natural Law, and whether it comes from a creator needs to be re-evaluated. [404]

It may seem like an insignificant part of the faith, but I believe that a re-investigation of the matter of angels and devils will bolster our new understanding of how god works in the world. Besides, with angels on many greeting cards, garden decorations, and TV shows, this may be more timely than we think. On the other hand, if you have had quite enough of angels and devils as I have, just skip this note.[405]

The idea of a final judgment needs to be re-formulated or perhaps dropped, at least in its cruder forms.[406] Obviously, a judgment can be made on the universe and maybe someone sometime in another universe will make such a judgment. If they somehow find a record of our universe, they would judge it according to their standards. If we initiate the existence of another universe and engineer it so that the inhabitants will someday be aware of their progenitors, they will make some judgment on us, for good or for ill. One can imagine a being in another universe sitting on his porch scratching his silicone hair under a crimson sky streaked with vermilion clouds and thinking that our blue and white color scheme was certainly ugly. I didn't say it was a profound judgment! Whether or not god makes such a personal assessment of our universe, perhaps from the myriad of universes existing, and likes our color scheme or our morals, or the individuals in it, picturing god doing so after putting an abrupt halt to the progress of the universe and gathering everyone in front of a throne so as to separate the sheep from the goats needs to be rethought. So does the more sophisticated view of a final judgment.

Our individual existence after death could use a little retooling as well. The pat answers to the bereaved have always been vague and unsatisfactory. The idea of a soul, which continues our existence as a natural function of its immortal make-up, doesn't even square with the Scripture, much less with our present understanding of the composition of the mind or will. According to the soul view, we are not at home in the universe, our true home is in heaven and we are either exiled from there (did we live there before being imprisoned in this body?) or on our way there. It is in our nature to exist as our true spiritual selves after having had this human experience. This is certainly not an apparent truth.

The resurrection of Jesus is also a doctrine that will require a different explanation, as it has many times in the history of the Faith.[407] The resuscitation of a physical body which is the understanding of many, has already been rejected by many of the theologians in the Church, but even their nuanced explanations will have to be rethought both in the light of what we mean by Jesus being the incarnation of the divinity and in regard to what it means to live a resurrected life. If god cannot make a human body, god cannot make a resurrected-divine-human body, either.[408]

The idea of the Assumption of Mary is in line for revision. Just what this doctrine is intended to convey has always been problematic. More so now.[409] The whole idea of contact of the dead with the living is not just a problem for mediums and séances.[410]

The role of the saints, if any, will depend on our understanding of existence after death. Their physical influence on the universe and its inhabitants will surely be the same as that of god, which is to say they have none. This means that we will not look for them to find lost articles, shower down rose petals, cure our cancer, or any of the other powers with which we have credited them. Our relationship to Christ and to the Saints will have to be based on a new understanding. Liturgical prayers to St. Blase to help us expel chicken bones and avoid strep throat have been transformed by more sensitive pastors to be prayer for the correct use of speech, for example, or other more meaningful petitions for assistance. This kind of revision will have to continue, that is if the Saints have any role at all.[411]

Obviously, heaven, purgatory and hell and whatever we mean by them will have to be expressed in terms of the Eschaton and existence after death and in a radically different manner than previously. The same will be true of "indulgences" and the effect the living have on the departed.[412]

Eschatology will have to similarly be examined and, if retained, expressed in new ways. Certainly if god has no effect on the fate of the universe then it cannot be said that it is proceeding according to god's plan to some preordained end. This does not mean that god may not have some plan in mind for us, however, and we will consider this in the final chapter.

The same will have to be done, and again has already been largely done, with such concepts as original justice and original sin Unfortunately, explanations of these concepts have sometimes been canonized by the Church and defended at all costs. As in the Scripture, there are often valid messages to be delivered about god and god's relationship with us that were conveyed to other cultures and times with some effectiveness by these concepts and words of official pronouncements, but which today are often obscured by their outdated theology. There are times when the whole idea must be thrown out, but there are many more when the vehicle of communication can be changed while the content can remain. Of course the content will be understood differently because of the new way of explaining it. One author says of original sin,

> "If [my] interpretation is admitted, original sin is not a unique catastrophe at the birth of our species; it is the continually perpetuated perversion of mankind, in which new sins are conditioned more or less by the preceding sins and carry on the existing disorder...Many modern exegetes refuse to see in the story of the Fall the story of an individual event. In their eyes it is only a kind of parable, illustrating the universal fact of sin..."[413]

This was not so acceptable to many theologians in 1964 when the book was written from which this quote was taken, but it certainly is accepted today. This re-examination which I am suggesting has already deepened our understanding of what the Church has been inspired to teach rather than corrupted it as some would fear.[414]

Our explanation of the Sacraments will need to be revised as I have pointed out previously. Transubstantiation in the Eucharist will have to be rephrased to explain "what" we understand god is doing in the Eucharist and not so much "how" god is doing it, at least in a physically manipulative manner. The alternative explanations current in theology do make some sense in terms of modern ontology, but more will have to be done. This same re-examination has also occurred in regard to the other Sacraments. We have seen much revision in the liturgy of the Sacraments in the past years and this had to be done because of a deeper understanding of the Scripture and the Tradition of the Church. Certainly our growth in appreciation for the Sacrament of Matrimony and the marriage covenant; our manner of responding to sin and Reconciliation; our care of the sick and the dying; a different approach to Baptism as initiation instead of a ticket to salvation and Confirmation as a sacrament for maturing in the wisdom of the Gospel rather than the infusing of amorphous gifts of the Spirit;[415] All of this and more has occurred in large part due to the findings of modern research into the functioning of real physical human beings with all their foibles instead of idealized potential saints. We have already rethought much of the faith, especially as it is actually lived in the world today, and the process will only be helped by the addition of the findings of physics, cosmology, biology and evolution.

SO WHAT KIND OF GOD DOES THIS LEAVE?

Let me be clear where my re-examination has led me. You may or may not agree, but this is where I am in answering these questions. Can our belief in the god we presumed exists be justified? In many aspects, the answer seems clear that such a god cannot. God need not be and in my opinion is not the creator of the universe, nor of anything else physical. God does not sustain these things in existence and if there were no god there could still be a universe and everything that it contains or will contain. God does not manipulate the universe on a physical level, not by making it rain or withholding the rain, not by curing, not by blessing with wealth, health or long life. God does not decide when "it is your time," physiology and chance decide. God does not decide the future movements of atoms and does not configure them for the benefit of god's subjects. Since everything that happens is the result of physical laws on some level, including the laws of chance and random action, god cannot be approached as one who has any control over our physical welfare or existence. This is not the same as a reductionist theist or the deists might contend, that if there is a god, or since there is a god, god has chosen not to intervene in the working out of god's creation. This *is* to say that god does not get involved because god *does not* control these things. To say otherwise is to fly in the face of every assumption of science, which has so far been proved right in its conjecture that everything that occurs has an explanation open to science's scrutiny. As has been said, medical personnel are sometimes heard to speak of miracle cures, a "miracle baby" and the like, but they know that they are expressing more about their ignorance of the factors involved than a conviction of fact. Some astronomers can be found who will speak of the vastness of the universe giving us some glimpse of the vastness of god, but in their normal routines scientists presume that, or at least they act as if, god is not involved, at least not in any miraculous manner, in the workings of the human body or the expansion of the universe, or any other subject of scientific investigation, and that presumption is proved out in every experiment. No need to have the angels pushing the celestial spheres to explain planetary or sidereal motion, nor to invoke a guardian angel to explain why this child recovers from the illness that killed another child. If the scientists or medicos are religious people they may wonder why god does not get involved in this case or that and attribute to god the power to do so, but they do not expect it. The answer to their query, and to Susan's, is not some contorted explanation of the ways of the Lord, it is the simple fact that god does not act in this arena. Unless we are willing to jump again into the argument of why god permits evil in god's creation, we are going to have to make again the assertion that god not only *does not* get involved in the workings of the universe, but god ***cannot*** direct the physical processes of the universe. We might fervently wish that god would "take charge" of a situation and make things work out for us. Perhaps god even wishes that it were possible. What person, divine or human, would not wish to have the power to rescue a child, or any other sufferer, from their pain?[416] This is the can of worms that is opened once again if we do not take the bit in our teeth and admit that god is powerless to change the physical ills that afflict us, the worms being the question of evil.

EVIL

Many attempts have been made to explain the presence of evil in our universe and god's relationship to it. We must speak of "our universe" since there is no evil we know of in any other "universe" (unless, of course, it is one that is also inhabited by beings who know good and evil). Walter Kasper sums up the problem succinctly and says,

> "No one has formulated this argument more forcefully than Epicurus. Either God wants to get rid of evil but cannot - and then he is helpless and not God; or he can get rid of it but chooses not to - then he himself is wicked and at bottom is really the devil; or he does not want to and is not able - then both of the previous conclusions follow; or he wants to and is able to - but then whence comes evil? A. Camus gives the argument in this form: `either we are not free and God the all-powerful is responsible for evil. Or we are free and responsible but God is not.' All the Scholastic subtleties have not made this paradox any more telling nor, on the other hand, have they been able to take the edge from it. After the dreadful horrors our century has seen, post-Auschwitz theology believes it now impossible to speak responsibly of a God who is both omnipotent and good."[417]

If we reached no conclusions resulting from our investigation here, the faith must still respond to this challenge in new ways than it has in the past. The traditional answers, which I review briefly in this endnote, must be rethought.[418]

The "free will argument" which I have alluded to previously, is explained very well by Richard Swinburne when he says,

> "The central core of any theodicy must, I believe, be the 'freewill defense', which deals - to start with - with moral evil, but can be extended to deal with much natural evil as well. The free-will defense claims that it is a great good that humans have a certain sort of free will which I shall call free and responsible choice, but that, if they do, then necessarily there will be the natural possibility of moral evil. (By the 'natural possibility' I mean that it will not be determined in advance whether or not the evil will occur.) A God who gives humans such free will necessarily brings about the possibility, and puts outside his own control whether or not that evil occurs. It is not logically possible - that is, it would be self-contradictory to suppose - that God could give us such free will and yet ensure that we always use it in the right way."[419]

John Polkinghorne agrees and says that

> "...there is only one broad strategy possible for any theodicy. It is to suggest that the world's suffering is not gratuitous but a necessary contribution to some greater good which could only be realized in this mysterious way. In relation to moral evil...this leads me to embrace the free-will defense: that despite the many disastrous choices...a world of freely choosing beings is better than a world of perfectly programmed automata."[420]

By that thinking, if god had made it possible to do even more evil than we are presently capable of, just think of the great choices we could make! If choosing not to bash your brains in with a flower pot makes it possible for me to be responsible for you and to you, just think how much greater a person I would be if I could choose whether or not to destroy your galaxy! I have been perhaps too sarcastic, but this *ad hoc* argument merely gives a possible explanation of why the situation is as it is. It does not answer whether it could be otherwise, and in my opinion has caused much suffering because no matter how you twist it, this explanation makes god responsible, even if the human with free will is also responsible. This position runs into the obvious question we have seen before. Will there be no free will in "heaven"? Will everyone be an automaton? Is there not another kind of free will that god could have given? Could we not, for example, have choice only between how much pleasure to give or receive or when to give or receive? To choose to not help you in a pleasant task is hardly the evil I choose when I decide not to help you survive. The free will defense has some holes in it. For a treatment of the usual argument, see the endnote.[421]

If we posit, only for the sake of argument, that in the traditional explanation god is not responsible for moral evil, for which humans will have to shoulder the blame, what about all the physical "evil" and suffering in the world. It looks like this at least is laid directly at god's feet, even by the Catechism. How much of human wrongdoing is more error or innate reactions induced by our evolutionary history, making it more natural evil than "moral evil"? Certainly much of it, quite possibly including the behavior of the husband of our long-suffering Susan. Richard Swinburne also claims to have an answer for the question of whether natural disasters and natural evil must be laid at god's feet. He believes that there should not be blame since god had no choice in the matter of natural evil either if we were to have meaningful free will. "Natural evil is not to be accounted for along the same lines as moral evil. Its main role rather, I suggest, is to make it possible for humans to have the kind of choice which the freewill defense extols, and to make available to humans specially worthwhile kinds of choice.[422] Maybe today if a volcano blows up in the Philippines I am faced with the choice of sending some help or taking a vacation. If I choose to help, the assistance can be there in a relatively short time. I have a moral, free-will choice to make. Then what about two hundred years ago when my help would take months to reach the afflicted? Was it necessary that the volcano blow up and kill hundreds or thousands of people to give me a choice then? That seems rather drastic. And what of all those disasters of which I am ignorant? Proponents of this thinking would of course say that the moral choices they have in mind are those that those close to the scene have, and I suppose that this is a valid answer, but the whole thing seems rather "*ad hoc*". If we lived in a world that did not contain wholesale destruction of peoples, would I really be so deprived of the opportunity to make meaningful moral choices? Most inhabitants of technologically advanced countries are not regularly confronted with such life and death situations and yet manage to be reasonably moral individuals. A little shallow sometimes in the wealthier neighborhoods, but not bad people. And how much evidence is there that those who do face natural "evil" are so much better people for it? Talk to the National Guard about looters. If we lived in a world in which these things did not exist, would we feel the loss or the necessity to invent them? Since we live in a world that does contain such things, and since we presume that god must somehow be responsible for them, we find ways to exonerate god of blame. However, once again, if it could have been otherwise without great loss of some goodness in the universe, and god did not make it otherwise, then god is to blame. Proponents are going to have to re-examine this argument. I have much more on opinions on the subject by others as well as my own reaction to them that I have loaded into yet another endnote if you have the time for it, and I hope you do since this is the crux of the question.[423]

If these previously accepted aspects of the faith must be examined in the light of the findings of science, where should we look for answers? I propose to look at the writings of first those who deny a role for god in the creation or the operations of the universe, many of whom would also question the existence of god. Most are honest enough to admit that science cannot prove that god does not exist, but all find no compelling reason to believe it.

Second, I will deal with people, scientists, theologians, and others, who try to claim a role for god in creation and/or the operation of the universe. I have not found anyone with faith in god who does not grant god such a role. If I had, I would possibly not have written this book. As you probably surmise, that is the direction I will go in the next chapter.

A SURVEY OF FAMOUS SCIENTISTS (AGNOSTIC PERSUASION
CARL SAGAN

Carl Sagan recently had to face his own death without religious faith. He had earlier written,

> "Nietzsche mourns the loss of 'man's belief in his dignity, his uniqueness, his irreplaceability in the scheme of existence.' For me, it is far better to grasp the Universe as it really is than to persist in delusion, however satisfying and reassuring. Which attitude is better geared for our long-term survival? Which gives us more leverage on our future? And if our naive self-confidence is a little undermined in the process, is that altogether such a loss? Is there not cause to welcome it as a maturing and character-building experience?"[424]
>
> "In a life short and uncertain, it seems heartless to do anything that might deprive people of the consolation of faith when science cannot remedy their anguish. Those who cannot bear the burden of science are free to ignore its precepts. But we cannot have science in bits and pieces, applying it where we feel safe and ignoring it where we feel threatened...because we are not wise enough to do so."[425]

True to his convictions, he wrote while much nearer death,

> "I would love to believe that when I die I will live again, that some thinking, feeling, remembering part of me will continue. But as much as I want to believe that, and despite the ancient and worldwide cultural traditions that assert an afterlife, I know of nothing to suggest that it is more than wishful thinking."[426]

Yet he believes in something, and continues,

> "...the world is so exquisite, with so much love and moral depth, that there is no reason to deceive ourselves with pretty stories for which there's little good evidence. Far better, it seems to me, in our vulnerability, is to look Death in the eye and to be grateful every day for the brief but magnificent opportunity that life provides."[427]

These two themes of honesty in the face of the findings of science (as well as an antipathy to believing in that for which there is no scientific proof), and an awe of the world that science has revealed are found in the reactions of most scientists who have responded to questions about their own faith. Sagan expressed joy in the discovery of something previously unknown about the universe and sounds almost mystical when he says that,

> "Nearly every scientist has experienced, in a moment of discovery or sudden understanding, a reverential astonishment. Science - pure science, science not for any practical application but for its own sake - is a deeply emotional matter for those who practice it..."[428]

Replace "science" with "God" or with "Wisdom" or with "Torah" and you have the monk, the philosopher, or the rabbi waxing eloquent.

STEPHEN JAY GOULD

Stephen Jay Gould claimed often to be an agnostic Jewish scientist. Yet he claimed to have respect for religion, especially since he allowed that there are two realms, the scientific and the religious, which should not trespass on each other. Indeed, his book, "Rocks of Ages" is specifically about his solution to the science/religion debate, a solution that he calls NOMA for "non-overlapping magisteria".[429] In reacting about the Pope's recent pronouncement in favor of evolution as a proven theory to be accepted he says,

> "Religion is too important to too many people for any dismissal or denigration of the comfort still sought by many folks from theology. I may, for example, privately suspect that papal insistence on divine infusion of the soul represents a sop to our fears, a device for maintaining a belief in human superiority

> within an evolutionary world offering no privileged position to any creature. But I also know that souls represent a subject outside the magisterium of science. My world cannot prove or disprove such a notion, and the concept of souls cannot threaten or impact my domain...I surely honor the metaphorical value of such a concept both for grounding moral discussion and for expressing what we most value about human potentiality."[430]

Talk about damning with faint praise. The Pope has no intention of speaking metaphorically (although in my opinion it is indeed a metaphor). In an editorial commentary published in many newspapers, Gould says that, "no scientific truth can pose any threat to religion rightly conceived as a search for moral order and spiritual meaning."[431] These may be the boundaries of religion for Gould, but certainly his "religion rightly conceived" is not that of John Paul II or his predecessors. In the article cited above Gould goes on to express sympathy for a person who would have to struggle with the concept of evolution instead of god as the engine behind the scenes to maintain any notion of a good and caring universe. He prefers the, "cold bath" theory,

> "...that nature can be truly `cruel' and `indifferent'...because nature was not constructed as our eventual abode, didn't know we were coming...and doesn't give a damn about us (speaking metaphorically). I regard such a position as liberating, not depressing, because we then become free to conduct moral discourse...in our own terms, spared from the delusion that we might read moral truth passively from nature's factuality. But I recognize that such a position frightens many people..."[432]

Not being a coward, Gould takes his reality in strong doses, not cut with the sugary flavor of the traditional folk faith! However, anyone who supposes that he gives religion any more status than a useful source of metaphor is deceiving themselves.

EDWARD WILSON

Edward Wilson makes the connection of our genetic tendency to hope and the "cold bath" of scientific knowledge and a commentator says that Wilson has,

> "...a straightforward solution: as long as we've got these orphaned religious impulses lying around, we might as well hitch them to a belief system that still has legitimacy - science, for instance. He would like to see people get their epiphanies the way he's gotten his - by, becoming engrossed in the `endless unfolding of new mysteries'; by investing their faith not in

> Genesis but in `the evolutionary epic.' We must cultivate a `scientific humanism' that taps the energy of our innate religious drive says Wilson."[433]

In his book, "*Consilience*", Wilson himself says,

> "For Centuries the writ of empiricism has been spreading into the ancient domain of transcendentalist belief, slowly at the start but quickening in the scientific age. The spirits our ancestors knew intimately first fled the rocks and trees, then the distant mountains. Now they are in the stars, where their final extinction is possible."[434]

However, he continues immediately, "*But we cannot live without them.*" This sounds like good news for the theists, but it is not. He proposes replacing faith in such spirits and,

> "...if the sacred narrative cannot be in the form of a religious cosmology, it will be taken from the material history of the universe and the human species. That trend is in no way debasing. The true evolutionary epic, retold as poetry, is as intrinsically ennobling as any religious epic. Material reality discovered by science already possesses more content and grandeur than all religious cosmologies combined."[435]

Wilson, like others, waxes eloquent about the depth of meaning of each human life in the context of the story of the universe, but the fact remains that any meaning is restricted to this universe. If our own descendants know anything about us and give us any respect, to say nothing of inhabitants of other universes, we will never know it and will be long gone. Our story may or may not be told to the accompaniment of trumpets and strings, but this is the best hope we have for the life we crave. Since the universe itself has only passing meaning, it is no hope at all.[436]

STUART KAUFFMAN

Stuart Kauffman holds,

> "...the hope that what some are calling the new sciences of complexity may help us find anew our place in the universe, that through this new science, we may recover our sense of worth, our sense of the sacred...[in order to]...reinvent the sacred - this sense of our own deep worth - and reinvest it at the core of the new civilization."[437]

Our own deep worth for him comes from our embodying the universe within ourselves, not in being the image of god or embodying Christ in ourselves.[438] Reporting on a sense of awe, even spirituality, at spectacular scenery spread below him, he nonetheless agrees with a friend who prayed, not to a god, but, "glanced at the sky and offered a prayer: `To the great nonlinear map in the sky.'"[439] We are, for Kauffman, *At Home in the Universe* and have only ourselves to rely on. He is under no illusion that we, or any other power, are up to the task. There is no guaranteed progress, such as Francis Bacon presumed in his *New Atlantis*,[440] and Kauffman exclaims, "Bacon, you were brilliant, but the world is more complex than your philosophy." Therefore,

> "We latter-day players are heritors of almost 4 billion years of biological unfolding. If profound participation in such a process is not worthy of awe and respect, if it is not sacred, then what might be? If science lost us our Western paradise, our place at the center of the world, children of God, with the sun cycling overhead and the birds of the air, beasts of the field, and fish of the waters placed there for our bounty, if we have been left adrift near the edge of just another humdrum galaxy, perhaps it is time to take heartened stock of our situation."[441]

For him, our situation is not that we are the darlings of a divine creator, but, "expressions of a deeper order inherent in all life."[442]

LEE SMOLIN

In the same vein, Lee Smolin in *The Life of the Cosmos* feels that we are indeed at home in the Cosmos and that we must find any meaning there. He concludes with a metaphor of the universe,

> "which I would like to set against the picture of the universe as a clock" which is, "an image of the universe as a city, as an endless negotiation, an endless construction of the new out of the old. No one made the city, there is no city-maker, as there is a clockmaker. If a city can make itself, without a maker, why can the same not be true of the universe?"[443]

He finds the possibility of meaning in working together with one another and with the rest of the universe in this ongoing "making." His conclusion is also that,

> "...there never was a God, no pilot who made the world by imposing order on chaos and who remains outside, watching and proscribing...There is nothing behind it, no absolute or platonic world to transcend to. All there is of Nature is what is around us. All there is of Being is relations among real, sensible things. All we have of natural law is a world that has made itself"

Thinking he has rejected the possibility of a transcendent God, but actually having rejected a controlling, ordering, creating "absolute" whether as a beginning point, a goal of philosophy, or an end point, what is his conclusion? "All that is possible of utopia is what we make with our own hands. Pray let it be enough."[444] Pray, indeed!

DAVID NOBLE

David F. Noble argues in *The Religion of Technology* for a decoupling of science and religion, at least Christian religion. He shows how science has long been at the service of religion in looking for a paradise above and beyond the universe in some way. First, literally above it and then a world built by technology that would far surpass and transcend the normal course of the universe. All this to be accomplished by a chosen people, first the Hebrews, then the Christians, and now the technicians. His conclusion endears him to environmentalists and lovers of humanity when he says that,

> "If dreams of technological escape from the burdens of mortality once translated into some relief of the human estate, the pursuit of technological transcendence has now perhaps outdistanced such earthly ends. If the religion of technology once fostered visions of social renovation, it also fueled fantasies of escaping society altogether."

His "at home in the universe" conclusions will not thrill theists when he says that we must, "...disabuse ourselves of our inherited other-worldly propensities in order to embrace anew our one and only earthly existence."[445]

KAFATOS AND NADEAU

As Menas Kafatos and Robert Nadeau see it, the universe is "conscious" and we, and any other conscious beings in the universe, are the personification of this quality.[446] We become the incarnation of "Being" and participate in that quality of being that defines reality. Of course any meaning and purpose and sense of awe at our position is entirely contained within the universe and the "Being" that we express is also restricted to the being that is possible to our universe. There is no sense that this is an omnipotent "other", for there seems to be no "other", the universe containing in itself and expressing in us, and perhaps even more clearly in our descendants be they human or machine, only that which has always been present but unexpressed.

DANIEL DENNETT

Daniel Dennett recognizes the tendency among non-believing scientists to try to replace the faith with something else, such as Kafatos and Nadeau have just expressed. Some, he says,

> "...ground their highest concerns in entirely secular philosophies, views of the meaning of life that stave off despair without the aid of any concept of a Supreme Being-other than the Universe itself. Something is sacred to these thinkers, but they do not call it God; they call it, perhaps, Life, or Love, or Goodness, or Intelligence, or Beauty, or Humanity. What both groups share, in spite of the differences in their deepest creeds, is a conviction that life does have meaning, that goodness matters."[447]

Dennett asks, as any believer such as Susan may ask, "Can any version of this attitude of wonder and purpose be sustained in the face of Darwinism?" His answer comes after much discussion, but it is in the same vein as the answers of the others.

> "Darwin offers us [something sacred]: it is in the distribution of Design throughout nature, creating, in the Tree of Life, an utterly unique and irreplaceable creation, an actual pattern in the immeasurable reaches of Design Space that could never be exactly duplicated in its many details. What is design work? It is that wonderful wedding of chance and necessity, happening in a trillion places at once, at a trillion different levels. And what miracle caused it? None. It just happened to happen, in the fullness of time. You could even say, in a way, that the Tree of Life created itself."

And, Dennett concludes,

> "...if [the universe] is not Anselm's `Being greater than which nothing can be conceived,' it is surely a being that is greater than anything any of us will ever conceive of in detail worthy of its detail. Is something sacred? Yes, say I with Nietzsche. I could not pray to it, but I can stand in affirmation of its magnificence. This world is sacred."[448]

Of course, this is not the sacred being of the theists, but it evidently is good enough for many modern people, or at least it seems the best we have. Any thought that we might live forever with this being is out of the question, but we can revel for a little while in this opportunity for existence.[449]

PAUL DAVIES

Also in "the best we have" category is a conclusion of Paul Davies who theorizes that heaven would be awfully boring if it meant an infinite repetition of activities, no matter how pleasurable, but immortality could be fine if it were combined with progress since humans like new experiences. Since there would be no end to the former state, and therefore it was not going to any destination, there could be no purpose and human beings do not, according to Davies, like to do things without a purpose. His conclusion is rather dreary, therefore, and the best he can do is to pose two choices.

> "If there is a purpose to the universe, and it achieves that purpose, then the universe must end, for its continued existence would be gratuitous and pointless. Conversely, if the universe endures forever, it is hard to imagine that there is any ultimate purpose to the universe at all. So cosmic death may be the price that has to be paid for cosmic success. Perhaps the most that we can hope for is that the purpose of the universe becomes known to our descendants before the end of the last three minutes."[450]

WEINBERG AND SMOOT, DAWKINS AND CRICK

On the subject of a purpose for the universe, Steven Weinberg writes that, "The more the universe seems incomprehensible, the more it also seems pointless...The effort to understand the universe is one of the very few things that lifts human life a little above the level of farce, and gives it some of the grace of tragedy."[451] Of course, there is no audience for the play, and it is much more enjoyable to be an actor playing the part of the tragic hero than to be the tragic hero who is doomed from the start.

George Smoot disagrees with Weinberg's assessment that the universe is pointless and finds that the universe is, "...quite the opposite. More and more, the universe appears to be as it is because it must be that way; its evolution was written in its beginnings - in its cosmic DNA, if you will"[452] Nevertheless he does not conclude that there was an intelligent designer to this "DNA" of the universe. It is just that given the initial conditions, "...there is an overall inevitability to the development of sophisticated complex systems. The development of beings capable of questioning and understanding the universe seems quite natural."[453] He must fall back on the sense of awe reported by many others and he concludes with,

> "The religious concept of creation flows from a sense of wonder at the existence of the universe and our place in it. The scientific concept of creation encompasses no less a sense of wonder: we are awed by the ultimate simplicity and power of the creativity of physical nature - and by its beauty on all scales."[454]

Richard Dawkins has written a book, *Unweaving the Rainbow*[455] in which he tries to uphold this view that the beauty and awe inspired by science is quite enough to replace that described by poets and mystics. At least one reviewer has found his efforts rather forced.[456] On the other hand, Francis Crick joins the chorus of praise for the beauty and complexity of the universe and also joins the attack on those immature persons who,

> "...prefer the myths of the past even when they are in clear contradiction to the science of the present. I do not share this view. It seems to me that the modern picture of the universe - far, far older and bigger than our ancestors imagined, and full of marvelous and unexpected objects, such as rapidly rotating neutron stars - makes our earlier picture of an earth-centered world seem far too cozy and provincial. This new knowledge has not diminished our sense of awe but increased it immeasurably."[457]

It is not theology that is needed to describe the origin of these wonders, says Crick, contrary to Dawkins, it is poetry.

Steven Weinberg, true to his convictions, puts a pox on both houses and says,

> "It would be wonderful to find in the laws of nature a plan prepared by a concerned creator in which human beings played some special role. I find sadness in doubting that we will. There are some among my scientific colleagues who say that the contemplation of nature gives them all the spiritual satisfaction that others have traditionally found in a belief in an interested God. Some of them may even really feel that way. I do not."[458]

He admits to the temptation to believe dishonestly, as when the character in the novel "1984" convinces himself that 2 + 2 = 5, but this he cannot do.[459] Still, "The honor of resisting this temptation is only a thin substitute for the consolations of religion, but it is not entirely without satisfactions of its own."[460]

RICHARD FEYNMAN

Richard Feynman would like to keep both worlds, but he is not sure how to do this. He sees a value in religion, at least the religion of western civilization, first in recognizing that there are mysteries that we do not understand and in encouraging the search for answers, and second in providing Christian ethics. However, he does not know how to maintain, "the real value of religion as a source of strength and of courage to most men while at the same time not requiring an absolute faith in the metaphysical system"[461] which he supposes religion must insist on keeping. He does not see how one can maintain a religious faith that will have consequences in one's life in the face of any kind of doubt about god, and he does not see any way of having a metaphysical system that is certain and will never be questioned by science. He should have consulted any honest believer who patterns their life on god and yet lives with doubt.

URSULA GOODENOUGH

Ursula Goodenough does claim to find a metaphysical system of a sort and though a biologist at Washington University in St. Louis and a professed agnostic if not atheist, also sings in the choir at a Presbyterian Church, very much like Thomas Hardy, the English poet of *God's Funeral* who remained taken by religious trappings even when faith had faded. She wants "an approach that can be called religious naturalism."[462] Professor Goodenough covers most of the topics of my work, beginning with the origin of the universe in the Big Bang. At the conclusion of her concise explanation of the events at the beginning of the cosmos she draws her first lesson,

> "I have come to understand that I can deflect the apparent pointlessness of it all by realizing that I don't have to seek a point. In any of it. Instead, I can see it as the locus of Mystery. The Mystery of why there is anything at all, rather than nothing. The Mystery of where the laws of physics came from. The mystery of why the universe seems so strange."[463]

I suspect that the physicists will not give up so easily and will respond, "spoken like a true biologist." The replacement for religious faith and faith in a personal, relational god which she offers will probably prompt the Presbyterians to respond, "spoken like a true atheist". Goodenough would most likely not have a problem with the latter, but she does try to replace the god of the theists, and certainly the god of the deists and still, "join the saints and the visionaries in their experience of what they called the Divine."[464] When she speaks of the emergence of life from non-life she rightly dismisses the Anthropic Principle and the call for a Designer, but wants to retain the category of "miracle" to apply to these events. Like the other scientists discussed above, she finds a sense of awe and majesty not in the person of god but in the working of nature. Her attempt to semi-personalize nature so that it can be related to at least in some manner as the "saints and visionaries" relate to their god is stretching the concept. One can be grateful to another personal being, for example, but saying, "when such gratitude flows from our beings, it matters little whether we offer it to God…or to Mystery or Coyote or Cosmic Evolution or Mother Earth."[465] Believers in a traditional personal god will beg to differ. Those who claim to know a divine being to whom they are grateful will find this unsatisfying. For Christians, her argument includes another problem. In saying,

> "For me, the existence of all this complexity and awareness and intent and beauty, and my ability to apprehend it, serves as the ultimate meaning and the ultimate value. The continuation of life reaches around, grabs its own tail, and forms a sacred circle that requires no further justification, no Creator, no superordinate meaning of meaning, no purpose other than that the continuation continue until the sun collapses or the final meteor collides. I confess a credo of continuation",[466]

Professor Goodenough stimulates more than just disagreement on whether or not there is a creator or a meaning to anything. Her "sacred circle" is inimical to the Christian faith that god is taking us somewhere. Like the others cited, any meaning and purpose she finds in the universe and in human life is relatively short-lived, lasting only until the earth is no more or, presumably, the universe itself can no longer sustain life in any form. Also like the others, her effort is praiseworthy and may be the best that we can do if there is no god, but that is our question.[467]

CHET RAYMO

In a book titled, *Skeptics and True Believers* with a subtitle of "The Exhilarating Connection Between Science and Religion", Chet Raymo could have saved me a lot of trouble in writing down many of the results of my reading in science. Brought up, like myself, as a Catholic with all the attitudes, explanations and assumptions I have previously examined, he became a skeptic after exposure to science. His book examines many of these assumptions as I have attempted to do and he does a masterful job of knocking down much of what the faith has assumed. His attempt to rescue something of a faith in a god is the same as we have seen in other scientist-humanists. He arrives at a god who is purified of the baggage of previous assumptions but who is also bereft of any personality and being apart from the universe. As he points out, in agreement with Steven Weinberg,

> "If God is the equations of particle physicists, then the word *God* has been gutted of its traditional meaning...In this respect Weinberg is right. The traditional, scriptural concept of God as a personal being, interested in our individual fates and capable of intervention at will in the workings of nature, does not rest easily with what science has discovered about the creation."

Departing, he says, from Weinberg, Raymo continues,

> "Weinberg's mistake is to assume that the only alternatives are the God of the scriptures, on the one hand, and the God of the GUT [Grand Unified Theory], on the other."[468]

Yet his, "God of the spiraling powers" whom he means to replace the god of the scriptures,

> "...resides in nature beyond all metaphors, beyond all scriptures, beyond all `final theories.' It is the ground and source of our sense of wonderment...He is indeed the `dread essence beyond logic'"[469] to whom, "we can then give ourselves *unreservedly* to spiritual union with creation and communal celebration of its mysteries."[470]
>
> "Natural and supernatural, immanent and transcendent, body and spirit will fuse in one God, revealed in his creation." [471]

This is not a god of the scriptures, personal, transcendent, to whom we can truly relate as a "thou," even though he cites Martin Buber and his "I-Thou" relationship theme. His god is Buber's "I-It," a god of awe and mystery but is certainly the power of the universe by another name. If this is the best we have, maybe it is good enough, but my question remains whether or not this pantheistic god, or panentheistic god, is the only one in which we can put our trust and faith.[472] Certainly we cannot trust Raymo's god in our personal lives and existence. We must, he says, put aside any idea of immortality and be content with being one with the universe. That is grand enough for him, awesome and mysterious. It is most certainly not the god he was taught about in his rejected schooling. Then again, neither is mine.

A.N. WILSON

Another author who writes of the effects of science on religious faith, A.N. Wilson, points out that the traditional faith was in trouble long before Darwin's appearance. On philosophical grounds questions had been raised and, "the God of the philosophers lay dead."[473] Wilson investigates many more philosophers and savants than I do and his work provides a much more interesting read and I recommend it. Yet, as Wilson points out, religious observance did not die out in the face of overwhelming logic and, "it seemed [in the 19th century] as if there were no good arguments left for religion. If, either for emotional reasons or because you believed in religion as a socially conservative cement, you wished to preserve the forms, you could only do so at the expense of the intellect."[474] There is an excellent summary of the effort to reconcile the two areas, especially the "Modernist" controversy in the Catholic Church.

However, Wilson can find no basis for a religious faith. One can act on a belief, of course. "It becomes a matter of life, and how individuals wish to lead it, and whether temperament or experience makes this `deep' kind of life something which appeals to them."[475] This is not a choice based on an "experience" of something or someone real, namely god. It appears that Wilson's conclusion is not just that, "it would be a bold philosopher in the late twentieth century who thought he could prove the existence of God."[476] With this, I would certainly agree. The continuation of a belief in god by millions, including a list of notables Wilson assembles, does not come from an experience of a real god according to Wilson but rather, "these world-changing men and women decided to ignore the death of God in the nineteenth century. They spoke in the name of a God who was First and Last. They put their trust in One who said, 'I was dead, and see, I am alive for evermore.'"[477] It would seem that for Wilson they just missed the funeral.

There are many others who could be treated here, since there seems to be a very real interest, if not to say a crying need to fill the void of a personal, humanlike God which science has left and there are many attempts to do so. I suppose mine is one more. I will spare the reader any more here. Consult the endnote for a few others.[478]

AND THEIR ANSWER IS...COLD COMFORT

All of the scientists mentioned, and many others, have faced the questions that their new knowledge of the universe has brought about. They have often also examined, or been exposed to at some point in their lives, the faith that used to provide answers to these questions in their old form and they have tried to respond.[479] Others, of course, like a good many in the human population throughout the ages, either never ask the questions or never question the answers from faith. Biologists who are concerned with living things, but are often looked down on especially by astrophysicists and cosmologists, are fond of saying that their could-be colleagues are too narrow to consider the wider, philosophical questions. This may or may not be true, but those scientists cited at least, and to their credit, have tried to respond to the questioning of the previously presupposed answers such as that there is a god who brought the universe about, endows it with purpose, governs its activities, and cares about its inhabitants. They are to be admired for their honestly accepting the skepticism or outright denial of the tenets of faith which science has brought into question and for their effort to provide some sort of replacement for the usual answers. Nevertheless, it must also be admitted, as they themselves do, that they can provide cold comfort to the questions of humanity. It may well be true that our hope that there is purpose and meaning to life and a reasonable expectation that our lives do not end in total annihilation at death are all the results of our evolutionary history and can be explained as necessary lies that our brains must make us believe or hope in in order to allow us to function once we became conscious. That does not remove the wanting or the hoping. Indeed, we have seen the scientists themselves admit to such hopes and fears. It is noble to accept what we can believe from the evidence of our senses and our instruments, and only that, and to find a sense of awe and sometimes a sense of the "sacred" in the mysteries as yet unexplained of the universe. Noble, but not very satisfying.[480]

IS THAT ALL THERE IS? SCIENTISTS AND OTHERS (BELIEVER PERSUASION)

Is that truly all we have to believe and hope in? Does modern scientific knowledge absolutely deny us any reason to hope in those things that the faith provided in the past? There has been some work done on integrating the faith with science and philosophy and we need to take a look at some of the results, for these individuals have done precisely what this work asks, that we examine the faith in the light of science.[481] This was not always thought necessary, of course. The pioneers of modern science were people of faith such as Galileo, Newton and others who held the same assumptions as the rest of their world and felt and hoped that their investigations would not only shed light on the operation of the world, but on the world's creator as well. One could "read the mind of God" by reading god's creation. It is only after centuries of development in the relationship between the practices and methodologies of science and philosophy that a breach was noted.[482] A 1961 treatment of the question of purpose and cause[483] gives a rather standard survey of the state of the question after what the philosophies of Hume and Kant had done to Aquinas. These and others are rejected, of course, in favor of Scholasticism. However, the author felt it necessary to try to enlist Darwin himself in the investigation, at least leaning to the side of purpose, even though Darwin obviously does not belong in this camp. Still, the challenge of science to the age-old philosophical belief in the presence of purpose, design and teleology was certainly recognized. John Courtney Murray, the famous Jesuit apologist, recognized the dynamic of Modernists whose,

> "...fundamental refusal had fallen on the 'God of explanation,' as he was called, though his proper Name is the Creator, the Craftsman of the Book of Wisdom (13:3) who is also He Who Is, without whom nothing is intelligible because without him nothing is."[484]

Because there is no need for a god in order to explain the world, therefore, they conclude, there is no god. Laplace's famous, if apocryphal, response to Napoleon's question of the place of god in the scientist's scheme of things,

"I have no need of that hypothesis", would have been shocking in Laplace's time and Einstein's references to god would not have bothered many of his fellow physicists in his.[485] By the time of the Second Vatican Council it was certainly clear to a contemporary author that

> "...the presentation of the Christian revelation through the traditional forms of theology, with their reliance upon Greek speculation on the timeless and the divine, is seriously suspect today. These theological forms, so useful in harmonizing the revelation with accepted strains of classical wisdom, are quite inadequate to interpret 20th-century man's preoccupation with his history, his space-time existence."[486]

The first reaction when it was realized that science did not always agree with the theorems of faith was to deny science. This continues to be done in fundamentalist and creationist circles, by disputing the findings of research as bad science or by changing the meaning of the results or the meaning of the words of Scripture. For example, the questioning of the accuracy of dating methods for fossils and geological structures is one way to dispute the data, and changing the meaning of the word "day" in Genesis to mean billions of years instead of twenty-four hours is still a favorite method for some preachers. This last can be crudely done or managed with some sophistication in the use of the idea of relativity to say that from our position one day is twenty four hours, but from another position, god's, it can be a much longer time, just as from our position an astronaut traveling near the speed of light might be gone for millennia while for the person moving at such speeds they have been gone for only some years. Another favorite tactic is to claim for religion a sphere of knowledge that is inaccessible to science, and sometimes graciously granting to science a sphere where religion cannot go. Again, this can be done crudely with a "you stay out of my playground and I'll stay out of yours and let's just not talk about it," kind of mentality, shared by theologians and scientists alike, or it can be a serious contribution to the discussion. In 1986, a conference on Religion and Science was held under the auspices of the American Catholic Bishop's "Committee on Human Values" of the National Conference of Catholic Bishops. As the introduction to the official proceedings makes clear, there was,

> "...a range of questions that the conference hardly touched...How does one explain the Incarnation or the Resurrection or the Real Presence in terms acceptable to science? Does the transcendent God of tradition truly touch our lives? If so, how are we to understand this intervention?"[487]

These are some rather large questions indeed, still, the presumption was and continues to be that there is a mutual ground on which the two sides can interact, and it is the ground of the physical world, not merely the ground of values and certainly not merely the ground of faith.[*488] My position here is that this is not possible, but we want to look at the work of others who have felt with the Bishops' committee that it is.

KAFATOS AND NADEAU

Menas Kafatos and Robert Nadeau refer positively to Wolfgang Pauli's call to recognize that science and religion are not antagonists and cite his words,

> "Contrary to the strict division of the activity of the human spirit into separate departments - a division prevailing since the nineteenth century - I consider the ambition of overcoming opposites, including also a synthesis embracing both rational understanding and the mystical experience of unity, to be the mythos, spoken and unspoken, of our present day and age." [489]

They call for the use of the concept of *complementarity*, meaning that both disciplines are real and legitimate but approach their subject from different angles. Their findings will, then, not contradict one another, though they may appear to do so, but will ultimately complement one another.

> "...the evolution of human consciousness in both scientific and religious thought is toward the affirmation of the existence of the single significant whole, and...these two versions of the ultimate truth exist in complementary relation. It is also important to note that the scientific world-view...simply cannot in itself satisfy our need to better understand the character of ultimate truth...It is time, we suggest, for the religious imagination and the religious experience to engage the complementary truths of science in filling that silence [about what really matters] with meaning."[490]

It is up for discussion whether the conclusions reached by these authors can really be used to describe anything like the conscious, personal god of classical faith, or whether the "single significant whole" is really the universe itself. If the latter, the insights of religion into deeper, and even transcendent, realities than the surface heretofore studied by science, could be put to service and seen as complementary to physics and the other sciences, but they would still only give insight into a deeper reality of the

---

* See the endnote for an extended treatment of the various presentations at this important conference.

universe and not into the reality of a god. Kafatos and Nadeau point out that,

> "...quantum field theory, like the rest of modern physics, discloses a profound new relationship between part and whole which is also completely non-classical. Physicists, in general, have not welcomed this new relationship primarily because it unambiguously suggests that the classical conception of the ability of physical theory to disclose the whole as a sum of its parts, or to 'see' reality-in-itself, can no longer be held as valid. We are now forced to realize that physics is no longer the business of disclosing pre-existent truths, or, if you like, that such truths are not, as we once supposed, revealed truths. There is no choice, in our view, but to view the truths of science as subjective, and useful to the extent that they help us coordinate greater ranges of experience with physical reality. Yet the loss of this old conception of scientific truths has its compensations in the form of the alternate framework of complementarity. The compensation...is more than adequate to cover this loss. Although the new epistemology of science indicates that we cannot, even in principle, know reality-in-itself in the old terms, it also provides a foundation for conceiving of our relationship to this reality which makes the business of being conscious human beings a good deal more meaningful than classical physics ever allowed."[491]

That this "reality-in-itself" may be grounded in god, as it is in classical theology and philosophy, is a good deal less certain than that they are referring to the universe, with all its quantum weirdness.[492]

CANDICE PERT

Candace Pert in "Molecules of Emotion" seen in the previous chapter, speaks of being forced to think about her own beliefs when participating in a "Wellness Conference" where the moderator makes the statement, "We are certainly spiritual beings in a physical body and not the other way around."[493] Pert tells of her own contribution to the discussion and freely uses words and phrases that sound "spiritual" and non-physical. She is really speaking of the emotions and says,

> "For me the key concept is that the emotions exist in the body as informational chemicals, the neuropeptides and receptors, and they also exist in another realm, the one we experience as feeling, inspiration, love - beyond the physical. The emotions

> move back and forth, flowing freely between both places, and, in that sense, they connect the physical and nonphysical."[494]

As I read her, she is not using the words in the same way they are used by most believers. Her "nonphysical" is more an emergent quality of the actions of the emotion-brain system. Then again she leaves much room for something, "out there" which,

> "...we can experience as emotion, the mind, the spirit - an *inforealm!* This is the term I prefer, because it has a scientific ring to it, but others mean the same thing when they say field of intelligence, innate intelligence, the wisdom of the body. Still others call it God."[495]

When asked if she is not really speaking about the "Holy Spirit" she, "had to admit that, yes, maybe I was." Still, she speaks of an "emotional resonance" in which our emotions are affected by those of others and even by an "extracorporeal peptide reaching" which is bigger than any one human. Whether this is the traditional god in her mind is doubtful, it may be the collection of all human intelligence affecting each individual. Pert herself seems to be still seeking just what she believes, but then aren't we all?

FRANK TIPLER

There is also the attempt by Frank Tipler in "The Physics of Immortality", to reconcile faith in some kind of god and belief in the resurrection from the dead with modern physics. He employs a final cause situation to bring about the construction of an almost universe-wide computer which will reconstruct each and every human being that has ever lived,[496] plus a good deal more that never previously lived lest we miss someone, and give them access to all the experiences of the universe. He explains that,

> "...if the universe is globally hyperbolic (deterministic), then the information contained in the whole of human history, including every detail of every human life, will be available for analysis by the collectivity of life in the far future. In principle at least (again ignoring the difficulty of extracting the relevant information from the overall background noise), it is possible for life in the far future to construct, using this information, a perfectly accurate simulation of these past lives:...[and] a sufficiently perfect simulation of a living being would be alive. I shall argue...that the Omega Point will choose to use His/Her power to do this simulation."[497]

This Omega Point, reminiscent of Chardin whom we will see in a moment, is the final cause working on the universe to inspire the creation of computer reproductions of each person. Since it is still contained within the universe, and since Tipler opts for the universe being a closed system heading for the Big Crunch, (a position held with less and less enthusiasm today), the persons resurrected and living in this paradise will not truly be eternally immortal since time itself will wink out of existence along with the universe, but the subjective, relative time will be well nigh eternal since one can conceivably fit an infinite amount of experience into an almost infinitely small amount of time. My treatment here does not do justice to Tipler's argument and he certainly has the mathematical background to make his case, so his work should be read before the reader makes a judgment for or against his thesis. However, speaking from a background of traditional faith, even though he allows for much of that faith to contain at least a grain of truth, his Omega Point is far removed from the god of Judeo-Christian experience. This is not cause for rejection of his thesis for I, too, am arguing that the traditional formulation of that experience of god is in need of some serious reconsideration. So the judgment must be based on his reception by other physicists and scientists, both about his science and in how he is accepted. In this there does not seem to be much enthusiasm for his work, so he is not getting much hearing from either camp.[498]

ERNEST STERNGLASS

Sternglass champions Lemaitre's ideas, Sternglass seems to be a respected scientist with rather different ideas about cosmology. No one seems to be following his ideas, however about a primordial electron-positron from which came the universe. The scientific hypothesis seems to be his focus in the book, but he does begin it by stating that,

> "...when there is no singularity in a theory for the origin of the universe, there is no point where the laws of nature break down. It is therefore possible to believe that the laws of nature were laid down by a Creator before the first pair came into being."[499]

He ends by saying that,

> "...the architect of this design and the energy required to bring this about remain a source of mystery, awe, and wonder beyond the ken of science."

He does not belabor the point but he certainly makes it clear that there is room for a designer and possible creator of the universe. He represents a number of scientists, some fundamentalists, some intelligent design advocates, but all open to belief in god. The assumptions are alive and well in most, however.

ATTEMPTS TO RECONCILE BY PHILOSOPHERS AND BELIEVERS

Leaving the scientists who try to include religion in their science, we turn to philosophers and believers who try to include science in their religion.

BERGSON

In earlier times, philosophers had worked from the other end of the spectrum, looking for ways to integrate the philosophical and theological assumptions of their culture to the facts being revealed by science in their time. Science had indicated an increasingly deterministic universe, working like clockwork according to the laws discovered by Newton and others. In response to determinism, some philosophers, such as Henri Bergson, "opts…for the concept of a continuously creative universe, in which wholly new things come into existence in a way that is completely independent of what went before, and which is not constrained by a predetermined goal."[500] There was for him a life force which animated, not each individual being such as a form or a soul, though he did not deny the soul, but rather his, "…*Êlan vital* [life force] was said to be an insubstantial essence that permeated and animated the inorganic molecules of cells and brought them to life."[501] Science could not study this force, just as it cannot study "real duration" of time, but only study that manifestation of real time that can be broken into discrete quantities. "Hence the philosophical method of studying both consciousness and the material world, in function of duration, is a distinctive *metaphysical intuition* which is irreducible to physical analysis into elementary parts and laws."[502] There is certainly a place in Bergson for scientific inquiry, since *"homo faber,"* the humans who must manufacture their own perception of reality in order to function in the world, function as though their perceptions are valid and complete, but the objects of scientific study do not constitute total reality. The object of metaphysics is to study this moving life force and, "Creative evolution is this ceaseless movement of life to reverse the process of matter, and to establish the values of freedom and consciousness."[503]

> "Life is like an immense wave, which spreads out in all directions. The ripples are stopped everywhere along the edge by material obstacles, and are converted into oscillation: in human awareness and freedom alone, does the vital movement break through the wall of matter and progress indefinitely. Thus the life-process finds its proper term in man's intuitive consciousness."[504]

Bergson recognized that his philosophical position had been heavily influenced by science, but we can also see the weight of prior philosophical assumptions at work in his metaphysics. He had no trouble in positing a force at work in the universe that was not amenable to empirical knowledge, but of course, science does have difficulty with this. If one is speaking to philosophers, such arguments can carry some weight, but not if one is speaking to a scientifically literate culture with different assumptions. One difficulty is that science today *is* reductionist and does not welcome true novelty into the universe. Paul Davies rather misreads Bergson, but his conclusion is one shared by many in science when he says, "Such a point of view can never be called scientific, for it is the purpose of science to provide rational universal principles for the explanation of *all* natural regularities."[505] Davies goes on to call for the steady organizing of the universe as a "fundamental property of nature" and, of course, he finds this property in a very different manner than Bergson, but the *Êlan vital* was itself a fundamental property of nature. Indeed, one of the assumptions that Bergson helped dispel was that life cannot arise from matter without the intervention of some outside source. His *Êlan vital* was not an outside source of influence but something intrinsic to the universe itself. Matter and the force were two different expressions of reality. Bergson was not very favorable to religion, as such, and so his hearing among religious philosophers was not always wholehearted. However, some listened, especially those whose own assumptions had begun to change in response to the new understanding of how the universe, and especially the evolution of life, worked.

CHARDIN

Among the more faithfully orthodox who listened, one of the first to attempt a synthesis, or at least a dialogue between religion and science was the priest and scientist Teilhard de Chardin, though the word "orthodox" was seldom connected to him. "He wanted both his fellow-scientists and his fellow-believers to share in his joy about the structures that he had discovered in the world of phenomena - the inner cohesion of the universe which is growing towards a divine fulfillment in Christ."[506] Influenced by Henri Bergson and by the evolutionary theory being debated in the decades after Darwin, Chardin gradually came to believe that there was an emerging "spirit" in the world that had brought evolution to its present point, but that it was capable of continuing to an even higher level now that human consciousness had appeared. However, that this would happen required,

> "...an actual, concrete, loving and lovable, and hence personal, center of convergence...already present and at work...And just such a person *does* in fact exist: the Logos incarnate in Jesus Christ, who has already passed through sacrificial death and who rose again as the universal Lord, who is present at the eucharistic

> heart of his body, the Church, and who is to come, `to unite all things in him, things in heaven and things on earth,' `so that God may be all in all' (Eph. 1:10; 1 Cor. 15:28)."[507]

Thus, Teilhard sought to unite the science of his day and the *Êlan vital* of Bergson and his followers with the whole sweep of salvation history and god's action in the universe. This action had been there since the beginning as the creative "spirit" of the universe and was directing the course of evolution to the "Omega Point" of Christ, as the Scriptures attested. In seeking a, "...hierarchy between past or present living forms, he took as his reference the progressive complexification of the nervous system, whose growth is linked with a growth of consciousness. Spirit and consciousness, however much they transcend matter, can manifest themselves only if the conditions of organic complexity are fulfilled."[508] There is a gradual "hominization"[509] through an irreversible ascent of life which is built into the warp and woof of the universe and so is open to scientific investigation, but which also expresses the action of god and so is open to the investigation of theology and faith. This approach was rejected by many in the Church because of its seeming pantheism or some related "ism" and has by now been rejected by science because there seems to be no evidence, despite the outward appearances, of an "ascent" in evolution, and certainly not of a spiritual component working through the nervous system. Chardin's approach struck a nerve among many of the faithful when it showed that the findings of modern science did not have to be rejected by a person of faith. They perhaps could be integrated, or at least reconciled, if not in Chardin's way, then through some other. He was very influential for many acquainted with biology, physics, cosmology and the other sciences that had made it increasingly more impossible to merely accept the assumptions through which the faith had been expressed.

BUT THEN WE HAVE NO ANSWERS.

The philosophers Karl Popper and John Eccles are also among those who have tried to find common ground with science and faith,[510] but with about the same results. Popper would like to see a place for god as the ultimate explanation of an infinite series of explanations, but even if there is an infinite series of explanations, there is little agreement that this calls for an "ultimate".

This desire for answers, which will not be available if the opposition is correct, inspired the Dominican, Raymond Nogar, to attempt an answer to the famous objections of Julian Huxley[511] to the religious ideas of his time, contrasted with the new theory of evolution. Huxley and the humanists cannot be correct, Nogar figures, because Huxley, "misses the necessity for transcendent being". In the matter of evil, for example, the evolutionary view espoused by Huxley cannot be correct, for,

> "Without a revelation from God, man cannot discover the moral and religious meaning of his destiny or raise a finger to extricate himself from the moral pathology into which he has fallen. Belief in the revelation of God and the redemption of Christ, accessible to the man who is open to the timeless and the divine, is the only release available to the despair of evolutionary humanism. No plan for the space-time future existence of man is realistic until it solves this fundamental moral and religious issue of absurdity. Huxley's instinct that the future of man must be thought out and acted out in the timely is an authentic one, but the meaning of that future can only be found in the timeless, in the gift of the divine."[512]

It does not seem to occur to Nogar that there might be no answers to his questions. There may be no meaning to human life and all may indeed be absurd. Huxley tries to paint the best possible picture of life and existence according to his understanding of reality. It is not good enough for Nogar, so he assumes that there must be some other answer. For Nogar, that can't be all there is!

At least Nogar and others did recognize that science was not giving them answers to some of their most human hopes and fears, but had hope that these discoveries could yet contribute to an improved understanding of god rather than merely undermine one.

MISSING THE MARK

Some Christians appreciated the effort of those who tried to reconcile the two, but others fell back on the oldest response to the challenge that we have seen before, i.e. denying the question. Joseph Ratzinger, in a 1970 work on faith, described it thus,

> "...to believe as a Christian means in fact entrusting oneself to the meaning which bears up me and the world; taking it as the firm ground on which I can stand fearlessly. Using rather more traditional language, we could say that to believe as a Christian means understanding our existence as answer to the word, the logos, that bears up and holds all things. It means affirming that the meaning which we do not make but can only receive is already granted to us, so that we have only to take it and entrust ourselves to it. Correspondingly, Christian belief is the option

> for the view that the receiving precedes the making - though this does not mean that making is reduced in value or proclaimed to be superfluous. It is only because we have received that we can also `make'. And further: Christian belief - as we have already said - means opting for the view that what cannot be seen is more real than what can be seen. It is an avowal of the primacy of the invisible as the truly real, which bears us up and hence enables us to face the visible in a calm and relaxed way - knowing that we are responsible before the invisible as the true ground of all things. To that extent it is undeniable that Christian belief is a double affront to the attitude which the present world situation seems to force us to adopt."[513]

In this view, no matter how much one protests the value of science and investigation, it is ultimately going to be useless in finding the truth. The truth is already revealed and we have but to assent to the "deposit of faith". The encyclical "*Humani Generis*" by Pius XII, according to one commentator,

> "...was a balanced synthesis of traditional neo-scholastic principles - that towards which faith is directed is always described as `truths', the concepts of which `have, in a work lasting many centuries,...been formulated and worked out with the most subtle shades of meaning, with the purpose of expressing the truths of faith even more accurately'. These terms could be `perfected and worked out even more subtly', but they could not be pushed on one side without making dogma `like a reed shaken by the wind'. `Whatever the human mind, in an authentic search, may be able to find in the way of truth cannot be in conflict with the truth that has already been acquired.' The Church's teaching office had been given the task of `preserving, guarding and interpreting the deposit of faith' by Christ, `in order to throw light on and explain in detail those things which are contained in an obscure form and so to speak implicitly in the deposit of faith'."[514]

For Pius and Ratzinger, all the old knowledge could not be wrong, it is claimed, though it could be better explained, which unfortunately meant in practice that the old assumptions could not be wrong, though they could be nuanced a bit if necessary. Otherwise the faith would be like the shaken reed, and one cannot have that. I certainly agree that faith gives a basis for hope, for meaning, for morality, and it does so as faith, not scientific knowledge. Nevertheless, there are many assumptions embedded in the "deposit of the faith" which must be discarded or re-examined, not maintained at any cost. One need not fault the Pope or Cardinals for taking this position at the time. It is, after all, their job to be conservative, and there was no overwhelming reason to shake the faith to its foundations or evidence that could not seemingly be contained within the traditional framework. However, the evidence was mounting and more theologians would feel the pinch to respond with something more than an appeal to blind faith. Science, after all, had been willing to kick the traces and bolt from established thinking, at least in its better moments, putting aside some long-cherished beliefs, some of which it had inherited from the religious faith itself. Granted scientists did it, and continue to do it, grudgingly and some not at all, but it had been done eventually. Why should the content and assumptions of the faith be any less open to question than those of science? They are not, of course, and so, many continue to endeavor to take science seriously and find a place for the faith within it, instead of insisting that science and faith cannot be taken on the same terms, and continue insisting that there is no warfare between the two.

KEEP ON TRYING

A little known scholar, the Jesuit M. B. Martin, though he maintained the ideas of god as the creator of everything physical and such concepts as an immortal soul, nevertheless recognized in 1972 that the entire basis for Western culture had to be questioned and, in his opinion, abandoned and he begins his study of the "cultural bondage" which afflicted the faith by saying,

> "While Platonism and its Christian adaptation, Neoplatonism, are highly developed thought patterns expressed in admirable literary style, they are, nevertheless, a primitive philosophy very much akin to the primitive Hebraic philosophy...Our understanding of reality is entirely different from that of early...culture and consequently we cannot accept the deductions drawn from their concepts...Platonism and Neoplatonism were useful as long as natural knowledge remained roughly within the confines of Plato's experience but they have little validity in the space age. It is amazing that in the waning decades of the 20th century the religious thought pattern of Western culture should be predominantly Hebraic and Platonic when our scientific knowledge of reality must hold these systems invalid."[515]

The handwriting was on the wall for traditional Thomistic explanations of the faith, but few were capable of seeing this at the time. We have already seen the attempt to recognize the process of evolution without a grand director, as long as god is allowed to set the tone. George Ellis says,

> "The more modern version [of the design argument]...would see God designing the Laws of Nature and the boundary conditions of the universe, in such a way that life (and eventually humanity) would then come into existence through the operation of those laws...through the process of evolution"[516]

From the theist's point of view, this is very restrictive of the god of Genesis especially when Ellis continues,

> "...from the standpoint of the physical sciences, there is no reason to accept this argument. Indeed from this viewpoint there is really no difference between design and chance, for they have not been shown to lead to different physical predictions."[517]

ARTHUR PEACOCKE

Arthur Peacocke has labored hard to reconcile the two areas of design and chance, but he is stuck with the position that god must have a creative/directive function and continues to hold that,

> "In the Christian understanding the world of matter has both a symbolic function of expressing His mind and the instrumental function of being the means whereby He effects his purpose."[518]

This is true for him on all levels of evolutionary development. The constraints of the genetic process are not the result of pure chance, according to Peacocke, but the means god uses as a matrix on which to build toward god's designed result.[519] His image of god as a composer, reminiscent of Augustine, encouraging certain potentialities in the universe in order to fashion god's masterpiece even extends to the evolution of human culture,[520] since it has developed along the lines that god intended because it had just those potentialities it needed to do so. Needless to say, this is the design argument dressed up in scientific terms and there is no compelling reason to embrace it.[521] Anyone who does not start out with a notion of a designer would have no reason to invent one. Once again this short dismissal hardly does justice to Arthur Peacocke's effort to integrate the two disciplines, and his work is certainly worth the investigation, but the fatal flaw is in the assumptions and presuppositions that continually creep into our thinking, assumptions that are often shared unconsciously with unbelievers in looking for cause and effect, since they seem to be almost built into our genetic makeup, and sometimes a good deal more than "almost" built in. None-the-less they are assumptions and certainly not everyone shares them any longer. Peacocke presumes that he must make some reconciliation between the beginning and the development of the universe and must give god some role in its direction and still be true to the findings, and the assumptions, of science.[522] He and others make an admirable attempt of fitting the two together, but I do not find them asking whether or not they need to be so joined.

WOLFHART PANNENBERG

Wolfhart Pannenberg is another who has taken up the challenge and he too honestly faces the questions, insisting for example that, "theology has to relate to the science that presently exists rather than invent a different form of science for its own use."[523] We are at the position that even though everyone must admit that the universe is contingent in the sense of not having had to be at all, still the notion of divine providence as a continuing "holding in existence" was "deeply challenged in the seventeenth century because of the introduction of the principle of inertia...a continuous conservation of what once was created becomes unnecessary."[524] But Pannenberg still assumes that contingency of this kind requires a cause and concludes, "the laws of nature appear to the theologian as contingent products of the creative freedom of God."[525] No unbeliever is going to fail to point out the two objections we have already seen to this position: that the universe did not *have* to be but it *can* indeed come to be by chance alone; and that god did not have to "be" either, the ontological argument of Anselm having been rejected, and that in this sense god is also contingent. Pannenberg tries to use the field theories of Faraday to make a connection with faith and says that a theologian is justified in feeling that,

> "...God has to be conceived as the unifying ground of the whole universe if he is to be conceived as creator and redeemer of the world. The field concept could be used in theology to make the effective presence of God in every single phenomenon intelligible."

Not that the theologian can take the field theory of physics and just apply it directly to god.

> "...Theological assertions of field structure of the divine spirit's activity in the cosmos will remain different from field theories in physics"

since after all, we are speaking of god and can therefore speak only by analogy, but he feels that it makes the faith intelligible and reasonable to the non-believer. Once again an admirable effort to be relevant, but his conclusion says it all,

> "...the contingency of nature's laws...cannot be accounted for apart from understanding the whole of nature as the creation of a free divine creation...we must go beyond what the sciences

> provide and include our understanding of God if we are properly to understand nature."[526]

If one is talking to believers whose faith needs to be enlightened and expanded by taking into account scientific knowledge in our time, these gentlemen, Peacocke and Pannenberg, should be on the top of anyone's speaker list. Both have much to offer in putting the faith in a new light and a new depth. They challenge some of the traditional explanations of the faith and make it relevant to a believer in the twenty-first century, but they do little to make it equally believable to the non-believers who do not begin with the same presumptions as the faithful.[527] Who says that we must conceive of god as creator and redeemer? And who says contingent means "not by chance alone?"

JOHN POLKINGHORNE

John Polkinghorne also tries to respond to science with faith and begins his treatment of creation with an account of the Big Bang and its consequences. He, too, attempts to show that there need be no contradiction between a mature, modern faith and scientific explanations of the universe, but he also begins with assumptions which he cannot shake and finally concludes that,

> "...clearly we cannot perform the ultimate experiment: remove the divine presence and see if the universe disappears. Belief in creation *ex nihilo* will always be a metaphysical belief, rooted in the theologically perceived necessity that God is the sole ground of all else that is."[528]

He goes on to treat the idea of "*creatio continua*" or continuous "holding in existence" of the universe along with a continued involvement in its existence, with or without involvement in its operation. Working from Chaos theory, Polkinghorne would like to find a need for a "top down causality", presumably from god, in addition to a "bottom-up causality" from physics and evolution. In holding out this hope for a place for religion in science he uses phrases such as, "developments of metaphysical thought along these lines are necessarily speculative", and "they hold out the possibility", and "the picture from [chaos theory] is not so tightly drawn as to exclude", concluding that there is, "a coherent possibility that this is how God interacts with creation."[529] To be consistent he would have to say that this, too, is ultimately a theological wish rather than a necessity in physics, and in treating it he leans to the belief that the universe is allowed to "make itself". He prefers this to saying, as Peacocke does say, "God the Creator explores in creation." Polkinghorne does not like the idea of god not already knowing all the possibilities of existence, "for God cannot be thought to need to use the universe as an analogue computer to explore possibility."[530] Two assumptions underlie this position, first that the universe needs to be "held in existence" and second that "god can have no limitations" but must be ahead of every game. While science can say nothing affirming or denying either assumption, it certainly would say that there is no necessity for positing the first about the universe. Though Polkinghorne recognizes that anything he says about god's under-girding the universe is a statement of faith and not of science, the title of his book, *The Faith of a Physicist,* leads one to think that he is reconciling science and faith. Certainly if one starts with the same assumptions that he does, his work is very enlightening, but if, as I propose, these assumptions can no longer be taken for granted and in fact are untrue, then his seemingly scientific statements and conjectures are really statements and conjectures of faith and religion.* He probably would agree with this assessment of his work. In another book, *Quarks, Chaos and Christianity*, he concludes that in his religious faith as in his science, there is room for an answer that provides, "the best available explanation of a great swathe of physical experience."[531] One of these explanations is to a question I have used in these pages. In Polkinghorne's words,

> "If the new creation [heaven presumably] is going to be so wonderful, why did God bother with the old? If *that* world will be free from pain, death, and sorrow, why did he create this one instead, which seems to have so much suffering in it?"

His answer is that this is a,

> "...creation allowed to make itself. Such a world must have death as the necessary cost of life...The new creation is something different. That means that it can work in a different way...The new creation is not a second attempt at what went before; it is the redemption and transformation of the old."[532]

I don't know about you, but that did not explain a thing to me, and I think Polkinghorne knows it, for his next words are not so sure of themselves, "These are deeply mysterious thoughts, but I think they are true thoughts." For him, god's, "existence makes sense of many aspects of our knowledge and experience" such as the presence of order in the universe, its creativity, and even the sense of hope experienced by human beings. Faith for him then becomes just as intellectually honest as the pursuit of knowledge in

---

* One helpful critic of these pages asked if the same cannot be said of my work. I hope that I did not begin with unexamined assumptions or preconceived conclusions. Being human, however, it is also to be expected that I have. The reader needs to be aware of any preconceptions on my or his or her part.

science. Yet as we have seen, there is, for many, no reason to propose a god as the author of order in the universe, nor for its fecundity, and hope is an evolutionarily expedient emotion for us to have. Consequently, his argument will probably not be persuasive to anyone not sharing the presumptions he exhibits and which I have been examining.

IAN BARBOUR

The reader of endnotes (and I hope you read them sometime if not in context) will find references to Ian Barbour's, *Religion in an Age of Science* throughout. He analyses the relationship of science and religion far more completely than I am doing here and is an excellent reference. His own conclusion seems to be that something akin to Process Theology, at least a "process model", "...seems to have fewer weaknesses than the other models".[533] Put briefly, I do not believe that he has avoided the assumption of god as creator but has tried to reformulate it to accommodate modern knowledge. I certainly agree with his assessment that, "Only in worship can we acknowledge the mystery of God and the pretensions of any system of thought claiming to have mapped out God's way."[534]

EUGENE MALLOVE

Pursuing the treatment of some who have sought to speak to the scientific mind with the viewpoint of faith, we look also at Eugene Mallove who in his book, *The Quickening Universe*, makes a distinction that appeals to some theists. He describes some of the speculations of physicists David Bohm and John Wheeler as pantheistic when they suggest that perhaps the meaning or names ascribed by conscious beings such as ourselves to the observable universe is a form of being akin to that which we have previously ascribed to god. Pantheism, Mallove points out, "...regards God as the indwelling presence of the divine in all things; that is, the divine is *immanent* in all of nature. Pantheism is thus distinguished from classical *theism* that considers God to be *transcendent* - beyond nature." And he adds, "Another variant of belief, called pan*en*theism, considers the divine to have both immanent and transcendent qualities."[535] Mallove points out that panentheism would be somewhat more acceptable to modern science, but mostly because it recognizes that there is an unknowable quality to the universe itself which science might grudgingly allow, especially in the light of quantum physics and chaos theory. Still, the theists' insertion of anything like the Christian god would find little acceptance since one could never distinguish between what is unknowable because divine and what is unknowable but very much of the universe. This leaves little or no room for a traditional faith and no compelling reason for any faith at all.

JOHN HAUGHT

John Haught, in *Science & Religion*, points out that there is much interest in chaos and complexity today since scientists are finding it in so much of what was once thought rather straight-forward deterministic physics. The famous "butterfly effect" in which the weather over Texas could perhaps be affected by the wind current set up by the flapping of an insect in Brazil is such a case, and, "even to predict the position of a billiard ball accurately after only one minute of movement would require that we take into account the gravitational attraction of an electron at the outer edge of our galaxy."[536] As Haught points out, pals of Paley would certainly find the guiding hand of a designer in all the order and complexity that surprisingly comes from the midst of chaos. Yet, Haught, who is certainly interested in finding points of conversation between science and religion, must admit that,

> "Most of us who are intellectually most at home with scientific skepticism find nothing whatsoever in the new scientific emphasis on chaos and complexity that would lead us to religion. In fact, the utter spontaneity of matter's self-organization, ...would seem to render more superfluous than ever the idea of an ordering deity."[537]

Haught does find points of contact between the two disciplines, however, one being the argument from contingency when he asks, "can scientists ask the very deep question as to why there is any patterning at all and pretend that they are not thereby steering perilously close to metaphysics...and can they pursue such inquiry to the very end without making contact with theology?"[538] Of course they cannot avoid such contact, but whether they must reach the same answers as theology is another matter. His contact leads not to new scientific insights, but to theological insights when we must reject a god of linear order in favor of a god of novelty.

> "The religions descended from Abraham, whose faith consists of openness to an indeterminate promise, think of God not just as an orderer, but as One who brings about an always new future. This means, however, that we expect the future to be always open to surprise - and therefore to the chaos that surprising new developments can cause."[539]

He sees the picture of a universe full of chaos and complexity as fitting much better with "an understanding of God's power as gentle and persuasive rather than coercive",[540] the god of the prophets. He also finds the universe as described by science, so self-sufficient, to be in agreement with what we would expect of a god who has "emptied himself" in the words of Philippians, a god who does not impose god's will on the universe but instead offers a self-emptying love, a *kenosis*.[541]

I agree that the *authentic* religion of Israel does indeed feature such a kenotic god, at least in its mature form, but most of the popular understanding of this god sees a creator in control. The Huguenots, Calvinists, Anabaptists, Pilgrims, the Boers, the Fundamentalists of all denominations, right wing militias and most Christians from Second Grade to Doctor of Theology have not seen god in such a seemingly diminished light. It is precisely my conclusion, along with Haught, that we must, "...think in a fresh way about any Creator we wish to associate with this world," [542] though I am not so sure that scientists see any more need of a god who is the "author of novelty" than they do of a god who is the author of order and design, and would find such a gentle, loving, self-effacing god to be as invisible as no god at all. Haught mentions Stuart Kauffman who argues for a self-organization principle in the universe operative even before natural selection, and though Haught does not argue for a role for god in placing this inherent creativity into nature and admits that Kauffman himself does not find one, he obviously hopes that there is room for such, though the most he will say is that science is going to have to at least change its reason for being skeptical of the claims of religion. Small consolation.

TIMOTHY McDERMOTT

Timothy McDermott tries to resurrect the Prime Mover argument in a novel manner. Instead of god being responsible for the make-up of beings, god is responsible for the big picture. "The displacing of God as engineer of the insides of things, has left their outsides intact." The environment, crafted by natural selection, makes it possible for beings to act in certain manners. Here he sees a regression of causality,

> "...and so on and so on, though not ad infinitum...We have straight-away stumbled on the seven-word essence of Thomas's five ways of proving God's existence. If the being of an organism is the doing of the environment which favors it, and the being of that environment is the doing of some further favoring environment, and so on, then to get the being of the organism at one end of the chain we must appeal eventually to some ultimate favoring environment at the other end that needs no further favoring by anything else, because that is just what it is, the favoring of existence. `And this is what all men call `God'.'"[543]

He is willing to grant that, "God is not primarily a planner or engineer preceding and manipulating nature; God is rather the ultimate favoring and doing which allows nature to be itself." Indeed, he leaves much room for the operation of chance, though he qualifies it severely. God already knows the outcome, even though chance is left to operate - somehow. One can guess that my opinion of this effort is that McDermott begins with the assumption that god must be involved in the existence and operation of the universe and is, somehow, directing its history, if only by "granting his favor" to some chance events and not to others. His non-infinite regression need not end in a Prime Mover. As we have seen, it can just as well end in Chance itself.

ANGELA TILBY

Angela Tilby also seeks to make science a handmaid of theology and spirituality and makes some interesting discoveries about the nature of the god as we have presumed god to be. An example is found in her application of quantum theory to the creator's role as she states that,

> "The classical [faith] world portrays creation as a task. God rests at the end of his creative work. In the quantum picture creation is a game in which chance plays a major part. It is as if God creates, not with any one end in view, but out of the sheer joy and delight of seeing what will happen. God is in labour with a trillion possibilities and brings them slowly and patiently to being through time and chance. Much is lost in the process, and much is suffered."[544]

In this one short section, she manages to work into the faith, quantum physics, chance, indeterminacy, recognition of the profligacy with life that evolution entails, and an answer to the presence of evil in the world. Having admitted to a predisposition to believe in a creative god; and having commendably investigated whether that creativity was expressed in the manner which she had been taught; and feeling that we must rethink some aspects of the traditional faith; she fails to prove rather than assume that any such creative god is in fact at work in the universe. The presumption is too strong in her. The only way to truly rethink the faith position is to remove any presence of god from the history and practice of the universe and see if it stands on its own. Some scientists have removed this presence and believe that the universe does stand alone. Few theologians or other believers have really been able to root out their assumptions, discard them, and face the consequences.

DIARMUID O'MURCHU

The Irish priest, Diarmuid O'Murchu, has a book, *Quantum Theology*[545] which should be must reading for anyone interested in the topics covered here. O'Murchu insists theology and faith take into account the discovery of the quantum nature of reality and thoroughly upsets the traditional faith while emerging with the awesome, creative, incarnated god of Christianity as well as the god known to other faiths. Since I recommend that the book be read, I will not attempt to summarize it here. He says so many things with which I agree that such would be entirely too lengthy, especially about the challenge to the Church presented by a modern understanding of god taking into account our present knowledge of reality. However, I believe that he also begins with the assumption that the universe is created by, "a superhuman, pulsating restlessness, a type of resonance vibrating throughout time and eternity."[546] While he makes it clear that god is not an external agent acting on the universe to bring it into being as if from outside, that god is a personal agent at all seems absent. Indeed, "Quantum theology is not particularly concerned about the nature of God."[547] We do not find a personal god as is present in traditional theology and in fact can only know god by immersing ourselves in the story of the universe and the dance of this "pulsating restlessness."

While it is pleasant to think in terms of such intimate contact with god - indeed O'Murchu does not shrink from pantheism - he presents no proofs or any reason to think that his ideas of god have any claim to credibility. Admittedly, it is not his purpose to argue people into faith by proving that there is a god, in fact he argues that this is entirely futile, and I agree. He begins with faith in god and explains god from the quantum perspective. There is much to be learned from this approach and I believe that if he were to discard some of the assumptions to which I have consistently objected, such as god as Creator or Sustainer, he would arrive somewhere near where I do, but he does have these assumptions and they color his conclusions. There is also no answer to the presence or evil, not even an entry in the index, except one that traditional theodicy would recognize as a variation of the argument that if there is to be good then there must be at least the possibility of evil. However, he has a twist from a quantum perspective when he speaks of chaos as a positive force in the universe that, while it may bring some things we humans consider negative, is all part of the creative, and very positive, process of evolution. Perhaps there is no answer to the question of evil except that god cannot exclude it but does not like it and did not cause it. I like my personal god better if I can exonerate god from being the cause of evil and negativity and feel there is just as much reason, and more, to believe in this god than his. Read on to see if I can.

TIMOTHY FERRIS

Timothy Ferris in *The Whole Shebang* is not trying to reconcile science and faith but to give an honest assessment of the relationship. He summarizes the arguments of the theists who would like to save some place for a creator god and finds them wanting and likewise the position of some atheists.

> "When the cosmological arguments propounded by atheists are subjected to reasoned criticism, they fare no better than the comparable arguments of believers...So we are left with - what? In my view, a situation in which we would be better off if we left God out of cosmology altogether."[548]

This is a far cry from the philosophy books used in many seminaries in the 60's in which god was the only subject and the Big Bang was left out of cosmology altogether, but neither does it answer the questions.

OTHERS

There are, of course, many theologians and believers who are struggling to use the findings of science, especially of cosmology and evolution, to illuminate theology. I have probably cited enough at this point to show that, in my opinion, most leave the fatal flaw of assuming god as creator, consciously or unconsciously intact in their exposition. You might want to see the endnote for more examples, including C.S. Lewis, Patrick Glynn, Walter Brueggemann, Karen Armstrong, Matthew Fox, Kenneth Miller, Denis Edwards, Jerry Korsmeyer, Russell Stannard and especially Steven C. Kuhl whose short article mentioned there certainly takes evolution and its challenge seriously and shows how Pauline theology may be a good response.[549]

SO WHERE DOES THAT LEAVE US?

To summarize the point we have now reached, I began with an assessment of the assumptions and presuppositions that underlie the traditional faith in god. This need not be the Hebrew's Yahweh or the Christian's Trinity. In fact the understanding of the god of the Israelites and the Christians was colored by these already established assumptions whose genesis is far back in the mists of prehistory, perhaps long before the appearance of *homo sapiens*. The ideas of how the universe works which developed in a very pre-scientific world-view were refined and evolved over the eons but were rarely questioned because there was no reason to doubt them. Very slowly the myriads of powers and spirits that were said to guide the world and the events of human history were whittled down, at least in the West, to a few and then to "One", and the assumptions were not abandoned but subsumed into this lone deity. It was not until very recently, with the advent of modern science, that there was any reason to question such universally accepted dicta as "*ex nihilo, nihil fit*", or that life, at least intelligent life, must have been a gift from god, or that there was some purpose to everything, not to mention a chain of causality going back to an unmoved mover. I have examined a number of these assumptions and found them unnecessary. I went on to ask about the historical faith that has grown up based on these assumptions and found that much will have to be re-thought and possibly abandoned or at least re-phrased. There are many more aspects of the faith, especially as it is popularly understood, and particularly in private devotions that we have not examined, but if the reader accepts the major points of doubt, it is rather obvious what must be changed.

At least these are my contentions, though I would love to be shown the error of my ways and why there must be a creating, designing, purpose-giving god who has a special affection for human beings rather than beetles, and makes the physical laws of the universe work to their advantage, at least when god is properly invoked, and who should be approached as such in worship and prayer. This work is meant to inspire criticism and stimulate a fruitful discussion, but those who have so far reviewed it have not as yet disproved its conclusions to the author's satisfaction. Therefore I proceed with the conviction that it is not necessarily true that the universe was caused by anything other than a chance event in the quantum vacuum, which itself is contingent in a manner of speaking; that there may well be other universes; that our universe, by chance again, happens to have the conditions which allow not only the evolution of stars and galaxies, some with solar systems of planets, and also some with the conditions conducive to the evolution of life; that consciousness is no more radical a step in the evolutionary process than any other, though it may be rare; that miracles are not only not necessary but do not occur; that there need not be any such thing as good or evil as we normally define them, but merely subjective assessments of the situation; nor that there need be any purpose or goal to the universe or to anything or anyone contained within it. I must once again repeat that all of these things could indeed be just as the traditional faith understands them, *but they need not be.* Therefore, anyone who does believe in god must take one of several possible positions. They may deny the findings of science that have brought religious faith into question. They may accept the facts of science, but not its implications, marveling at the vastness of the cosmos and insisting that god is even more glorious than once thought when the cosmos was envisioned as a globe around the earth, for example, even though science sees no reason to posit such a glorious overlord. Or they may accept the implications and ask what this can tell us of the god in whom we believe, the god we believe we have met.[*550]

This last is the course I have set for the final chapter of this work. It seems to me that this is not any different after all from what humans have done over the millennia. When new information was available, a new understanding of god was attained. Because the basic assumptions were not under consideration or questioned until recently, it was not all that noticeable when the relationship with god was understood in a new light. The god understood by Abraham is certainly the same as the god of Thomas Aquinas, but how much more deeply is this god's covenant love understood because of the growth taking place over the centuries.[551] The same is true of human beings. You may be the same person that I married or befriended twenty-five years ago, but I know so much more about you that you are also a "different person". Then I assumed I knew you, but as more facts came to light and I saw you from different aspects, my understanding was deepened. I hope that the road we are about to travel will have the same effect in our understanding of god, or at least plants a seed.

---

[*]The reader may want to refer to this endnote to check where you are now in relation to where you were in the choices in chapter 2.

5

We began with a question from Susan. A seemingly straightforward question, but one that was actually full of assumptions and presumptions. There is nothing wrong with making assumptions. All of human knowledge, including modern science, is full of them. They are necessary to the development of knowledge and serve as postulates that can then be held up to testing. If they agree with observations then they are held to be true unless and until other observations are made to negate them. In science or any area of human knowledge that attempts to explain physical phenomena, the assumptions, and the conclusions to which they lead, are tested against observations of the physical universe. In religion, attempting to make metaphysical explanations, the assumptions must agree with both the physical universe and the faith tradition of its adherents. This can be done honestly or dishonestly.[552] It is dishonest if it does not really face the challenges to its assumptions that observations have made but attempts to make reality fit its doctrines. This has, of course, been done on too many occasions. Certainly, claiming that the theory of evolution cannot be true because god made the world in seven days and with species fixed in their places was dishonest after the theory had reached a certain level of acceptance and proof. Claiming that it cannot apply to humans because of our ability to reason and that that reasoning must be explained through non-physical means would by now be dishonest. There is just too much evidence that our reasoning processes have evolved like everything else. With the development of algorithmic complexity theory in which there is a fundamental order to chaos, some believers seek to integrate this insight into the faith, as Arthur Peacocke does, holding that, "God uses chance which `offers the potential creator many advantages...God chose to make a world of chance because it would have the properties necessary for producing beings fit for fellowship with himself '."[553] His is an honest effort, and certainly at our level of knowledge today, saying that chance cannot produce anything at all would be dishonest. Speculating that god uses chance is the kind of statement that can help believers but unfortunately is considered rather gratuitous by nonbelievers

True religion wants to be honest. If knowledge of the physical universe changes, then religion must and will change with it, and if the community of the faithful reinterprets its own faith tradition and reaches other conclusions than its adherents did formerly, then the doctrines of the religion will also change. Chaim Potok has one of his characters, a rabbi, saying, "...if the

Torah cannot go out into [the] world of scholarship and return stronger, then we are all fools and charlatans. I have faith in the Torah. I am not afraid of truth."[554] Certainly in Judaism and in Roman Catholicism this has often been true over the centuries. While the Catechism maintains that, "there will be no further revelation"[555], it also says, "yet, even if revelation is already complete, it has not been made completely explicit; it remains for Christian faith gradually to grasp its full significance over the course of the centuries."[556] The Spirit of god works through this "*sensus fidelium*" or the understanding of the faithful people, to lead the church into an ever deeper understanding of revelation. As I said, it is like a spouse who might know their loved one well on their wedding day, but also grows over the years in their appreciation of the goodness of the other as well as in their ability to articulate these qualities and to respond to them. This is not something new in the Church and as we see in the reforms over the centuries and in the working out of dogmas and doctrines, there has always been a willingness to examine the faith and its practice from new perspectives. In our own time Karl Rahner,

> "...dared to criticize the [old] approach and to say that in the Catholic interpretation...dogmas had a temporal relativity, that there was not only a history of how dogma evolved, but of how it was forgotten, that even solemn Conciliar definitions of the person of Christ...could be not merely an end, but at the same time a beginning of theological thought. For that reason dogmas should not only be repeated, but also given a new `understanding' in a new age."[557]

So in chapter 3 I asked whether the assumptions made by religion could stand the test of comparison to the knowledge provided by the physical sciences and in those cases where they could not, I asked where changes seem to be needed in the doctrines of the faith in response to this new information (chapter 4). Now it is time to begin the process of restating the faith. This will not be done with any completeness in this work, though it is hoped that a contribution to a beginning will be made here. It will be done in the years ahead as the recipients of the revelation of god integrate new knowledge into their metaphysics, into their daily living and understanding of the gospel and revelation, and into their catechism and apologetics.*

---

*If you are reading this chapter, it is because no one who has read chapters 3 & 4 has yet to my satisfaction proved my speculations unfounded or wrong. I am still open to such a possibility and would actually welcome it since we could all happily go back to our old ways of understanding God, of praying to God, of teaching and preaching about God. If you, gentle reader, have reached this point and feel that you have reason to differ, please do not hesitate to join the discussion.

The question is, then, when Susan comes to her pastor today, what can and should she be told of god in response? Before she can appreciate who and what god is, of course, she must first recognize for what they are, namely assumptions, those beliefs she has been taught. She will then have to be led into accepting and finding some answers to the questions discussed in chapter 4. Only then can she begin to appreciate the reality of god as god should perhaps be understood and make choices about her own faith and subsequent faith response.

STEP ONE

In my experience, one way to begin to do this is to draw out the implications of what Susan is saying. She is having problems because of the abusive behavior of her husband. Perhaps she has sought professional help for herself, since he probably would not go with her, and it has not helped much. She was most likely told that she is going to have to remove herself from the situation and leave home or remove him somehow and insist that he get some treatment before the relationship can be resumed. She might have done so, possibly several times. Maybe he did go with her for counseling or attended some meetings on his own and made all kinds of promises that were not kept. Maybe she is out of the marriage now or may still be with him. No matter the circumstances, she has, whether or not she went through any other process to change the situation, prayed that god would change it and change has not occurred. So she is asking the question with which we began, "why is god doing this to me?" Other pastors or lay Christians may have given her all the stock answers about the relationship of god and evil and why she is just going to have to accept it, or grow from it, or repent of something in her life before god will lift the punishment. However, it happens that the pastor she consults does not believe these usual answers and wants to challenge her to rethink her position. And so he asks her just what it is that she really thinks god is doing to her and how god is doing it. Does she mean to say that god is deliberately punishing her? Is god trying to teach her a lesson? Trying to make her appreciate something that she has forgotten or learn something she has never learned? She may well answer yes to one or more of these questions, or like Job, she may protest that she is innocent of anything that god might be punishing her for. Or she may say she does not believe in the kind of god who would do such things to others. Somewhere along the line, though, god is felt to be responsible. Well then, the pastor continues, HOW is god doing this? Is it god who decided to make her husband abusive? How did he get that way? When did this start? Did god pick parents for him who would teach him to be this way? Did god in god's eternal wisdom intend that he should have the experiences that he did in his life so that some day he would marry Susan and make her life miserable? Male chimpanzees inherited the same aggressive tendencies towards females from the ancestor that they share with us that have now manifested themselves in her husband. If god was designing these ancestors of humans and chimps, did god know that putting this tendency in would some day result in the treatment that Susan suffers? By now Susan will usually see the folly of pushing the causality of god back so far. Certainly god has not micro-managed her life back millions of years before her conception. From this position the pastor can probably get her to agree that god has not been plotting against her at all, and certainly not from all eternity. It is true that bad things happen in this world, and some of them happened to her, but to explain this we do not need to feel that god has singled us out. This can go a long way in helping her feel, if not better, then at least differently about what has occurred in her life. Does she really believe god punishes in this way? Maybe she needs a course in Scripture, especially in the New Testament, where god's will is said to be the welfare of humanity, not its destruction. Or Jesus' teaching about those who were killed in the falling tower of Siloam[558] who were obviously not killed as punishment for their sins since many others in Jerusalem were worse sinners. Eventually she may understand that her problems are not a punishment for some unknown sin for which she needs to feel somehow guilty. Maybe god is not angry with her and maybe god has not forgotten her, and this may satisfy her questioning for a while. I have made this an easy situation, of course. No one is going to buy this argument as soon as they hear it, particularly if we have a very troubled person. He or she may very well have been told all their lives that god is involved in many if not all of the daily events of our lives, and certainly in the important ones. This kind of thinking begins on the positive side when god is given credit for "blessing" us whenever something good happens. When it is an unmitigated good there is no problem, such as when someone wins the lottery or some "prize patrol" comes to the house, presents the winner with a check and films them dancing in the yard shouting "Praise the Lord". It is a little more problematic when the blessing is along the lines of, "god saved me from my burning home when the smoke parted for a moment and I saw the way out." I am so grateful to god until I begin to wonder why my home burned in the first place. Couldn't god have prevented that? If god could, then why didn't god? Still, these negative thoughts are often kept to oneself or even deliberately ignored so that god can continue to accept the praise and not be in line for any blame, or more likely the "curse" is turned into a "blessing" somehow. A lifetime of this kind of thinking is not going to be reversed in one session with the pastor who tries to carry her thinking to its limits and show her that she, and everyone who ever taught her about god, have been wrong. However, in time she can admit that things may not be as simple as she thought.

STEP TWO

In our scenario, Susan has now come to accept that god has not singled her out for special treatment. Sooner or later, however, it is going to occur to her that even though god may not have a personal dislike for her, god is not off the hook yet. Why is it that bad things happen to *anybody* To this question she sees no way to remove god from the answer. All right, maybe god doesn't sit around planning what hard situations to put individual people in, but couldn't god have made a world that did not have all the violence, troubles, misfortunes, illness and death and just plain orneriness in it? Her question has returned, though it may now be phrased, "why is god doing this to *us?*" She may agree that it would be unreasonable to expect god to step in and change her situation. What is god supposed to do, change her husband's DNA and remove the offending abusive tendency or even aggressive capability? Still, why is violence there in the first place? If god controls the universe, even if doing so through the instrumentation of evolution, then surely god could have directed it down other paths. We have already asked if there is evidence of design in the evolutionary process and answered that, well, it could be god controlling the workings of the universe, or maybe it is little green men from another galaxy, but there is no credible evidence that there has been any tinkering with the works and certainly no need for there to have been any. The universe is perfectly capable of reaching the condition in which it now exists on its own, warts and all. In fact it would be better for god if it did reach its sorry state without god's assistance, for then god would not have to answer for the shape it is in. Susan may begin to doubt the usual answers to theodicy, even if she does not understand the world.

STEP THREE

All her life, Susan has heard prayers addressed to the "all powerful God" or "almighty God". God is called, "Lord of the Universe", "mighty King", "Author of Life", "Giver of the gifts we receive". Doesn't god have to have all these qualities and attributes in order to be god? She has always been told that god must contain all power and potentiality, as well as all actuality, in godself. Now she may wonder, was god constrained to "allow evil" so that we might have free will? Must there be a proclivity for violence if we are to be conscious? In order to do great good, must we be not only capable, but seemingly very likely, to do great evil? If god is to be really god, must god be in control of all the details of the operation of the universe? Must god be in control of *anything* about the operation of the universe in order to be god? She is in a quandary over these questions. On the one hand she believes that god must be in complete control of everything or else god is not god. Since she feels that *someone* must be in control of everything then it must be god, but if god is in control of everything, then god is ultimately responsible for the presence of evil in the world, up to and including the evil that has affected her. Maybe god only allows it, or uses it to bring about good, but if god could have done things differently, then god is responsible, as I have maintained elsewhere. Now, however, it seems so ridiculous to make god responsible for things that will happen millions or billions of years after god's initiatory actions, yet if god is all-knowing then the results should have been foreseen, and if foreseen, then god is responsible. It has become the thesis of this work that the only way to remove the responsibility for evil in the universe from god is to remove any divine responsibility for the universe. If god did not make the universe like it is, then god is not responsible for what it contains, including evil. Lo and behold, this is exactly the finding of science, that there is no need and no evidence that god or anything or anyone else has any accountability for the state of the universe. Not in its inception, not in setting its initial conditions, and certainly not in its day-to-day operation.

STEP FOUR

Since god's responsibility for evil has always been the biggest question for theists, a question they have strived mightily but ineffectually to answer for millennia, why do they not embrace this position that god did not create the universe according to god's grand design and declare god innocent? If god was on the scene at the inception of the universe, according to my scenario, it was as a spectator, and not only that, a spectator who could not interfere in the evolving of the universe even if god wanted. They do not take this position because they have always felt that they must preserve god's prerogatives as all-powerful creator, and god must be the creator because someone must have been, and if that someone was not god (or at least one of god's henchmen) then whoever it was must be the real god.

To get Susan over this hurdle, the pastor is going to have to turn scientist as well as theologian, first disabusing her of her presumptions about the necessity for a creator, sustainer, or director of evolution, as was done in chapter 3. This could be a long, slow process, but let us say that it has been done and that Susan is convinced that the universe does not require a creator, even though she still believes that it has one in god. If she can take one more step and understand that we can allow god to be whoever god is and that we do not have to insist that god is what we have presumed god to be for reasons that have little to do with the real message of revelation, unencumbered by presuppositions, she is ready to take this most important step. Susan may come closer to letting god tell us who and what god is, rather than humans telling god how god ought to be to fit philosophical and theological suppositions.

As I have said, the only way that makes sense to me that god is not responsible for the presence of evil is if god *did not* plan and create the universe. And I must say more, since it is in theory possible that god does have the power to plan and create but does not in fact use it, preferring to let things go in their own direction. This would be the position of the Deists and, in my opinion, often that of what is called "Process Theology".[559] Maybe there are other universes or even other solar systems in this universe in which there are conscious beings with free will to whom it has never occurred to act violently, or for whom there is no need to do so, or who have evolved mechanisms to keep their behavior always positive, like the Star Trek Vulcans. That is not our universe, or at least not our world, and if god is to be found innocent of afflicting Susan or any other inhabitant of our universe, then it is not enough to say that god did not in fact plan and make our universe as it is. What we must say is that god *cannot*. I have made this conjecture before but I wish to make it very clear now. This indeed is my proposal, that GOD CANNOT. If god has any power to physically change anything in the universe then god is responsible. If god could have directed the process of evolution in a more positive direction and did not do so, god is responsible. If god could do something about the major directions of evolution, or even if god can only tinker around the edges and is able to remove genetic disposition for greed or selfishness from someone *and does not do so*, then god is responsible. If god *cannot* physically change who and what we are, or what kind of universe we live in, then god *is not* responsible.[560] Susan and others can rant and rave about god's injustice in not solving their problem; they can feel guilty of something for which they feel god is punishing them; they can stop going to church because their god never listens; or any of the other usual reactions when people come personally up against the question of evil and god. They might as well blame it on a witch doctor in Brazil or the witch next door, or the machinations of Lady Luck as blame it on god, for none of these is guilty. They do not have the power to act and so are not liable.

Let me be clear. Theists have been pushed to the wall to explain the role of god in the existence and operation of the universe precisely because there is no overwhelming evidence for such activity. So we bite the bullet and grant, if only for the sake of argument for some, that god is powerless to effect any direction or change on anything physical in our cosmos. And what results? We remove all the difficulties scientists, and we are all becoming scientific, have had with the whole notion of god as god was presented in religion. We also do away with all the questions that philosophers and theologians have had with the problem of evil. But do we still have a god? And if we still have a god, is god relevant at all to the world or only to believers?

LEADING TO GOD AS GOD IS

If god is not to be blamed for everything that happens, and certainly if god cannot be said to be responsible for *anything* that happens in the course of physical events; if god is not the creator nor the sustainer of the world in which Susan must live, why does she need god at all? Perhaps she should just resign herself to her fate of living a life without purpose, or even escape her fate by removing herself from the scene. If there is no real purpose to life, if everything that occurs is really meaningless, and if there is no hope of change, why participate in the charade? If she, like many scientists quoted, can find joy in this life, awe in the power of the universe, not only the raw power of a supernova, for example, but also in the power of evolution to produce the myriad kinds of life around her, not to mention the appearance of her and her conscious kin; if she can forego the thought that her actions and her life can have infinite meaning and eternal effect; if she can put aside the need for hope in some nebulous life after death;[561] and if she can stop hoping that some savior god can come to her rescue like a knight in shining armor; then there is hope that she can face her problems realistically, enjoy the good times to the hilt, and lead the best kind of life that chance makes possible for her.

If she can do this, the kind of god she may then find will be infinitely more accessible than the god she leaves behind.[562] For god is not like a human ruler, making decisions about the boundaries in which her life will be lived. God is one who, existing beyond any boundaries, wishes to lead her beyond as well. God is not like any human guide or counselor who, with the best of intentions will always lead her to a life similar to their own since this is all the guide can know. God will lead her and encourage her to live a life uniquely her own. How she reacts to the suffering in her life, for example, can be a liberating experience, a growth experience, just as many who have not concentrated on god being the cause of suffering have suggested. The Contemplative, Charles Rich, deals with *The Meaning of Suffering* in a very positive way, showing that,

> "...suffering detaches us from the accidental things of this life so that thus detached we may get the grace to cling and cleave to what is permanent and essential to it. To put it in the words of the golden-mouthed Doctor of the Church, St. John Chrysostom, 'suffering destroys in us the sympathy we have for the present life.'...Long ago the Apostle to the Gentiles told those around him that it is not enough to perform the various deeds and acts usually effected by those who love Christ; he laid down as an indisputable principle that they must also partake of the Cross of Christ if they would hope to share in the glory of His risen state...We cannot know what the divine sweetness of God is like unless we ask for the grace to bear whatever we have to in the right Christian way."[563]

The author does not bring up the question of why this suffering is present in the life of the Christian and there is no need to do so. The universe is full of suffering and destruction, but not because god wishes it so. That is the way god found it. What god teaches us is how to overcome it or even use it and grow from facing it. God is also confronted with suffering in god's experience, maybe not by personal suffering or threats to god's well-being, but certainly the pain of watching god's loved ones suffer and being powerless to remove it. Anyone who has sat by the sickbed of a loved one knows the pain of this that can be more severe than what the patient is suffering. How does god respond to suffering? With empathy, sympathy and the never-ending effort to lead us away from inflicting suffering on others as well as dealing with our own in as healthy a manner as possible just as John Paul envisions god, though without the emptying or *kenosis* of powers that god never had. Susan would do well to imitate this god rather than rant against the god she once thought caused the problem.

SO WHO IS GOD? UNCAUSED, BY CHANCE?

Who is this god and what is god's relationship with the universe, and therefore with Susan? There is no reason why there must be a god. It could have happened that there was not. In our examination of the argument that nothing can come about "*ex nihilo*," we have already seen that the presumption is that the universe arose as a fluctuation in the quantum vacuum. The universe needs no cause, certainly not a conscious, rational causer or prime mover, and in fact need not exist at all. The presence of the universe does not require the presence of god; that either god or the universe exist is not necessary, nor does one depend upon the other. Both are the "result" of chance. It may be that, given things as they are, universes will always spring from the vacuum, but things do not have to be as they are. There may be penultimate reasons why our universe exists and exists in the conditions it has, and it is the duty of the human spirit to inquire into them, but it will remain true that there is no necessity that these "almost ultimate" situations need pertain. There could have been no quantum vacuum, the penultimate cause of the universe. The ultimate reasons will always elude us for the simple reality that there are no ultimate reasons. It could always have been otherwise. With god's being and existence there is also no penultimate reason; nothing at all causes god to exist or to exist as god is. Neither are there ultimate reasons. It could be otherwise, but by chance god is as god is, "I am who am."

It could be said that "chance" is the ultimate reason. "Chance" however is not the chance of the scientist or mathematician. Their chance is the chance of our universe about which "laws of chance" can be formulated. If it were the scientists' chance, then chance would be a "real" force or cause which could be investigated and which is ultimately responsible for what occurs. Chance is used here to mean simply, "it could be otherwise", without ascribing any causality. We must use some word to describe what we mean and "chance" is as good as any, but it does not have to correspond to any causative agent. And so it just "happened" that god exists, and that god exists as god is. It also just "happened" that anything else exists. Since it happened to be, there can be causality of subsequent events. Forces can exist which bring about change in the reality.

TRANSCENDENT

We have always claimed that god exists outside of time, but never could really appreciate what that might mean since our whole universe is enmeshed in space-time and to imagine something outside of our universe is impossible for us. Now we can understand in a theoretical but fuller way that such is possible. If there is another universe, or many universes, which exist alongside of ours, they are outside of our space-time. We are not aware of other universes, but we know that they could exist. In fact since we exist, the probability that others do as well is much higher. We should not visualize such a thing as seeing from the outside a number of universes, perhaps like so many soap bubbles floating in a room. The best we could manage would be the same kind of universe that we inhabit, though immensely far from us, even though to put it in those terms would be to include it in our own space-time. The word "far", used in this way, cannot include distance or time but merely that it is inaccessible. Our tendency to visualize ourselves as existing in some space-time larger than the universe in which other universes might exist as well, is proof that we cannot imagine such a thing, but we can admit it. If other universes can exist outside of our universe and us, certainly god could as well. The faith invites us to experience a being outside of our universe which is not another universe but a different kind of being altogether, one not sprung from the vacuum. A universe in which the laws of physics are slightly different than ours, or radically different than ours, is another kind of being in one sense, but as far as the theory of emergence from a vacuum knows, it is still made of "stuff" of some sort and still liable to the laws of evolution. It is still "our kind of town".

Since we cannot prove that something does not exist, we must admit the possibility that a universe could spring into existence with consciousness already a quality of it, a truly "conscious universe", but since we are the only example of a universe with consciousness contained within it that we know of, and it took eons for it to develop here, our tendency is to declare such a possibility highly unlikely. On the other hand, if universes are continually popping into existence, most for nanoseconds but some for much longer periods, then the probability of a fully conscious universe springing full-blown from the vacuum is higher. (Of course if they have "always" done so, it is a certainty. Note that I am not restricting consciousness to our kind. There may be many ways to be conscious). If we can admit that a conscious physical entity *could* arise from the quantum vacuum, and exist "alongside" us, and if we can admit that communication between such an entity and ourselves is not beyond the realm of possibility (some hold that quarks flit off into another universe during a quantum leap, others that another universe might exist on a "brane" a quarter inch away from ours), then would we have to deny that the existence of a conscious entity, able to be aware of our universe and able to be aware of us, but that *did not arise from a quantum vacuum*, could exist. The testimony of many millions of people who claim to have been in contact with just such a being (what the heck, let's call it "god" or something like that,) cannot be summarily dismissed.

EVOLVING AND GROWING

This god could maybe bring about change in godself, perhaps "internally" by self-growth, but also by exposing godself to other forms of existence. We must also explore whether god can affect other beings and if so how this is done. It could have be otherwise, of course. God could have neither of these capabilities, but then we would know nothing of god and what we think we know would all be a figment of our collective imagination. Or god could have all the powers we have traditionally assigned to god, but then we run into the problem of evil again. It could be otherwise, but it seems that it is not otherwise. It is as it is. And as it is, "being" wishes to exist and strives to continue to exist. Any being of our universe that does not would long ago have vanished as a result of the selection process of evolution. Any god that did not want to change would be irrelevant to any but itself, since it would make no effort to communicate or to learn, but that is not our experience of god. We know god as one who revels in god's own existence and what is more important to us, revels in the existence of anything and everything else.

OMNI-EVERYTHING?

If a being is god and exists, does it have to exist as everything that could possibly exist in order to be god? Is it a necessary attribute of god that god contains all experiences and powers and "actuality" in itself or else it cannot be god? In the past we have maintained that god must possess everything in godself so as to be the source of everything that exists in a contingent universe. Since "nothing can come from nothing", if something exists then it must arise from a source, indeed an ultimate source. As I have asked before, "whence does it come from the ultimate source?", but now this is a moot question since there seems no need for ultimate sources.

So it is not necessary that god know what it is to be a stone, or a lizard or else is not perfect and hence is not god. God is who god is, and a lizard is what a lizard is. Neither is diminished in their being by not being the other. Certainly the lizard would be different than it is if it had all the attributes of god, but it would also not be a lizard, and being a lizard is a form of being and therefore good in itself. It will not be more of a lizard by being more like god. However, this is not true of the other direction. God will be enhanced by experiencing the lizard since being a lizard is being and that is what god is about. For that matter, god is enhanced and perfected by exposure to any other being, as are humans. Far from existing as an idea in the mind of god first and then coming into being, out of the myriad things that can exist, some, by chance, begin to be, opening for the universe that contains it a new way of existing and opening for god a new understanding of the possibilities of existence. God is neither a singularity nor a nascent universe that evolves into complex structures. God is god, a way of being. And now here is another way of being that is not god. It is something other, this universe. It too, is itself. And god is drawn to it for its similarities to godself, for the potentials that god can see in it, but also for the dissimilarities which reveal new ways of being, and therefore for the potentials within god that it might reveal. But let us not start with a lizard. How interesting can a quark be? The first time it is experienced, yes, but once you have experienced it (if you could meet a quark for a drink?), how often do you need to experience it again? And how engaging is its twin, and their twin? How they may react with one another and the new ways of being that they may become have not yet occurred. And if they have not occurred, they do not yet exist. Is it impossible to be god without knowing all the potentials that will become existent? I see no reason to insist on this. Is it even possible for god to know all the potentials that will be actualized? Not according to quantum indeterminism. And so the observing god learns from the quark the lessons of its existence as it is, but not as it will be. The astute observer, as god presumably is, may see the possibilities, but will not know if these can be actualized until they in fact are. Logically one can know whether or not a lizard can exist with five legs instead of four, but one does not know a five-legged lizard until one actually exists. Engrossing as it may be to speculate on what could maybe be, one is not instructed in what can really be until it is. And so god observes the quarks as they begin to interact to form hadrons (and god had better look quick because this happened "between one-millionth and one-hundred-thousandth of a second after the outburst from an initial singularity."[564]) and these acting to form atoms and then the simplest molecules, followed by all the great things molecules can make. Interesting stuff that has enraptured many a human observer and would have the same effect on god.

God is not, however, involved in directing the process. Some have speculated that perhaps god does not wish to be involved, preferring to observe and learn without interfering, much as Jean Luc Piccard will follow the prime directive of Star Fleet and not intervene in the evolution of other cultures. Piccard does intervene of course, since he decides not to do so would make him responsible for evil consequences. My speculation is that god *cannot* intervene, but exists in an entirely different manner than the materials of the universe. This inability does not diminish god's being god. We may arbitrarily believe only in a god who can and does have the power to intervene in the movements of a quark or molecule. We may prefer that god be able to control the universe, but once again, it is not necessary that this kind of god is the god who actually chanced to be. It is probably true that the (unconscious) material of our universe is not able to respond to any other stimuli than those of its own universe and the "laws" of its own reality. It can be influenced by what classical physics knows as the natural laws of the universe, and human consciousness may well affect it if it is true that observation by a conscious observer "collapses the wave function" of quanta, making them behave as particles. Human consciousness exists as a result of these same properties of the universe. Human consciousness is a consciousness of the universe, not a consciousness as a gift from god, and god's consciousness is of god, not of the universe. God's consciousness is not that of a brain nor of a computer nor of any other manner of consciousness that the universe may devise, and so perhaps has no effect on the wave function of the quark or of the universe as a whole. The quark or the molecule, or for that matter maybe the lizard, cannot respond to the consciousness of god. God can be aware of them without affecting their being. Humans, of course cannot be aware of others in the universe without affecting them in some manner. Our very awareness of being may determine what the being is. According to the so-called "Copenhagen Interpretation" of quantum effects, the physical states of particles such as momentum, energy, spin, color and others do not exist until it is "observed" or measured.

> "Measurements in quantum mechanics...are very active processes. They aren't processes of merely learning something; they are invariably processes which drastically change the measured system." "Measuring the color of a hard electron...isn't a matter of ascertaining what the color of that hard electron is; rather, it is a matter of first changing the state of the measured electron into one to which the color predicate applies, and to which the hardness predicate cannot apply...and then of ascertaining the color of that newly created, color-applicable state."[565]

Some have imposed the requirement that the observer be a conscious being,[566] and some have gone farther and conjectured that a conscious god would fill the bill quite nicely, but there is a problem here, especially with human consciousness. As we noted in our investigation on the soul, there is no reason to hold that consciousness is anything other than a manifestation of the human brain, which is certainly a work of the universe and a physical object, no matter how we may think we experience that there is something non-material about it. If the human brain, which is matter, can have an effect on other material objects, transforming a photon into a wave or a particle for example, then do not other physical objects have a similar effect? Or perhaps it is more a matter of our brains extracting a part of the information contained in each object so that we can make some sense for ourselves of the universe around us. In any case, there is no evidence that any kind of consciousness other than a matter-based consciousness has any effect on physical objects. And since it is my conjecture that god cannot affect the physical universe, this conclusion fits in quite nicely.

BUT INTERESTED

So throughout the history of the universe god observes its progress and its experiments in new ways of being. God can learn from its experiences as new realities come into existence and god is certainly interested. Whether god can grow in godself without relation to other beings we may perhaps never know, but god can grow in godself through exposure to others, just as the only other conscious beings we know, ourselves, can grow. At the risk of anthropomorphism, one can imagine god as an interested observer of the progress of the universe, and perhaps of myriads of universes, some of which actualize a seeming potential and some of which show what being cannot be. An interested, curious, perhaps absorbed observer, learning ever more of the possibilities of being. However, again at risk of putting our own emotions on god, perhaps also a "lonely" observer for while god can evidently respond to the universe, the universe at first cannot respond to god. And god has so much to offer. There are so many ways of existing that the universe never produces. It is locked into its own history. If the strong force which binds the nucleus of the atom had been just a wee bit different, chemical reactions of various atoms and molecules could be just a little different and, as we saw in the butterfly effect, there is no telling what might have developed. Or another example, life on earth may have started around undersea volcanic vents, living primarily on sulphur and with no dependence on oxygen, indeed O_2 is poisonous to those forms that survive in the oceans today. Because of the vagaries of history the atmosphere is not conducive to this kind of life and so it exists underwater (or in a backwater)[567] but who knows what it might have evolved into if things had been a little different. (It did, in fact, evolve into us.) This universe could have produced so many more ways of being than it has and it could be otherwise than it is but it will not be since the universe has gone in another direction. Constrained by its own history, the universe changes, sometimes quickly and sometimes ever so slowly. Though god may, like aliens from another galaxy, check in from time to time or send a message just as our species now does hoping to make some contact with something that can respond; nothing does. Like the apocryphal Star Trek saying, "Beam me up, Scotty, there's no intelligent life down here."

EUREKA!

Then the universe produces something that is more than just of passing interesting to god. It begins to produce, by chance, a new way of being, a consciousness that begins over the millennia to resemble, however remotely, that of god. Not through the same process, of course. God is evidently conscious in god's being. The universe becomes conscious through the interaction of nervous tissue, at least in humans and on our world, and perhaps through other materials in computers one day. This is a different manner of being conscious, but conscious nonetheless. Now god is not alone in being as god is, aware. There is something of the universe that will respond to god's invitation. There is something that can understand its own being, however imperfectly, and revel in the fact that it is. There is something that can share itself and communicate itself. And god is more than interested, god is FASCINATED.

True, this "something" began on such a primitive level as to be almost non-existent. How much response is there in the mind of an amoeba? A lizard? A beagle? How much in the brain of the first primates? More than in a stone, certainly. The human might be fascinated by a stone, of course. Someone made a fortune several decades ago with a mood ring that was certainly interesting to those who saw it. Interesting for about ten minutes. Its charms wore thin rather quickly. A diamond is certainly engaging, but more for what it can purchase or what it means than for its intrinsic being. The geologist can spend a lifetime with rocks and stones, but it is their history and what they tell of our history that is so fascinating, not the individual lump of material. And if a stone is lost, unless it has a monetary, nostalgic or scientific value, there is little regret. The same may not quite be true of a plant. A plant develops and changes over time. To know the plant in some detail today is not to know it tomorrow. The stone will lay there unless subject to some external influence but the plant will unfold in a new configuration. The botanist or gardener may develop some attachment to a particular plant, but as with the stone it will probably be more for the personal investment in the item rather than the plant's intrinsic lovableness. Even here there may be gradients since one is perhaps more likely to be attached to an African violet if that is one's particular passion, or to an orchid, and even more to a century plant in which one has invested many years in waiting for the bloom, but most gardeners do not grow sentimental over a carrot. To use the word "fascinating" of a plant would be to misuse the word.

Animals are another matter. Humans can form much deeper relationships with animals and at times must be careful lest they do. Never let the farm children develop an attachment with tomorrow's porterhouse, or pork chops. The more the animal resembles us, the more we can relate to it. A slug is a little hard to connect with, but even a fence lizard can be the object of our affections, until it eats a fly and grosses us out since eating flies is a little strange to us. The closer we come to the order of mammals, however, the more identification we make with ourselves. Among mammals the connections become stronger the more the animal responds to us and seems to understand us, or even resemble us physically. Many an environmentalist is chagrined when they must defend from encroachment and possible extinction an animal, even a mammal, with which we do not readily identify. Pandas are easy, as are lemurs and orangutans because of their looks; dolphins and whales are interesting because they live in an environment alien to most mammals so we root for the underdog, plus they seem to respond to us and perhaps even care about us, saving us from drowning perhaps. Rats are harder, since even though their relatives, the hamsters and gerbils, are cute, no one has caught the plague from the latter, but the former are deemed enemies of humanity. And bats are a nightmare to drum up sympathy for since they are so unlike us and seemingly care nothing for us. Dogs are our favorites since we perceive them as not only "almost human", but able to understand our speech and respond to us. And the young of almost any mammal, with their flat faces and big eyes, are enough like our own babies as to inspire a riot if you club a young seal.

Even the mammals can keep our interest for just so long. If it is not our pet we can enjoy its antics for a while, and if it is an exotic specimen, for a little longer. If it knows a "stupid pet trick" we can maintain our curiosity for an extended time. If it is our pet the attachment can last for years, but not for too long at a time. One can be happy to see one's dog after an absence and feel flattered that it would jump up to greet us, but our tolerance of this behavior pales rather quickly. However, we are exposed to them in many circumstances and are treated to the full repertoire of their behavior and so can observe more of their capabilities, hence the prolonged interest. We make allowances for their lack of ability to respond. Though a dog may sense our mood and become depressed along with us, and we may even tell them what is troubling us, we certainly do not expect them to give any good advice or empathize with us in a rational way. When the situation changes and we regain our good spirits, the faithful dog will rejoice with us, but we do not expect it to know why we are happy.

The human being who will quaff our celebratory beer, "drinks all around", without knowing the occasion is welcome to it, but is certainly not classified a friend unless they know what has prompted our good cheer. The human is expected to respond beyond the ability of the pet and we can be much more interested in one of ourselves than one of another species. This is not merely human chauvinism. No other species can respond to us in the same way and with the same depth as another human. True, the infant's cries can be a particularly annoying form of communication and response, but we endure them for the joy of knowing the person the infant will become. For the parents even the crying can be endearing, except in the middle of the night. As the parents spend more time with the child and see the developing personality, it beomces more accurate to use the word "fascinated". This will continue throughout the life of the person, or at least it can. True, some personalities have been abdicated by their possessors. They are so predictable or so shallow as to be boring, but if this has not occurred, even the very limited in personality can be fascinating. Herman Melville observes in "Moby Dick",

> "Seat thyself sultanically among the moons of Saturn, and take high abstracted man alone; and he seems a wonder, a grandeur, and a woe. But from the same point, take mankind in mass, and for the most part, they seem a mob of unnecessary duplicates, both contemporary and hereditary. But most humble though he was, and far from furnishing an example of the high, humane abstraction; the Pequod's carpenter was no duplicate."[568]

Nor is any other individual person. We do not hold it against the person of retarded brain function and can be genuinely fascinated with the Down's Syndrome person if we sense that they are responding to us or to themselves with some *élan.*

THE FASCINATED GOD

In just this way is god fascinated when a being arises in this universe that can respond.[569] The means by which the being is conscious may vary tremendously from the manner of god's consciousness. The depth of human consciousness and self-awareness as well as the depth of response may be limited, but nonetheless it is consciousness responding to consciousness, and god is enthralled. The ancient Babylonians may have thought that the gods would flock around a burnt offering like flies around honey, but it is not the lure of food that will beckon god but the lure of relationship that cannot be had with a stone nor a plant nor even an animal, much less a quark or a neutrino. We may not be the best that will ever be, but we are the only game in town at the moment, at least in our corner of this universe. Beyond that, we are also not too bad. Our potential is, if not unlimited, quite substantial. This is true of every human who has not abdicated his or her humanity, an act which we will explore in the section on morality. We tend to sell each other short, usually the other person, but even ourselves.

Parents, counselors and psychiatrists spend much time in encouraging their charges to recognize their own goodness and develop it since there is sometimes a tendency to deny it. Yet most, if not all, humans are fascinating once they are known in depth and allowances are made for conditions beyond their control. Of course one of the "greats" is quite interesting and complex. Who could not spend much time, even a lifetime, with one such as Napoleon? Yet there are literally millions of others just as capable who could have accomplished what he did, given the opportunity. They may seem commonplace, but once you see them in their element they show their potentiality. This is likewise true of people with more modest capabilities than the "greats" or potential greats. In their own way they are just as fascinating in how they respond to life as one of more headline-making talents. Many a pastor has been surprised at the ingenuity and often the morality and spirituality even of one they took to be a total nonentity appearing at their door for a handout.

FASCINATION TURNS TO LOVE

This idea of god, not as the creator and ruler of the universe but as a lover of our universe and especially its conscious inhabitants is the key to understanding the divinity, and it always has been. True, the attributes of god have been explored by philosophers and theologians for millennia. Those of omnipotence, transcendence, omniscience and omni-everything-else receive top billing, but no one prays to such a god with anything like affection. With awe, perhaps, or in deference to an exalted position, and certainly to manipulate such a being if possible. We relate to this god because he/she/it exists and has some power over us and there is certainly plenty of this in the Scriptures and in the liturgy. Still, in our better moments we recognize that this is not the way in which god relates to us. We may address god as the creator and sustainer, even as Aristotle's prime mover. We call god "Lord" and "Baal" and "Almighty God" in the mistaken conviction that god can and might use power to intervene, but we also take the advice of Jesus and call god "Father". And in our really playful moments we use the same terms of endearment as Jesus really counseled and use "Abba", or even a pet name we have for god. We do not do this in response to the teaching of the philosophers who have not reasoned to a fascinated god, and though many a theologian and preacher has led in this direction they would not have dared if god had not set the tone for the relationship. When we speak of god, at least in the Judeo-Christian tradition, we are speaking of the god who covenanted with us, who adopted us as family, who treats us as a mother or loving father would. Like a parent, god does not frown on the children, even when they are in error, even when they rebel, but smiles on them in the love of fascination.

HOW DO I SAY I LOVE YOU

All of this presupposes that god can do what I have said all along god cannot do, interact with the physical world. As we have seen previously, it makes the most sense to say that god is not a part of our physical reality and cannot have an effect on it, nor does god's awareness of the universe change the reality. Nor can this reality of our universe, which has sprung unbidden into existence, be physically manipulated by god. The universe does not need god as a cause of any kind and we see no evidence that god intervenes in its operation. There are anecdotes galore of god doing so, of course, and a plethora of films and books, especially recently, purporting to show god intervening through the medium of angels. The skies have not been so full of angels since the Middle Ages. Our demons, however, seem to come more in the form of aliens today, though the occasional ghost needs to be busted or devil exorcised. Despite all this, few expect god to intervene in these ways in their lives and science presumes that every phenomenon has an explanation. The best explanation of where our knowledge has led us is the conclusion that god not only does not intervene in the physical universe, but cannot. Perhaps we will try to say that god can indeed interact but that god does not since this would certainly defeat god's purpose of learning from the existence of other beings about the possibilities of life and consciousness. If god could intervene and chose not to do so this leaves god open to the charges addressed under the relationship with evil. If god could do something about misfortunes happening to people and did not, then god is partially responsible and even culpable. This, along with the evidence, or rather lack of it, from science in some intervening causative agent, has led us, or me at least, to the position espoused here that god cannot intervene.

If god expects a response, however, god must reveal god's presence. And if that presence is not to be found in the handiwork of creation, nor in changing the course of world events, or the course of an illness, then how can god make this presence known and invite a response This will lead us to the only seemingly testable hypothesis of the theological position of this book. There must be some manner in which our human brain and the consciousness to which it gives rise can be aware of a reality apart from this universe. Certainly those who believe in ESP will have no difficulty in believing that this is possible since they already believe in emanations and auras that are able to impinge upon our consciousness. Whether or not this is part of the explanation is, of course, open to further investigation and questioning as to whether it is even possible for one human person to communicate with another in this manner. If it is, and this seems to be unproven, then there would be some indication that it is also possible with god. However, it is presumed that these avenues of communication between ourselves, if they exist, would have a physical basis in electromagnetic waves or some other means and that there is a physical receptivity in our senses or in the functioning of the brain that is open to them.[570]

Since god seems not to have a physical basis, or else god could intervene in physical events, this could not be the manner of communication, but it suggests that the brain is open to much more subtle influences than large-scale sensory input. Is it possible that our brain is open to the perception of larger realities that are only hinted at by their physical manifestations? Is it possible that we are aware of realities that have no physical manifestations? Certainly we are not aware of anything in the universe that does not have a physical manifestation, but we are aware that there could be physical manifestations of which we are not aware. Until the proper instruments were developed we had no idea and no way of sensing ultraviolet, though many insects sense it, or gamma rays or X-rays, even though they emanate from very physical objects. By chance, our senses developed which pick up the clues we need to make sense of the environment that we normally experience, and by chance, we did not develop a means to detect other clues.[571] We have not evolved a sense organ to perceive some of these things, but we have developed an organ that can convert them into something that we can sense, as when the human brain developed a Geiger counter to change radiation into audible clicks and the deflection of a dial that we can see. Is it possible that by chance our nervous system is capable of perceiving a signal from a non-physical being? Maybe something about our system for intuition, a function of the brain which is still not well understood but which seems to be much more important than previously recognized.[572] It makes sense that intuition, as a sort of hazy but quick judgment, would evolve in a species which for much of it's existence has been potential prey of other animals. Acting on a hunch that there is something large and furry and hungry who is stalking me is certainly preferable to not knowing it is there at all until it pounces. Maybe a hunch is based on some physical sensation or some trigger to a memory of a similar situation that turned out to be dangerous. I am not suggesting that the working of our unconscious mind is somehow not physical, but I am suggesting that while evolution was providing us with the kind of brain that could perceive very subtle physical clues and extrapolate them into the existence of the presence of an enemy, it might have, by chance, set up a mechanism that can perceive a non-physical existence as well. This would be one of those "emergent properties" of the brain, like consciousness itself, which is an unplanned but welcome by-product of evolution of the brain.

Another hint at the possibility of our perceiving a non-physical being comes also from our experience of physical persons. Denis Edwards talks about a "pre-conceptual experience" of others, that extra perception of the other in a way that is hard to explain. He uses this familiar apprehension to explain human beings' experience of god and says that,

> "...when I speak of experience of God I will always mean pre-conceptual experience. God always transcends our senses and our intellects. We do not have access to the inner being of God. However, I will argue, this always transcending God has come close to us in love, and we can experience this presence in our lives in an obscure way. The concept of pre-conceptual experience allows us to speak of a real human awareness of God who yet remains always incomprehensible to our intellects."[573]

Indeed, the practice of meditation so highly recommended by the spiritual masters calls for emptying ourselves of intellectual arguments and even of our own thoughts so that god may be heard in Edwards' "obscure way." What is this except removing the distractions so that god's presence may be experienced?[574] It is true that scientific investigation of the workings of the brain have now mapped out those portions of the brain that either become more active or less active during such religious practices as meditation. The authors of *Why God Won't Go Away* point out how in meditation the brain is deliberately deprived of sensory input as it focuses on a mantra, a crucifix, etc. This causes the parts of the brain that normally give us the experience of being an isolated individual no data to work with and so we feel what the conscious brain perceives as our unity with all beings. Other religious experiences have also been mapped on the physiology of the brain, including the presence of a loving god. Still, the authors point out that our perception of anything is provided by our brain, and the fact that my brain is providing me with an experience of bread does not mean that there is no bread present. There may be a part of the brain which when stimulated gives us an experience of what we may well term "god", and indeed some researchers have dubbed it the "god spot". Just so, our brain being responsible for the experience of the divine or other mystical phenomena does not mean that there is nothing real about it external to my brain.[575]

PROOF OR ENCOUNTER

I would love to have someone devise an experiment to show that perception of a non-physical being is indeed within the capability of our brains, but since I suspect that there is only one non-physical being in existence and we have no way of sensing this being except perhaps through this fortuitous and serendipitous capacity of our neural system, I am afraid no such experiment will be conceived. [576] But then I am not trying to prove the existence of this non-physical being, god. Conversely, I do contend that there is absolutely no way to prove god's existence, *except the subjective and personal experience of god's presence*, and this will "prove" it only to myself. The testimony of others who have sensed god's presence can introduce me to the possibility, can teach me how to be open to the experience, and can reinforce my faith, and these are not inconsiderable things. The villain in "Pale Rider" wants the preacher, Clint Eastwood, removed because, "one ounce of faith, they'll be dug in deeper than ticks on a hog." Yet, just as the theologians have been insisting,[577] though they also persist in trying to prove themselves wrong, one cannot prove anything about god in the way that science can "prove" something about the universe. This should not be so surprising, since we cannot prove who another human person is to a third person no matter how much explanation of his or her parts, physical attributes, habits and quirks, experiences and memories, values and biases. We can prepare them to experience this third person by telling them all this information, but they will not know that person to be real and will not know them as a person until they experience them.

INTERPERSONAL COMMUNICATION

And so god encounters a being who can somehow experience god's presence and whose presence god can experience. God is fascinated by these beings, and would be interested even if they were not aware of god and/or could not respond, but they are, and they can, and god is overjoyed in this people, just as the Scriptures say.[578] And so what does god do? Observe them from afar? Perhaps, since god could grow within godself from learning alone. Yet, there seems to be a quality of god that wants to share godself and god's knowledge of being with others. Not so unusual since we ourselves have these same noble tendencies when we allow ourselves to have them. So what does god do? Why, god communicates! Saunters right up and says "howdy".

Before we continue on how and what god communicates, we need to be clear on how god does not communicate, and that is through messages embedded in nature. Certainly if god is not the creator of the universe then we cannot know god through god's creation as has been asserted in the "two books" hypothesis and is implicit in the assumption that god created everything. We cannot isolate qualities from the universe such as "beauty" or "being" and say that these same qualities must be present in god or else they could not be present in god's creation. Nor can we extrapolate from god's actions on physical events as recorded in the Scripture and make assertions about the attributes of god. It is not permissible, for example, to take the story of god stopping the flow of the Jordan River or bringing about the plagues in Egypt or Jesus calming a storm on the Sea of Galilee and see these as proof that god is a god with power over the monsters of the deep. Indeed, when we read these stories we are going to have to extract the premises based on prior assumptions about god's power and, denying that these events could have happened because of god's intervention, look for the theological and social implications of these stories and myths rather then on the physical event described. God cannot be known by the study of the material universe, though god can be known at times by how we are inspired to recognize some attribute or presence of god even if we express this awareness through the wrong symbols.[579]

DIAMONDS IN THE ROUGH

We are not the goal that a creator-god had in mind from the beginning. The rather discredited Anthropic Principle[580] is welcome to hold that we, or someone like us, are the goal of the universe that cannot exist without conscious observation. That is certainly not what the traditional faith meant when it speaks of humanity as the darling of the creator's eye. We are not the end product of the creator's activity, and since the actual creator is the principle of evolution and not god, we are not even an end product. Surely something else will evolve from us, both genetically and culturally. Our genes and our memes will conspire to make from us another species, perhaps, or at least another variety of homo sapiens, or we will join all the others whose extinctions we mourn. That does not diminish our importance to god. We may not be the result of the sweat of god's brow or the fruit of god's loins, but we are a treasure to god anyway. The diamond miner did not create the diamond, but rejoices in its beauty nevertheless, and then seeks to manipulate it to make it even more beautiful. So it is with god.

WHO DO WE THINK YOU ARE?

Granting that god does not communicate through creation, I do assert that god converses in some way with us. And what would god communicate to us? The first would probably be god's presence. Maybe this is what is underlying animism, shamanism and totemism that perceives some "presence" beyond what is immediately sensible in objects.[581] I hasten to say that the experience of the supernatural reported by many will certainly be explainable by scientific means.[582] God may be able to use mental constructs that evolved as by-products of the brain's development but there will be no "proof" that anything other than natural processes are occurring when people claim to have these revelations. If god is able to communicate in this way and this presence is processed in thought, it would become less diffuse and more analyzed over time. The analysis will also certainly be primitive at first and slowly develop as the humans compare notes, the very process we see at work in the scriptures and in the theologizing of later Judaism and Christianity, as well as in the other religions of the world. Authentic qualities and traits of god would be discerned, but others would be ascribed to this presence by the wishes and desires of those experiencing it. That humans would like this presence to be powerful, in charge, and rather like them, but more so, does not mean that this is in fact a characteristic of the presence. It is also true that hoping that this presence is beneficial and even personal and loving does not mean that this is the reality of that presence. Indeed the presence is by and large not perceived at "first" as all that favorable to humanity, and that "first" may be for some millions of years. It may be perceived as oppressive and quixotic or rather unconcerned with humanity, since most of the other "presences" in the universe are this way. This is the same pattern that we still see in many human reactions to new information, and certainly to the first contact with new cultures and ways of life. First, we see something that may be an enemy, then we see something that may be exploitable, and only later do we relate to the other as a person like ourselves. Too much like ourselves at times, since we insist that we are superior and it is our ways and customs that must prevail and "they" must be more like "us". Humans always seem to try to mold the others into themselves, or to deny them human status or allow them to have their own personality. Only after long exposure do we begin to see the goodness of the other and develop a willingness to learn as well as teach, and to take lessons as well as give them. And is not this the pattern that we see at work in the scripture. It is a rather primitive god that Abraham calls his own, and one given the same qualities that Abraham has, or would like to have, and Abraham is within the period of oral history and not near as primitive as when we began to be human. However, god never ceases trying to communicate the truth of godself to this species and gradually, over millions of years, we begin to get a clearer picture. Give two people fifty years of wedded bliss and they might begin to understand each other a little better than they did at their engagement. [583]

COVENANT

What is the first clear development in the understanding of this Judeo-Christian god? Is it not the covenant of Sinai? Whether or not the Exodus story is written backwards into Israel's history from a later period, it is certainly apparent that covenant is the cornerstone of Israel's relationship with god and of their understanding of what this god is like. The first really clear understanding of god is that god loves them and wishes to relate to them in the analogy of a human family.[584] There has always been a question of why god would wait two million years before making this desire known in an effective manner. If you allow that the older Confucian tradition is a legitimate revelation from god, it is only minutely older than the Judeo-Christian roots compared to the history of conscious humans. Perhaps it was not god who waited, perhaps it was our own awareness of the presence of god to us that had to be developed over the millennia. The presence of god seems to have only a very subtle influence and perceptibility to our brains, else we would have no atheists today. How long does it take for a species with our capabilities to become aware of such a subtle presence in our minds? It took thousands of years to become aware of the benefits of fire and the skills to control it. Evidently, it can take a long time to realize that we are perceiving god. Or at least a long time before we started keeping any records of it. There are, to be sure, records of gods and goddesses from Mesopotamia, Egypt, the Hittites, Chaldeans, and all the other inhabitants of the area that gave rise to the Hebrews. There are cave paintings and rock art from around the world that predate the Old Testament by tens of thousands of years. All of these seem to preserve a record of the perception, at least in the minds of the artists, of something or someone other than ourselves. We are not going to investigate these because they seem to have remained stuck in their own assumptions and did not lead to new insights into god, but they certainly indicate that god did not just begin to make godself known at the time of Moses, or the time of Abraham.

LIVED IN COMMUNITY

The Hebrew scripture is witness to another aspect of this awareness of the presence of god. It is perceived in community. In spite of the stories of revelation to individuals such as Moses alone in the desert before a burning bush,[585] it seems that the revelation is always for the community and not for the individual. Witness the call of Moses to go back to the incipient community in Egypt and bring them into the desert to encounter the god of the covenant.[586] The prophets who experience god in the desert or on the mountaintop or even in their own personal life experiences and insights, were themselves formed by the faith of the community and took their insights back to it. "We are not to think of the prophets as people who had an entirely different source of information from other people. Rather, they saw ordinary things more clearly than others did."[587] In fact they are often called precisely to bring the community back to its own authentic faith in which the prophet was formed.[588] God is present to all people, but in such a subtle manner that it takes the process of community development of a body of experience of this presence to make it intelligible and to weed out the false impressions that have found root in individual analyses of these experiences. We find this same process in the biblical stories of the kings being confronted by the prophets, representing the community. Certainly the Law, while it is attributed to a sudden burst of revelation to Moses on the mountaintop, is really the distillation of the wisdom of the community of Israel throughout the centuries as influenced by their awareness of this divine presence. For example, in a later time there was a movement among some prophetic circles to say that god should only be approached in Jerusalem in the Temple, as god had revealed to Solomon. This was the royal party line,[589] and evidently not from god since it was never really observed by the whole community of Israel/Judah. What the prophets eventually realized to be genuinely from god was that it was the pure heart that was the acceptable sacrifice, not the bullocks. This was not the invention of one person, but the distillation of the experience of the community of this god. Indeed, the replacement of sacrifice in the Temple by the Judaism of the synagogue is the result of the community's experience in the Exile and here there is also no claim to a sudden revelation through one person's experience. It is true that there was a movement after the Exile in the time of Nehemiah and Ezra to rebuild the temple and it was in fact replaced, complete with sacrifice. Herod built one of the wonders of the ancient world as his temple, succeeding the earlier replacement. Nevertheless, something new had occurred and for many of the people, if not most, their faith was expressed in the synagogue much more regularly than in the temple and after the destruction of the temple it was the Torah scroll which represented the divine.[590]

The early Christian Church does claim such a revelation in the one person of Jesus, but it is clear that it was in the life of the early Christian community acted out in long periods of living the faith in community, that the authentic experience of the community was articulated. From the beginning this is true. The experience of the risen Christ on the road to Emmaus is obviously for Luke a parable of the faith journey of every Christian. After their personal encounter with Christ,

> "Luke tells of the return of the disciples to Jerusalem not merely to locate this one anecdote in the totality of the apparition tradition...More important may be his wish to emphasize the Christian situation, what must happen to all those to whom it has been given to recognize Christ in faith. The demand of this situation may be summarized in two words: *community* and *testimony*. The disciples do not experience what has happened to them as a merely individual concern: straight away they return to the community in Jerusalem. The message they have received is destined for all...It is only through the convergence of complementary experiences and apparitions that, in the very young Church, there emerged a dawn of recognizing certitude."[591]

Over the centuries, this growth in recognition and understanding, often expressed in dogmatic form was frequently in response to heresy, and again it is the community that responds. It is on the level of dialogue, whether amicable or not, that the truth is hammered out. Jesus himself was not an isolated individual giving voice to a totally new revelation. He was firmly rooted in his own community and what he said and did sprang from a very Jewish faith.[592]

THE LISTENING COMMUNITY

It takes long periods of "comparing notes" to gain some understanding of the meaning of the subtle influences of the presence of god to the human brain. It takes a process of winnowing to discover and remove or lessen the wishful thinking and the assumptions that have been imposed on this experience to understand what is authentic. Maya Angelou, though speaking of individuals, could well be speaking of the whole human experience of god when she says,

> "I'm startled and taken aback when people walk up to me and tell me they are Christians. My first response is the question, `Already?' It seems to me a lifelong endeavor to try to live the life of a Christian...It is in the search itself that one finds the ecstasy."[593]

This is precisely the understanding behind the Church's idea of Tradition".[594] While there is certainly a tendency among the hierarchy to restrict this communitarian avenue of revelation to themselves,[595] it has always recognized that it is the whole community of the Church which is actually active. "Even as authoritative voices within the church proclaim that unless you believe *this* and reject *that*, you cannot consider yourself a Catholic"[596] as a popular explanation of the faith, aptly titled "The Good

Enough Catholic", points out, there is always the other side of the story. The Catechism admits that, "The apostles entrusted the `Sacred Deposit' of the faith...contained in Sacred Scripture and Tradition, to the whole of the Church."[597] It is considered the role of the leaders to interpret and legitimize what the community as a whole has come to understand, as well as to root out those human interjections that will inevitably creep in, even through the bishops and authoritative voices themselves. And in practice if not in the theory that some authorities would like to have consecrated, this is the way it has always worked. The bishops follow the lead of the laity until it is time to lead. Then in another long communitarian process the pronouncements are hammered out, not always in a spirit of love and patience and humility as they should be, but sometimes forged in the heat of battle complete with denouncements and excommunications until what has been there all along is recognized. And what has been there all along is not a "deposit of faith" but the presence of a living god.

The convocation of the Second Vatican Council can serve as an example. John XXIII was certainly not born a Pope. He grew up in a family and in a culture that was trying to be attuned to god's presence. None of the conclusions of the Council, from a recognition that god works in *all* the "People of God", the laity as well as the clergy,[598] through the revisions of the liturgy that this brought about, was done in a vacuum. All of these things had been growing and fermenting in the consciousness of the community for a long period prior to the Council's pronouncements. And not only the community of Italy, but the whole Church; from Taize in France to Hellreigel in Baden, Missouri; from Scriptural ideas of Source Criticism to reform of Canon Law. John XXIII was at least unconsciously aware of most of these and consciously aware of many. Yet he encouraged the myth that the calling of the Council was in response to a special intervention of god the Holy Spirit. It was indeed an act inspired by the Spirit, but it was in the usual manner of human growth in understanding, that is, gradually, through the process of the whole community coming to a recognition of its rightness.

THE QUESTIONING COMMUNITY

True, it is often one person who gets the credit, and sometimes it is even one person who actually distills the growing awareness into a usable form. Galileo was hardly the only person with the facts about a heliocentric system, but he does seem to be the person through whom the facts came together. And despite the opposition of some, the community readily accepted his conclusions, including those that led to a different perspective on god, because the community had already come to the same conclusions.[599] Not in a functioning form, perhaps, until Galileo put words to the feelings, but by and large the human community was ready to accept a wider view of god than previously. Francis of Assisi[600] was not just a bird lover. He was a new force in the life of the Church of his day and his revitalizing of the monastic tradition had far-reaching consequences. Yet he succeeded, not just because of his own charisms, but because the community was ready to accept what he had to say. This has been true of reformers through the ages.[601] Darwin certainly ran into much opposition, but his diminished place for humanity was not intolerable since the community had already sensed it. This is not to say that every human being was ready, nor that the whole of humanity is ready yet to accept all the implications. Nor should we say that Darwin's findings were the result of revelation from god, though we would not want to deny that possibility, either. It was the result of a greater understanding of the universe of which we are members, quite independent of god. The medieval concept of the universe was constructed because of their ideas of god. If god is a creator in the literal sense of Genesis, then the world must be constructed in a certain way and we must interpret the findings of geology and paleontology in a manner that is explained by god's creative activity. If god is perfect, then the heavens, which are in the realm of the divine, must be perfect, which is to say circular. If god is a sustainer and involved in the most minute operations of the physical world, then there need be no evolutionary explanation for why we are shaped the way we are, why our backs have trouble supporting an erect posture, why we have an appendix and a tail. They are this way because god wanted them so. However, if the community has come, more or less consciously, to the realization that god is other than was thought, then other explanations for these phenomena are acceptable, at least in time. And it has been hardly any time at all since these alternatives such as a less-than-perfect universe, a universe in evolutionary process, and humanity as a development in a small section of the cosmos rather then its center, were proposed and they are already accepted in the main. The readiness to believe that there could well be a superior race of beings that will arrive in their saucer-shaped craft to either enlighten us or eat us would certainly not be found in a medieval audience. It would not only not fit in with their world-picture, it would not fit with their god-picture. Today it does, to the delight of the movie makers and the chagrin of the scientists when it is taken uncritically. The unbeliever will say that humanity has just bowed to the facts in making these changes in attitude, though reluctantly. The believer will say that is probably true, but god has also been subtly revealing godself and we have come to recognize it more deeply, or rather that god has been showing godself all along and maybe helping us understand our own universe as well, but we are slow to pick up on it and quick to add our own interpretations. The believer is only briefly cowed by our new knowledge of our insignificance to the universe because god has not only been experienced as more than a king sitting on the throne above the seventh crystalline sphere, god has been experienced as one who loves us. True, if god loves us, god also loves the extra-terrestrial with a taste for us. God is also revealing godself to them, if they exist, and is showing them a better way to exist and relate just as god is doing with us.

GROWTH

This is indeed the goal of god's revelation. Not merely to reveal god's existence, but to reveal the potential in our existence as well. C.S. Lewis has a very high devil explain it well: "He did not become one of them and die among them by torture - in order to produce candidates for Limbo, `failed' humans. He wanted to make Saints; gods; things like Himself."[602] We grow as a species by exposure to god, just as we grow as individuals by exposure to others. Psychologists explain that, "A person's use of moral principles to define the self is called the person's moral identity...[and] moral identity drives behavior."[603] The moral influence of parents first, then peers, then other significant models encourages this self identity as a moral person who makes moral choices. God's influence likewise is not authoritarian from on high but encouraging and enlightening, just as a good parent. This is the basis for moral laws based on god's revelation, not to restrict our humanity but to expand it in the right directions.[604]

GROWTH GOES BOTH WAYS

Of course we know nothing of how god knows our response, nor for that matter how god is aware of the existence of the universe. This is certainly beyond our investigation and maybe beyond our understanding in any case. And if we are able to perceive the influence of god in our brains, it is certainly possible that god can perceive the action of our brains in god's own consciousness. This seems to be another reason for god's revelation of godself, to initiate a dialogue in which we can grow, but also in which god can grow. God, too, can benefit from exposure to other forms of being, and even more from being in relationship with these beings, not just a spectator. All the references to god taking delight in god's people are more than poetic license. We may have been startled by the image in the Psalms of god making merry, as one does over booty, or dancing with joy over god's people. This is not the god of the scholastics who, being perfect in godself, has no way to learn anything new, no way to realize something about being that god did not know before. The conclusion here is that we can allow this fascinated god to be godself, and not have to contain all perfection and being already. There is room for this god to discover, to nuance, and to take delight. God can take such happiness from relating to our universe, and others, not only by observing from afar, though surely god had to be content with that before there was anything here capable of responding. How much more fun the child is when it has reached an age when it can relate to us as a personality. Then we can teach him or her, but we can also be instructed by what they do with our teaching. In the same way, god is not only the teacher but also a partner in the process of discovery about the possibilities of being in the universe and in our case in the possibilities of human existence. We can take on something of god, which we have always hoped to do, and god can take on something of us.

FAITH, NOT PROOF

Can any of this be proved? Are we dealing here with a phenomenon that belongs to this universe and hence is open to investigation by human intelligence (or any other kind of intelligence which is the product of the universe itself)? Obviously not. What are the effects of god's action on the universe that can be tested? If they are physical and use the same physical laws of interaction of phenomena, they would be indistinguishable from causation by natural effects. If they are genuine "miracles" they may be noticeable and distinguishable, but few today really believe in the existence of miraculous events despite the vocabulary used by some, particularly in the medical field, the movies and among the fundamentalists. To speak of a miracle cure is understood to mean that something has occurred that is beyond our understanding, either because we are not aware of the causes or because the causes are governed by chance and are intrinsically unknowable. Even if there were genuine miracles, they are unusable as demonstrations of god's activity to those who do not accept them as such, which is the majority of educated people today. No, the presumption is, and ought to be, that god does not intervene, and as I have said, cannot intervene, through influencing the physical world in any manner open to investigation through the means presently available to us, which is not to say that it does not occur through means that we have not and maybe cannot investigate.

If god has been influencing human consciousness in a "non-physical" way throughout the ages, would there not be some detectable result of that influence? If god has been leading us for millennium after millennium to live more and more in a covenant relationship with each other, should we not be farther along than we would have been without such guidance from god? Yes, we should be and likely, though arguably, are, but how could this ever be measured? At what level would we be if left on our own? Still clubbing our women and carrying them off to the nearest cave as the popular myth has it?[605] Altruism springing from the totally natural processes of evolution certainly seems capable of having brought us to our current level of morality, such as it is. In fact this has always been a telling argument against the existence of divine activity in our world. If god were acting on us with supernatural power, why are we not much farther advanced than we are, which often seems little advanced from the caveman and woman except in the technology of the eight pound wonder vacuum cleaner with which we clear the debris from the cave floor. God seems singularly ineffective in bringing peace to our world and inspiring advancement in the level of love that god seems intent on making possible for us. We so easily slip into primitive behavior of which the mountain gorillas would be ashamed. What is the benchmark we could set of progress through human capabilities alone and what would have to be attributed to the action of god? If god's action is on human consciousness it would seem like this should be open to measurement. If god's presence has an effect in the human brain, would it not be similar to the effect another presence might have? If there is such a thing as ESP, though to date there is no credible evidence that such exists, it would be an indication that such communication is possible. If such evidence were found, how could the influence of god be distinguished from the influence of some other agent or agents unknown quite removed from the recipient? If we could prove that a person sitting in the next room can influence our thoughts at a distance, but was not now doing so even though we were being influenced by some other external agent, how would we know that second agent was god and not someone in the next building, or the next city, or the next planet? We would not.

No, trying to prove the existence of god and god's influence on us has not worked by investigating the physical universe, contra Paley and the new Catechism, and trying to prove god's presence through other means is not going to provide any answers either. We are left with what the Scripture says we will be left with, Faith. In 1968 W.H.Van de Pol in *The End of Conventional Christianity* saw this direction when he said,

> "Christian faith in God has always been understood by the Christian as an answer to the way in which God has revealed himself in Jesus Christ. Without this revelation belief in God in a Christian sense would be impossible...The Christian has

> always been convinced that faith in the Christian sense is not a human achievement but a gift of God."[606]

He then proceeds to examine philosophical attempts to explain god, ending with the conviction that in our modern understanding, the traditional god is dead, but concludes,

> "Since the death of God, within the limits of the human condition, Jesus takes the place of God. But what must be kept in mind…is that Jesus as God's representative also keeps the place of God open. For belief in Jesus involves both death *and* resurrection…However, this resurrection in which Christians of all times have believed is not…a historical event…The resurrection is an eschatological event…The resurrection must be believed. It cannot be ascertained or proved."[607]

We have seen that looking to the physical world for some kind of proof of faith is futile as is the converse, trying to say that the content of the faith is somehow provable using the scientific method, that the realities the faith describes are somehow "facts" in the empirical world.[608] Paul had perhaps seen this way before our times and said to the Christians in Rome, "In hope we are saved. But hope is not hope if its object is seen; how is it possible for one to hope for what is seen? And hoping for what we cannot see means awaiting it with patient endurance."[609] "Faith is a gift of God," says the Catechism, though it goes on to ruin its own insight by adding "a supernatural virtue infused by him."[610] God did no infusing into our nature on any ontological level and it is an error to phrase it in this way, though just how it ought to be phrased is a good question.

The believer is going to have to give up all hope of proving our case except in the one way that the Scripture asks us to, by the lives we lead, but then this may be the most effective way. We have tried to command belief through coercion, but saying one must believe in a loving god or we will stretch you on the rack is hardly a good method, though it may be compelling. We have tried to argue people into belief and have called them dolts and worse for not believing the evidence of their own eyes, in not believing the arguments from design and the necessity of a prime mover. Now that it is perhaps only the dolts who believe that these *are* good evidence, we are reduced to complying with the Lord's request that we show that we are disciples by our love for one another.[611] We are forced to prove that god changes human history by allowing god to change *us*. No one needs to prove to one who has experienced god that they are a better person for it, or at least a different person, and by extrapolation that the human race is better for it, too. If we are not farther along the road to reaching our potential, though I believe that we are, it is not because god has the world but because we have not allowed god to act within us.

DISCIPLESHIP

This may mean that the Church will be a lot fewer in numbers if it is comprised only of those who believe they have been called to be disciples and are willing to live by the gospel values, notoriously difficult to live no matter how satisfying the results in the long run. The numbers may be fewer who are willing to commit to a life of seeking growth in the ability to love and respond to love. No longer will we have the presence of those who have been argued into membership through specious logic and bad science. No longer will we have the membership of those who are trying to bribe god into exercising some of god's fabulous power to override the normal course of things. We have not been a great gathering place for business people for some time, intent on making contacts with other members of the business community behind the pillars of the cathedral, nor of those seeking votes from a monolithic bloc of Catholic voters, but there will be even less from the business community when the Church is composed of those who are probably not powerful in the economic sense and whose political persuasions are not those of self-seeking Pac's. Then, just as Paul said, "his power is shown in our weakness."[612] The Scripture also points out that it is just such people, like Jesus of Nazareth, who fascinate god the most. And let us not sell short the majority of human beings. It is also possible that the more the Church is what it is called to be, the more attractive it will become, not less. We often remark about the hunger for the transcendent in our society today and bemoan that people look for it in X-Files and crystals and gurus from the hinterlands. Timothy Ferris laments that, "With the decline of organized religion...millions of people began investing their religious impulses in science fiction fantasies about superior alien beings aboard flying saucers."[613] When it becomes apparent that god is not to be found in magic and the paranormal but in the human heart and mind, just where god always said to look, who knows what will be the result.

GOD'S PLAN

Perhaps we are deceiving ourselves and there is no influence on us from a loving and covenanting god, just the dictates of our own brains. The people of faith must answer that question honestly, first to themselves and then to others and the answer must be "possibly". Any other answer is dishonest and is not an answer of faith. The non-believer must also be honest in their assessment of the possibility that believers are deluded and the answer must be "possibly not", but the evidence that could convince them will be the lives of the faithful, not how different we are from the rest of the animals in the composition of our brain and its emotions.[614] People often ask, "what does god want from me?", or "what is god's plan for me?" This may well be coming from a belief that god has an already-worked-out plan for everything, including each human life, but someone who does not subscribe to this thinking may well ask the same questions. What we have always been told that god wants for each person is that they be authentically themselves. Finding that self is a function of the brain and is often directed if not determined by genetic constitution. Yet many believing questioners find god involved in the process as well. Some, of course, feel that god always intended them to be a nurse or a mother and some say it is the creator-god who implanted that tendency within them.[615] However, most report a more general direction, leading them to use a talent they have discovered within themselves, or even an area which they just enjoy doing, "to the best of my ability", or "not only for myself but for others", or "to repay the debt of what others have done for me." All of these generalized feelings we might use in choosing a direction in life may be and probably are genetically influenced by both genes and memes, but they are also open to the influence of god, believers propose, an influence to which they try to open themselves.

MORAL GUIDANCE

Earlier I spoke of the basis for morality. While the theologians have spoken about the moral laws written into our essences by a creative god through a natural law, a position that can no longer be maintained, much of popular morality has been based not on such a philosophical grounding but on revelation. The judgments made on human actions in the Scriptures and in the Tradition of the Church were understood as edicts from god. Of course there were some problems with this approach when god was seen doing something seemingly "immoral". Since god could, it is presumed, affect the course of history by direct intervention and the manipulation of physical powers, then when god sent serpents to bite those who grumbled against the conditions in the wilderness,[616] it followed that this "historical" incident revealed something of god's will and god's concept of correct human activity. When the Israelites were ordered to massacre every citizen of a town, god's will got a little murky since god was ordering for god's purposes what would have been grievous sin if done for human purposes. Of course commentators would make exactly that point, that god's purposes were not human and that god's ways were far above and different from human ways, and that the author of life, who makes decisions every day about who will die and who will live, can decide on the death of a whole town, along with all their cattle and dogs, and not be liable to a judgment of sin. How much more sense it makes to read the Scriptures as exactly what they purport to be, the human reflections on a divine influence that has worked throughout their history. It then becomes understandable that, since this influence is so subtle, there will be misunderstandings as well and many "commands from God" were actually our own desires put in god's mouth.[617] It was not god's wish that every enemy would be killed, but ours, and one that we have hardly outgrown. Natural and historical occurrences can be interpreted any way we wish, and have been. The defeat of Jerusalem by the Babylonians, an event I propose was beyond the power of god to bring about, was interpreted as a message from god and it was considered by some to be a message of divorce. God had abandoned the people as the prophets had foretold god might if they did not cease their infidelity. Others, after reflection, considered the event a result of god's activity and interpreted it as a corrective discipline designed to shock them back to their senses. What then, would god do if a people that had been rather open to god's influences and looked sometimes as if they might abandon some of their all too human and negative ways and truly live in covenant with each other began to backslide? God would try to make them aware of the consequences of such actions, just as the pre-exilic prophets said that god was doing through these events, though this required an intervention. God could try to make them more aware through the natural consequences of bad actions. And when disaster did befall them, god would try to influence them to pick up the pieces. And now aware, with god's prompting, of some of the negatives in their culture that had brought them to such a sad end, they would perhaps respond to god with hearts and minds renewed, just as the exilic prophets counseled. It is true that the same disaster may have happened if they had been the best of people. The fall to Babylon was, after all, a normal event in human history and not something brought about by god, but it is also possible that if they had played by god's wisdom rather than human, the invasion and destruction would not have occurred in the first place because they would have been the enemy of no one, including Babylon. In any case, though the Scripture records that god is responsible for the destruction of Jerusalem by Babylon, which the early Christian apologists would also say was true when the Romans repeated the conquest in 70 AD, today we can absolve god of the deed and still admit that the lessons learned were learned in part because of the action of god. God does not cause these things, just as god did not cause Susan's problems, but god can help us respond to them in more positive ways.

MORAL CRITERIA

What then is to be the basis for our morality if we cannot look to some moral code written into us by a creator or taught by a frequently stern teacher? If we cannot see in the events of human history a record of god's intervention in punishing the evil and rewarding the good, to what can we look for our morality? Certainly we can look to our own make up as directive and call this a "natural law" if we wish, but this has nothing to do with god. For example, if we have evolved as social beings then unsocial behavior is not going to be in our best interest. The example of history has always been a guide to wise behavior. Repeat what works and do not repeat what brought disaster. This may have to be the basis for our view of government as a guardian of social and individual rights since other forms of government have not promoted the natural order that has evolved. Jefferson would have to be rewritten to say, "that the human genome has evolved endowed with certain natural rights." Certainly this does not have the same flair, but endowment by a Creator is out of the picture.

Beyond this there is a model of behavior that can guide our own and it is indeed provided by god. In fact it is revealed in the Scripture though not through events, but through the human reactions to those events by those attuned to god's presence. For example, in a previous note we looked at the morality of our actions in regard to the environment and especially to the preservation of other species. If we have evolved in company with these species and there is some benefit for ourselves in keeping this company, then by all means let us do so for our own good. The lessons of history would seem to bear this out. Native Americans in the midwest, the southwest, and in Central and South America, and African civilizations in central and southern parts of that continent have all left impressive remains of their occupancy. Europeans who found them often made up elaborate stories of wandering lost tribes of Israel or Caucasian groups who were given credit for these societies since it was thought that the natives, who were probably being exploited by these newcomers, could not have had the wit to build them.[618] Our present knowledge shows that it was merely a case of the original inhabitants, whose descendants often remained in the area in an impoverished state, having overtaxed their environment to the extent that it could no longer support their population. So the wisest course is to live in balance with the environment. We need no divine revelation to recognize the wisdom of this course. This does not solve the problem, however, since today we could perhaps substitute something else to supply the needs that only the biosphere could provide previously. Could we not then use up the old environment with impunity? What difference if we kill off most or all of the animals that compete with us, or just get in the way of another fast food restaurant or housing development? Let's see what difference.

We are a different species from god, as it were. How does god respond to us? This can be a model for our own treatment of and reaction to other species, to other ways of being alive. God is drawn to us and is fascinated by us. God always treats us as a beloved "other". Granted, god may not be *able* to do anything to exploit us or even harm us, but god could just ignore us. Instead, god is drawn to us for godself, since god then grows in understanding of the possibilities of existence. God not only learns of life from us but seeks to teach us of life as well. God respects us as living beings other than godself. More, there is another element or flavor of god's attraction to us that we can only feel and not define. It is such a complex attraction arising from many motives that the best we can do is recognize some glimmerings of it in ourselves and give it a name, though a name can never truly explain what it is that it designates. And the name of this movement toward the other is "love". That god loves is the deepest lesson of the scripture and the most profound aspect of revelation. So much does this love affect us that we can experience god not as a lover but as love itself.[619] To put it another way, god is drawn to life and life's goodness. John Updike phrases it as,

> "All joy belongs to the Lord. Wherever in the filth and confusion and misery, a soul felt joy, there the Lord came and claimed it as his own; into barrooms and brothels and classrooms and alleys slippery with spittle, no matter how dark and scabbed and remote...wherever a moment of joy was felt, there the Lord stole and added to His enduring domain. And all the rest, all that was not joy, fell away, precipitated, dross that had never been."[620]

This combination of interest, fascination, respect and caring characterizes the most perfect being we know of "in the universe." Never mind if that being is not necessarily the omni-everything that the theologians maintain god must be to be god. Just take god as god is, or at least as humans have experienced this presence. God is still the most perfect being we have ever met. And is that not what we ourselves wish to be, perfect? What better model than this god? So our behavior needs to be based on this archetype of perfection. We too need to be interested in other lives than our own, allow ourselves to be fascinated by other ways of being, up to but not limited to other ways of being human. We need to respect the other and on some level to love the other. We are evidently not capable of loving every other to the same extent that god can, but we are able to love some deeply and other perhaps less deeply, but really. This sounds very much like "anyone who loves themselves will lose themselves while those who puts themselves aside will find themselves."[621] And so god, finding rudiments of the ability to love enmeshed in our innate tendencies to take care of our descendants (genes), encourages us to expand this into genuine care and concern, into love, as god knows it can be.

FOR EXAMPLE, THE ENVIRONMENT

We consider the example of the environment introduced above. At the moment it seems that we have a need for the plants as the basis of the food chain on which we depend as well as for the production of oxygen, control of erosion, as a source of chemicals for drugs, as sources of beauty, perhaps for our psychological well-being (at least as a way to smooth over disturbances in our relationships by giving flowers as a sign of contrition), and for many other purposes. And so we also have a need for insects that pollinate them. Since we do not know how to keep the beneficial insects and destroy those we consider pests, we are stuck with the whole kingdom. We value birds and bats if only because they keep the insects in check. What if we reached a point where we no longer needed them? You rarely see potted plants on the starship *Enterprise*. Would we feel free to do away with birds, bats, and especially insects? Or does god's fascination with all life give us a motive for our own actions? Should we not respect every form of life and even grant it a measure of caring (read love), and certainly the right to exist? If we do not, can we lay any claim to wanting to be perfect as god is perfect?[622] Of course how this should be done will still remain a matter of compromise and setting of priorities, just as it is today, but if we start where god starts, with care for the other, our deliberations will keep on a positive tone.

FOR EXAMPLE, LIFE

There is certainly an application here for the issue of abortion, which has hinged so much on the question of when human life begins.[623] There is not an obvious connection since we have been speaking of respect for and protection of species, not necessarily of individuals, but our respect for the potential of a species as well as the actuality is grounded in the example of god and the potential of a species is based on each individual, not to mention that god's care seems to be primarily for the individuals. I would have to conclude that if we base our morality on imitating god's fascination with each individual, abortion would certainly be questionable. The question of when a fetus becomes human would follow the same path as deciding when on the continuum from non-human primate to human primate some threshold is crossed and I am afraid I have no answer to this.[624]

Birth control has been a problem in Catholic moral teaching but the morality has never been judged on the potential child, though some people almost take the position that there is a soul somewhere waiting for the production of a body to house it. This is not held by the Church or justified by science. The question has much to do with the relationship of the couple having sex and their motivations and morality, not on the rights of an unconceived child.

Population control is also a question that needs some input from science. Just how many people can inhabit the planet and maintain a quality of life that we can agree we need? The means used to accomplish this goal, which is almost certainly going to become an ever more pressing issue, are certainly moral questions to which the Church needs to give continuing attention. Abortion has been declared not a moral means and I would have to agree. Contraception (and possibly the abortion of a fetus prior to some as yet unidentified age) are a different question.[625] My own opinion would certainly be colored by my ideas of sex explained in a moment.

RESPECT FOR ALL (IF NOT FASCINATION)

This proposed basis for morality does not solve all the questions, but then neither does the basis of commandment or revelation of laws, natural or casuistic. Yet if we allow ourselves to view all life as god views it, as a furthering of our own understanding of life, is there not reason to maintain everything and everyone even from a selfish motive? And more, is there not reason to respect as god does each manifestation of the possibilities of existence and even to care for it, whether or not we profit or benefit? The same is true of our relationships with one another. We have all experienced the discovery that another human being has far more to them than meets the eye. Not everyone other than ourselves is as shallow as they seem.[626] They have their own memories, talents and opportunities and they respond to them in such varied ways that there is always something for us to learn about human life from one another. Some, of course, live their lives unproductively by making selfish decisions and our fascination with them quickly wanes, but there is usually much more to the person in front of me than meets the eye. If I wish to live my own life to the fullest, if I wish to understand more about the potentialities of being human, I must be attentive to these other life histories. Indeed, the more I can allow myself to respect the other, perhaps even to care for the other, the more fascinating they become and the more I can learn. I asked a man preparing to celebrate with his wife their sixtieth wedding anniversary if he usually knew what she would say in a given situation or how she would react, having observed her for so long. His answer was, "yes I know her pretty well, but the old gal surprises me about once a week". This surely shows his own fascination with this other person and also gives a hint of how he himself had benefited from this openness for many decades.

FOR EXAMPLE, SEX

The questions considered in previous notes about proper sexual morality can be considered now, but really as just the pattern of all human relationships. If I bring to my own behavior the qualities of god's behavior and apply them to the area of my sexuality, how will I deport myself? I will understand those drives that are a part of my own makeup, having evolved through my genes. And I will respect the genetic drives and motives of the other. For example, while the male seeks many partners, the female usually does not but rather seeks some depth to the relationship with her sexual partner. This tendency in the female may be just a feeling that evolution has placed there so that she will act in a way designed to keep the male around and defend the young, but no matter how it got there and for whatever original reason it is part of her make-up, it is there. For the male to ignore this reality in the woman and treat her as an object with no feelings and needs, or place his own desires above hers is to show a lack of care for her. It also deprives him, of course, of what he could learn from her about the possibilities of human relationship and he will try to flit from flower to flower without ever experiencing the beauty of this flower. And the beauty of this flower may be well nigh inexhaustible, especially since much of it does not even exist as yet but is only potentially present. Much of the fascination will be in watching the person develop and respond to new challenges in new ways. The male who sees, and wants, only the surface of the person is hardly acting as god acts, and therefore hardly acting morally. Is not the sex that he enjoys without really seeing the person, just as sinful as it is when there is a law spelling out what is morally correct and what is not? It is being closed to the person of the other while professing openness that is morally repugnant because of the dishonesty, not because the sexual pleasure is wrong. This approach will obviously allow for a great range of choices and moral decisions in each relationship. What is the minimum relationship required for honesty in the exchange of a handshake? In a kiss? In a hug? In a romp in the hay? What is the minimum in this relationship with this person under these circumstances below which it would be dishonest to engage in this physical expression of unity? And it is not only a question of what I am seeking to express in this physical action but also what the other person is understanding and receiving. This is the eternal battle of the sexes. The male may, of course, not be trying to relate to the female at all but merely be responding to physiological urges to reproduce,[627] and not even be aware of that but be seeking only physical pleasure. This is the level of relationship in prostitution, and is considered rather suspect in most cultures, though its ubiquitousness in human cultures militates against its automatic condemnation under all circumstances.[628] Apart from this kind of encounter, it is expected that there be some deeper relationship with the woman before sexual expression is appropriate. For the male, the relationship may be one of respect alone, or it may be an expression of friendship, or a declaration of a desire to spend some time getting to know the female, or a profession of undying love for all eternity. Or some step in between any of these. For the female, while her sexual activity with the male may also be an expression of the same kind of relationship that the male understands, she may well receive his attentions as an expression of something deeper than he intends. And the male who does not care what message is being received nor what is being sent is not acting responsibly.

GUIDANCE, NOT CONCRETE LAW

However, in the absence of a law inscribed in our nature by god and designed to fit all occasions, we must be prepared for differences in behavior from one person to the next and one couple to the next. We cannot necessarily condemn in another person what might be dishonest in ourselves. We must also be prepared for making mistakes. I may well convince myself that this particular activity with this particular person expresses this level of relationship, only to find myself going farther in my expectations of the relationship than I intended, or than they intended. One of us is going to get hurt, but if it is an honest mistake, let me learn from it and not repeat it, but let me not get mired in guilt over it. Laws are useful, of course, since as much as I think myself, or ourselves, to be unique in all the world, we are not going to be that different from other human beings. Laws merely express the hard-earned wisdom of others in the same situation, and also the wisdom of god influencing the human mind, but they do not reflect an arbitrary edict of the almighty. However, they can be darned handy and timely, whether working with sexuality or with high voltage electricity, which is about the same thing. And of course the female must understand the male and not use him for her personal advantage nor read in more to the relationship than is being expressed. For her to seek to fulfill emotional needs at the expense of the other is just as reprehensible as the other way around. If it is true that the female is generally more interested in the depth of the relationship than the male, this is a blessing for the race, for she can teach him something that his own genetic disposition has not impressed upon him to the same extent. But who says that our genetic dispositions are perfect? They are merely there, and we can nurture them and extrapolate from them to something even better. This is, in fact, the way that god works with us, pointing out the good that we have inherited through the evolutionary process and making it better, and pointing out the harmful that is our legacy and teaching us to counteract it. We can reach some rather sophisticated insights on our own, of course, and the history of sexual relationships of man and woman in various cultures shows the possibilities. Anthropologists find many levels, from the sublime to the commercial, but at its best human intelligence can find quite deep meaning in the addition of sexuality to human relationships as Abel Jeanierre does, saying,

> "The reciprocity of being-for-another is achieved through perfection of being-within, lived in the mode of intimacy. And sexuality is revealed in this dimension as a universal structure of the human being in genesis. Its development discloses what it is in principle: transsubjectivity, the real and concrete basis of all interdependence. Thus sexuality is at the source to the degree that each can only be given to himself as responsible for the other, each is himself only through the other...The person transcends himself fundamentally, and this is revealed by sexual differentiation."[629]

Though a religious person wrote this, it is a conclusion that human beings can reach independent of a faith perspective. It is also a place from which god can lead us to something even deeper, (as if the citation needed deepening). A recent catechism makes the same point when it says,

> "How far from merely avoiding sin is the free and joyful consecration of two Christians who are already totally committed to Christ! They see each other as complement one to the other. Love is the glad fulfilling of another's needs, and sexuality is an obvious sign of human incompleteness. Human life does not begin or continue through one person. Life, and the beginning of life, demands com-munion."[630]

Evidently god is getting through. This is not the kind of treatment of sexuality that one would find in the past.

FOR EXAMPLE, RACISM

Racism is another topic of moral concern as well as political concern and since we can no longer talk about any human being as being made "in the image of god",[631] on what basis do we relate to one another? Certainly science has something to teach us about this since the genetic makeup of human beings is so similar that it is obvious that we are all closely related to one another. There is more genetic difference between a man and a woman than there is between two men of different "races". Race is, of course, not a proper word for the differences among humans for it has a specific meaning for scientists to distinguish differences within a species and we are not nearly dissimilar enough to correctly use the word "race" to describe us. God can certainly inspire us to use this information and put aside our prejudices.

On the other hand, there certainly are cultural differences that often coincide with the so-called races because the races developed through geographical dispersion, the same way that cultural differences develop. Culture is values, among other things. Would it not be to our advantage, not merely to our amusement, to open us to these values by exposure to the way these values are transmitted in that culture. The Mexican Day of the Dead is certainly interesting to those from other countries, and the value of respect for family and continuity with the community is worth reinforcing and remembering. If god learns from and values all these human differences and the individuals who embody them, can we do better than imitating god?

The same pattern can be used for every moral question. How should I relate to this person begging for a handout on the street? As a person, not an object, of course, just as god relates to them. How to relate to this mentally impaired person? To this dying person? In the same way. And how should they be encouraged to relate to themselves? With the same fascination in how they are responding to the challenges of life, and of death, with which god regards them. "Love your neighbor." And while you are at it, love yourself as well.

HEAVEN AND HELL

This may be a good time to speak about the question of heaven and hell, since one or the other has been considered the consequence of our moral decisions. This relates to the question of continued existence after death, but let that ride for the moment. If morality is taken as response to divine laws or revelations, then the image of judgment enters in. Some decision will be made, by god or by the life of the people themselves, as to whether he or she deserves reward or punishment. However, if god is not a lawgiver, either of the laws of physics or of morality, but rather is an observer and advisor, there is not a question of judgment on how we have or have not responded to god's edicts. Let's reverse our order of consideration from understanding how god responds to us as an indicator of how we should respond to one another and observe how we respond to one another and see if perhaps this tells something of how god might respond to us. If I am in fact "fascinated" with another person it can be through interest in their goodness or in their baseness. The James Bond stories are fascinating not only because of the hero but also because of the villains, some of the most creative and bizarre in the genre. People sometimes profess that they would rather go to hell than to heaven because the really interesting people will be there. Still, if we had to pick our neighbors or our spouses from a list of heroes and villains we would surely pick the good. We might be perversely interested in the scoundrels, but would hardly want to spend our lives with them. Perhaps god reacts the same way. The good, god would want to preserve; while the evil can be allowed to disappear.[632] Just as we may attend the funeral of crabby aunt Hortense, we are not going to miss her and would hardly want her back. [633] On the other hand, aunt Hermione, whose company we always enjoyed, will be missed and if we could keep her with us we certainly would. If god does somehow preserve us in existence, god, like us, would want to keep those who are good. Since neither god nor we are attracted to the evil and negation of life and goodness that is possible in human existence, then hell may well be non-existence rather than continued existence in some bad place. On the other hand, people can grow out of their negativism. "Jaws", the man with the stainless steel teeth, not the shark, changed from an enemy of James Bond to an ally. Could not his villainous boss, Goldfinger, change after death if given the opportunity? Their potential for life and goodness is fascinating, maybe more so in contrast to the actuality, even if their state at death is not positive. In recent decades, probably because of a renewed interest and therefore deeper understanding of the Bible, the actually non-scriptural emphasis on god as a god of punishment that many were taught has been less believed in at least in the more liberal circles. There is less thought that more people will end up in hell than heaven, in favor of the idea that a loving god would find a way to redeem most if not all rather than losing them, even if they fought god pretty hard. There is also the finding that much of what we do is much less under our conscious control than we thought. If so, there is much less culpability. I am saying here that hell understood as non-existence makes more sense than god constructing some place to send the bad guys. God's fascination with life, even life not so well spent, may be more deep than ours and perhaps god finds a way. The whole idea of Purgatory in the Catholic tradition seems to testify to this faith that god finds a way, even after death, to help us grow toward our potential.

So what is heaven and what is hell in this new scheme of things? Both are going to remain what the faith has always said they were, either perfect union with god or perfect separation from god. However, all the simplistic ideas about heaven, and hell as well, are going to be discarded. If we continue to exist as ourselves, with all our memories and our consciousness and our own personal individuality, but now contained in the being of god, then two things are liable to be true; we will have our own personalities and all the potential of a human being to experience existence around us and within us; and since our environment has changed, we will have the potential to experience also the existence of god and all that god experiences. We will be caught up in the activity of god, which is to live and develop one's own existence and to expose oneself to any other form of existence possible. God, of course, has already experienced humanity and its history as well as all the other existence that came to be in our universe and all that is now available to us as well. In addition, for all we know, god has experienced innumerable other universes and all the ways of existing that came to be in those universes, likely to be quite different than our own, and these too are available for us to experience as well. There is no reason to suspect that ours will be the last universe to spring into existence from the quantum vacuum. Those "following" us (judged from our own internal perception of past and future), which god would surely want to investigate and experience, and perhaps influence as god did our own universe, we will experience as well. How active we might be in regard to these others we have no way of knowing, but in such a condition we may well be perfectly content to just "be" and grow to "be more" and will not feel the need to be involved and active in the affairs of others. In our own time and universe, humans do not seem content unless they are changing things (although even then we are rarely content), but perhaps by that stage of our development we will find other pleasures. In addition, we have always insisted that god is not contained or constrained by time. How we would experience this in our own personality is anyone's guess, but time is certainly a limiting factor in human life. However, the Church believes in eternal life, not everlasting life. Everlasting has an element of time about it, but eternal means to exist as god exists, without time. Its absence should be interesting.[634]

On the other hand, hell is missing out on all these things because we no longer exist. Of course the person that once existed will not suffer from this lack, having no existence to experience anything. It is only from the perspective of those who exist that they lack anything. Besides, this would be the normal course of our existence anyway, so it is not that they would be missing out on something to which they had a right. The reason we want everlasting existence is because this is the means that evolution, whether of genes or memes, has hit upon to help us deal with the awareness of mortality that consciousness brought along with it. It is obvious that we could have gotten along without it since many an ancient Hebrew or modern Jew, and many a scientist or agnostic or atheist has lived quite well without counting on life after death.[635] To find that continued existence is something that god has in mind for us will be a very pleasant surprise.

RESURRECTION

What of resurrection from the dead? Without a soul, much of the popular myth is no longer possible. This myth had it that when a person died their "life force" or essence soul or something that is vitally "them" continued in existence, an idea that seems to again be popular in our TV culture, if it was ever unpopular. This "something" seemed to function in all the ways that a living person did and was spoken about and spoken to in the same terms. This presented something of a problem to orthodoxy because, even though it admitted of a soul which was a necessary constituent of a human being, the authentic teaching of the Church was that the physical was also essential to being human. The Catechism says, in Thomistic terms,

> "The unity of soul and body is so profound that one has to consider the soul to be the `form' of the body: i.e., it is because of its spiritual soul that the body made of matter becomes a living, human body; spirit and matter, in man, are not two natures united, but rather their union forms a single nature."[636]

So what are we to make of the existence of a soul without a body? Mostly the question is ignored and the soul is treated as capable of thought without a brain, sensation without senses. However, some theologians, unofficially, taking their cue from the scripture and from a recognition that the concept of "soul" has more to do with Greek philosophy than biblical concepts, hold that the person does indeed cease to exist until resurrected by god, body and soul together. The official position is that the soul is immortal[637] but the idea that escaping death would be a result of god's intervention rather than inherent in human nature is tacitly approved. Another way some approach the problem is to point out that since time is irrelevant with god (and relative with us), who knows if there is a "period of time" from our relative view from the death of the person to the resurrection when the soul is not in existence, but I, and I hope we, are left from this study with no necessity of positing a soul.[638] We can fall back on the relative-time answer or we can just state that god raises the whole person at the resurrection, and that prior to that they just do not exist, but why god would raise us, and how, are certainly questions worth pursuing. The "why" is not hard to figure. Shakespeare's Hamlet is struck by the incongruity of death when he makes his famous, "Alas, poor Yorick! I knew him well" speech and forms an image of the remains of Alexander the Great plugging a bunghole or Julius Caesar as chinking in a wall, but he knew Yorick and one gets the impression that he would really like to have him back. Just as we can be fascinated by another person and wish to have their company, so is god. While death will take the person from us and we can do nothing about it, though we would wish to, so god wishes and god can do something. But how? There is no problem in believing that god can keep in existence a being that god brought into existence in the first place, but as we have seen there is every indication that god had no role in bringing us or anything else into physical existence and probably has no ability to do so. So how can god keep us in existence? The final answer is going to be that we do not know. For that matter, it is an article of faith even in the traditional view that this will occur and that god has promised it, but that the how or when is of faith and faith means that there is no proof. Could it be that this belief really is nothing more than our projected wish to continue in existence onto our experience of god. If this is true, however, then the faith is truly useless and outmoded. A reviewer of James Wood's book, *The Broken Estate*[639] points out that,

> "...as Wood notes so devastatingly, unless Christianity is preached as `true,' as God-given salvation, without which we will be lost forever, it can't win souls or claim any authority in the world. `We need eternal life, or we are lost...Theologies that deconstruct this need eliminate Christianity's uniqueness, hence its power, hence its existence."[640]

However, who says that we are not "lost" and that Christianity is not without authority? Is that the conclusion to which we must come from the present investigation? If we base our faith in resurrection on the existence of a soul, then it is. If we base our faith on god's ability to keep our physical bodies in existence beyond death, then Christianity has nothing to say, but our faith is that god has promised us life after our death, and we must ask how could god bring this about without soul or physical body.

One theory on the possibility of resurrection is Frank Tipler's, which we have seen. He looks at the possibility of our engineering our own immortality through programming a universe-wide computer to recreate each individual person who has ever lived. If this is possible, something that many physicists would dispute, will we have the far-sightedness to bring it about? Will we allocate the resources that will be needed to make it happen? Given our history to the present, the likelihood of our doing so is slim indeed. Many people today believe that if they live a certain way, and not live in other ways, they will "go to heaven" but are unwilling to so live. To get a significant part of the human race, or even a majority of politicians in one country to spend money now on something that will not bear fruit for several billion years or longer would seem hopeless. If Tipler's scenario is indeed possible, the god of traditional theology has no role in it. However, might the god to whom our reflections have led us be able to capitalize on this? Tipler's theory itself, or a development of it, may be the result of god's guidance of individuals, including Tipler, to discover this possibility for resurrection. It is just this kind of thing, this inspiring of new ideas and possibilities leading to new realities, that the believer finds god bringing about in the history of Israel and the Church. Would anyone expect such an insignificant people as Israel to have the immense influence on human history that they had? Would anyone expect it from the Christians of any age? Or the Buddhists or Muslims, for that matter. Yet, with the influence and guidance of god these things have occurred against all odds. God would not even have to know ahead of time how we could keep ourselves in existence, or rather recreate ourselves. God need not be an expert in cybernetics to inspire us to seek life, though presumably god could figure it out much more easily than we have. Or maybe god has figured it out and is leading us to build Tipler's "Resurrection City". Or perhaps god has had experience from another universe so as to know that such a thing is possible for these other beings than godself. Our belief that god is leading us to life everlasting is satisfied no matter how god does it, even if it is by inspiring us to do it ourselves.

REAL RESURRECTION

There is another possibility that is much closer to traditional faith and avoids a life after death that is not truly eternal, but only "everlasting". In Tipler's scenario, each individual is "reconstituted" through the re-creation of the physical memories and patterns of every individual who ever lived, and to be on the safe side, of every individual who ever could have lived. Since our identity is completely bound up in the physical and there is no spiritual component of a human, and since nothing of the physical remains except the atoms, every one of which may also have been used in the body of another human being, the theory does not envision anything of the recreated person being physically identical to the original. However, since the memories of the once flesh and blood person and this computer double are identical, it is the same person. Tipler has all this occurring in the memory of a vast computer, but his point is that if my memories are contained in a computer, this is really "me" since "I" am nothing but the memories and experiences I have contained in the computer of my brain.[641] Change my brain with a stroke or by driving an iron tamping rod through it and "I" am not "Myself" anymore. On the other hand, god has consciousness and knowledge and learns and therefore has memories, though the process is undoubtedly entirely different from our evolved, physical, process. Still, memories are memories. Is it possible that god keeps us in existence and "in the presence of god" by allocating god's own memory process to our configuration of memories? If something of god knows what I knew when I died, remembers the same experiences as I did, and more, and is capable of learning even more, and if this pattern of memories is conscious of an individual identity, then it is me. There have always been philosophies and theologies that suspected that we continued our existence as a part of god. Hans Kûng concludes that,

> "Jesus did not die into nothingness. In death and from death he died into that incomprehensible and comprehensive absolutely final and absolutely first reality, was accepted by that reality, which we designate by the name of God. When man attains his Eschaton, the absolutely final point in his life, what awaits him there? Not nothing, even believers in nirvana would say that much. But that All which for Jews, Christians and Muslims is the one true God. Death is a passing into God, is a homecoming into God's mystery, is assumption into his glory."[642]

This and similar conclusions were flawed because they presumed that this was so because we had begun as god or a part of god or a spark of the divine or, more orthodox, from god's creative and purpose-giving action. Kûng tipped off his presumptions with the word "homecoming", assuming that the choice must be that god created and is the ground of all being or that there is no god. He says in his concluding section,

> "In terms of natural science, the evolutionary process as such neither includes nor excludes a first origin (an alpha) and an ultimate meaning and goal (an omega). But even for the natural scientist and the medical expert, for the historian and the social scientist, there arises the existential question of the origin and meaning and goal of the whole process, from which he cannot escape. It is my decision of trust or mistrust, my decision of faith, as to whether…I want to accept an ultimate groundlessness and meaninglessness or…a primordial ground and primordial meaning of everything, in fact, a God who is Creator and Finisher of the cosmic process, as the Christian proclamation assumes."[643]

Kûng would admit, I am sure, that just because I decide to trust does not make the object of my trust real. So, I must point out that we did not arise from god or god's action. We are our own being and a part of the universe that is independent of god. We are our own Alpha, but this does not disqualify our being a "part of god" in our resurrected form. We are certainly not our own Omega.

And this is what the faith has always said: "somehow", god keeps in existence those whom god favors; that existence is of the human personality; that it is a truly human existence and not that of an angel or some other life form; that it is outside of time; and that it is the result of god's gratuitous action and not an innate quality and consequence of human life, actions or nature. Of course we do not like this, wanting to be in control of our own destiny and personally and inherently immortal, and so other explanations of the process of salvation from this life and the beginning of another life have been set forth, from reincarnation to Tipler's *The Physics of Immortality*. A new-old wrinkle was forcefully shown in 1997 with the suicide of 39 members of the "Heaven's Gate" cult in California and the subsequent addition of more deaths later. Salvation for these cultists consists in liberation from the confines of this body and this planet, shades of Plato. The "away team" leaves spaceship earth for another level of human existence among the stars. The instruments of this redemption are not divine, however, but a higher life form of the universe cruising in a mother ship behind a comet, though why they would have to hide themselves from the puny humans is anyone's guess. Presumably this salvation is not eternal since the redeemers and givers of new life are themselves products of the universe and therefore eternal life is not theirs to give, but when you are talking about an existence thousands or millions of times longer than the human life span who is going to quibble? The same is true of the search for continued life through cryogenics, freezing oneself, or at least one's head, until you can be brought back to life by an advanced civilization, presumably one's own centuries hence. Shooting your body off into space alá the wealthy businessman in Sagan's "Cosmos" is another manifestation of this desire for some approximation of eternal life without the intervention of a god. Of course, some of this is possible. If the human brain and personality can be contained in a computer and if the human body can be replaced by bionic components, then human life can be extended indefinitely. Not eternally, certainly, for there is no practical eternity in the universe. The traditional faith that ultimately there can be only one source of true eternal life and that source is god, is the only satisfying response to the question of death. Dom Sebastian Moore points out that the death of Jesus did not have the desired effect of reducing his "kingdom" to ashes,

> "...for the Kingdom had not owed its life as an idea to the man, in the same way that it would have done if he'd *said* 'I'm the King' and had fallen at the head of the defeated insurrection. Jesus had not identified the Kingdom with himself: he had identified himself with it."[644]

Jesus did not rely on his own power to bring about the kingdom, much less to bring about his resurrection. And Jesus did not rely on a continued existence in the kingdom of this universe, but on existing in the Kingdom of God. Jesus relied on living in god. Envisioning god as keeping us alive by, in a sense, allocating a "part" of god's memory to us, which thereby becomes "me" is just as good as putting a new garden of Eden above the seventh celestial sphere and letting me romp eternally through the Elysian Fields. Maybe better! True, "I" would not be a physical being anymore, made of the same stuff of the universe, though I may not know this and may well experience myself and others as physical beings, but if not, that seems a small price to pay.

JUDGEMENT DAY

What, then, is to be the judgment on our lives? How well we adhered to a moral code imposed by a creator? Or worse, arbitrarily imposed by a tester? Is there a certain level of moral responsibility that we must reach or any standard to which we must measure up? Many have thought so and were very willing to relegate to eternal punishment those who did not meet their criteria, often involving avoiding anything that might look like fun. If my assessment of god and god's relationship to sentient beings is correct, then there is good news and bad news. The bad news is that it *is* an arbitrary judgment on god's part. There is not, and there cannot be, a code of conduct for every conscious life form in the universe or in every universe since our evolutionary histories will be so varied, our challenges so different, and our starting points so diverse. What is innate, and hence mundane, in one universe may be quite remarkable if reached in another universe. Indeed, this is so between any two human beings. One who has all the advantages and goes nowhere is much less interesting than one who starts in the sewer and manages to claw their way up to the gutter. Conversely, one with advantages who makes even more of them is more interesting than one who stays in the sewer. And who is god more likely to wish to keep in existence?

The good news is that god seems to have a low threshold of fascination, or maybe there is more goodness even in the worst of us than we realize. Maya Angelou records that she,

> "...began to sense that there might be truth in the statement, that there was a possibility that God really did love me. Me, Maya Angelou. I suddenly began to cry at the grandness of it all. I knew that if God loved me, then I could do wonderful things, I could try great things, learn anything, achieve anything. For what could stand against me with God, since one person, any person with God, constitutes a majority? That knowledge humbles me, melts my bones, closes my ears, and makes my teeth rock loosely in their gums. And it also liberates me. I am a big bird winging over high mountains, down into serene valleys. I am ripples of waves on silver seas. I'm a spring leaf trembling in anticipation.[645]

While it is possible to live a life of no use to myself or anyone else, including god, god's reaction to us, individually and collectively, seems to be not only very understanding and patient but also liberating, allowing us to become even more what fascinates god.

WILL THE REAL GOD STAND UP

So we have a god who is physically powerless, but in every other respect seems to be the same god attested in the Scriptures: one who is in a covenant relationship with us, not because of our own worth but because of god's initiative; one who guides and directs us as best as god can; who leads us in developing a moral code which begins with our selfish genes and their induced but limited feelings of altruism and love. Then god leads us to surpasses them and leads us to new definitions of these basically self-serving tendencies. God seeks to share god's own fascination with life and consciousness and lead us to treasure our own life and all its potentialities. We find anew a god who does not want to let us go upon our death but raises us, one way or another, to another form of existence. How does one pray to such a god?

PRAYER

Prayer has been seen as having four "ends": thanksgiving, adoration, petition and contrition. The Catechism takes this approach[646], devoting one whole part of its four divisions to prayer but only about a page to the prayer of petition, and it is reluctant to explain the prayer of petition in crass terms, rather maintaining that the prayer of petition is "...already a turning back..." and that it is centered on trust in the coming of the Kingdom, saying, "we pray first for the Kingdom, then for what is necessary to welcome it and cooperate with its coming." Only in one small section is there much about turning to god for every little thing.[647] It is obvious that the Church wishes to lead us to a higher understanding of prayer. In most Celebrations of the Eucharist the closest we come to begging god to intervene in the workings of the universe is "give us this day our daily bread", apart from the "Prayers of the Faithful" which can sometimes get much more detailed in their expectation that god will intervene in the workings of the world. Mostly, the prayers of the Eucharist do not ask god to change anything but ourselves, and that radically. "The whole movement of the liturgy is a call from the Church to go to God not only with the keenness of our intelligence...but with all the powers of feeling or joy within us."[648] These feelings are directed towards opening ourselves to god, not manipulating god to do our will. This is not restricted to Christian prayer, and Joseph Campbell points out that in all religions, "prayer is relating to and meditating on a mystery."[649] Indeed, the more mature authors on prayer stress this opening of the person to the wideness of god's love and the wideness of the universe as the primary result of prayer. St. Therese of Lisieux speaks of her inability to use "some beautiful formula" and defines prayer as, "an aspiration of the heart, it is a simple glance directed to heaven, it is a cry of gratitude and love in the midst of trial as well as joy; finally, it is something great, supernatural, which expands my soul and unites me to Jesus."[650] Her autobiography is full of prayer for god to change her or other persons but has very little petition to change circumstances or physical objects. Maurice Nedoncelle says that,

> "God answers us by giving us a spirit of dissatisfaction...[This] allows god to effect our deliverance from that narrow possessiveness which hinders us from accepting the inter-elationship of the realities of the universe and from perceiving what they reveal."[651]

This includes the prayer of petition, which he likens to our petitioning of other human beings for attention and favors. The result may go much beyond merely having the other give us "things."

> "Self-development demands that we should petition others, and this petitioning is the cause of self-development...It follows that if God wants us to pray to him, this is in order that we may become aware of him and of ourselves, i.e. that we may achieve our own real being, and this implies an openness to all reality...Apart from it man remains a mere *thing*, and might be caught up into the moving belt of unconscious causes."[652]

By petitioning the other we acknowledge their existence, as well as the (limited) circumstances of our own; we are forced to give some thought to our own priorities and resources; and we acknowledge the need for bringing conscious, efficacious action to bear on the problem about which we make our request.[653] Just as we may ask something of another human person without really expecting a problem-solving solution, so can we ask of god without assuming god will respond by manipulating the situation. We would be just as amazed if god answered the prayer of the grieving mother, "please don't let my child be dead", by raising the child back to life, as we would be if the police officer on whose shoulder she also cries this request were to turn to the child and raise him or her from the sidewalk, alive and well. The narrator in "Sula" by Toni Morrison realizes something she had not understood about,

> "...the women who shrieked over the bier and at the lip of the open grave. What she had regarded since as unbecoming behavior seemed fitting to her now; they were screaming at the neck of god, his giant nape, the vast back-of-the-head that he had turned on them in death. But it seemed to her now that it was not a fist-shaking grief they were keening but rather a simple obligation to say something, do something, feel something about the dead."[654]

They expect nothing from god except that god would hear their grief, even though the grief may be expressed in words asking for a reversal of the dying. Prayer has the power to open us to the reality of life, even though it is phrased in such a way that, if heard as voiced, would close us to the way things really work, or more to the point, the way god really works with us. Karen Armstrong suggests that, "all the world's faiths do not see the sacred as simply Something `out there' but as a Reality that is also encountered in the depths of our own beings", where god is truly to be found. She points out another motive for prayer in that, "it is also true that people who pray are addressing deep personal needs and fears." In prayer people have dealt with these terrors, and, "have invoked them, described them to themselves in prayer (as well as in art-a related activity), and in so doing have managed to reach beyond them." This same effect can be achieved with a good psychiatrist, of course, but we have seen throughout that god does not do things for us that we cannot achieve by other means eventually. That we have discovered in our day how to do what god made possible in other days by other means is in the same pattern. So god continues to inspire god's people to use prayer to, as Armstrong points out,

> "...acknowledge our vulnerability, our frailty, our failures, and our sins. By putting our unutterable weaknesses into words, we

> make them more real to ourselves but also make them more manageable...This daily discipline helps us to break through the defensive carapaces that we all form around ourselves, thus allowing the Benevolence and Rightness for which we long to penetrate the prisons of our cautionary being."[655]

None of this requires that god be as we have assumed god to be. These same motives for prayer exist for the scientifically literate Christian as for the most primitive.

Annie Dillard reaches the conclusion that,

> "...there is not a guarantee in the world. Oh your *needs* are guaranteed, your needs are absolutely guaranteed by the most stringent of warranties, in the plainest, truest words: knock; seek; ask. But you must read the fine print. 'Not as the world giveth, give I unto you.' That's the catch...Did you think...that you needed, say, life? Do you think you will keep your life, or anything else you love? But no. Your needs are all met. But not as the world giveth."[656]

Some, of course, have used this line of reasoning to excuse god from responding as we wish, but Dillard really does see it not as a "catch-22" but as an appropriate response from god to the petitioner, opening one up to what is truly valuable, in contrast to the trash that the world can offer under the guise of value. Dillard reaches much the same conclusion as I have, at least I think she does. In a somewhat confusing book whose "form is unusual" in her words, she says,

> "God is no more blinding people with glaucoma, or testing them with diabetes, or purifying them with spinal pain [and a series of other human afflictions]...than he is pitching lightning bolts at pedestrians...The very least likely things for which God might be responsible are what insurers call 'acts of God.' Then what, if anything, does he do? If God does not cause everything to happen, does God cause anything that happens? Is God completely out of the loop?"[657]

She asks the right questions about, "Does God stick a finger in, if only now and then...Or is prayer eudaemonistically - praying for things and events, for rain and healing - delusional?" She gives various approaches from Tillich, Chardin and others, but her personal conclusion, as best I can tell, is, "I don't know. I don't know beans about God." But she certainly knows enough to ask the right question and, I believe, to lean in the right direction.

These traditional and primary reasons to pray other than in the grosser forms of petition need not change in the light of what we now know about how the universe works. If one believes in god, then these motives for prayer are just as true as they ever were, including the petition that god would inspire us to be involved in bringing about a change.

In spite of these more mature emphases on the purpose of prayer, the Church still feels it must insist that everything comes from god. Pope John Paul points out a lessening of prayer because,

> "...modern people have an increasingly less religious view of the world and life. The secularization process seems to have persuaded them that the course of events can be sufficiently explained by the interplay of this world's immanent forces, independent of higher intervention. The achievements of science and technology have also fostered their conviction that they already have, and will continue to increase, their ability to dominate situations, directing them according to their own desires...In reality...because they are creatures and of themselves incomplete and needy, human beings spontaneously turn to Him who is the source of every gift..."[658]

Well said, except that seeking anything physical from god is going to be a waste of time. Asking god, or thanking god, for a change in the weather is futile since god has no control over the weather. It has always been problematic whether this petionary form of prayer was proper because one's desire for a certain kind of weather, or anything else is very subjective. The farmers may want rain so badly that they are praying for it on the same day, and maybe in the very same church, that others are praying for clear weather for the parish picnic. What is god to do? General Patton asking his chaplain to pray for good weather so that his army can go on to kill other soldiers places god in another difficult position. Certainly even if Patton's cause is just there are many innocent combatants on the other side, most of whom have mothers earnestly praying that god do everything possible to bring their child home alive. If god is not able to bring about these conditions, for either side, of what good is the prayer? Commentators have always pointed out that the prayer of petition often has the effect of changing the petitioner rather than changing god, and certainly this is a salutary result, but it can be obtained in other less dishonest phrasing of the prayer. "Give us this day our daily bread" has a nice ring to it, but if it really means that we are asking god to put food on our table it is looking at the wrong source. God's food pantry is empty. It makes sense to clean up our language and go ahead and pray not for a miraculous recovery but for the ability to respond well to

Uncle Harry's illness, for example, or to be of help to Uncle Harry in this difficult time, or to not be so overwhelmed by the reminder of our own mortality that we forget about Uncle Harry. Still, a little short-hand in our prayer is alright and we can just pray "for Uncle Harry" as long as we remember on some level what we are really praying for. Thanking god for the food we are about to eat "which we have received from your bounty" is also not an accurate prayer, though it is only by implication a petitionary prayer. A book called *Prayers from the Ark*[659] is very cute with a poem from each animal, but almost all are asking god for something; dew on the flowers for the bird, plenty of little slugs for the duckling (I am not sure what the slugs would pray for), or the hedgehog thanking god for quills. Cute but misdirected. Many of the prayers by Michel Quoist[660] ask god for nothing physical. His illustrations remind the bottom of a wall reminds the pray-er that all that matters is faithfulness. His prayer does not even thank god for inspiring such a thought, it just tells god what is being thought. Rick Hamlin in *Finding God on the A Train*[661] works his way through all the familiar prayer patterns in a very amusing but mature manner, leaving room for all manner of prayer, including finding the Lover god in his own lover wife, and concluding that "Prayer is the only earthly endeavor where trying is enough."[662] The Benedictine Macrina Wiederkehr subtitles her book, *Seeing the holy in the ordinary*, which would lead one to expect finding god in the beauty of a tree or some other phenomenon, which she does, but more in the manner of the prophets who were led to god not as creator of these objects but as known by analogy to their qualities. Though she has the same assumption of god as creator, this enters very little into her meditations.[663] Desmond Tutu, though he includes a section on the prayer of supplication which he begins with the common assumption, "All we are, all we have, is a gift. God is always more ready to hear and to give than we are to ask and receive"[664] still has three other sections of selected African prayers which are not so petitionary. Teresa of Avila[665] is certainly a woman of her times and speaks of many things in her treatment of prayer that we may question today, including devils and angels. A good psychologist would have a field day finding underlying motivations and analyzing some of her visions, but her work is a classic in spirituality and for good reason. The possibilities of growing in our openness and relationship to god are evident in her life and, if we can translate her experiences to our own day, in ours as well. She certainly does see god as the creator and the Majesty who possesses all power and she relates many prayers for physical healing for others. She records that, "I saw fully and clearly the favor the Lord granted me; the next day this person was completely cured of that affliction."[666] Still, the bulk of her book, and of her prayer life, was not in beseeching god for physical favors but in nurturing her openness to god and opening herself to god's guidance and wisdom. Underlying it all, she senses that god is just as interested in her as she is in god. In one of her Soliloquies she exclaims,

> "O my Hope, my Father, my Creator, and my true Lord and Brother! When I consider how You say that Your delights are with the children of the earth, my soul rejoices greatly. O Lord of heaven and earth, what words these are that no sinner might be wanting in trust! Are You, Lord, perhaps lacking someone with whom to delight that You seek such a foul-smelling little worm like myself?...Well, what need is there for my love? Why do You want it, my God, or what do You gain?"[667]

If she accepted the argument of this book, she would have to change much of her language and the titles she gives to god, but she would not have to change her appreciation that god is fascinated with her. Her question of why god would bother with such worms as us would remain a good one!

Another Teresa, Mother Teresa of Calcutta and the founder of the Missionaries of Charity had much to say about prayer and she certainly believed that god was very much involved in the workings of the world. Yet, the editor of a book of her thoughts included in a collection of her writings a chapter on prayer that does not once mention begging god to intervene. For her, in the silence of prayer, "we will find new energy and true unity. The energy of God will be ours to do all things well...Silence gives us a new outlook on everything."[668]

Another writer on prayer says that god, "...does not torture, [but] he certainly loves to tinker...the disasters...are not added on for the sake of Christian progress...But neither are the circumstances of life impersonal visitations with which he simply helps us cope."[669] I obviously disagree but if this were not presumed and the author did not have to face the question of god's response to or responsibility for evil from a traditional perspective she would surely have reached other conclusions, for her thesis is that the god to whom we pray, "is a God who laughs". Teresa of Avila, Mother Teresa and others' approach is basically just where we would expect to find the god envisioned by this book, changing our patterns of thinking, helping us to see new horizons and realities. Louis Evely sees this opening up to god's correction as fundamental.

> "Prayer alone can wear down our frightful resistance to God. Praying is exposing ourselves to His influence, placing ourselves under His command so that He may do in us for once what He'd want to do forever, giving Him, at last, time and

> opportunity...Praying is letting Him kill in us that boorish, loud-mouthed, egotistic character whose bellowing keeps us from conversing with God."[670]

There are much better patterns for prayer and conversation with god than all the thanking for the wonders of a creation god did not create, too often ignoring what god has indeed created, a little more peace and harmony in our universe, and the possibility of a loving relationship with every human person. A little creativity in meal prayers, thanking god that we know how to sit down in peace with one another, for example, and a little more celebration of life from the animals we eat, would be better phrased. Thanksgiving Day needs to thank god for the right things, instead of keeping the agricultural god happy so that next year's crop will also be bountiful. Maybe thanking god for the wisdom to take the day off to spend with family, or even to watch football, is not too off the mark, either.

WHAT ABOUT GOD HEALING?

Much of petitionary prayer is in search of god's healing, physically, mentally or emotionally. One might expect me to be against this prayer, and if the prayer is that god would manipulate bacteria, viruses, or parasites, or knit broken spinal cords or remove a cancerous growth, I certainly believe that this is not proper prayer. In the Catholic tradition much individual prayer and much prayer in groups is directed to achieving this end, but most liturgies do not center on such healing. The Sacrament of the Anointing of the Sick is celebrated with individuals in their home or hospital bed and sometimes with a larger group in the church, but physical healing has not been stressed, except by the charismatic movement. Other denominations may put much more emphasis on god's physical healing and certainly much tent ministry and TV evangelism is about the healing of physical maladies. I once received in the mail a glove that the cover letter instructed me to put on and hold aloft during my prayer. Then return the glove, along with $39.95, and Reverend Ike would put the glove on and pray for my intention. Results almost guaranteed! I presume such ministers would not say that god can be bought for $39.95, but then why would god heal if two pray, but not one? I suppose Ike is holier than his correspondents, so god will listen more closely.

However, there are other approaches to healing that I feel are made even more effective if we accept this new vision of the reality of god. An article in the Jesuit magazine, *Company* talks of a group that seeks to heal whose, "message is grounded in Ignatian spirituality and almost always relates to healing: healing one's image of God, healing past hurts, healing the loss of a loved one, forgiving one who has hurt you."[671] In fact, says one member, "one of the easiest ways to receive healing is to change your image of God." One explanation of the effectiveness of their ministry is that, "we assume...that imagination and memory are as real as `real' events...When you imagine something, your physiology changes just as it does during a real event." This is precisely the way in which I envision god working. In addition, while there is much individual reflection, it is through and with other people that god's healing love is most effective. Again, precisely the manner in which I see god and the Christ working through and in the Church.

OTHER PRAYER FORMS

Much of the prayer of adoration, and much spirituality, is based on praise of god the creator. Many lessons are drawn from this role for god that are going to have to be rethought. "All powerful God" and "almighty God" or "Lord of heaven and earth" are all wonderful-sounding but have no basis in reality in the tone in which they have customarily been set. Praising god for the results of purely physical and probably random forces credits god where credit is not due. Worse, it keeps us from looking where god is really acting. The prayer of petition may be legitimate if we phrase it to ask god for inspiration and guidance, the areas in which god is indeed active among us. It has the disadvantage of intimating that if we do not pray, god will not act, and this must be wrong. God is continually active in making god's inspiration available to us. Why pray that god will give us guidance in some activity or about some question when god is already active in doing just that and probably has been for millions of years. It is ourselves that we must change and not god, just as we have long suspected, and this change can perhaps be brought about in much more effective ways. Using the traditional phrasing as more a shortcut to our actual meaning can be done, but better to phrase it right the first time.

If we take the petition out of the prayer of Scripture and the Church, and if we no longer praise god for actions that were none of god's doing, what is left? How about what the scripture and the church have also constantly counseled, the only form of prayer in which the primary motive is not that god would do something for us? This is the prayer which thanks god for what god has already done. True, the motive for remembering the past is to apply it to ourselves in the present, but this is not the reason that this is done in the form of prayer and praise of god. The reason is purely to show gratitude and the rest is a welcome byproduct. For the Church to assemble on Sunday, or at any time, to study and to open themselves to the wisdom of god that we have come to understand and to thank god for this inspiration and counsel throughout the ages is a worthy motive of prayer. The fact that this opens us to this same wisdom from the past and to the wisdom that god is trying to help us understand today is welcome and sought and should not be muddied by inappropriate prayer forms.

The prayer to god for assistance is often said to have been answered by god sending a savior in the form of another person who solves our problem for us. A character in one of Joseph Conrad's stories surprises himself when he prays in dire circumstances. He is a little ashamed, perhaps, to admit to another man that he has become so desperate and says,

> "I prayed like a child, of course. I believe in children praying— well, women, too, but I rather think God expects men to be more self-reliant. I don't hold with a man everlastingly bothering the Almighty with his silly troubles. It seems such cheek. Anyhow, this morning I - I have never done any harm to any God's creature knowingly - I prayed. A sudden impulse - I went flop on my knees..."[672]

When someone offers the means to escape his difficulty, the character is convinced that his prayer has been answered. "I may be unworthy, but I have been heard. I know it. I feel it."[673] Did his prayer cause god is inspire his benefactor to make the offer of assistance? Or did it make him open to accepting it once made? After all, if he had not prayed, given his feelings about a real man not "bothering the Almighty" like a child or a woman, he may well have refused on the basis of his pride. The prayer opened him up to accepting the help. On the other hand, there is no point in god broaching the subject to his friend if he is not open to accept the help, so in a manner of speaking, perhaps his prayer did move god.

The Israelites bring up the exodus quite frequently. Whatever the historicity of the stories of Moses, the pharaoh, plagues, crossing of the Red Sea, the encounter of god on Sinai and the years of wandering in the desert, they certainly attest that god somehow gave them a basis for community identity and inspired them to live by a life of freedom, respect for other covenanters and for the alien as well, and gave the basis for a moral code advanced over all others. The story of the Exodus was most popular during and after the Exile in Babylon and in the Diaspora as a promise for the future when god would manipulate history and bring them home and, once home, make them a great nation again. It of course is problematic as to whether or not this is in god's power. If it means that god will use god's power to change the course of physical events, then it is not going to happen. If it means that god can and does "inspire" Cyrus or some other agent to do the right thing and let them go home (II Chronicles 36), or inspire the Jews to take advantage of the opportunity to return home and reconstitute a covenant community and/or to remain where they are and develop a new understanding of god and god's covenant, these things god was surely doing, just as in Conrad's story. When the Exodus is mentioned in prayer, which is frequently, it is in thanksgiving for what god did for the people in the past, but it is also recognized that "remembrance" has a result for today as well. The experience of the ancestors is made available to the descendants. The action of god in the past continues today, so much so that the prayer makes the person involved in it a participant in the past event as well as the present situation. God brought "me" out of Egypt, and brought "us" out of the Exile as well. Today this same realization is brought to bear on the Eucharist. The participant in the Mass today is present at the Lord's Supper just as surely as the disciples in the upper room. This kind of prayer need have nothing to do with petition but has a great deal to do with becoming aware of god.

RECONCILIATION

And what of the prayer of contrition or sorrow for sin? If morality is based on response to laws set down by a creator, whether they are arbitrary laws designed to test us or laws built into creation by god and so "natural" and really a part of wise living, then an offense against the law is also an offense against the lawgiver. This has been recognized in many ways in civil and religious society. If you break a law against speeding you are not just offending against some written words but against the people behind the words. The police officer has a right to feel personally affected, and indeed some experience an emotional response to the violation of the law as if it were a violation of their own person. The alderpersons would have a right to feel the same since the stop sign on the corner is merely an inanimate substitute and symbol of them personally standing there and commanding a halt. Most of us have experienced the same feelings upon being passed on the road "as if we were going backwards", not only because the speeder has put us in a real danger, but also because they have thumbed their nose at the laws of the society of which we are a member. If god has personally issued the laws of the universe and, more to the point, the laws of moral behavior, and I violate them, then god has a right to feel personally offended as well. So if there is going to be a repair of the damage I have caused an apology is in order. The practice of "Confession" emphasized this by demanding a special apology to god and sometimes to the human person or persons offended, but still declaring that the reconciliation had occurred without the apology to the humans offended, since god was obviously the most offended party. It is to be hoped that the modern practice of the whole process of "Reconciliation" brings out much better the need for dealing with the brother or sister and healing the breach there as well as any breach with god. However, if god is not the author of the moral laws, for what am I apologizing? I have not offended the lawgiver by breaking the law. I have not flaunted my own will in the face of god's, for god has expressed a word of guidance, not a demand based on a right to be obeyed. I have merely rejected the guidance. So what is the point of the prayer of contrition?

It has lately been recognized anew that there is in sin much more about relationships than there is about infraction of law.[674] Sin has been called ingratitude to god who has reached out to us out of love and been rejected. It is true that while I may have a perfect right to reject unsolicited, or even solicited, advice from another, I run the risk of hurting that person if I do so, especially without explanation. If I hurt myself in the process, which I will have done if my action was truly sinful, I can expect the other person to be personally hurt as well since they gave the advice out of concern for me. It may be that they have no "right" to be hurt, particularly if I did not ask their advice, but hurt they may be anyway. If the relationship is to be repaired, this aspect will have to be dealt with. Not only that, sin is always something hurtful to a person, and if I harm someone the damage has been done not only to them but to all who love them. How wide is the circle of people who are affected when the drunk driver runs someone down, not to mention the family of the person with the alcohol problem? The others love these harmed and harming individuals and the circle of those to whom he or she needs to be reconciled widens. One of these loving persons is god.

The prayer of contrition opens me to my error and my sin. Once again this can be better phrased than it sometimes is. To pray that god would forgive me, and not incidentally not punish me, for breaking one of god's commandments, may help bring about the reconciliation and change within myself and my values that is needed. Still, if god has issued no commandments through a natural law, the prayer is mistaken; they are not god's commandments through a natural law. On the other hand, to pray in thanksgiving that god, who loved me enough to teach me and guide me in the ways of love from which I have strayed and will now continue to guide me back, such a prayer is going to accomplish the same thing, but perhaps also open me even more to the introspection, the consultation with others, and the intention to be reconciled with the human beings I have offended that is necessary. It will do so much more readily than concentrating on sin as the effect my actions have had on god as the most offended, ignoring the needed guidance to change, and the need to reconcile with the other human person.

ADORATION

Adoration is the other frequently mentioned "end" of prayer, and it fits well with these conclusions. If god can be fascinated and "in love with" such specimens as us, certainly we stand in awe of the person of god. A human can be awed by the fact that another human they respect can love their old sinful self. I have quoted scientists who are in awe of the intricacy of the universe and turn it into what can be considered their own brand of god. Certainly we can be in awe of another being at least as complex as our universe.

AM I DREAMING?

Rick Hamlin brings up one aspect of his prayer life that needs to be pursued a bit. He speaks in several places of dreams and daydreams. The Bible, too, along with the mystical experiences reported in many other religions, mentions the presence of god in dreams and trances. Since we are not aware of god speaking to our conscious minds, unless we are speaking of apparitions of the Sacred Heart to Margaret Mary Alacoque or directives to build a university or a cathedral delivered to Fundamentalists or Mexican peasants, all of which are suspect, then god may do god's best work in communicating to our unconscious minds. Since ninety percent or more of our decisions and value-forming operations are performed by the unconscious mind, this would be the best place to have an effect anyway. I do not know if this dream business is or is not a pointer to how we can be open to god and how god can communicate with us. I do not usually remember dreams, nor do most males, unless I have a fever or a hangover and these are certainly to be mistrusted, as are, maybe, those induced by hallucinogens, fasting, and other mystical or contemplative methods, but then maybe not. A cautious mistrust does not preclude some truth being found in this area of dreams. However, I will leave that to wiser heads.

NEW PRAYERS, AND PRAY-ERS, NEEDED

Taking these things into consideration will mean the abandonment of many of our favorite prayers and hymns since they are contaminated with false notions of god and god's activity. "*He Knows Just How Much You Can Bear*" will have to be changed radically, as will the verse of "*Let All Things Now Living*" which speaks of the stars in their courses and the sun in his orbit obediently shining, and certainly the song "*Give Thanks to the Lord*", which has god mastering the winds and seas and spreading out the winds and the golden plains. Add the song that thanks god because "*he woke me up this morning and sent me on my way*" which always was a bit of a problem. If I did not wake up this morning I might have even more to thank god for, but whether I wake or sleep in death or coma has nothing to do with god's design or decision. Some prayers, songs or verses will have to change or will have to be removed, but this will open the way to a more honest prayer life, not to mention the employment of generations of psalmists and musicians. Any reference to god as creator will have to be seen in the light of the creative god of Israel, creative of a people rather than a universe, but this change will probably go on internally in each believer for a while since the establishment will undoubtedly be slow in making changes. As with petitionary prayer, it is possible to sing the song with mental reservation about its factuality, but better to sing what we really believe.

When the bread and wine are presented at the table during the Eucharist, there is a prayer in the *berakoth* style that says, "Blessed are you, Lord, God of all creation. Through your goodness we have this bread to offer, which earth has given and human hands have made." It really needs to be changed entirely to praise god for something that god really has done, such as gathering us around this table. Until it is changed, how does one pray this prayer that is so obviously at variance with what one believes? You change it internally. For a while when we hear "creator" we will just have to think "creative", even though this is obviously not the intention of the author of the prayer. It is indeed through the goodness of god that we have the faith to bring this bread and wine to offer, and earth has certainly given it and human hands have made it. It takes some mental gymnastics and may strike one as vaguely dishonest, and there will be times when one has to say things that are certainly not what one believes; "I believe in God the Father Almighty, creator of heaven and earth" from the Nicene Creed comes to mind.

In the same way, for a while we still have to hear "God" when we sing "Father", at least until the Bishops and liturgists get around to changing sexist language. For individuals who reach this new understanding, the reference to "almighty God" will indicate a different "might" than before, but that god is mighty is nevertheless true if rightly understood. God is mighty in love and wisdom and goodness if not in the power to change my physical surroundings or situation. Still, this saying one thing aloud in prayer but thinking another is not a comfortable situation. What is one to do? Remain silent while the rest of the congregation says or sings these things? Say something different while others are saying the traditional formula? Dissent openly and nail your thesis to the cathedral door? No, but sometimes a little gentle persuasion may work. Many congregations have just quietly dropped one word from the Creed we say on Sunday and proclaim our faith in Jesus who, "for us [*`men' omitted*] and for our salvation," came down from heaven. Having the choir sing new words a few times can change refrains of songs sung by the people. Unfortunately, this often cannot conveniently be done. If you are the priest celebrant, it is even more of a problem. Still, one must remember that we are members of a faith community and the most loving thing is often going to be a little less than the ideal. What family could survive if everyone constantly made a federal case out of every quirk in the others' behavior, speech or ideas? The history of the Church has enough schism and division. There is room enough for many ideas and if sometimes we have to use a little mental reservation so as not to startle the faith of others, so be it. While it is true that "*lex orandi, lex credendi*" or "the way people pray is the way they will believe", it is also true that when the Church comes to believe in a different manner than it does now, its manner of worship will change as well. It has happened before, even recently. Even though there is a movement on the right to bring back adoration of the Blessed Sacrament and dust off the monstrances, as well as a tendency to make the Eucharist as regulated as it was prior to the `60's, it is a long shot that we will go back to the practices of the pre-Vatican II days. Why did we change after the Council? Because we came to understand, through the patient work of liturgists, theologians and many of the laity and clergy that the Christ wanted to be seen as the brother gathering us around the table rather than the king on the throne of the tabernacle. These pioneers plowed the field and eventually the crop came in. The same will be true for this new understanding as well.

GOD FACE TO FACE

All this is how we will respond to god if god is considered in a manner consistent with the findings of science and the reinterpretation of the scriptures and traditions that report on contact with god. In Judaism and Islam there are reports of contact with god almost directly, but there is also the understanding that god is not sensible in the same manner as another human being is. True, there are the Old Testament stories of god speaking with Moses face to face, not to mention dropping in on Abraham and Sarah for dinner, but even those who would maintain that these things really happened would admit that the being perceived by these people was not the true image of god but only something that could be experienced by human senses. Certainly god is not a human being and has no face to encounter. Muhammad saw the face of Gabriel astride the horizon[675] but even when he ascended to the heavens he did not report on "seeing" god. Since most Christians and Muslims understand that god is not to be experienced in any physical manifestation, not a lot of change is needed here.

FACE TO FACE WITH JESUS

For Christians there is another matter to be considered, since they believe that in the man Jesus we encounter the physical presence of god.[676] Paul Wilkes speaks for the orthodox understanding when he says that, "with the birth of Jesus Christ, God stepped beyond using intermediaries to bring his message, and...came to earth to live as a human being.[677] Not only that, when you meet this physical presence of god in Jesus you are meeting the fullness of god. There has always been controversy over how and if this is true. Most of the heresies addressed in the various Councils from the Fifth century through the 18th were Christological, questioning whether Jesus was truly god, or questioning his true humanity. If not truly god, then Jesus is not the physical presence of god but merely an ambassador, such as the prophets. If not really human, then was Jesus really physical or was he a "ghost" or "spirit"? Both of these positions have been rejected over the ages in favor of Jesus being both divine and human. I have proposed that god seems not to have control of matter. God not only does not but cannot manipulate the universe or its physical parts for god's purposes. If god could but did not, then god is responsible for the presence of what we term evil in the universe. However, the usual explanation of the incarnation of god in matter in the person of Jesus maintains that god can invest godself with matter, that the personality of the Second Person of the Trinity can express itself in the same way that the human personality finds expression in the body. (Actually we would be more correct to say that the body expresses itself in the human personality since the brain uses personality as a survival tool.) In addition, this idea of the incarnation of the second Person in flesh was easily understood when we thought in terms of matter and form in the manner of Aristotle and Thomas Aquinas, which the Church has certainly done. Then it is not difficult to understand that while in the human person we have a human form or nature expressing itself through matter, in Jesus we have both the human and the divine forms or natures expressing themselves through matter, but only one person. If there is no such thing as a human "form" or "soul" and the body alone is what constitutes a human being, a human person, then we are left with no good explanation or conception of how the divine person or hypostasis[678] can impress itself on the physical. Since there is no doubt that the faith of the Church is that

Jesus was truly a human being and not god masquerading as one, and the normal presumption is that Jesus was a human being, then in what way can Jesus be said to be god?

Other humans who are not Christians experience the presence of god in a non-physical manner and do not claim that god becomes incarnated in the world, and Christians themselves claim to experience the presence of the "Spirit" in a non-physical way. Indeed this would seem to be the only way in which god can be experienced if god cannot express godself through matter. How would Jesus experience god's presence within himself in a different way? How would Jesus experience god present on his constitutive level? How could god even be present to Jesus or through Jesus or in Jesus on some constitutive level? To admit that god could be physically present in Jesus and physically present through Jesus to his disciples is to claim that god has some influence on matter, that god can manipulate it or change it or use it to god's purposes, and then we are back to Susan being correct when she blames god for the evil in her life. No, we are going to have to understand Jesus in some different manner than heretofore. The Gospel of Matthew, for example, does just fine in understanding Jesus, even though, "[Matthew] does not say Jesus is God."[679] The author of the Epistle to the Hebrews, possibly around the time of Matthew, understands Jesus as the Christ, but there is no mention of Jesus as god as the later Church would believe.[680] So, we may actually have to get back to the roots. On the other hand, Tradition has concluded that Jesus is both a human and divine presence. Certainly, this could be a total mistake, or it could be a question of semantics, but perhaps not. Just how we can understand Jesus may become a little clearer if we look first at another problem.

MEETING IN THE SACRAMENTS

The Sacraments are also said to be physical avenues of god's presence, specifically by being a continuation of the physical presence of Jesus. In the traditional explanation, Jesus is himself a Sacrament of the encounter with god[681] by being a physical presence through which god works and which god manipulates and by being a sensible presence of god which can be perceived by humans senses. Since Jesus died, there is a problem if Jesus is present to us in a resurrected body rather than in the body easily experienced by his disciples. True, in the Gospel stories even the resurrected body is sensible in somewhat the same way that the pre-resurrection body had been, though the historicity of these stories is surely questioned by many scholars. Still, no one today expects to be able to shake hands with, or put their finger into the side of, the risen Christ. And yet in the theology of the sacraments the believers are encouraged to experience through their senses the presence of Christ by sensing the material objects of water, wine, bread, oil and so forth and experiencing the rituals of the Church in which they are used. It is not only the Church that one can see pouring the waters of baptism, it is Jesus physically present. Indeed it is not only the water that is cleansing, it is Jesus cleansing presence being poured over us. It is not only the hands of the priest on the head of the penitent or the hands of the bishop on the head of the *ordinandi*, it is the hands of Jesus. This is most stressed in the Eucharist in which the bread and wine are so much seen as the physical presence of Christ that for a long time the bread could only be taken on the tongue and any small crumbs were carefully consumed while even microscopic remnants could only be handled by the priest and were washed into the ground hallowed by the spilling of the blood of Christ rather than into the sewer which was profaned by human presence. (Though why Jesus would be offended by contact with unconsecrated laity with whom he had spent his life was hard to say, and why the human alimentary canal would be worthy to receive his presence but not the hands was not explained.) We still have the same problem for the Sacraments as we had for god manipulating matter to become physically present to us in the body of Jesus. God, it seems, has nothing to say about what this physical object is or what it will do. The usual explanation of the sacraments, especially of the Eucharist, again involved the presence of a "form" in every object. In the Eucharist the "forms" of bread and of wine, along with the prime matter, supported the "accidents", which were expressed through physical matter, the philosophers making a distinction here between prime matter and matter. Mere humans can make a change here by manipulating the matter or accidents until the form changed, but not vice versa. One can change the matter of bread by burning it to ashes, at which point the form changes from "bread" to "wheat ash", or some such, but we cannot change the form except by manipulating the accidents. However, it is said, god could change the form and leave the accidents intact. In other words, for us to change something into something else, we must change its properties and characteristics. You cannot call a tree a table until you have done a fair amount of work on the shape, texture, and perhaps color. No philosopher or theologian ever claimed that god could change a tree into a table without changing its outward appearance. Normally, however, it was only in explaining the sacraments, and especially the Eucharist, that this ability of god's to change the form without changing the accidents was invoked. For more (a lot more) on the orthodox explanation of the Sacraments, see the note.[682]

However, my position remains that the manipulation of matter in any way is something of which god is not capable, not to mention that the philosophical postulation of such a thing as a "form" is no longer taken seriously by most. There are, however, a number of other explanations than "transubstantiation" for the reality of the Sacraments, and one is of special interest. It is called "transignification" The premise is that we must take seriously the view of science that there is no "form" underlying the object. In fact, the object itself has no real identity on one level since it has been formed from pre-existing things and will dissolve into its constituent parts which are the only objects that have any kind of long term identity, the quarks or possibly some irreducible components underlying even quarks. To say that an object is a "table" has meaning only to the humans who use it as such.[683] It is not a table to a dog but merely another flat surface from which to snatch a crumb. It may not even be a table to another human. If I design something as a pedestal for a statue and you use it for your breakfast nook we will both call it by different names. If pressed I may agree it is a table, but I will always see it as a pedestal and feel a little funny putting my cup of coffee on it. Similarly, I may call my pancake turner a spatula but a chef would not, yet to me that is what the item truly is, though this is obviously a matter of semantics. And so, transignification points out that it is the reality that the users wish to express that gives the item its identity, barring mistaken identification by an individual. A telephone pole really *is* a telephone pole, and not just a piece of wood, if we all agree that this is so. In the Eucharist, we have the user, god, and the receiving believers, with both agents agreeing that this item is not bread but is in fact the physical body of Christ. Of course it is what god says that most matters in this line of theology, since the creator-god can say what the creature is. If the "believers" do not believe, that makes the bread no less the body of Christ, and if the believers believe but god does not, then it is not the presence of Christ. This is important when a group of lay believers wish to celebrate the Eucharist without the presence of the ordained priest. The Church's belief is that this is not the Eucharist because god has determined that the priest representing the bishop, and hence the rest of the Church, must be present. So, while this is handy in setting up a hierarchy and officials, the Church may be responding to a felt need that the community agree on what the reality is.

MEETING GOD IN THE PHYSICAL

This theological position of transignification cannot be used in its entirety in our current investigation, since it insists that god has some real control over matter, but it may point us in the right direction. If we allow that it is the conscious mind that gives the object its meaning and purpose, at least for other conscious minds, then we can allow god some input into the meaning of physical objects. This bread can truly be expressive of god, a sacrament of god, and in some sense a physical presence of god if god can inspire other conscious minds to accept it as such. There is no miracle here since this is the same manner in which humans proceed. If the inventor of the spoon had made the archetype as a launcher for peas (and called it a spauncher, perhaps) then that would be how the rest of us would continue to see it, even if we subsequently found another use for it in conveying soup to our mouths. We would consider this latter a secondary use and maybe a little uncouth, just as in the present dispensation we find launching foodstuffs a secondary and gauche use for an elegant silver spoon. It all depends on what we say it is. If both god and the believers consider that this is not bread and wine but the body and blood of Christ, then it *is* the body and blood of Christ in no less real a way than this collection of molecules and ultimately of quarks *is* bread or wine.[684] Once again, we have god having an effect on the physical universe not as creator or manipulator, but as a loving being communicating with other beings.

It is worth mentioning that the significance of the Eucharistic meal in the Gospel of John is the action of eating it with the community that is the body of Christ, not the "real presence" in the bread and wine. As one scholar observes,

> "The core of his argument is not `this bread = Christ's flesh,' but rather: you must participate in the community's eucharistic meal by literally eating and drinking if you hope to receive the life of the Risen Jesus. John's primary theological concern is the meaning of the actions, not the meaning of the elements."[685]

Paul says that anyone who eats and drinks unworthily drinks condemnation for not recognizing the body.[686] However, in context it is the body of the Church that is meant to be recognized. All the emphasis on the manner of the physical presence of Christ in the bread and wine while ignoring the physical presence in the community sharing the meal may be not only misplaced scientifically but scripturally as well.

GOD INCARNATE, IN THE FLESH

This may be a way to think about the Incarnation, god becoming flesh in Jesus. If god intends to express godself through some human person and both the person themself and others who believe that he or she has this identity agree that this is so, then this human person, who is after all no less a collection of molecules and quarks than the bread and wine or a table or telephone pole, ***is*** in fact, for us, the physical expression of god. Whether Jesus was aware or not of this identity during his historical lifetime, he could still be god incarnate because of this process of transignification. Who and what Jesus is is not determined by any form or soul or substance but by the relationship with god and others which makes him who and what he is, the same process as in ourselves, though, I have no doubt, with much deeper results in Jesus than in myself.[687] Jesus of Nazareth is a human product of the evolution of the universe just like every other human, but, Denis Edwards points out, he,

> "...is unlike anyone else...in that in him we find a radical and complete openness to God's self-giving in grace. In this one product of evolutionary history, the cosmos accepts God in a definitive and absolute way. The whole process of the world's self-transcendence into God reaches its climax in the `yes' to God embodied in Jesus' person and life."[688]

This is not a quantitative or essential difference between Jesus and other humans but a qualitative difference. His response is much deeper than others but still totally human.[689] Jesus can be the physical presence of god in the same way in which the Sacraments are the presence of Jesus through this process of transignificance. This does not do justice to the carefully worked out dogma of the two natures in Christ and the presence therefore of both the physical universe through the human nature of Jesus and the non-universe-god in the divine nature, both in the same physical manifestation. This would mean that god had some direct effect on something physical and controlled it in some real way.[690] The traditional viewpoint maintains that every act of Jesus is a totally human act, and therefore an act of a resident of the universe, with all the constraints that that imposes. It also maintains that the same action is fully an act of god the Second Person of the Trinity. Especially evident in miracles and the show of mastery over the powers of this world, his actions are both fully human and fully divine. In Jesus, a human being is Lord of the Universe (and his mother will be declared Queen of the same universe, though what effect the Jesus event had on a denizen of a planet circling a star in a galaxy 300 million light years away from our own has always been problematic)[691]. This in turn means something we have always assumed, that as its creator, god is Lord of the Universe, and creator is precisely what my investigation has denied that god is. The sticking point in many of the heresies about Jesus was that a human could be addressed as Lord, when all the time the impossibility seems to have been that god "himself" should be addressed as "Lord and Master" of physical reality. It now seems more possible that humans can bend the entire universe to our will, given a few billion years, than that god can control it. This is going to leave the usual understanding of the Incarnation and other interventions in the physical world through historical events untenable. However, with our understanding that reality-for-us, the only reality we care about, is dependent on the interpretation that our brains and sensory equipment, including our measuring instruments, extract from the quantum probabilities presented to us, we can, and I do, make a claim in faith that god has an input on the reality of what we experience. This is not done through the creative, ruling power of god, but through god's influence on our conscious and unconscious brain, and the reality that we bring into being for ourselves. This is the same conclusion I reached in the case of the Sacraments, which are themselves the real presence of Christ. Christ can be the real presence of god in the same manner, which is just as real as any which has been claimed in the past.

WHO IS JESUS

Our faith in and understanding of Jesus is radically changed by these findings and yet remains peculiarly the same. Some Christians, including many Catholic Christians, are practical monotheists[692] and there is for them one god, the Father, who has a Son, Jesus, but this son is not considered to be also god unless someone asks them a direct question. "Is Jesus God?" will be answered in the affirmative, but their prayer for intervention is directed to the Father since Jesus is viewed more as a great teacher and prophet than god.

Others are practical polytheists and end up with three gods, Father, Son or Jesus, and the Spirit. They may even speak of Jesus creating the world or possibly of Jesus talking with Abraham and Moses as contemporaries, as if Jesus is eternally Jesus, just as the Father is eternally the Father. The earliest Church fathers spoke of there being a "Son" only when the "Logos" or Word was spoken at the moment of creation, but they are not very clear what they mean, not taking up the question of an eternal Trinity. They only knew that they believed in one god, not three, but that in the one god there were three "somethings", variously described and explained. In the faith of the Church since early on there was no Jesus before the conception of a human person by Mary. There was the Son, the Logos, but the Son had not yet become human, and since Jesus was both god and human, prior to this "hypostatic union" there was no Jesus. This "union" was of the divine nature and a human nature. The "*hypostasis*" which is the second person of the Trinity is the "something" which makes the Son god, every bit as much as the Father is god. They form one being which has one nature or substance. This is united with a "human nature" to form the person of Jesus who is then both god and human, to the detriment of neither. All of this is based on an explanation[693], which is no longer used in philosophical circles, except among the Thomists, and certainly not used in scientific circles. The scientist of Aristotle's day may have explained the cosmos in terms of matter and form, but none do today. Because of the conclusion that god cannot manipulate matter, there can be no use of matter by god to introduce godself into the universe. Let me be clear, once again. For most people Jesus is either an irrelevant figure from the past or perhaps a prophet, a teacher, a model for revolutionaries against the prevailing system, but not different from any other human person except in terms of accomplishments or influence. However, for Christians, Jesus the person is both "of" this universe and "not of" it. Jesus is the ultimate example of god's control over what exists and what does not and of god's manipulation of the make-up of the physical universe. This is precisely what I am denying god is capable of doing. Therefore, my conclusion must be that Jesus is a human person in precisely the manner of every other human person and every human person is no more nor any less than Jesus. The difference is not ontological but in the vocation from god to be a messenger of god's presence and love. Every (baptized?) person is called to the same kind of vocation but perhaps not to the same degree. It is probably also true that Jesus responded to this call from god to a greater degree than any other human being, but these are qualitative differences, not ontological. This certainly upsets the traditional understanding that our redemption must be accomplished by god personally since humans are incapable of saving themselves from the futility of the accidental universe, and that this was in fact done by god when god became human in Jesus. However, it does not deny the basic premise that salvation must be done by god, but done now "through" Jesus, not "in" Jesus in the manner it has been formerly understood. It also means that Jesus is not the divine savior we have said him to be, again in an ontological manner, but there is still plenty of truth in saying that our salvation has been brought about in Christ Jesus.[694] I must insist that this denial of the traditional understanding of the nature of Jesus does not mean a denial of the reality of Jesus as the presence of god, indeed as "the" presence of god, but understood from a quantum ontological view. However, I must also admit that this conclusion on the surface is not compatible with the Councils of Nicaea and Chalcedon who call Jesus, "true God of true God"[695] and obviously do not mean it in a quantum physics manner. I am also afraid that Raymond Brown, who was one of the first to question Jesus' knowledge of his divinity during his lifetime, is correct when he says,

> "If Jesus is not 'true God of true God,' then we do not know God in human terms. Even if Jesus is the most perfect creature far above all other, he can tell us only at second hand about a God who really remains almost as distant as the Unmoved Mover of Aristotle. This God may have been so thoughtful and loving as to take an interest in history and to send a Savior, but then it cost Him nothing in a personal way. Only if Jesus is of God do we know that God's love was so real that He gave Himself for us...Only if Jesus is of God do we know what God is like, for in Jesus we see God translated into terms that we can understand."[696]

I admit that in my scenario we would lose an important component of the explanation of the faith of the Church. While it is true that god, having made godself vulnerable through loving, and especially at risk in what was surely a special relationship with the man Jesus, would suffer in and through the rejection and suffering of Jesus, and while god, like ourselves, could have spared godself this suffering merely by remaining as a spectator, I question, however, why love has to cost god something. If the theology of the Church did not conclude that Jesus' death was suffered by a divine as well as human person, I wonder if we would have thought this is so important. On the other hand, with this new view, even in the orthodox understanding of Jesus as an avenue of understanding god, one must admit that whatever we say in human terms must still be said by analogy. Understanding god by understanding Jesus is still done in these human terms alone. Having a real relationship with god by having a real relationship with Jesus remains. It will be much more helpful to abandon all of the language of substance, form and nature altogether and explain the person of Jesus without recourse to Greek and Thomistic terminology, no matter how ingrained it is in our theological thinking and doctrine. The early Church had no problem living with the Risen Jesus as THE (unique but not only) Son of God with god as his Father. In fact, this may be a moot point anyway since it is not the historical Jesus but the resurrected Jesus, the Christ, to whom we look to be the incarnation of god *for us*. The Resurrected Christ is experienced as present through Word, Sacrament and Church, the Church being the underlying sacrament of Christ's presence. For the Second Vatican Council, "the church is an 'analogous sacrament': Jesus Christ is the primary sacrament, the church is the root sacrament, and the seven sacraments are the unfolding and development of the church's life."[697] All three of these take their reality from what the persons involved say they are, especially, but not restricted to, the divine person. The Church explains that the Scripture would be the word of god, and that Sacrament would be the presence of Christ, and that Church would be the bond among all those called, even if no member of the Church believed in them any longer. In this view, it is the person of god, or god in Christ, who gives these expressions of god their reality and not the persons of the Church, following from the authentic "*ex opere operato*" position. However, even in this theological position the agreement of the members of the Church is still necessary for these things to be effective in making god sensible to them, so it is also their understanding and belief that makes Word and Sacrament "real" for them. In my view, the understanding of the people of the Church is as necessary for transignification as god's intent. Most ordinary Christians understand that, "Vatican II clarified that the sacraments are not some act of magic performed by a priest or bishop *upon* us. Sacraments are potentially blinding moments of God's grace, which work because we who participate *in* them bring the interior disposition, the desire, the faith *to* them.[698] Quantum physics may well indicate just how deep the role we, and god, play in making the sacraments and the presence of the Risen Christ real by forcing us to re-examine our definition of "reality." However, this need not stop the Church from holding what it has always believed, that Jesus is "The" real presence of god for us and that this presence has as much reality as any other presence or being. We must try to understand it in quantum physic's terms rather than scholastic, but the result will be the same. This person of Jesus is a human being like ourselves and yet is also god-for-us. His presence in the Church and in the Scripture and Sacraments as a new form of presence which was formerly experienced in the body of the historical Jesus is also just as real as the presence of a brick wall in a quantum universe where the presence of a brick wall is also only real for us if and when we, or someone, experience it. Jesus can be approached as a real person, a brother, a friend, a Savior, the Messiah, the "physician of our souls" or any of the titles by which he has been addressed, short of "all powerful". We are still invited to form a personal relationship with him as with any other human being and, because of his identity with god, to form therefore a personal relationship with the Father, and the Spirit and uniquely the Son. None of this requires god to meddle in the physical composition of the matter or energy of the cosmos. Susan can still look to Jesus for understanding, guidance, encouragement, forgiveness, and love even though she will no longer look to Jesus to physically change her situation, to win the lottery for her so she can move out, or to break her husband's leg if she could just talk Jesus into doing it. She can rely on Jesus, and therefore god, to be constantly working to inspire in her husband, and in herself, a desire to change, learn, and grow. Her best prayer will be for herself, of course, since this will open her more and more to Jesus' presence to her, and if she slips and prays that god will change her husband, as if without the prayer Jesus might not work with the husband or might just forget to do so, this is alright, too, since it will help her change in her own attitude to her husband as an object for prayer and perhaps pity and understanding rather than hatred. It makes no difference if she knows the quantum mechanical explanation of how Jesus could really be the presence of god, just as it made no difference if she ever understood any of the philosophical or Thomistic theological explanations. What is important, as all the Saints have attested, is if she trusts Jesus to be a loving and faithful person and through this trust believes in god as well. It would be better if she had at least the understanding that such an explanation is possible, and even better if she knew what it was, but many a Christian has found salvation with even a faulty, not just uninformed, explanation. You don't have to be a rocket scientist to be saved, just as you never needed to be a St. Thomas of Aquinas. Thomas himself, for all his learning and struggling to make sense of the faith before the questions of his day, understood that the person, "who does not use his reason will never get to that boundary beyond which reason really fails." One author who cites this quote from Aquinas also records the oft-told story about Aquinas' realization that rational understanding is not a substitute for faith and the experience of god.

> "On the feast of St. Nicholas, in the year 1273, as Thomas turned back to his work after Holy Mass, he was strangely altered. He remained steadily silent; he did not write; he dictated nothing. He laid aside the *Summa Theologica* on which he had been working. Abruptly, in the middle of the treatise on the Sacrament of Penance, he stopped writing," explaining later, "All that I have hitherto written seems to me nothing but straw...compared to what I have seen and what has been revealed to me."

This author points out that,

> "...the last teaching of St. Thomas concerns [the Canticle of Canticles], that mystical book of nuptial love for God...All our knowledge can only be the cause of new questions, and every finding only the start of a new search."[699]

Though there may well be Trinitarian thinking in the New Testament, at least by the time of the Gospel of John, it was not well worked out as it would later be and even so this meant that perhaps seventy years had passed after his death with no definitive thought of Jesus being a Second Person of a Trinity. There is certainly an insight contained in the doctrine which says something to us of the meaning of the person and message of Jesus and his intimate connection with god and vice versa, which will be lost. Yet, claiming that in Jesus we meet the real presence of god in the same manner in which we meet this presence in the Sacraments, ie. in terms of quantum reality, may well be enough. In addition, claiming that in understanding the real potential of the primitive human movement-towards-others that we call "love" we have understood and experienced something of god, which we can call Sophia or Spirit, may be as far as we can go. Karl Rahner has set the pace for modern Trinitarian theology with his statement that, "the Trinity of the economy of salvation ***is*** the immanent Trinity and vice versa." [Italics his][700] Most modern authors on the Trinity follow this dictum that one cannot reason to the inner life of god, the immanent Trinity, but must accept the revelation from god about godself as it is contained in god's works, the "economy", or history and method of god's acting. I am obviously not going to explain what the ancient doctrine of the Trinity may look like today after reviewing the findings of science, but I am suggesting that here especially we must put aside our assumptions, especially the assumption that god is the creator of the universe, and the assumption that god can become a human person, since after all "god can do anything". If it is true that god did not and that god cannot control the physical universe except by influencing the consciousness of those who do have such control, namely ourselves, (perhaps among others,) then we are going to have to interpret the inner life of god in vastly different terms. On the other hand, if we follow Rahner it is obvious that we did NOT experience god creating the universe, nor did we experience Jesus being the physical presence of god. Even those who did experience the physical presence of Jesus did not experience him as god.[701] What we DO experience is the guiding, sanctifying, creative presence of the Spirit of the Risen Christ. Perhaps there is a Trinity there. Elizabeth Johnson begins her discussion of the Trinity with the Spirit-Sophia and not with the creator, (be this Almighty Father or Mother-Sophia,) nor even with the incarnate person, because that is what we experience of the action of god in our lives. She may be on to something.

To expand on the Trinity, Jesus may make god physical, sensible and therefore understandable on our level, but he does not express the totality of god in at least one important sense. This is the doctrine of the Trinity which states in the words of the Catechism that, "we do not confess three Gods, but one God in three persons, the `consubstantial Trinity.' The divine persons do not share the one divinity among themselves but each of them is God whole and entire."[702] And yet, "the divine persons are really distinct from one another. `God is one but not solitary.'"[703] We have no experience of such a thing in our universe, you cannot have three human persons sharing the same "humanity", for example, and so the Church has had to hammer out its own theological terminology in order to explain how such a thing is possible. There would be no idea of Trinity, of course, unless it was revealed. There is no way to know anything about god unless god communicates it to us. We are surely not above inventing our own gods, but this is something we fervently hope we are not doing. In our better moments, we realize that we embellish the revelation with our own conceits and must make a concerted and constant effort to root out what we have read in. For an obvious example, our understanding is that god has been revealing godself throughout the ages and we experienced this god as creative, though we read in "creator." We experienced this god as familial, though we read in "paternal" rather than "maternal", called this god "Father", and then proceeded to lay on god elements of human paternity including discipline, authority, even threat and, sometimes, capricious behavior. Still, having understood that god invited a familial, covenantal relationship, we realized also not only our son-and-daughtership but our brother-and-sistership as well. So for Jesus to express the belief that he was A son of god was not unusual. Whether and/or to what extent Jesus experienced himself as THE son is for the theologians and Scripture people to determine if they can, but certainly the later Church began to experience Jesus as such, thus presenting themselves a dilemma. THE son was on a par with THE father and thus was somehow divine. Not only that, but Jesus spoke of THE spirit as a person distinct not only from himself but from the Father. Thus the idea of the Trinity was born (and/or revealed) and subsequently developed. This is only mentioned here because it is part of Christian doctrine, not because our re-examination of the faith based on the findings of science has anything to say about it except for the cautions mentioned above on the identity of Jesus. The Trinity is strictly a revelation from god, (unless of course the whole thing was a misunderstanding by Jesus or more likely by the early Church,) and no light is thrown on it by anything of the universe, though it has been a favorite pastime of theologians to discover "images of the Trinity" in the world. Augustine finds one especially in the human mind and its operation. In it, "you have a certain image of the trinity, the mind itself and its knowledge, which is its offspring and its word about itself, and love as the third element, and these three are one...and are one substance."[704] One suspects, however, that if there were four Persons, Augustine could as easily have found an image of four in human epistemology.

THE FATHER AS CREATOR?

We must refrain from making a distinction among the persons of god according to who has what role as the creative principle of the physical universe, since none of the divine persons are responsible. Each has been assigned a role in the creation of the universe from the earliest times in the Church. Justin Martyr explained the Son as an "idea" of the Father which idea is "uttered" in creation. Just where the Spirit was during this process is unclear until Irenaeus speaks of the Word and Spirit being the "two hands" of the Father in the work of creation, though it is clear from a study of their writings that both of these early theologians and their successors up to the present day are assuming that god is the creator of the cosmos rather than reasoning to this conclusion from the evidence. If we remove this creative work of bringing about the physical existence of the universe or of anything physical in it, this does not mean that god has not been creative. Remember that Israel experienced god as creative of their covenant community long before they developed the idea of a creator *ex nihilo*, and the distinctive actions of the Father, Word and Spirit may well be valid when speaking of the creation of bonds and relationships, though some rightly call for the use of other language as more appropriate, such as Spirit-Sophia, Jesus-Sophia and Mother-Sophia.[705] Still, Elizabeth Johnson, who calls for the creator to be Mother-Sophia rather than Father still assumes that this,

> "Holy Wisdom is the mother of the universe, the unoriginate, living source of all that exists. This unimaginable livingness generates the life of all creatures, being herself, in the beginning and continuously, the power of being within all being...Her creative, maternal love is the generating matrix of the universe, matter, spirit, and embodied spirit alike. This is true not only in the case of human persons but of all living and inanimate creatures and the complex interrelationships between them that constitute `the world.'...All creatures are siblings from the same womb, the brood of the one Mother of the universe who dwells in bright darkness. In her, as once literally in our own mother, we live and move and have our being, being indeed her offspring. (Acts 17:28)."[706]

It is well that she cites Acts, for in it Paul is quoting from a Greek poet, Epimenides, and has just referred to a shrine to an unknown god. Paul claimed that this was Israel's Yahweh, Christianity's God the Father. Johnson claims that this is Mother-Sophia. I submit that today we know that the mother-of-us-all from whom we have all sprung is the universe itself, and its process is called Evolution. None of this detracts from the idea of a Trinity; it will just force us to explain it anew.

Another recent study of the Trinity by Catherine LaCugna says that,

> "There is an entirely different way of approaching the doctrine of the Trinity, one that is more consistent with the Bible, creeds, and the liturgy, and also one that makes it possible for theology of God to be intimately related to ecclesiology, sacramental theology, grace, ethics, spirituality, and anthropology. It requires that we root all speculation about the triune nature of God in the economy of salvation (*oikonomia*), in the self-communication of God in the person of Christ and the activity of the Holy Spirit...The guiding principle in this book is that for Christian theology, the mystery of God can be thought of only in terms of the mystery of grace and redemption. We can make true statements about God - particularly when the assertions are about the triune nature of God - only on the basis of the economy, corroborated by God's self-revelation in Christ and the Spirit."[707]

This is a far cry from the Thomistic practice of investigating the nature of god independent of god's creative activity. Aquinas ends up with a god who is utterly unchanged by being the creator of the universe. It is the heart of Rahner, Johnson and LaCugna's approach to the Trinity that,

> "God truly comes to be God in creation which is united with God and also diverse from God...The mysteries of God's Covenant with Israel, of the cross and resurrection of Jesus, of new life in the Spirit, form the only solid basis for pondering the nature of God. That is, theology is the contemplation of the divine *oikonomia*."[708]

All three of these authors, and others who follow this approach, continue with the presumption that ultimately god is the ground of all being, though their explanations of just how god creates and what this means is quite far from the traditional explanation. I am questioning that presumption, of course, but their suggestion that we understand god from the relationship that god has forged with the universe is valid and helpful whether or not god is the author of creation. God remains the author of these relationships, the author of the covenant, the author and guarantor of the promises, even though god had nothing to do with bringing our universe or us into existence.

SUSAN, WOMAN OF FAITH

Susan may have come a long way by this time or she may be thoroughly confused. She possibly began with a belief in a creator god who is in charge of everything that happens in the universe since its creation from nothing at the word of god; a god who manages the lives and experiences of each of its creatures, theoretically for their own good, but often in rather strange ways; who can be petitioned to intervene or to change god's mind about things that are happening in our lives; who showed care and concern by becoming one of us in the person of Jesus; and who is now accessible in the sacraments of the church in a quasi-magic manner.

Now she is presented with a god who had nothing to do with the coming-into-being of the universe; has no control over its physical evolution; and should not be petitioned in prayer to change the physical circumstances of her life because there is nothing god can directly do about them. Her understanding of Jesus as the physical presence of god will have to be radically altered from the relatively simple "union of two natures" to a complex theory of transignification based on reality as defined by the laws of quantum mechanics. And nothing, absolutely nothing, can be empirically proven, but is all a matter of faith. For this the believers do not apologize for they are still convinced with as much certitude that there is a god as that there is a universe matching the description that their brain develops. They have experienced sensory information and developed an understanding of their environment that works for them. Presumably, since our brains are all about the same, our perception of the universe is about the same, though as we know from the effects of illness or injury to the brain, this is not universally true. It is also true that different cultures may inspire its members to experience things differently than those in another culture. Still, there is much about the universe that we do not perceive that is nevertheless true and much about the universe that we think we perceive that is not true. It is all based on what has impacted us and what we have absorbed from these encounters. This is true whether we are talking of the universe or of god since I conjecture that there is some means whereby god can be perceived by our consciousness. This point of contact is so subtle and so tenuous and so liable to be misinterpreted by the still-primitive equipment inside our heads that it is a far cry from the god we thought we knew was obviously the ground and reason for all other being and whose presence can be experienced in a blade of grass. Though we might wish for the god we used to have, after further inquiry we discover that the god we are "left with" is far greater and far more valuable to us than the god we lost. The former faith inevitably led us to the contradiction of a god who loves and yet is ultimately responsible for the evil that we also experience in our lives. There is nothing negative about the god that this reinvestigation leads us to. While this god may seem rather puny and relatively impotent compared with the god who can stop or start a volcano, it turns out that, as the faith really always said, this god is effective in the one way which we really need. We do not need to be protected from life and from the universe but to fulfill our role in it to the maximum, both individually and collectively. This is precisely what god is interested in helping us to do, for the more we become, the more we learn, and the more god learns as well, as the parameters of "being" are expanded. True, many die with little chance to grow, and no human dies perfect, but these things are not god's will. While the best of all possible worlds would be one in which god took this role of guide and counselor as well as the role of controller, fashioning a universe that fitted our needs, and in which everyone could develop all their potential, this is not what we have and we must live with it. Still, with a god who is a sojourner with us, we can enjoy the journey as best the circumstances allow. Further, and this is particularly important for those of us who are not fortunate enough to live a life formerly called "blessed", there is still the hope/knowledge that god is in sympathy with us and is still/always in love with us and will make it possible to live a life worth living when this sometimes miserable but always potentially wonderful one is over. Then we will live an existence that god does indeed create in "heaven," the "reign of god," the "new creation."

THE CHURCH, THE BODY OF THE RISEN JESUS

The faith proclaims that this Jesus, raised from the dead, continues to live, and to live not as he did, and not somewhere distant, but to live a resurrected life in the Church.[709] The believer can reach the same conclusion with this new understanding of the reality of god's working with us, but now with much more intensity. While we are used to saying that the Church is the body of Christ, we often do not take this literally. The Body of Christ, it is felt, is really in the Eucharist, or in Heaven, or at the right hand of the Father, but is not literally the Church (and certainly not this fellow next to me.) However, the faith holds that Jesus continues to be a physical reality and this precludes Jesus being anywhere but in this universe, composed of the quarks and molecules of any other human being. This cannot be said to occur because of the resurrection. I hope to be resurrected as was Jesus, but do not expect to remain a part of this universe. If I exist then, I exist "in the mind of god", whether it feels any different or not. I will not be composed of quarks and atoms since god has no way to constitute me from these elements. So if Jesus really is a "flesh and blood" person, then through what elements of the universe does he exist? It must be that it is through the people of the Church. We have been through the quantum physics explanation of how this is possible, so I do not believe we need to go through it again, so just apply the same reasoning to how this is possible.

The result, however, will be a much-deepened respect for the Church, meaning the people who compose it and the life in the Spirit that enervates it. Certainly, god is at work influencing all people at all times. Certainly religious traditions other than Christian have a long tradition and much accumulated wisdom. Most certainly all the Christian Churches come from the same tradition. It is these traditions that put us in touch most surely with what god has actually been trying to teach us. I certainly believe that the Christian tradition has been the most open to god and has allowed the least amount of human misinformation to intrude, though god knows that the amount is significant. This book ought to be sufficient proof of that by pointing out all the assumptions that have colored the Church's thinking but were mistaken. For someone else, perhaps god's wisdom is most accessible in another tradition, but for me it is Christian and Catholic that allows me to find god.

This understanding that it is in the community of the Church where god can most effectively work brings certain consequences. It means that I must know what the tradition is, and what the "T" radition is. It means that I must be present when the Church gathers, listening when the Church expresses itself, praying when the Church prays and worshipping when the Church worships. I may delude myself that I can find god exclusively on my own, in the comfort of my own home or the serenity of the woods, and certainly god is working with me there. However, god's influence on me is subtle, mostly because of the limitations of my own brain. It is far less subtle when it has been heard, investigated, probed and lived by millions of brains over the course of millennia. How better than "at church" can I expose myself to this collective understanding in the community of the Church, with real people, people I know, people whose works I respect, people whose lives I can emulate and people whose life in Christ I share?

There are many Christian denominations that invite their adherents to "have a personal relationship with Jesus" which is good advice in the main. However, if Jesus is envisioned as a resuscitated human person sitting on a throne above the skies, or as some kind of spectral presence who can be related to in the same way as Peter and John could relate during Jesus' life, this is not where Jesus is to be found. The presence of Jesus is to be found in the Church, which in actuality cannot be experienced as a mere concept, and certainly not with a relationship to a distant hierarchical figure such as the Pope which is not a true interpersonal relationship. It must be experienced in the local Church, the parish or its equivalent, with a relationship with real people. On a personal note, I do not find the posture of prayer in the Assembly kneeling with head down, hand often hooding the eyes, oblivious to the neighbor to denote a personal relationship with Jesus. Such a posture may be helpful in private prayer, but when Jesus is present to me in a personal way through the personal presence of the people around me, cutting them off means cutting off Jesus as well. The Roman tradition of withdrawing within oneself after receiving Communion fostered this false meeting, as do some of the other denominations practices. All need to keep in mind where Jesus is really to be found.

Christ's presence in the Church also means that I must understand that god is working through human beings who frustrate the work of the Spirit at every turn. Not on purpose, as a rule, but nevertheless just as effectively as if it were done deliberately. Every Christian must be a part of a faith community and open to the wisdom of that community and yet critical of it at the same time. Every teacher (from Bishops to parents) must realize that when they preach the gospel they are instruments of god and at the same time vectors of misinformation and must help their charges to ferret out the difference through their own experience of life in faith. They must, for example, speak as people of faith, not claiming evidence from science. They must speak from experience, not from theology books, as helpful as these are if they are recording the faith of the author and not just an intellectual exercise.

The Church must speak to the world by its life, not by its doctrine, especially since as we have seen so much of its doctrine is based on assumptions that are no longer valid. The language of this message, interpreting the experience of god in their lives, must not be cluttered by outdated theological interpretation. Recently I spoke with someone who suffered multiple gunshot wounds during a robbery but survived. "God spared me for something" was the interpretation of the meaning of this. How often we have been treated to an interview on TV of the one person in the neighborhood whose house was not destroyed in a tornado and the message was, "God blessed us." This is not the way to speak of our faith or our experience of god. When I pointed out to the shooting victim that god had nothing to do with his being shot or his surviving but much to do with how he reacted to this traumatic event, he readily agreed. He had just been trained to express it as he did. We need to train ourselves and one another to explain god's working in accord with how and when god really acts, not in terms of any false assumptions.

The role of power and authority in this must be minimal, though it must be present. When there are multiple ways to react to god's truth, several ways for each individual at times, there must be a "testing of the spirit" which is actually done by the community, but under the direction and final authorization of the community's officials who themselves must recognize the inadequacy of their own formulation. The development of new ways of receiving and understanding god's wisdom has always been through the faithful living together. There must be more faith-sharing and religious education and investigation rather than sermonizing and teaching from on high alone. "One alone is your teacher, the Father."

Personally, I sometimes feel that I could readily be a hermit and forsake human company, especially on days when the humans ringing bells for me, phone and door, are only a source of aggravation. However, were I to disappear into the wilderness I would take along inside of me a whole community of faith that has formed me. And I would not stay long alone for I realize that each of those sources of aggravation were also sources of blessing. Each one taught me something of god, or helped me discover new depths in understanding as I try to share my own faith with them.

REVELATION

Let's continue with our questioning of a few other areas where the results of our scientific knowledge of the universe have led us. Another topic is "Revelation" which is a word used in many ways,[710] but all have the common meaning of "knowledge". We can have a revelation of something new to us or unknown to us so that "that's a revelation to me" could have a meaning very similar to "that's news to me", except that it implies that because of this news other factors are also clearer. Or a person can reveal themselves to another in the sense of allowing them knowledge of a hidden facet of their personality. This may be as simple and crass as opening one's raincoat and exposing oneself to others or it may be as profound as an intimate discussion where lovers bare their innermost being. The doctrine of the Trinity is considered a revelation not only because it tells us something we did not know before, and could not know, but also because it reveals something of god's inner being. Aside from this form of revelation there is another that must be re-examined in the light of our premise. Revelation has also been taken as the communication of some detail of god's plan for the universe, a universe that was created by god according to a certain teleology. Most often these claims have to do with revelation through natural law and that, of course, we must reject. God does not reveal godself through physical creation. There is no built-in plan according to which the universe is proceeding, not even the Deists' wind-up toy which god set in motion eons ago to follow its own course. There is no teleology from the beginning, though god may have subsequently developed a plan that would be optimal for us and feasible given our make-up, towards which god tries to guide us.

NO WOMEN PRIESTS. A REVELATION?

One of the latest and most controversial examples of claiming revelation from god is the declaration by the Vatican that it is a revealed truth from god that women are not to be ordained to the priesthood. We can use this to re-examine our understanding of at least one kind of revelation. This claim presupposes that the institution of priesthood was done at god's direction and to fulfill god's plans and will. It is possible that this is the case, of course, since the Church is considered more than a human institution which god merely passively observes, but that it is an instrument that god has inspired to be a device for disseminating the acquired knowledge of the wisdom of god communicated by god through the ages to diverse people. However, the Church is also a human invention which god observes as it evolves as another example of the human life and psyche with which god is so fascinated. We must be careful, therefore, that what we claim as revealed is really telling us something of god and not merely something about ourselves and is not the result of our own reading in of cultural biases. In this example of the ordination of women, the reasons given in an apostolic letter, "*Ordinatio Sacerdotalis*" from Pope John Paul II are mainly the example of Christ in "ordaining" only men and the history of the Church that has always followed that example.[711] The ideas that we have been following here about how god works in our universe cannot add or subtract anything about this argument, but this would not be the case if the document had argued that god the creator always intended that the restriction of the priesthood to males be so because they were created in the image of god and only the male gives the image in this case that god intended. Some authors on this topic have made such claims in a round-about way by quoting Scripture to the effect that the Christ is likened to a bridegroom and the Church to the bride, and so only males like Jesus can be called to the priesthood or deaconate[712] to represent the male groom to the female Church bride. Since Christ is considered the basis and model for all creation, it would follow that the eternal plan of god *ab initio* intended that only men be ordained to this symbolic role. If this is the argument of the Vatican, I would have a problem with this line of thought since god did not create any beings, nor have a plan for anything from the beginning of creation, including the composition of the Church. God does however seem intent on creating the image of god within us now that god has "discovered" us, as well as creating our image in godself to some extent. The document does not make the claim that the exclusion of women has been god's intention ever since the plan of creation was formed. The Pope's letters on ordination and another letter to women as well as the opinion of the Congregation of the Faith that the teaching is definitive, do state that this exclusion is not an insult to women, nor does the Church "see this teaching as in any way denying the fundamental equality of women and men, *both created in the image of God.*"(Italics added)[713] As helpful as the "image of god" concept has been in issues of human rights, we are forced to abandon it as it is usually stated, namely that there is something innate in each human person, usually considered the soul, that, being in the image of god, demands respect. However, we have replaced this concept with several other ideas. For the Christian, baptism makes one an image of Jesus, something that should grow in reality in the person's life. This image is specified by the sacraments. This, of course, only makes sense to Christians, if to them. Another replacement concept is that each person is another evolutionary experiment in what it means to be a member of this species and since in our species there is so much complexity in each person that each version is a valuable example and presence to the other members of the species, each should be preserved and encouraged to develop. We should be as fascinated with each other as god is with each of us. Another approach is to remember what it is that is fascinating in the worthwhile human being. Judgments on who is expendable and who is valuable to society are too often made on the basis of external factors such as wealth, race, intelligence, the privileges of birth or political or class hierarchy. These things, when investigated, are not the measure of each individual. As we have seen when discussing morality, the person born to privilege or who merely falls into an important business or political position or as a media personality can be a very shallow person indeed and of little value to others. The same can be said of those with real abilities and talents who use them to attain these positions but through "evil" means, while a person with few resources who nonetheless uses them well will be a "fascinating person", though they may attain to no high rank or notoriety because of their lagging starting point. Certainly, anyone who starts with advantages and uses them for the benefit of society rather than only themselves, will be a fascinating and worthwhile person. Often the most fascinating are those who start out from humble beginnings and limited resources but with significant talents and use them well and achieve great things. Think of Abraham Lincoln. Certainly, the two most obvious alternatives in being human, male and female, are equally valuable and fascinating ways of being, though we humans have not always appreciated this, but god seems equally fascinated with both. Indeed god has guided those who would listen to make no differentiation of worth based on these characteristics: "there is no more distinctions between Jew and Greek, slave and free, male and female, but all of you are one in Christ Jesus."[714] So the documents on the role of women in the Church are correct in stating that god has revealed that the best course of action for the human race is if the contributions of both men and women to the life of the Church be equally prized and equally encouraged. This is not because god created us intending that we should do so, for god did no creating. Nor has god revealed that an exclusively male priesthood has been a part of god's plan all along, since there has been no plan on which creation is based. If the plan that is claimed is one that god developed "after" "discovering" our existence (pardon the clumsy phrasing, but you know what I mean), and if the authorities in the Church have some special gift for discovering what this is, then fine, let them teach it. It does seem passing strange that the same god who has guided us to put aside our divisiveness and expand our genetically influenced vision of who is my family, would then take it back and deny to women some role that they could obviously fulfill and in fact possibly expand beyond what males have done with it. It would seem that god would rather encourage any new development of any human institution since it is another way to live life, and this is what is fascinating to god. As Elizabeth Johnson points out,

> "...all persons are constituted by a number of anthropological constants, essential elements that are intrinsic to their identity. These include bodiliness and hence sex and race; relation to the earth, other persons and social groupings; economic, political, and cultural location, and the like. These constants mutually condition one another, and in their endless combinations are constitutive of the humanity of every person. Significantly change any one of them, and a different person results. It is shortsighted to single out sexuality as always and everywhere more fundamental to concrete historical existence than any of the other constants."[715]

Nowhere do we find god singling out any one human characteristic to be considered more interesting or valuable than another. We have had to relearn this repeatedly about questions of race, economic position, the privileges of birth, and the differences from sexuality. It would seem that god is leading us in another direction than the Congregation for Doctrine wants to follow.[716] Many theologians have pointed out that though the scriptures are products of their culture and often denigrate women, overtly or through neglect, there are many indications that god was leading those who were open to god's influence in a different direction when it came to personal relationships, including that of man to woman, and when it came to the worth of the individual, man or woman. The women's liberation movement looked for evidence of this in the Bible and had little trouble finding it.[717]

IS THERE A PLAN?

Once again, let me be clear about what I have just said about teleology or a plan. Teleology is usually defined as a final cause that shapes the development of things. If I want to reach a certain end result then I must begin at one of a limited number of ways and proceed in a direction designed to reach that goal. It has long been assumed that god has had some goal in mind "from all eternity", in fact that goal is the Word that is uttered in creation, "in the beginning was the Word."[718] In order to reach that goal, which we chauvinistically understand as the gathering of the human species into the kingdom of god, god created a world which could support such creatures and then directed it along lines that would lead to us, either in six days or in billions of years of evolution. This kind of a teleological cause is now obviously excluded from consideration. It is certainly possible that, once we had arrived on the scene, god could see the possibilities in us and could then form a goal for us to reach that would maximize our beings. Seeing how far we could go if we pursued certain courses, god would naturally try to guide us, subtly but effectively, over the centuries and millennia, in this direction. In this case it may make some sense to believe that god would reveal to those in the Church charged with pointing out god's authentic wisdom, at least the broad outlines, and maybe even details such as, "you will go farther faster if you don't let women become priests." Seems a little strange, but then the universe is full of mysteries like why does buttered bread always fall butter-side-down, so why not some mystery to god's ways? Still, as mentioned above, it is a little unbelievable that this is something god would conclude is good for the human race. On the other hand, the "goal" or "design" of god may be the same as above, to help us maximize our existence, but since god has never encountered critters just like us before, god has no way to develop an all-encompassing plan and is content to just point out the obvious, such as "you have tried putting down women or other races, or anyone different from yourself for more than two million years, why don't you try the opposite?" This seems to be where the majority of enlightened males are today, thanks to god and no thanks to our superior intellect. This is not the teleology of the Scholastic textbooks, so carefully worked out instead of "ad hoc", but then we are starting from different assumptions than the Thomists and traditional Catholic theologians.

ESCHATOLOGY

What then of eschatology? As discussed in chapter four, the subject of a final "kingdom of god" which is the destination that god wants humans to reach, is a bit out of science's line. Whether there will be a "big crunch" or a slow fizzle it is sure that the universe is not tending toward the establishment of a "new creation", and since god cannot manufacture a new or a renewed physical universe such as we now inhabit, the traditional picture must be abandoned.[719]

There has been, in fact, an abandonment of belief in the future, and part of the reason is the success of the standard model of the universe and the discovery of the time scales of evolution. If the universe is some 15 billion years old, if it has taken hundreds of millions of years of life to reach our level, if we can expect the sun to continue burning for another two billion years, then the old expectations of an immanent new world or a kingdom of god gets harder to believe. Surely there are always the chiliasts predicting the end of the world. It happened again in the year 2000. For most educated westerners, at least, there is little expectation of "rapture"[720] but there can be an eschatology without a teleology, by which I mean that it is perfectly possible that god, or anyone with half a brain for that matter, can see the possibilities in human life and have a fair estimate of how they could be achieved, "if only". If only there were no aggression directed in negative paths, Susan's life, and everyone else's would be so much better. If only there were real love and harmony, care for the neighbor, valuing of differences, physical and mental health, challenge without dire consequences for failure, and so on, life would be fine. And if there were no end to it, how sweet indeed! God can see that this kind of existence would be good for a people god has come to love and god would like to see them have it. Maybe, if Tipler is on to something, god could inspire us to build it within the universe, something that eschatology in the traditional teaching encourages us to do anyway, though not in the same way. If not, then god will provide it in godself, something that the doctrine of the Beatific Vision has always proclaimed. The promise and all that god inspires in us brings up all or at least most of the topics of traditional eschatology.[721] This does not mean that god had this plan as a teleology in mind before the universe began, or even before human consciousness began. It could well have developed "ad hoc". Some of the ways we talk about it will have to change. We should no longer say things like "from all eternity god intended us to be with god", but what do I care if god meant that I should exist and took steps to bring this about or if god just discovered me on the doorstep and took me in? We have always talked about ourselves as adopted sons and daughters, and being a foundling is quite all right.

REVELATION. THE SCRIPTURES

Another form of revelation is the Scriptures. It is customary to treat this form first, certainly before revelation through the teachings of the Bishops, and the reason is that it has been assumed that the revelation by god to the authors of Scripture was of a different order than that to people of today, including the teaching authority of the Church. Can a case be made for this? It would seem not since the ability of god to impact the human consciousness appears to be effective but subtle. To insist on a word-by-word, or even concept-by-concept, revelation to a human mind through dictation by god would seem to be stretching things a bit. Indeed it has become problematic to define just what inspiration of the Bible means since we have discovered, pretty much only in the 20th century, that the Torah was not written down by Moses. What we have received, even the latest compositions of the New Testament, has gone through such a convoluted history of transmission and editing and redaction, much of which changes the meaning of the material with which they began, that it is impossible to say who was inspired. For example, many of the patriarchal stories have indications of being composed, in part at least, during the times of which they speak; were passed down orally for some centuries; were perhaps written down during monarchical times; were worked into a continuous narrative during or after the Exile; were sometimes reworked by someone recognized now as "inspired"; were copied, sometimes with the additions of marginalia or glosses or even downright errors; and are read now in translations which themselves can change meaning from one language to another. Just who was inspired by the Spirit so as to make the work truly the "word of God?"[722] The first author or storyteller? The last? Every-one?[723] In Exodus, the god who contacts the people on Sinai comes down on the mountain in a very physical manner, in a cloud, in smoke and thunder, very much like any of the storm gods of the surrounding peoples. In Deuteronomy, god remains in heaven (and the mountain is Horeb, not Sinai). In a more primitive view in Kings, god resides in Solomon's Temple and is physically present as on Mount Sinai. In the later, more refined view, the Temple is, "the locus where his name was placed...Unlike Exodus,

Deuteronomy never refers to the Temple as 'the House of God,' but repeatedly calls it 'the place where Yahweh chose to cause his name to dwell'. God was resident in heaven and not in the Temple."[724] How does this reconcile with inspiration? Was the first idea in need of correction? Are both ideas correct? We also see that, "many passages from the Book of Proverbs, especially 22:17-24:22, seem to be paraphrased from the *Instructions of Amenemope*, a hieratic, or cursive hieroglyphic, papyrus probably dating to between the seventh and sixth centuries B.C.E., which contains the advice of an Egyptian wise man."[725] Was the Egyptian inspired, the Israelite who used his material, the various scribes who passed it on, or everyone involved? Some commentators seem to be migrating to the last choice, that everyone involved in the construction of the Scripture was inspired by god, though including the non-Hebrew authors from whom they borrowed is pushing it for some. The category gets wider when we inquire into the formation of the Canon, the official list of inspired books. There are other works in the same vein as the Scriptures, the book of Enoch or the Gospel of Thomas for example, which might, with just a little different twist of history, have ended up in the Canon. There are others, or parts of accepted works, the Apocrypha and deuterocanonical, which are accepted as inspired by some Christian denominations and not by others or by Christians but not by Jews.[726] It is the understanding of the Catholic Church that a work is inspired by god if the Church says it is. This brings in the question of whether those who decided on the Canon were inspired or not. Were they a part of the process, or did inspiration end before their decision? Does this mean that the entire Christian community was inspired by god with this special guidance, since after all it was the usage of the early Church that influenced the decisions of the authorities who drew up the final list? Then what kind of guidance or revelation would this be that is any different than what god normally does? For god to "take over" a human mind (brain) and make it do precisely what god intends means that god has more power over physical events than seems to be true and in any case most scholars of the Bible do not claim that this is how inspiration works. No one but the literalists and fundamentalists claim that the inspired authors, whoever they may be and at what point they impacted the formation of the Scriptures, were so overwhelmingly influenced by god that what they wrote was not their own words and ideas. Somehow influenced by god, yes, but still their own. It may well be that there is no different manner in which god can influence the human consciousness than that in which god can do it today. This would make the accumulated wisdom of the faithful, which the Bible contains, open both to the wisdom of god and the misinterpretation of humans, precisely the situation in which we find it. True, it would do away with the Church's claim that there is some kind of "special" inspiration responsible for the Scripture that makes it the word of god in a manner different than could be written today. I will personally leave this question open, claiming no divine inspiration, though I would lean to saying that the writing down of the accumulation of two million years of experience of god rather shortly after the invention of writing itself does have some special claim to our attention. It is also certainly possible that the particular human person who wrote this or that part of the Scripture was more open to receiving the subtle influence of god than were his or her neighbors. This certainly seems to be the case with Jesus, who was so open that for all purposes he was a presence of god. The prophets evidently believed that they had some special handle on what god wanted to communicate, or at least their disciples thought so. And even though the classical prophets developed from madmen or religious fanatics or out-and-out charlatans who claimed they could find your lost donkey for you,[727] it seems to have been a real development. Classical prophecy was something new, or at least it was the first written record of just how open to god a human being can be. So the Scripture may be more than just the accumulated wisdom of the humans who preceded its writing; it may be a new communication from god received by individuals who were genuinely more attuned to god. That their insights should be the last word on every controversy is more problematic since this is not the way of the universe that the oldest is the wisest. Alexander Graham Bell's ideas on how to run a phone company would be hopelessly outdated today. The decisions of Church Councils and Synods have hopefully been advances over their predecessors so that we have a clearer idea of god today than did Matthew, Mark or Luke. Still, the seminal experiences of god by the human race over eons of evolution are probably not totally wrong in spite of our discovery that we, and presumably they before us, have burdened ourselves with a number of unwarranted assumptions, now discredited. The Scripture can also be a corrective for those areas where we went astray after the composition of the Bible. While some ideas found in it are surely in need of correction, others are genuine insights influenced by god but whose development veered to error in later times. Going back to the roots can be a remedial activity to these errors. This has arguably been the case with our modern conception of god as a very forgiving god who is infinitely patient with us. It was not that long ago that god was seen by most Christians, certainly the Calvinist tradition but including most other denominations as well, as a very "jealous" god who brooked no deviation from god's inspired laws, and who approved of the power of the state being used to assure compliance. Indeed, correction of the wrongdoer by the harshest of measures was often seen as an act of kindness. Better to fall into the hands of man for punishment which might lead to repentance than to fall into the hands of god who knew no mercy for transgressors or heretics. In the not too distant past it would have been the Christians clamoring for the execution of the sinners instead of holding vigil outside the execution chambers of states allowing capital punishment. What caused this new understanding of god's mercy, patience and forgiveness but a return to the Scriptures with a new approach allowing us to correct the Old Testament witness (which itself had been wrongly interpreted) on which so much of the puritanical ideas had been based, with the ideas of the New Testament (which turned out to be a logical development of the Old). There may or may not be a "special" inspiration to the Scriptures, but either way they are the word of god in human history. Any inspiration is helpful. The sacred writings of other human traditions almost surely also contain truths inspired by god through the long history of humankind's experience of god's presence. This is surely true of the Qur'an and the Hadith of Islam[728], the Tipitaka of Buddhism[729], the Upanisads, the Mahabharata and Purunas and other writings of Hinduism[730] and other written sources of the world's religions, as well as the oral traditions of many peoples. American Indians have made efforts to preserve their mostly oral histories and wisdom in such books as *Black Elk Speaks* and others.[731] Collections have been made of Afro-American folk wisdom, which even in tall tales distill the wisdom of generations and perhaps much older traditions.[732] The same has been true of other peoples such as the Indians of the Amazon basin, indigenous peoples of the Philippines, New Guinea, Australia, the Kaonde of Africa[733] and other peoples. Inevitably much of the same hard-won wisdom is contained in sources from very disparate cultures. Since humankind is pretty much the same, this is to be expected, but it is also to be expected if god has been working throughout our history to teach us.

How then to read the Scriptures? In the past everyone read them as coming from the hand of god with virtually no human interference. For some it was and is as if the Bible was brought down from the mountain by Moses instead of just the Ten Commandments. Indeed, the way some react to any change in biblical translation they seem to believe that Moses brought the King James version down from the heights. Today, of course, we must read them for what they are, a collection of human writings but with the caveat that these humans were inspired by god in the manner we have discussed. Does this mean that whenever the name of Adam or Eve comes up we must first point out that these are not historical people but figures in a myth? Must we presage every mention of Israel or Esau with the reminder that they may or may not be real persons, or that they might be eponymous ancestors? This will depend on the audience. Until the people in the pews, and those in the pulpit for that matter, understand these things we probably should be careful of contributing to misunderstanding and remind listeners of the provenance of the Scripture being read. However, it is certainly permissible with an audience that does not need these warnings to speak *as if* the characters and events are real even if they are not. After all, they are at least characters in a story whose author invites speculation as to their motivation and to god's motivation in dealing with them. To speak of Abraham's willingness to go forth from Haran may or may not need to be prefaced with a reminder that he was in the middle of a mass migration from the area and may or may not need to mention the question whether he is a real character or not, in order to speak of his willingness to relocate at the express command and direction of god. In the story, and it is this which is inspired after all, he is certainly aware of his motivation. The point is that god deals with us as god did with our ancestors and that we, too, must be open to god's direction. This same approach can and should be used for any of the Scripture. Whether or not the story of the Transfiguration of Jesus is a "relocated resurrection event" is beside the point when we are attempting to apply the story to the reality of our lives. This is true in preaching but also in private reading and meditation.

REVELATION. TEACHERS

We also need to say a word about another avenue of revelation that has been claimed for god, the teaching authorities in the Church. If we are tempted to conclude that the Scriptures are both the result of millions of years of influence by god in the human consciousness, plus perhaps that same influence as perceived by some persons more finely tuned to perceive god's message, like a critical situation which finally crystallizes all at once, then where do the authorities in the Church fit in? In a perfect world, only those who had demonstrated that they were more sensitive to the work of the Spirit would become the bishops and Popes of the ages (no one looks to the grunts of the clergy, the presbyters, for any special holiness). Certainly some have been so and rightly claim a special understanding of god's revelation, and this is usually recognized by the community of the Church, although sometimes only after their deaths. More frequently, men become bishops (only men, but that is another story) because of their administrative skills, their orthodoxy, or their unlikeliness to rock any boats, not because of their evident openness to the Spirit, especially since the Spirit seems to be so progressive. So what claim can be made that they have some special avenue to the wisdom and guidance of god? On the other hand, few bishops aside from the bishop of Rome would lay claim to special inspiration as individuals, though as a college they do make this claim, and this collegiality includes the Pope since his office is seen as only exercised in union with the other bishops. It is possible that god is somehow able to be perceived better by the group of bishops, just as god seems to be more discernable in the collective Church than in the rest of society. The process, whatever it is, may or may not be the same and the tentativeness of the pronouncements of the Pope and bishops should also be the same as the "sensus fidelium" which often takes centuries before the truth is sorted out. It would help if teachings from the hierarchy were couched in less apodictic terms given the provisional nature of their collective wisdom.

These, then, are major topics of reinvestigation and my approach to explaining the faith as it has been explained in a new way, taking into account the new information presented earlier.

ANGELS AND DEVILS. AGAIN?

There are obviously any number of other topics that could be covered, but we must stop somewhere. Before we summarize our findings, however, there is another which, though it is perhaps inconsequential, we have addressed in previous chapters and might as well say a few things about at this point. What about angels and devils? Obviously we are not going to talk about beings that god created having some physical influence in the universe. If we do, then we are right back to the problem of finding any real evidence of such intervention, and to the problem of god being responsible for evil in the world. If god cannot intervene, neither can any of god's creatures, if god has any. However, to speak of an angel one need not think of one out of Hollywood. An angel is nothing but a messenger, and a messenger is nothing but the message. True, some messengers have been executed for bringing bad news, while others have merely had their beard shaved off, or half shaved, and probably none-too-gently. Still, it is the sender with whom the recipient is angry, not the messenger. It is the content that has brought down the wrath, not the person bearing it. So to speak of an angel is merely a mythical or poetic way to express that god has somehow communicated, it is not to insist that there was a real being as an intermediary. It would be legitimate, then, to keep some of the language used in psalmody and spiritual reading, especially the more poetic, as long as the reader keeps in mind that this is just a figure of speech. In our day, with New Agers willing to find spirits in every prism or cactus, and with every manufacturer of trinkets willing to put angels everywhere, we would be better off by not aiding and abetting false ideas of the angels' existence or of god's intervention in human affairs in some physical or physiological manner.

Devils, of course, were invented to explain evil and temptation as well as absolve god from some responsibility for evil. It does seem sometimes that there is a force (ask Darth Vader) that propels us towards negative actions, and indeed there is. However, this force need not be personified, it is the influence of evolutionary behavior which, according to it's own definition of the good, makes perfect sense. Or sometimes it did make perfect sense but does so no longer. When humans lived in small tribes it made survival sense to know and care about those with whom one lived and to distrust, and even seek to selfishly use, those outside the group. It also made sense for the tribe to develop a culture that set it apart from others. Even a simple thing such as bodily decoration could be useful if it defined this assembly of people as "family", to be protected. And if it also served to identify at a distance friends from foes, so much the better. How natural, then, racism, which did not require painting the body to set one apart, the "paint" was already present. However, in our own society when skin color is not a good identifier of friend or enemy, our evolutionarily induced tendency to make these distinctions is no longer useful or "good". Whence comes the evil of racism, then? From the devil? Hardly. It comes from ignorance of our roots and failure to adjust fast enough to changing conditions. There is no need to invoke an external agent to explain these temptations and tendencies, we have met the enemy and "they is us" in the immortal words of Pogo. Enough said.

SUMMARY

As we conclude this assessment of the challenges presented to the faith from the new information gathered by the human scientific endeavor, I want to put in this one place what Susan should be told. It might be even more helpful to start over and talk about what Susan should have been told in the first place, when she was young and just being exposed to the faith. Then consider how she would then react to the problems that prompted her to ask her questions full of so many unexamined assumptions. What do we want to tell the children and the people just coming to the faith?

We must begin with a thorough grounding in the universe as we know it; how it began, evolved, functions, and where it is going along with our place in it. While from the viewpoint of those who were brought up under the old worldview this is a step down for humanity, this is not so when we begin anew. The universe is a truly marvelous occurrence, no less so because it is the result of mere chance. It arose out of a quantum fluctuation like an infinite number of other fluctuations that occur around us and within us all the time. Because of processes we have reviewed, this one did not flip out of existence at the moment of its birth. Instead it persisted and remains today on the knife-edge of non-existence. Indeed, it can be said not to exist if you add up the sum total of energy contained within it. How much "stuff" really exists with no danger of being annihilated by negating forces? Evidently nothing. The sum total of the universe is exactly zero. Now one can look at this as many philosophers have done and conclude that it is all futile and absurd, or one can look at it as wonderful and marvelous. Once it exists, the universe could have gone in any of an infinite number of directions, but the direction it took led to the development of "life", a particular kind of existence which, though not so different than "non-life" physically, being composed of everyday molecules, nevertheless is something new. Evolution is the driving force of the whole process, but especially of life. Galaxies and solar systems evolve or devolve, just as the whole physical universe does, but "evolution" takes on a special meaning when applied to life. Again, there are an infinite number of paths that evolution could have pursued, but the one it did pursue led to the development of sentient life, then to conscious life. Where on this continuum one would put any one animal species is often arbitrary. Is a mongoose conscious while a paramecium is merely sentient? Or must we reserve consciousness to humans only? The distinctions become somewhat meaningless. We are what we are. And what is it that we are? We are accidents of the universe. We are the universe conscious of itself. We are wonders of improbability and that we exist at all, or that anything at all exists, is a cause for wonder. We have barely scratched the surface of the wonder that is our existence or the wonder of existence itself. And this is what we need to teach one another and to learn as a species, that the universe is a wondrous place in its own right. There need not be, and indeed there is not, a grand plan for it all. Most of the wonder is in the coming-to-be of all that is through the mechanism of evolution. Never mind that the process is unbelievably "wasteful" and "cruel" if looked at from a negative perspective, and it is. In order for the bird to soar, how many gliders landed just a little short and ended up a meal for the pursuer? There may be a necessity for pressure for survival for evolution of life to work or else there would be nothing but eucaryotes drifting in the sea. It was not until one ancestor developed into a predator on the others, seizing their sunlight, eating their young, or enveloping them completely, that there was any reason to change. Without the possibility of extinction, the random mutations that occur would survive, it is true, as long as they did not kill their host, but the pace of "advancement" would be substantially slower. How wonderful it is that such a mindless (and heartless) process can lead to the rich diversity of beings that have populated the world. Susan could spend a lifetime merely learning these lessons.

Does it make any difference that in the long run it all comes to nothing? Many people do not feel a need to be downcast about this. The alternative to living briefly, or in the case of the whole universe to existing briefly is not to live or exist at all. Why not rejoice in the brief existence rather than rail about it not being longer? "Better to have loved and lost than not to have loved at all" is also true of existence. It is only when we *assume from the beginning* that eternal existence is possible that we feel bad about not having it. And it is only when we *assume from the beginning* that everything has a cause and that everything is already mapped out that I feel put upon when I learn that I am only one of many specks on the fringe of a run-of-the-mill galaxy or if I find myself cast in a secondary role in human society. If whether I am healthy or not, rich or poor, live a long life or a short one, is all the result of chance and circumstance and not someone's deliberate choice, then I will take what I get and make the most of it. Jealousy and the desire to do better and have more seem to be built into the human brain, and will sometimes give me a momentary pang, but understanding these feelings and where they come from will mean that I can learn to handle them and not let them direct my life or rob me of my own joy in being me. Of course this is easy to say when I live in a privileged culture and a privileged time. Being the victim at age ten of a murderous, genocidal mob armed with machetes, or being an uneducated, brutish peasant of an uneducated, brutish lord; being the species vanquished by a superior group; being one of the victims of the process of evolution; these lives may seem to be wasted, but the possessors of those lives would still rather have a chance to make the most of what they have rather than never exist at all. If this is true of the individual, it is most assuredly true of the species. Is it a tragedy that the dinosaurs no longer exist, or is it rather a wonder that they ever existed? We ourselves are surely only the ancestors of a much different kind of human, if not a new species that, unless we manage to bring to a halt by our thoughtless destruction the process of evolution, will look back on us with pity that we never experienced what they experience. Yet, we do not pity ourselves, we revel in our existence (with only momentary pangs of self pity that we will never get to wrestle with a Klingon or dock at a space station).

Certainly there are all the negatives and shortcomings of existence in this universe. We do not want to teach young Susan to concentrate on these, but rather on the amazing fact of her being. No matter that we rightly concentrate on the positive, Susan will still experience the negative. She will not live forever; she will not be able to reach even a fraction of her own potential; she will never know all there is to be known; and whatever she puts her hand to will come out incomplete. Indeed, whatever conscious life manages to achieve throughout the whole course of its history will come to nothing, either in a Big Crunch that wipes out all trace of previous accomplishment, or in a Big Whimper when nothing can ever react with anything else again. True, she has not missed out on *existing*, and that is vastly superior to the alternative, but she, and the entire universe with her, will have missed out on so much of *existence*. So much candy to eat and so little time! And it is precisely then that the faith can bring, in an understandable way, the "Good News" of god.

THE GOOD NEWS IS STILL GOOD
What good news can we give her? What we have told her is that there is a god who created her, has control of the universe, and, if she was lucky enough to be taught by the right group of Christians, that this god cares about her. What we should tell her now is not that there is a god but that we believe that there is a god. We believe it because we have experienced god's presence. Nor can we tell her that god created the universe and controls it and if she can influence god, physical events in the universe can be altered to her benefit. However, we can tell her that god has changed the universe through influencing those conscious beings who responded to god's initiative. And we can tell her that god knows her personally and wants to be known personally by her. In fact, given any sort of response from Susan, god is fascinated by her, or in the terms of the Scripture, god loves her.

Susan has some problems that brought her to her pastor, and along with her own, there were other problems in her family. All of them were "blamed" on god. Obviously, Susan would not be at the pastor's door with the same questions if she were brought up without the assumptions that we must now abandon. Would she be there at all? Or should she be at a competent counselor or psychologist? Does her pastor have anything to offer that Ann Landers couldn't? Well, yes and no. She may well benefit more from a counselor, especially one who can help her see why she is reacting in the way she is; why she has the feelings she has; and why she has not take some actions that others might think are obvious. "Leave the bum," for example. To the extent that the mental health people recognize our motivations as being greatly influenced by our evolutionary past, and to the extent that they can help Susan understand which of some possibly conflicting tendencies are healthiest for her, they can aid her in making some decisions. Which of these inherited tendencies or something else is going to lead in the best direction? This is where her pastor definitely has something to say from the perspective of the faith, for this is precisely where god has been active in the universe. Long before god encountered anything like us, god found creatures that were capable of reacting to stimuli in rather sophisticated ways, but they were not really "free." Observers have seen birds that will try to keep a hoard of food secret from their neighbors by not calling, but who raise the hue and cry when there is a cat present who can be intimidated or at least confused by large numbers of birds. Obviously the bird, whose brain is the size of a marble at best, can assess the situation and make decisions about appropriate activity. Yet there is in the bird no personality such as we possess which could look over all the possible solutions presented by its brain and invent another entirely. It is too hard-wired, even though there is great complexity. So, if Susan were the bird, there would be a limited number of possible responses to the problem of an abusive husband. Susan herself has many more and is "free" to choose any one of them for any number of reasons. This is not to invoke something spiritual or incorporeal in Susan that makes the real decisions, but it is to recognize in the human brain something that can be *thought of* as spiritual. No one feels it necessary to invent a "bird soul". In any case, god could do nothing with the beasts except observe. They have no values that can be redirected, no morals that can be refocused, but Susan does, and these god can affect. What animal (saving the effects of altruism discussed previously) would die on a cross for the good of all? What beast can take steps with their lives directed at reinforcing the "memes" of its culture? "Dying on the cross" of her marriage is probably not the best course for Susan to take, but she well might remain in the marriage as an example to others of the permanence and importance of the marriage bond. She cannot, of course, without making some changes in the relationship, but this is possible since her husband, too, can make the same choices or learn the same lessons. Both of them can change, and both can change under the direction and guidance of god, often as mediated by the Church in some way. If they do, they have accomplished something that no other species on earth could do. Can they do it without god? Of course, for they are doing nothing that is not "natural" to human beings. God does not change human nature except by guiding our choices. Would they do it without god? That is another matter.

The rabbits of Australia have no choice. They must eat every single blade of grass they can get to, and if this dooms their species to extinction there is nothing they can do about it, for they are entirely unaware of the consequences of their actions. Many species have eaten or reproduced themselves into extinction. Humans are quite evidently capable of doing the same and are well on our way to doing so, but we are also capable of the unselfishness that will enable our descendants to have a good life, too. Without god we are perfectly capable of making the right decisions, but with god we are more likely to do so. Indeed, influencing the memes of human culture would prove to be perhaps the most effective way for god to communicate to us. Memes are those values that guide human cultural activity, including individual behavior. What is considered valuable, including but also surpassing the transmission of genetic DNA to succeeding generations, is what is prized by the society and inculcated in its members. Through memes we pass on to our inheritors not only our own physical traits but also our wisdom, our experiences, our art, religion and values. What better way could god have for influencing human history than through influencing these memes? Just as in genetic evolution, the whole is greater than the parts. The physical evolution of our brain has opened possibilities that cannot be deduced from the brain of a crustacean. Human society can be led to heights undreamed of through the 'normal', i.e. without any intervention by god, course of the memetic change of society. Such influence, which may require no physical intervention, would be very subtle, a quality that I postulate would be present in the manner of god's operation. The more we open ourselves to the wisdom of god the better off we and our species will be. The conflict comes when our species is not open to god's wisdom and the individual is. And here we come again to Jesus.

Centuries ago the Monophysites were condemned by the Council of Chalcedon in 451AD for claiming that Jesus was not really a human being. Yet, for much of the history of Christianity Jesus has been definitely seen as more divine than human, not surprising since in many cultures he was replacing gods who were more human than divine. It is only in recent times that the recognition of the effects of Jesus' humanity have been investigated theologically and questions such as "did Jesus know that he was god?" have been asked with any seriousness. In what way Jesus can be said to be god was discussed above, but certainly his humanity is recognized today. Certainly if Jesus was a human being, but one entirely and completely open to the wisdom of god, he would run into opposition from those who were only reacting through their inherited capabilities without heeding the guidance from god as to which actions were the wisest. Indeed, he did meet such opposition. Because of his openness to god, he could set aside his own innate tendency to self-preservation and do what would be best for others.

He was willing to be faithful to following the guidance of god even to death, "death on a cross". Was this the wisest course of action? Was god correct in the direction in which god led him? Judging by his positive effect on human history and on the lives of countless individuals, it unquestionably was. Jesus was certainly directed and influenced by the memes of his own culture, and yet he also changed these in a profound manner, as is evident, writ large, in the subsequent history of humanity, at least in "Christendom".

Is this not how evolution works? By the mutation of a gene a new physical characteristic and/or a new behavior can be manifested, and by the mutation of a meme human history can be redirected. The life and example of Jesus caused a re-evaluation of the then-current memetic composition of society in many fields from human relationships between master and slave, male and female, gentile and Jew to the conduct of business, science and the arts. He did nothing very radical himself, just as most genetic mutants may be virtually indistinguishable from their parents, but in this manner, working in the man Jesus and his disciples, god was able to change the course of human history. The Cross and Resurrection retain their power to save and remain the central Christian message. "The Cross" in Jesus' life need not have been the instrument of execution. It could have been "the Noose" or "the Sword", either of which would have made excellent symbols. Or, as in the life of most of Jesus' disciples, the Cross could have been an extremely unpleasant situation or even the challenge to radically change one's life style with the ensuing painful process of actually doing so, but historically it was "the Cross and Resurrection" through which god was most powerfully able to speak a Word of guidance, direction and wisdom through Jesus and the effect on the world reverberates through it today.

Of course interpreting what god's guidance is in any situation is sometimes problematic. God's influence is usually much too subtle, even though some evangelical types are fond of saying "the Lord told me", or "the Lord revealed to me" that I should found a university, or build a cathedral, or in some way get into your wallet, but the Church has always maintained that god is to be found in the wisdom of the believing community, not in isolation. Even reading the Bible for instruction can be misleading unless it is read with the guidance of the Church. Though the message of the life of Jesus seems very individualized, Jesus is very much a man of his community. Though the gospels present a picture of one who stood in opposition to Judaism, this seems to have not been the case. Though he is at odds with the Pharisees, so were the Sadducees and the Essenes and the Zealots and who knows how many other splinter groups, all at variance with one another but all recognized as being a part of god's people. Jesus rarely, if ever, said anything that his contemporaries had not said. What was unique to Jesus was his unbending trust in god and his unflinching obedience to serving god. It is his example that, if Susan will follow it, will help her face her problems. Though all the gospels, but especially John, envision Jesus working out some pre-conceived plan of god in which he suffers a redemptive death predicted by the Old Testament, we need not agree that there was such a plan, and certainly not "from eternity", since god cannot direct the course of human events on such a scale. What we find in Jesus is a man who was thoroughly immersed in the community of his people, a people which had consciously sought to be open to the influence of god. Certainly they had often failed to be open, or being open they had failed to understand, or understanding they had allowed misinterpretation, or avoiding this had just failed to heed it. Still, theirs was a tradition in which god had found a voice, sometimes tenuous, sometimes strong. It is in this community that Jesus is formed and it is out of this tradition that he always acts. There need not be a comprehensive plan which directs his actions, but there is a community wisdom which gives him guidance. If Susan immerses herself in the community of the Church and thus in the accumulated knowledge of the wisdom of god as distilled in the life and teaching of Christ, she too will have this guidance. It is certainly not a specific guidance. Though, being human, the Church has often tried to codify the wisdom and to make a casuistic law of it, this is doomed to failure. The Pharisees tried to do this and Jesus rejected it, yet found the strong guidance of god in his life. Susan will do best not to seek a specific answer to her particular question but a general tendency to guide human activities. Following the direction of a cult leader or a pastor who says, "this is what you must do and this is the way you must do it" may bring her some relief from her problem, but another problem will come along soon enough. No, she and her pastor and the other Christians must examine the many ramifications of her problem and submit them to the wisdom of god as experienced by the community. This will mean that she must consider the other as well as herself. Her husband is one of the "other"s and just because he is the offending party does not remove him from her concern. So are children, in-laws, and friends. So is the Christian Community before whom they married and set themselves up as exemplars of married life and love. Still, she is also a person in her own right who deserves better treatment and a better life and not to be abused by someone else. All these factors have to be considered with a view to doing what is best for all. All this is a far cry from appealing to some code of morality or "natural law" laid down by the creator. It admits to much more subtlety because there is nothing permanent in the universe and no standard of behavior except that of reality itself. And every situation has its own reality. True, every society needs laws and regulations which everyone is expected to observe because they are almost always correct. Religion, which codifies the wisdom of god into commandments, is helpful to most people most of the time. Nothing that has been said about how Susan needs to approach her problems differs from the advice given by a pastor who believes in a law giving, boundary setting creator, except the motivation. It is no longer to please a "jealous" god, or to show proper subservience to the will of a "master" god. She needs to find the best solution to her questions for her own good, and god will help her with the gift of god's wisdom. This, too, is nothing new. Many a theologian has reached the conclusion that the will of god is the welfare of the human person, and this is not a conclusion based on logic following from the notion of god as an all-powerful creator, who could care not a whit for the creatures. No, it comes from the experience of god recorded in the scripture and lived in the church. The difference is that Susan does not now have to go through all the questioning and wondering about why god has caused her the problems in the first place, but can just be aware of god's helpful presence and concern.

The believer will also point to Easter and the resurrection as an indication, at least to themselves, that opening oneself to god's values was also the wisest course for Jesus as an individual. Since our study has not reduced the hope of resurrection for those who live, even loosely, as Jesus did, being his disciple is also the wisest course for each of us as well. If Susan allows god to guide her she, too, is much more likely to make the correct choice. She may make a mistake, but she will always know that she tried her best and god loves her even more for it.

CONCLUSION

I began with a pastoral question. When someone like Susan comes to me and asks, "Why is God doing this to me?", what should I say? In my early years in the priesthood I took the company line. Citing Job was my favorite, with its conclusion that asks who do you think you are to ask such a question?[734] Trust that god knows best. Then there was that question Jesus asked, "those eighteen on whom the tower at Siloam fell and killed them? Do you suppose that they were more guilty than all the other people living in Jerusalem? They were not, I tell you."[735] Certainly misfortune had nothing to do with one's moral position. I was a Catholic, after all, not a Calvinist. Maybe misfortune was just an accident, at least sometimes. Perhaps Susan just had some bad luck, or a string of bad luck, which had nothing to do with god. And then there was all the science, which interested me since my Boy Scout days. Some scientists, at least, seemed to feel that they could explain how the universe came to be as it is without bringing in the presence of a god, and the more I read the more I agreed with them. Surely they had to be wrong somewhere along the line, but the more I looked into it, the farther back along the line that was. When we reached the Planck time of 10^{-43} seconds after the Big Bang, I was at a decision time. Such a remote god was not the god of my experience. Either the scientists were wrong, who would not even put god in before this time in the great unknown beginnings of the universe, or the apologists for god were wrong.

What if it were true that god was not involved in the physical workings of the universe, beginning with creation and right down to Sustenance and Providence? What if we were not contingent beings, owing god our existence? Why then god is absolved of all responsibility for evil and for bringing Susan to grief, for whatever reason, good or bad. But then god would not be worth bothering about, would god? And why would all these people who claimed an experience of god in their lives be saying the same thing, that god did indeed intervene in the universe? And that for me was the key. They were all saying the same thing. We had gotten into a manner of speaking, a paradigm of reality which included god's governance that had remained fairly constant for at least three thousand years, and probably for eons before that. They believed that the mysteries of the universe around us had their basic explanation in a divine actor, period. Nothing could come from nothing; period. If something is, it had a cause; not "period" since you had to allow for at least one uncaused cause, another thing that had always bothered me. Believers rarely questioned these things, and non-believing scientists were not interested in the theological implications. These were assumptions that had not changed because there was no reason to question them, but now they could be examined with a fresh light.

When this was done, there I was where I hope you are now. And there was God, standing forth on the ashes of ancient presumptions just as creative and loving and present as God ever was before.

There is much that remains to be done, that is presuming that I am not found to be an ignorant dolt who has missed the point entirely. If not, then much more needs to be done with morality, with eschatology, with the Trinity and Incarnation, and with many other additional topics of the traditional faith that I have not even touched on, especially spirituality and one's personal relationship with god. I have no doubt that radical changes will have to be made, but I also believe that there is truth in most of the traditional positions and that when they are examined even deeper truths will emerge about the fascinated God and the people who are fascinated by this God.

Bibliography

Abbott, Edwin, *Flatland*, New York: Dover Publications, 1952

Afro-American Folktales, Ed. by Roger D. Abrahams, New York: Pantheon Books, 1985

Albert, David Z., *Quantum Mechanics and Experience*, Cambridge: Harvard University Press, 1992

Alcock, John, *The Triumph of Sociobiology*, New York: Oxford University Press, 2001

An African Prayer Book, Selected by Desmond Tutu. New York: Doubleday, 1995

The Anchor Bible: Genesis, Trans. by E.A. Speiser, Garden City: Doubleday & Company, Inc. 1987

The Anchor Bible: Psalms I, 1-50, Trans. & Notes by Mitchelle Dahood, S.J., New York: Doubleday, 1966

Angelou, Maya, *Wouldn't Take Nothing for My Journey Now*, New York: Random House, 1993,

Aristotle, *On the Heavens*, Trans. W.K.C. Guthrie, The Loeb Classical Library, Cambridge: Harvard University Press, 1926

Aristotle: *On Man in the Universe*, Ed. by Louise Ropes Loomis, New York: Walter J. Black, 1943

Aristotle's Physics, Trans. by Richard Hope, Lincoln: University of Nebraska Press, 1961.

Aristotle: Selected Works, Trans. by Hippocrates G. Apostle and Lloyd P. Gerson, Grinnel: The Peripatetic Press, 1983

Armstrong, Karen, The Battle for God, New York: Alfred A. Knopf, 2000.

Armstrong, Armstrong, in the foreword to Every Eye Beholds You, Ed. by Thomas J. Craughwell, New York: Book of the Month Club, 1998

Armstrong, Karen, *A History of God*, New York: Alfred A. Knopf, 1993,

Asimov, Isaac, *Foundation and Earth*, Garden City: Doubleday, 1986

Augustine Day by Day, Ed. by John E. Rotelle, O.S.A., New York: Catholic book Publishing Co., 1986

Augustine, St., "De Libero Arbitrio", *in St. Augustine, The Problem of Free Choice*, Trans. by Dom Mark Pontifex, Westminster: The Newman Press, 1955

Augustine, St., *The Trinity*, Trans. by Edmund Hill, O.P., Brooklyn: New City Press, 1991,

Aurelio, John R., *Mosquitoes in Paradise*, New York: Crossroads, 1985

The Autobiography of St. Teresa of Avila, Trans. by Kieran Kavanaugh, O.C.D. and Otilio Rodriguez, O.C.D., New York: Book of the Month Club, 1987

Bacon, Francis, *New Atlantis*, New York: Walter J. Black, 1942

Baldwin, James, *Nobody Knows My Name*, New York: Dell Publishing Co., Inc., 1961

Ball, Edward, *Slaves in the Family*, New York: Farrar, Straus and Giroux, 1998

Barbour, Ian, *Religion in an Age of Science*, New York: HarperSanFrancisco, 1990

Barbour, Ian, *The End of Time*, New York: Oxford University Press, 2000

Barclay, William, *Ethics in a Permissive Society*, New York: Harper & Row Publishers, 1971

Barrow, John D., *The Origin of the Universe*, New York: BasicBooks, 1994

Barrow, John D., and Frank J. Tipler, *The Anthropic Cosmological Principle*, New York: Oxford University Press, 1986

Barton, George A., *The Religion of Ancient Israel*, New York: A.S. Barnes & Company, Inc., 1961

Basham, Don, *Deliver Us From Evil*, Old Tappan: Chosen Books, 1972

Benoit, Pierre, O.P., "Inspiration and Revelation", in *The Human Reality of Sacred Scripture*, Concilium, Volume 10, New York: Paulist Press, 1965

Benson, Herbert, *Timeless Healing*, New York: Scribner

Black Elk Speaks: Being the Life Story of a Holy Man of the Oglala Sioux as told through John G. Neihardt, Lincoln: University of Nebraska Press, 1961

Blackburn, Simon, *Think*, Oxford: Oxford University Press, 1999

Blackmore, Susan, *The Meme Machine*, Oxford: University Press, 1999

Bodanis, David, *E=mc2*, New York: Walker & company, 2000

Bonhoeffer, Dietrich, *The Cost of Discipleship*, New York: Macmillan Publishing Co., Inc. 1959

The Book Of Confessions, Philadelphia: Office of the General Assembly of the United Presbyterian Church in the United States of America, #6.024

Bosler, Raymond, *What a Modern Catholic Believes About Moral Problems*, Chicago: The Thomas More Press, 1971

Boslough, John, *Stephen Hawkings Universe*, New York: Quill/William Morrow, 1985

Bouillard, Henri, S.J., "Human Experience as the Starting Point of Fundamental Theology" in Concilium: The Church And the World, Volume 6. New York: Paulist Press. 1965

Bowen, Catherine Drinker, *Miracle at Philadelphia*, New York: Book of the Month Club Inc., 1986

Bradbury, Ray, *Something Wicked This Way Comes*, New York: Bantam Books, 1962

Brady, James, *Holy Wars*, New York: Simon and Schuster, 1983

Brown S.S., Raymond E., *Jesus God and Man*, New York: Macmillan Publishing Co., Inc., 1967

Brueggemann, Walter, *Texts Under Negotiation*, Philadelphia: Fortress Press, 1994.

Brungs, Robert A., S.J. *You See Lights Breaking Upon Us*, St. Louis: by the author, 1989

Buddhism, Edited. by Richard A. Gard, New York: George Braziller, 1962

Bulst, Werner, S.J., *Revelation*, Trans. by Bruce Vawter, C.M., New York: Sheed and Ward, 1965

Burtt, Edwin Arthur, *The Metaphysical Foundations of Modern Science*, Garden City:

Doubleday & Company, 1954

Burton, Diane, *Many Moons*, New York: Prentice Hall Press, 1991

Butterworth, Robert, *The Theology of Creation*, Notre Dame: Fides Publishers, Inc., 1969

Campbell, Allan, "Transposons and Their Evolutionary Significance" in *Evolution of Genes and Proteins*, Ed. by Masatoshi Nei and Richard K. Koehn, Sunderland: Sinauer Associates Inc., 1983

Campbell, Joseph, *The Power of Myth*, New York: Doubleday, 1988

Capra, Fritjof, and David Steindl-Rast, *Belonging to the Universe*, New York: HarperSanFrancisco, 1991

Capra, Fritjof, *The Web of Life*, New York: Anchor Books, 1996, p. 228.

Carretto, Carlo, *Why O Lord?*, Maryknoll, New York: Orbis Books, 1986

Carrighar, Sally, *Wild Heritage*, New York: Ballantine Books, 1965

Carson, Rachel, *Silent Spring*, Greenwich: Fawcett Publications, Inc., 1962

Casson, Lionel, "Ancient Egypt", *Time-Life Great Ages of Man series*, Time, Inc. 1965

Castelot, S.S., John J., *Meet the Bible!*, Baltimore: Helicon Press, 1960

Catechism of the Catholic Church, Liguori: Liguori Publications, 1994

Catton, Bruce, *This Hallowed Ground*, New York: Pocket Books, Inc. 1956

Calkins, Hugh, *Living with Illness*, Chicago: Thomas More Press, 1973

Chesterton, G.K., *St. Thomas Aquinas*, New York: Image Books, 1956

Chardon, Louis, O.P., *The Cross of Jesus*, Trans. by Richard T. Murphy, O.P., St. Louis: B. Herder Book Co., 1957

Chopra, Deepak, *How To Know God: The Soul's Journey into the Mystery of Mysteries*, New York: Harmony Books, 2000

Clancy, Tom, *The Bear and the Dragon*, New York: G.P. Putnam's Sons, 2000,
Clark, William R., & Michael Grunstein, *Are We Hardwired?*, Oxford: Oxford University Press, 2000
Clarke, Breena, *River, Cross My Heart*, Boston: Little, Brown Company, 1999
Classics of Western Thought: The Ancient World, Ed. by Stebelton H. Nulle, New York: Harcourt Brace Jovanovich, Inc., 1968
Cohen, Jack, & Ian Stewart, *The Collapse of Chaos*, New York: Viking, 1994
Cole, K.C., *A Hole In the Universe*, New York: Harcourt, Inc., 2001
Collins, James, *A History of Modern European Philosophy*, Milwaukee: The Bruce Publishing Company, 1954
Colson, Charles, *Loving God*, Grand Rapids: Zondervan Publishing House, 1983
Connell, Evan S., *Son of the Morning Star*, New York: Harper & Row, 1984
Conrad, Joseph, *Victory*, Garden City: Doubleday Anchor Books, 1915
The Constitution of the United Presbyterian Church in the United States of America, Part I Book of Confessions Philadelphia, 1966
Cooke, Bernard, *Christian Sacraments and Christian Personality*, New York: Holt, Rinehart & Winston, 1965
Cooper-Lewter, Nicholas C. and Mitchell, Henry H., *Soul Theology: The Heart of American Black Culture*, San Francisco: Harper & Row Publishers, 1986
Copernicus, Book I of "De Revolutionibus Orbium Caelestium" (1543), cited by Thomas S. Kuhn, *The Copernican Revolution*, Cambridge: Harvard University Press, 1985
Copleston, Frederick, S.J., *A History of Philosophy*, Vol. II, "Mediaeval Philosophy", Part II, Garden City: Image Books, 1950
Crick, Francis, *The Astonishing Hypothesis*, New York: Charles Scribner's Sons, 1994,
Crick, Francis, *What Mad Pursuit*, New York: Basic Books, Inc. Publishers, 1988
Cross, F.L., *The Early Christian Fathers*, London: Gerald Duckworth & Co. Ltd. 1960
Cross, Frank Moore, *From Epic to Canon*, Baltimore: Johns Hopkins Univ. Press, 1998.]
Crossan, John Dominic, *Jesus, a Revolutionary Biography*, New York: HarperSanFrancisco, 1994

Cyril, St., "On the Maker of Heaven and Earth", in *Orations from Homer to William McKinley*, Volume III, Ed. by Mayo W. Hazeltine, A.M., New York: P.F. Collier and Son, 1902

Cziko Gary, *Without Miracles*, 1997

Damasio, Antonio R., *Descartes' Error*, New York: G.P. Putnam's Sons, 1994

Damasio, Antonio, *The Feeling of What Happens*, New York: Harcourt Brace, 1999

Damon, William, *The Moral Child: Nurturing Children's Natural Moral Growth,* New York: Free Press, 1990

Danielou, *In the Beginning...Genesis I-III*, Baltimore: Helicon, 1965

Darling, David, *Soul Search*, New York: Villard Books, 1995

Davies, Paul, *Abouttime*, New York: Simon & Schuster, 1995

Davies, Paul, *The Cosmic Blueprint*, New York: Simon and Schuster, 1988

Davies, Paul, *The Last Three Minutes*, New York: BasicBooks, 1994,

Davies, Paul, and John Gribbin, *The Matter Myth*, New York: Simon & Schuster, 1992,

Davies, Paul, *The Fifth Miracle*, New York: Simon and Schuster, 1999

Dawkins, Marian Stamp, *Through Our Eyes Only?*, Oxford: W.H. Freeman & Co. Ltd., 1993

Dawkins, Richard, *The Blind Watchmaker*, London: Longmans, 1986

Dawkins, Richard, *Climbing Mount Improbable*, New York: W.W. Norton & Company, 1996

Dawkins, Richard, *River Out Of Eden*, New York: BasicBooks, 1995

Dawkins, Richard, *The Selfish Gene*, New York: The Oxford University Press, 1990

Dawkin, Richard, *Unweaving the Rainbow: Science, Delusion, and the Appetite for Wonder*, New York: Houghton Mifflin, 1998

Dennett, Daniel C., *Darwin's Dangerous Idea*, New York: Simon & Schuster, 1995

Descartes, Rene, "Meditations on the First Philosophy" in *The Rationalist*, Translated by John Veitch, New York: Doubleday & Co. Inc. 1960

De Chardin, Pierre Teilhard, *The Future of Man*, New York: Harper & Row, Publishers, 1964

De Chardin, Pierre Teilhard, *The Phenomenon of Man*, New York: Harper & Row. 1959

De Gasztold, Carmen Bernos, *Prayers from the Ark*, Trans. by Rumer Godden, New York: Penguin Books, 1976

de Montfort, St. Louis Mary, *The Secret of the Rosary*, Trans. by Mary Barbout, T.O.P., Bay Shore: Montfort Publications, 1964

de Vaux, Roland, O.P., *Ancient Israel*, New York: McGraw-Hill Book Co., 1961
de Waal, Frans, *The Ape and the Sushi Master: Cultural Reflections of a Primatologist*, New York: Basic Books, 2001
de Waal, Frans, *Good Natured*, Cambridge: Harvard University Press, 1996
Diamond, Jared, *Why Is Sex Fun?*, New York: Basic Books, 1997
Diekmann, Godfrey, O.S.B., "Two Approaches to Understanding the Sacraments" in *Readings in Sacramental Theology*, Edited by C. Stephen Sullivan, F.S.C., Englewood Cliffs: Prentice-Hall, 1964
Dillard, Annie, *For the Time Being*, New York: Alfred A. Knopf, 1999
Dillard, Annie, "Pilgrim At Tinker Creek" in the Book-of-the-Month Club edition *Annie Dillard*, New York: Harper & Row, 1990
Dillenschneider, Clement, C.Ss.R., *The Dynamic Power of Our Sacraments*, Trans. by Sr. M. Renelle, S.S.N.D., St. Louis: B Herder Book Co., 1966
Doctorow, E.L., *City of God*, New York: Random House, 2000
Dolan, John P., *History of the Reformation*, New York: Desclee Company, 1965
Dossey, Larry, M.D., *Healing Words*, New York: HarperSanFrancisco, 1993
Dubarle, Andre-Marie, O.P. *The Biblical Doctrine of Original Sin*, New York: Herder and Herder, 1964
The Dynamism of Biblical Tradition, in Concilium, Vol. 20, New York: Paulist Press, 1967
Dyson, Dr. Freeman J., in "Religion, Science, and the Search for Wisdom", Ed. by David M. Byers, Bishops' Committee on Human Values, NCCB
Edelman, Gerald M., and Giulio Tononi, *A Universe of Consciousness*, New York: Basic Books, 2000
Edwards, Denis, *The God of Evolution*, New York: Paulist Press, 1999
Edwards, Denis, *The Human Experience of God*, New York: Paulist Press, 1983
Edwards, Denis, *Jesus and the Cosmos*, New York: Paulist Press, 1991
Eisenman, Robert H., and Michael Wise, *The Dead Sea Scrolls Uncovered*, New York: Penguin Books, 1992
Eissfeldt, Otto, *The Old Testament*, Trans. by Peter R. Ackroyd, New York: Harper and Row, Publishers, 1965
Eldredge, Niles, *The Triumph of Evolution dna eht eruliaF fo msiniotaerC*, New York: W.H. Freeman and Company, 2000
Elizondo, Virgil, *The Human Quest*, Huntington: Our Sunday Visitor, Inc., 1978

Ellis, George F.R., *Before the Beginning: Cosmology Explained*, New York: Boyars/Bowerdean, 1993

Erwin, Douglas H., *The Great Paleozoic Crisis: Life and Death in the Permian*, New York: Columbia University Press, 1993

Estermann, Claus, *Creation,* Trans. by John J. Scullion, S.J., Philadelphia: Fortress Press, 1974

Evely, Louis, *That Man Is You*, Westminster: The Newman Press, 1965

Eyes On the Modern World, Ed. by John G. Deedy, Jr., New York: P.J. Kennedy & Sons, 1965

Faulkner, William, *The Wild Palms*, New York: Vintage Books, 1939

Faulkner, William, "Nobel Prize Acceptance Speech" in *Essays, Old and New*, Edited by Robert U. Jameson, New York: Harcourt, Brace and Company, 1955

Ferris, Timothy, *The Mind's Eye*, New York: Bantam Books, 1992

Ferris, Timothy, *The Whole Shebang*, New York: Simon & Schuster, 1997

Feynman, Richard P. *The Meaning of It All*, Reading: Helix Books: 1998

Feynman, Richard P., *Six Not So Easy Pieces*, Reading: Helix Books, 1997

Fine, Steven, *This Holy Place: On the Sanctity of the Synagogue During the Greco-Roman Period*, Notre Dame: University of Notre Dame Press, 1997

The Five Gospels, New York: Macmillan Publishing Company, 1993

Five Great Dialogues: Plato, Trans. by B. Jowett, Ed. by Louise Ropes Loomis, New York: Walter J. Black, 1942

Foley, Leonard, O.F.M., *From Eden to Nazareth*, St. Anthony Messenger Press, 1983

Foley, Leonard, O.F.M., *Believing in Jesus: A Popular Overview of the Catholic Faith*, Cincinnati: St. Anthony Messenger Press, 1981

Fortey, Richard, *Life*, New York: Alfred A. Knopf, 1998

Fox, Matthew, thinks he has common ground for science and religion, especially in *The Cosmic Christ*

Fox, Matthew, *Original Blessing*, Santa Fe: Bear & Company, 1983

Fromm, Erich, *You Shall Be As Gods*, Greenwich: Fawcett Publications, Inc. 1966

Fuller, Steve, Thomas Kuhn: *A Philosophical History for Our Times*, Chicago, University of Chicago Press, 2000

Galligan, Michael, *God and Evil*, New York: Paulist Press, 1976

Ganly, John C., M.M., *Kaonde Proverbs*, private printing

Gebara, Ivone, *Longing for Running Water*, Minneapolis: Fortress press, 1999

Genz, Henning, *nothingness*, Reading: Helix Books, 1999

Gell-Mann, Murray, *The Quark and the Jaguar*, New York: W.H. Freeman and Company, 1994

George, Francis Cardinal, OMI, "Biotechnology and Some Thought on the Body", *Proceedings of The Genome, Plant, Animal, Human*, Ed. by Robert Brungs, SJ & Marianne Postiglione, RSM, St. Louis: ITEST Faith/Science Press, 2000,

Gerhart, John, and Marc Kirschner, *Cells, Embryos and Evolution*, Malden: Blackwell Science, 1997

Giblet, Jean, "Baptism- The Sacrament of incorporation into the Church, according to St. Paul", in *Baptism in the New Testament*, Trans. by David Askew, Baltimore: Helicon, 1964

Gilmore, Robert, *Alice in Quantumland*, New York: Springer-Verlag, 1995

Gladwell, Malcolm, *The Tipping Point*, New York: Little, Brown, 2000

Glynn, Patrick, *God: The Evidence*, Rocklin: Prima Publishing FORUM, 1999

Goldhagen, Daniel Jonah, *Hitler's Willing Executioners*, New York: Alfred A. Knopf, 1996

Goodenough, Ursula, *The Sacred Depths of Nature*, New York: Oxford University Press, 1998

Gould, Stephen J., *Dinosaur in a Haystack*, New York: Harmony Books, 1995

Gould, Stephen Jay, *The Lying Stones of Marrakech*, New York: Harmony Books, 2000

Gould, Stephen J., *Rocks of Ages*, New York: The Ballantine Publishing Group, 1999

"Great Religions of Modern Man", *Catholicism*, ed. by George Brantl, New York: George Braziller, 1962

Greeley, Andrew M., *What a Modern Catholic Believes About God*, Chicago: The Thomas More Press, 1971

Green, Roger Lancelyn, *The Tale of Troy*, Harmondsworth: Puffin Books, 1958

Greene, Brian, *The Elegant Universe*, New York: W.W.Norton & Company, 1999

Greene, Graham, *The Heart of the Matter*, Harmondsworth: Penguin Books, 1948

Greene, Graham, *Monsignor Quixote*, New York: Simon and Schuster, 1982

Greene, Graham, *The Power and the Glory*, New York: Bantam Books, 1940

Greenstein, George, *Portraits of Discovery*, New York: Knopf, 1999

Greenstein, George, *The Symbiotic Universe*, William Morrow and Company, Inc., 1988

Gribbin, John, *Companion to the Cosmos*, Boston: Little, Brown & Co. 1996

Gribbin, John, *Schrodinger's Kittens and the Search for Reality*, Boston: Little, Brown and Company, 1995

Groeschel, Father Benedict J., C.F.R. and James Monti, *In the Presence of Our Lord*, Huntington: Our Sunday Visitor Publishing Division, 1997

Gross, Paul R., and Norman Levitt, *Higher Superstition: The Academic Left and Its Quarrels with Science*, Baltimore: The Johns Hopkins University Press, 1994

Haag, Herbert, *Is Original Sin in Scripture?*, Trans. by Dorothy Thompson, New York: Sheed and Ward, 1969

Hamlin, Rick, *Finding God on the A Train*, New York: HarperSanFrancisco, 1997

Haring, Bernard, C.Ss.R., *Shalom: Peace*, New York: Farrar, Straus and Giroux, 1967

Harper's Encyclopedia of Bible Life, Madeleine S. & J. Lane Miller, Harper & Row, New York 1978

Harrington, Daniel J., S.J., *why do we suffer?*, Franklin: Sheed & Ward, 2000

Harrison, Edward *Masks of The Universe*, New York: Macmillan Publishing Company, 1985

Hart, Charles A. *Thomistic Metaphysics*, New Jersey: Prentice-Hall, Inc. 1959

Haught, John F., *Science and Religion*, New York: Paulist Press, 1995

Haughton, Rosemary, *The Passionate God*, New York: Paulist Press, 1981

Hauser, Marc, *Wild Minds: What Animals Really Think*, New York: Henry Holt, 2000

Hawking, Stephen, *A Brief History of Time*, New York: Bantam Books, 1988

Hawking, Stephen, *Black Holes and Baby Universes*, New York: Bantam Books, 1993

Hawkins, Michael, *Hunting Down the Universe: The Missing Mass, Primordial Black Holes, and Other Dark Matters*, London: Addison-Wesley, 1997

Hellwig, Monika, *The Meaning of the Sacraments*, Cincinnati: Pflaum/Standard, 1972,

Hermann, Ingo, "Total Humanism" in *Is God Dead*, Concilium Vol. 16, New York: Paulist Press, 1966

Hey, Tony, and Patrick Walters, *Einstein's Mirror*, New York: Cambridge University Press, 1997

Hinduism, Edited by Louis Renou, New York: George Braziller, 1962
Howard, Ted, and Jeremy Rifkin, *Who Should Play God?*, New York: Dell Publishing Co. Inc., 1977
Hoy, Michael, "A Theology of the Body: Body, Genes and Culture, Who's Holding the Leash?" in *Christianity and the Human Body: Proceedings of the ITEST Workshop, October, 2000*, St. Louis: ITEST FAITH/SCIENCE PRESS, 2001
Hurnard, Hannah, *Hinds' Feet on High Places*, Old Tappan: Fleming H. Revell Company, 1973
Hurston, Zora Neal, *Their Eyes Were Watching God*, Urbana: University of Illinois Press, 1937
Huxley, Julian, *Essays of a Humanist*, New York: Harper and Row, 1964
Huxley, Julian, *Man Stands Alone*, New York: Harper & Brothers, 1941
Huxley, Julian, *Man in the Modern World*, New York: Mentor Books, 1962
Irenaeus, *Adversus Haeres*, 4, 20, 7:PG 7/1, 1037
Islam, Edited by John Alden Williams, New York: George Braziller, 1962
Jammer, Max, *Einstein and Religion*, Princeton: Princeton University Press, 1999
Jastrow, Robert, *The Enchanted Loom: Mind in the Universe*, New York: Simon and Schuster, 1981
Jeanniere, Abel, *The Anthropology of Sex*, New York: Harper & Row, Publishers, 1967
The Jerome Biblical Commentary, Ed. by Raymond Brown, et al., Englewood Cliffs: Prentice-Hall, 1968
John Paul II, Pope, *Crossing the Threshold of Hope*, New York: Alfred A. Knopf, 1994
John Paul II, Pope, *Celebrate 2000*, Selected by Paul Thigpen, Ann Arbor: Servant Publications, 1996
John Paul II, Pope, "To All the Priests of the Church On the occasion of Holy Thursday, 1979", in *Set Apart for Service*, Boston: St. Paul Editions, 1982
Johnson, Elizabeth A., *She Who Is*, New York: Crossroad, 1997
Johnson, George, *Fire in the Mind*, New York: Alfred A. Knopf, 1995
Johnston, Victor S., *Why We Feel*, Cambridge: Perseus Books, 1999
Jones, Howard, *Mutiny on the Amistad*, New York: Oxford University Press, 1987
Jorgensen, Johannes, *Saint Francis of Assisi*, Trans. by T. O'Conor Sloane, Garden City: Image Books, 1955
Kafatos, Menas, and Robert Nadeau, *The Conscious Universe*, New York: Springer-Verlag New York, Inc., 1990

Kaku, Michio, & Jennifer Thompson, *Beyond Einstein*, New York: Anchor Books, 1987

Kasper, Walter, *The God of Jesus Christ*, Trans. by Matthew J. O'Connell, New York: Crossroad, 1984

Kauffman, Stuart, *At Home In the Universe*, New York: Oxford University Press, 1995

Kayzer, Wim, *A Glorious Accident*, New York: W.H. Freeman and Company, 1997

Kelly, William L. S.J., & Andrew Tallon, *Readings in the Philosophy of Man*, New York: McGraw-Hill Book Company, 1967

King, Stephen, *The Drawing of the Three*, New York: New American Library, 1987

Kitcher, Phillip, *Abusing Science*, Cambridge: The MIT Press, 1982

Klein, Etienne, *Conversations with the Sphinx*, Trans. by David LeVay, London: Souvenir Press, 1996

Korsmeyer, Jerry D., *Evolution & Eden*, New York: Paulist Press, 1998

Kowalski, Gary, *The Souls of Animals*, Wallpole: Stillpoint, 1991

Krauss, Lawrence, *Beyond Star Trek*, New York: Basic-Books, 1997

Kubler-Ross, Elizabeth, *Death: The Final Stage of Growth*, Englewood Cliffs: Prentice-Hall, Inc., 1975

Kubler-Ross, Elizabeth, *On Death and Dying*, New York: Macmillan Publishing Company, 1969

Kubler-Ross, Elizabeth, *Questions and Answers on Death and Dying*, New York: Macmillan Publishing Co., Inc. 1974

Kuhn, Thomas S., *The Copernican Revolution*, Cambridge: Harvard University Press, 1985

Kuhn, Thomas, *A Philosophical History for Our Times*, Chicago, University of Chicago Press, 2000

Kung, Hans, *Eternal Life*, Trans. by Edward Quinn, Garden City: Doubleday & Company, Inc., 1984

Kung, Hans, *On Being a Christian*, Translated by Edward Quinn, Garden City: Doubleday & Company, Inc., 1976

Kung, Hans, *Theology for the Third Millenium*, Trans. by Peter Heinegg, New York: Doubleday, 1988

LaCugna, Catherine Mowry, *God For Us*, New York: HarperSanFrancisco, 1991

Lakoff, George, and Mark Johnson, *Philosophy in the Flesh*, New York: Basic Books, 1999

Lane, Dermot, *The Reality of Jesus*, New York: Paulist Press, 1975

Laplace, Jean, S.J. *Prayer According to the Scriptures*, Brightons: Alpine Press

LaVerdiere, Eugene A., S.S.S., *The Presence of Christ in the Eucharist in Bread From Heaven*, Ed. by Paul J. Bernier, S.S.S., New York: Paulist Press, 1977
Leakey, Richard, *The Origin of Humankind*, New York: BasicBooks, 1994
Lee, Harper, *To Kill a Mockingbird*, New York: Warner Books, Inc. 1960
Leslie, John, *Universes*, New York: Routledge, 1989
Lewis, C.S., *Miracles: A Preliminary Study*, New York: The Macmillan Company, 1947
Lewis, C.S., *The Problem of Pain*, New York: Macmillan Publishing Co., Inc. 1962
Lewis, C.S., *The Screwtape Letters*, New York: Macmillan Publishing Co., Inc. 1961
Lightman, Alan, *Time for the Stars*, New York: Viking, 1992
Ligneul, Andre, *Teilhard and Personalism*, New York: Paulist Press Deus Books, 1968
The Living Faith, New York: Herder and Herder, 1959
Locke, John, *Essay Concerning Human Understanding*, 1690
Lucretius, *On The Order Of Things*, Book I, Translated by Charles E. Bennett, New York: Walter J. Black, 1946
Lukefahr, Oscar, *A Catholic Guide to the Bible*, Ligouri: Ligouri Publications, 1992
Lukefahr, Oscar, *We Believe*, Liguori: Liguori Publications, 1990
Lyddon, Sister Eileen, *Door Through Darkness: John of the Cross and mysticism in everyday life*, Hyde Park: New City Press, 1995
Maccoby, Eleanor E., *The Two Sexes: Growing up Apart, Coming Together*, CambridgeBelknap Press, 1998
Mackenzie, R.A.F., S.J., *Faith and History in the Old Testament*, New York: The Macmillan Company, 1963
Maclagan, David, *creation myths*, London: Thames and Hudson, 1977
Maertens, Thierry, O.S.B. *Bible Themes*, Vol. 2, Notre Dame: Fides/Claretian, 1964
Magee, Bryan, *Confessions of a Philosopher*, New York: Random House, 1997
Maguire, Daniel C., *Moral Absolutes and the Magisterium in Absolutes in Moral Theology?*, Edited by Charles E. Curran, Washington: Corpus Books, 1968
Maguire, Daniel C., *Sacred Choices: The Right to Contraception and Abortion in Ten World Religions*, New York: Fortress Press, 2001
Mallove, Eugene F., *The Quickening Universe*, New York: St. Martin's Press, 1987

Maritain, Jacques, *A Preface to Metaphysics*, New York: New American Library, 1962

Marlowe, Christopher, *The Tragedy of Doctor Faustus*, Edit. by Louis B. Wright, New York: Washington Square Press, Inc., 1959

Martin, M. B., S.J., *Christianity and Its Cultural Bondage*, St. Louis: Robert Bleile & Assoc., 1972

Masson, Jeffrey Moussaieff, and Susan McCarthy, *When Elephants Weep*, New York: Delacorte Press, 1995

Maurer, Armand, *St. Thomas Aquinas, The Division and methods of the Sciences*, Toronto: Pontifical Institute of Mediaeval Studies, 1986

May, Rollo, *Love and Will*, New York: Dell Publishing Co., Inc., 1969

McBrien, Richard P., *Who is a Catholic*, Denville: Dimension Books, 1971

McKenzie, John, S.J., *Dictionary of the Bible*, Milwaukee: The Bruce Publishing Co, 1965

McKenzie, John, *Did I Say That?*, Chicago: The Thomas More Press, 1973

McMullin, Rev. Ernan, "Natural Science and Belief in a Creator", in *Proceedings of a Conference on Religion and Science Conference on Religion and Science September 1986*, Ed. by David M. Byers, Washington: United States Catholic Conference, Inc., 1987

Medwick, Cathleen, *Teresa of Avila: The Progress of a Soul*, New York: Alfred A. Knopp, 1999

Melville, Herman, *Moby Dick*, New York: Airmont Publishing Co. Inc., 1964

Melvin, Edward J., C.M. *A Nation Built on God*, Huntington: Our Sunday Visitor, Inc., 1975

Miller, Kenneth, *Finding Darwin's God,* New York: Cliff Street Books, 1999

Michener, James A., *The Covenant*, New York: Random House, 1980

Monden, Louis, S.J., *Faith: Can Man Still Believe?*, Trans. by Joseph Donceel, S.J., New York: Sheed & Ward, 1970

Mother Teresa, *No Greater Love*, Ed. by Becky Benenate & Joseph Durepos, Novato: New World Library, 1997

Moore, Sebastian, *The Crucified Jesus Is No Stranger*, New York: The Seabury Press, 1981

Moore, Sebastian, *No Exit*, New York: Newman Press, 1968

Morrison, Toni, *Sula*, New York: Plume, 1973

Morse, Melvin, M.D., *Parting Visions*, New York: Villard Books, 1994

Mowinckel, Sigmund, *The Psalms in Israel's Worship*, Trans. by D.R. Ap-Thomas, New York: Abingdon Press, 1962

Moyers, Bill, *Genesis: A Living Conversation*, New York: Doubleday, 1996

Murphy, Dennis J., M.S.C., *His Servants the Prophets*, Collegeville: The Liturgical Press, 1965
Murphy, Roland E., O.Carm., in "The Tree of Life", in the *Anchor Bible Reference Library*, New York: Doubleday, 1990
Murray, John Courtney, S.J., *The Problem of God*, New Haven: Yale University Press, 1964
Murray, Michael H., *The Thought of Teilhard de Chardin*, New York: The Seabury Press, 1966
Nedoncelle, Maurice, *God's Encounter With Man*, New York: Sheed and Ward, 1964
Newman, John Henry, *The Idea of a University*, New York: American Press, 1941
Noble, David F., *The Religion of Technology*, New York: Alfred A. Knopf, 1997
Nogar, Raymond J., O.P., "Evolutionary Humanism and the Faith", in *Death and Dying*, Concilium, Vol. 16, New York: Paulist Press, 1966
Nogar, Raymond J., O.P., *The Lord of the Absurd*, Notre Dame: University of Notre Dame Press, 1966
North, John, *The Norton History of Astronomy and Cosmology*, Ed. by Roy Porter, New York: W.W. Norton & Co., 1995
Nowell, Robert, *What a modern Catholic believes about Death*, Chicago: The Thomas More Press, 1972
O'Connell, Timothy, *What a Modern Catholic Believes About Suffering and Evil*, Chicago: The Thomas More Press, 1972
O'Murchu, Diarmuid, *Quantum Theology*, New York: The Crossroads Publishing Company, 1997
O'Neil, Robert P., and Michael A. Donovan, *Sexuality & Moral Responsibility*, Washington: Corpus Books, 1968
One Lord One Baptism, World Council of Churches Commission on Faith and Order, Minneapolis, Augsburg Publishing House, 1960
The Ontological Argument, Ed. by Alvin Plantinga, Garden City: Doubleday & Company, Inc. 1965
Oraison, Marc, *Morality for our Time*, Garden City: Image Books, 1968
Orwell, George, *1984*, New York: New American Library 1949
Overbye, Dennis, *Lonely Hearts of the Cosmos*, New York: HarperCollins Publishers, 1991
Paley, William, *Natural Theology—or Evidences of the Existence and Attributes of the Deity Collected from the Appearances of Nature*, Oxford: Clarendon Press, 1938

Pannenberg, Wolfhart, "Redemptive Event and History", in *Essays on Old Testament Hermeneutics*, Ed. by Claus Westermann, Atlanta: John Knox Press, 1963

Pannenberg, Wolfhart, *Toward A Theology of Nature*, Louisville: Westminster/John Knox Press, 1993

Peacocke, Arthur, *God and the New Biology*, San Francisco: Harper & Row, 1986

Peck, M. Scott, M.D., *Further Along The Road Less Traveled*, New York: Simon and Schuster, 1993

Pellegrino, Charles, *Return to Sodom and Gomorrah*, New York: Random House, 1994

Pennock, Michael, *Prayer & You*, Notre Dame: Ave Maria Press, 1985

Pennock, Robert T., *Tower of Babel: The Evidence against the New Creationism*, Cambridge: The MIT Press, 1999

Pert, Candace B., Ph.D., *Molecules of Emotion*, New York: Scribner, 1997

Pieper, Josef, *The Silence of St. Thomas*, Trans. by John Murray and Daniel O'Connor, Chicago: Henry Regnery Co., 1957

Steven Pinker, *How the Mind Works*, New York: W.W. Norton & Company, 1997

Pius XII, Pope, *Divino Afflante Spiritu*, 1943

Plastaras, James, C.M., *The God of Exodus*, Milwaukee: The Bruce Publishing Company, 1966

The Pope Speaks: Dialogues of Paul VI with Jean Guitton, New York: Meredith Press, 1968

Popper, Karl, and John Eccles, *The Self and Its Brain*, Berlin: Springer International, 1977

Potok, Chaim, *In the Beginning*, New York: Fawcett Crest, 1975

Powers, Joseph, S.J. "Faith and the Eucharist", in *Bread From Heaven*, Ed. by Paul J. Bernier, S.S.S., New York: Paulist Press, 1977

Powers, J. F., *Prince of Darkness and Other Stories*, New York: Image Books, 1947

Polkinghorne, John, *The Faith of a Physicist*, Minneapolis: Fortress Press, 1996

Polkinghorne, John, "The Modern Interaction of Science and Theology", in *The Great Ideas Today*, Encyclopaedia Britannica, Inc., 1995

Polkinghorne, John, *Quarks, Chaos & Christianity*, New York: Crossroad Publishing Co., 1996

Pollard, Miriam, O.C.S.O., *The Laughter of God*, Wilmington: Michael Glazier, 1986

Powers, Joseph, *Eucharistic Theology*, New York: Herder & Herder, 1967

Price, Huw, *Time's Arrow & Archimedes' Point*, New York: Oxford University Press, 1996

Process Theology, Ed. by Ewert H. Cousins, New York: Newman Press, 1971

Quantum Questions, Ed. by Ken Wilber, Shambhala: New Science Library, 1985

Quoist, Michel, *Prayers*, Trans. by Agnes M. Forsyth & Anne Marie de Commaille, New York: Sheed and Ward, 1963

Rahner, Karl, *Do You Believe in God*, Trans. by Richard Strachan, New York: Newman Press 1969

Rahner, Karl, *Meditations on the Sacraments*, New York: The Seabury Press, 1977

Rahner, Karl, *Nature and Grace*, New York: Sheed and Ward, 1964

Rahner, Karl, *Theological Investigations*, Volume IV, Trans. by Kevin Smyth, Baltimore: Helicon Press

Ratzinger, Joseph, *In The Beginning...*, Huntington: Our Sunday Visitor, 1990

Ratzinger Joseph, *Introduction to Christianity*, Trans. by J. R. Foster, New York: Herder and Herder, 1970

Raup, David M., *The Role of Extinction in Evolution in Tempo and Mode in Evolution*, Ed. by Walter M. Fitch and Francisco J. Ayala, Washington, D.C., National Academy Press, 1995

Raymo, Chet, *Skeptics and True Believers*, New York: Walker and Company, 1998

Rees, Martin, *Just Six Numbers*, New York: Basic Books, 2000

Reformed Confessions of the 16th Century, Ed. by Arthur C. Cochrane, Philadelphia: The Westminster Press 1966

Religion, *Science, and the Search for Wisdom*: Proceedings of a Conference on Religion and Science September 1986, Ed. by David M. Byers, Washington: United States Catholic Conference, Inc., 1987

Remus, Harold, *Jesus as Healer*, Cambridge: Cambridge University Press, 1997

Jean-Francois Revel, Jean-Francois, and Matthieu Ricard, *The Monk and the Philosopher*, Trans. by John Canti, New York: Schocken Books, 1998

Rhodes, Jewell Parker, *Voodoo Dreams*, New York: St. Martin's Press, 1993

Rice, Anne, *Merrick*, New York: Alfred A Knopf, 2000

Rich, Charles, *Reflections from an Inner Eye*, Huntington: Our Sunday Visitor, Inc. 1977

Ridley, Mark, *Evolution,* 2nd Edition, Cambridge: Blackwell Science, Inc., 1996

Ridley, Matt, *Genome*, New York: HarperCollins, 1999

Rohr, Richard, *The Good News According to Luke*, New York: The Crossroads Publishing Company, 1997

Rose Eileen, Sr. M., C.S.C., *The Spirit of the 'Anawim, in Contemporary New Testament Studies*, Collegeville, The Liturgical Press, 1965

Rosse, Gerard, *The Cry of Jesus on the Cross*, Trans. by Stephen Wentworth Arndt, New York: Paulist Press, 1987

Rostand, Edmond, *Cyrano de Bergerac*, New York: Bantam Books, 1923

Ruffle, John, *The Egyptians*, Ithica: Cornell University Press, 1977

Ruse, Michael, *Mystery of Mysteries: Is Evolution a Social Construction?*, Cambridge: Harvard University Press, 1998

Ryan, William, & Walter Pitman, *Noah's Flood*, New York: Simon and Schuster

Sabourin, Leopold, S.J., *The Psalms*, New York: Alba House, 1970

Sacks, Oliver, *The Man Who Mistook His Wife for a Hat*, New York: Summit Books 1985

Sacramentum Mundi, Vol. 2 Ed. by Karl Rahner S.J. New York: Herder and Herder, 1968

Sagan, Carl, *Billions and Billions*, New York: Random House, 1997

Sagan, Carl, *Demon Haunted World*, New York: Random House, 1995

St. Joseph New American Catechism #3, arranged by Rev. Lawrence G. Lovasik, S.V.D., New York: Catholic Book Publishing Co., 1985

Scanzoni, Letha, and Nancy Hardesty, *All We're Meant to Be*, Waco: Word Books, Publisher, 1974

Schillebeeckx, Edward, *Christ the Sacrament of the Encounter With God*, New York: Sheed and Ward, 1963

Schillebeeckx, Edward, O.P., *Marriage, Human Reality and Saving Mystery*, Trans. by N.D. Smith, New York: Sheed and Ward, 1965

Schlier, Heinrich, *The Relevance of the New Testament*, New York: Herder and Herder, 1968

Schlink, M. Basilea, *Realities*, Grand Rapids: Zondervan Publishing House, 1966

Schokel, Luis Alonso, S.J., *The Inspired Word*, New York: Herder and Herder, 1965

Schoof, Mark, O.P., *A Survey of Catholic Theology 1800-1970*, Paramus: Paulist Newman Press, 1970

Schoonenberg, Piet, S.J., "TRANSUBSTANTIATION: How Far Is This Doctrine Historically Determined?", in *The Sacraments: An*

Ecumenical Dilemma, Concilium, Vol. 24, New York: Paulist Press, 1966

Schroeder, Gerald L., Ph.D. *Genesis and the Big Bang*, New York: Bantam Books. 1990

Schwartz, Jeffrey H., *Sudden Origins*, New York: John Wiley and Sons, Inc, 1999

Searle, John R., Mind, *Language, and Society*, New York: Basic Books, 1998

Segerstrale, Ursula, *Defenders of the Truth*, Oxford: Oxford University Press, 2000

Shea, John, *What a Modern Catholic Believes About Heaven and Hell*, Chicago: The Thomas More Press, 1972

Sheed, F. J., *God and the Human Condition*, Volume One, New York: Sheed and Ward, 1966

Sheen, Fulton J., *From the Angel's Blackboard*, Ligouri: Triumph Books, 1995

Sheen, Most Rev. Fulton J., Ph.D., D.D., *The World's First Love*, New York: McGraw-Hill Book Company, Inc., 1952

Shindeler, Emerson W., *Taking the Death of God Seriously in The Meaning of the Death of God*, Edit...by Bernard Murchland. New York: Vintage Books. 1967

Smart, Ninian, *Philosophers and Religious Truth*, New York: The Macmillan Company, 1969

Smith, John Maynard and Eors Szathmary, *The Origins of Life*, New York: Oxford University Press, 1999

Smolin, Lee, *The Life of the Cosmos*, New York: Oxford University Press, 1997

Smoot, George, *Wrinkles in Time*, New York: William Morrow and Co., Inc. 1993

Sobel, Dava, *Galileo's Daughter*, New York: Walker and Company, 1999

Solzhenitsyn, Alexander, *Cancer Ward*, Trans. by Nicholas Bethell and David Burg, New York: Bantam Books, 1968

Speyer, Edward, *Six Roads from Newton*, New York: John Wiley & Sons, Inc., 1994

Spinoza, Benedict de, "Ethics", Part V. Prop. XXIII Translated by R.H.M. Elwes, in *The Rationalists*, New York: Doubleday & Co. In. 1960

Staguhn, Gerhard, *God's Laughter*, New York: HarperCollins, 1992

Stannard, Russell, *The God Experiment*, Mahwah: HiddenSpring, 1999

Stapp, Henry P., *Mind, Matter and Quantum Mechanics*, New York: Springer-Verlag New York, 1994

Steck, John H., "What Says the Scripture", in *Portraits of Creation*, Grand Rapids: William B. Eerdmans Publishing Co. 1990

Steinbeck, John, *The Grapes of Wrath*, New York: Bantam Books, 1939

Sternglass, Ernest J., *Before the Big Bang*, New York: Four Walls Eight Windows, 1997

Story of a Soul, The Autobiography of St. Therese of Lisieux, Trans. by John Clarke, O.C.D., Washington: ICS Publications, 1976

Stoutzenberger, Joseph, *The Christian Call to Justice and Peace*, Winona: Saint Mary's Press, 1987

Stuhlmueller, Carol, C.P., *Thirsting for the Lord*, Ed. by Sr. M. Romanus Penrose, O.S.B. New York: Alba House, 1977

Swimme, Brian, *The Hidden Heart of the Cosmos*, New York: Orbis Books, 1996

Swimme, Brian, *The Universe is a Green Dragon*, Santa Fe: Bear & Company, 1984

Swinburne, Richard, Is There A *God?*, New York: Oxford University Press, 1996

Swizdor, Matthew, O.F.M. Conv. *Lay Hands on the Sick*, Granby: Franciscan Friars, St. Hyacinth College, 1983

Tattersall, Ian, *Becoming Human: Evolution and Human Uniqueness*, New York: Harcourt Brace & Co., 1998

Tekippe, Terry J., *What Is Lonergan Up to in Insight?*, Collegeville: The Liturgical Press, 1996

Thuan, Trinh Xuan, *The Secret Melody*, New York: Oxford University Press, 1995

Tilby, Angela, *Soul: God, Self and the New Cosmology*, New York: Doubleday 1992

Tipler, Frank J., *The Physics of Immortality*, New York: Doubleday 1994

Touching the Risen Christ: Wisdom from the Fathers, Ed. by Patricia Mitchell, Ijamsville: The Word Among Us Press, 1999

Trefil, James, *Are We Unique?*, New York: John Wiley & Sons, Inc., 1997

Trefil, James, *Reading the Mind of God*, New York: Charles Scribner's Sons 1989

Troyat, Henri, *Tolstoy*, Garden City: Doubleday & Company, Inc. 1967

Twain, Mark, *A Connecticut Yankee in King Arthur's Court*, New York: Airmont Books

Twain, Mark, *The Mysterious Stranger (and other Stories*, New York: New American Library, 1962

Updike, John, *The Centaur*, Greenwich: Fawcett Publications, 1962

Van de Pol, W.H., *The End of Conventional Christianity*, Trans. by Theodore Zuydwijk, S.J., New York: Newman Press, 1968 van Kaam, Adrian, C.S.Sp., *On Being Yourself*, Denville: Dimension Books, Inc., 1972

Van Riet, Georges, Ph.D, *Thomistic Epistemology*, Vol. 2, Trans. by Donald G. McCarthy, St. Louis: B. Herder Book Co., 1965

Van Till, Howard J., *The Fourth Day*, Grand Rapids: William B. Eerdmans Publishing Company, 1986

Van Till, Howard, "The Scientific Investigation of Cosmic History", in *Portraits of Creation*, Grand Rapids: William B. Eerdmans Publishing Co., 1990

Vanier, Jean, *Jesus, the Gift of Love*, New York: Crossroad, 1994

Vergote, Antoine, *The Religious Man*, Trans. by Sister Marie-Bernard Said, O.S.B., Dayton: Pflaum Press, 1969

Vermes, Geza, *Jesus the Jew*, New York: Fortress Press, 1990

von Balthasar, Hans Urs, "Meeting God in today's World", in *The Church and the World,* Concilium, New York: Paulist Press, 1965

von Balthasar, Hans Urs, *Prayer*, Trans. by A. V. Littledale, New York: Sheed & Ward, 1961

Walsch, Neale Donald, *Conversations with God: Book 1*, New York: G. P. Putnam's Sons, 1995

Walsch, Neale Donald, *Conversations with God Book 3*, Charlottesville: Hampton Roads Publishing Company, Inc., 1998

Ward, C.S.C., Leo R., *God and World Order*, St. Louis: B. Herder Book Co., 1961

Watson, Lyall, *Dark Nature*, New York: HarperCollins Publishers, 1995

Waugh, Evelyn, *The Loved One*, Boston: Little, Brown & Company, 1948

We Are But A Moment's Sunlight, Ed. by Charles Adler and Sheila Morrissey Adler, New York: Pocket Books, 1976

Weinberg, Steven, *Dreams of a Final Theory*, New York: Pantheon Books, 1992

Weiner, Jonathan, *Time, Love, Memory*, New York: lfred A. Knopf, 1999

Weisheiply, James, *The Development of Physical Theory In The Middle Ages*, Ann Arbor: The University of Michigan Press, 1971

Westerman, Claus, *Creation*, Trans. by John J. Scullion, S.J., Philadelphia: Fortress Press, 1974

Wetheim, Margaret, *The Pearly Gates of Cyberspace*, New York: W.W. Norton, 1999

Wiederkehr, Macrina, *A Tree Full of Angels*, San Francisco: Harper & Row, Publisher, 1988

Wiesel's, Elie, *Night*, New York: Bantam Books, 1960

Wijngaards, John, *How to Make Sense of God*, Kansas City: Sheed & Ward, 1995

Wilbur, Ken, *Quantum Questions*, Shambhala: New Science Library

Wilkes, Paul, *The Good Enough Catholic*, New York: Ballantine Books, 1996

Will, Clifford M., *Was Einstein Right?*, New York: Basic Books, Inc., Publishers, 1986

Wilson, A.N., *God's Funeral*, New York: W.W. Norton & Co., 1999

Wilson, A.N., *Jesus: A Life*, New York: Fawcett Columbine, 1992

Wilson, Edward O., *Consilience*, New York: Alfred A. Knopf, 1998

Wilson, Edward O., *Naturalist*, Washington: Island Press/Shearwater Books, 1994

Wilson, Edward O., *On Human Nature*, Cambridge, Mass: Harvard University Press, 1978

Wimber, John with Kevin Springer, *Power Healing*, San Francisco: Harper & Row, Publishers, 1987

Wood, James, *The Broken Estate*, New York: Random House, 1999

The World Treasury of Physics, Astronomy, and Mathematics, Edited by Timothy Ferris, Boston: Little, Brown and Co., 1991

Wranggham, Richard, and Dale Peterson, *Demonic Males*, Boston: Houghton Mifflin Company, 1996

Wright, Robert, *The Moral Anima,* New York: Pantheon Books, 1994,

Wright, Robert, *Nonzero*, New York: Pantheon Books, 2000,

Wright, Robert, *Three Scientists and Their Gods*, New York: Times Books, 1988

Zantig, Thomas, *Jesus of History, Christ of Faith*, Winona: St. Mary's Press, 1982

Endnotes

1. Introduction to Mark Schoof, O.P.'s A Survey of Catholic Theology 1800-1970", Trans. by N.D. Smith, New York: Paulist Newman Press, 1970, p.2

2. Job 5:17-18 See also Proverbs 3:11, "My son, do not scorn correction from Yahweh, do not resent his rebuke; for Yahweh reproves the man he loves, as a father checks a well-loved son."

A monk is dying in J. F. Powers, "Prince of Darkness and Other Stories", New York: Image Books, 1947, in the story "*Lions, Harts, Leaping Does*" p. 49 and we read,

> "Inevitably it occurred to him his plight might well be an act of God. Why this punishment, though, he asked himself..."

Still, he knows that it is his pride which has caused god to afflict him and he accepts it.

3. An example of this is found in, Mike Ferraren, "Why me, Lord", *Columban Mission*, Vol 80, No. 7, November 1997, p. 26, who relates,

> "Since the accident I have been discovering many things. My family has become my priority now, not money or prestige. I have become closer to my wife and children. I have stopped making plans for them. Instead I let God's spirit take over. As a result we have become closer as family. My deepest realization is about my wife. I have come to realize that Linda is my most valuable treasure in this world. It has been 28 years since the accident happened. They have been 28 years of *self-surrender to God's will.* I believe that I have a mission to do for God; that's why He spared my life. I have become closer to His people - to fisherman, tricycle drivers, businessmen, politicians, friends, strangers." (Italics added)

4. "20/20" of October 25, 1998

5. "Macbeth", Act IV, Sc. III, l. 261f

6. Alexander Solzhenitsyn, "Cancer Ward", Trans. by Nicholas Bethell and David Burg, New York: Bantam Books, 1968 writes this conversation between two characters,

"`Why is it,' Dymonka would ask Aunt Styofa, `that there's such injustice in fortune itself? There are people whose lives run smooth as silk from beginning to end, I know there are, while other's are a complete louse-up. And they say a man's life depends on himself. It doesn't depend on him a bit.'
`It depends on God,' said Aunt Styofa soothingly. `God sees everything. You should submit to him Dyomusha.'
`Well, if it's from God it's even worse. If he can see everything, why does he load it all on one person? I think he ought to try to spread it around a bit.'"

Elie Wiesel's first account of his experiences during the Holocaust, "Night", New York: Bantam Books, 1960 tells of a man, "Akiba Drumer said:

> `God is testing us. He wants to find out whether we can dominate our base instincts and kill the Satan within us. We have no right to despair. And if he punishes us relentlessly, it's a sign that He loves us all the more." [p, 42]

One answer to where is God during such suffering, which we will see elsewhere, is voiced at the hanging of a young boy when he relates that,

> "I heard a voice within me...`Where is He? Here He is—He is hanging here on this gallows...'." [p. 62]

Yet his faith was slipping until,

> "...I had ceased to plead. I was no longer capable of lamentation. On the contrary, I felt very strong. I was the accuser, God the accused. My eyes were open and I was alone—terribly alone in a world without God and without man."

At no time does he doubt that if there is a god, that god has some responsibility for the events of that time.

7. St. Joseph New American Catechism #3 arranged by Rev. Lawrence G. Lovasik, S.V.D., New York: Catholic Book Publishing Co., 1985, p. 21

8. "The Living Faith", New York: Herder and Herder, 1959, p.24

9. "The Living Faith", p.24

The idea of god punishing in order to bring us to our senses is very strong in the message of Fatima, an alleged apparition of Mary in Portugal. *The Fatima Crusader* of Autumn 2002, Issue 71 is full of arguments that the clerical pedophilia problem is such a punishment, as well as the cause for other negative events.

10. "Catechism of the Catholic Church", Liguori: Liguori Publications, 1994 United States Catholic Conference, #309.

11. "Catechism of the Catholic Church", #310

12. "Catechism of the Catholic Church", #312

13. See "Catechism of the Catholic Church", #313 on Catherine of Sienna, Thomas More and Dame Julian of Norwich

14. "Catechism of the Catholic Church", #314

15 . A Religious Sister puts the whole usual answer to the question of her cancerous condition when she asks,

> "Why is there suffering in the world? There is no response apart from the adoration of this mystery in the silence of love."
> [Sister Mary Clarita Shalom of the Annunciation, "To Suffer With Love" Trans. by Father Peter M.J. Stravinskas, *The Catholic Answer*, Vol. 12, No. 5, November/December 1998, p. 40]

Sister is admirable for her faith and in the course of her article she covers all the bases, mentioning meeting many other sufferers who were able to smile because they,

> "...had given their 'yes' to the Cross; because in spite of numerous difficulties, faith illumines every event and gives strength to confront the Passion in view of the Resurrection...Sickness teaches us abandonment to God, helps us understand those who suffer...and we see how many marvels are accomplished by the grace of the Lord."

She claims to know a man who was miraculously cured after seeing the face of Jesus. She speaks of the "plan of the Father" and expresses the faith that,

> "we are in the hands of God; for Him, nothing is impossible, and nothing is lost of what we suffer with love. The Lord makes every test and every suffering work for the good, and I am certain that He will never give us a cross that we cannot carry."

The whole standard answer in a two-page article.

16. Some maintain that this is not necessarily true, that beyond the light cone of our section of the universe other parts of the universe may be quite

different. Since we will never penetrate this event horizon, we can safely ignore this possibility.

See a discussion of this point in Wim Kayzer, "A Glorious Accident", New York: W.H. Freeman and Company, 1997, p. 218f

17. People have not always believed this. Aristotle (384-322 b.c.e) claimed that it is self-evident to all people, "both barbarians and Hellenes" that the heavens are composed of perfect, unchanging material quite different from that of the terrestrial realm.

[Aristotle, "On the Heavens", Trans. W.K.C. Guthrie, The Loeb Classical Library, Cambridge: Harvard University Press, 1926, pp.23-25 (270b1-24)]

Eighteen hundred years later, Copernicus (1473-1543) may have changed the perception of how the heavens moved, but not what they were. "Perhaps the contiguous air contains an admixture of earthy or watery matter and so follows the same natural law as the earth", he writes, to explain why the atmosphere moves when it would appear to be of the same nature as the heavens, heretofore regarded as being of a different substance than the earth. There must be something of earth mixed in.

[Copernicus, Book I of "De Revolutionibus Orbium Caelestium" (1543) cited by Thomas S. Kuhn, "The Copernican Revolution", Cambridge: Harvard University Press, 1985, p.152]

Galileo (1564-1642), the darling of many scientists because of his mistreatment by some in the Church, cast some doubt on the assumption that heaven and earth are different when viewing the rings of Saturn showed material suspiciously like that of earth, but he still presumed that the circle was the best expression of the divine and that it therefore, rather than the ellipticals the planets actually follow, had the place of honor since he took it for granted that the heavens and earth followed different laws. Isaac Newton (1642-1727), showed that the moon followed the same laws of gravity as prevailed on earth, something never realized before. The idea that the heavens were different was eroding. In our day there is good reason to believe that the presumption that the far reaches of the universe are the same as the known universe is true, but it will remain unproved empirically until a probe invades these regions. However, all of these classical physicists did have to act as if "reality" could be expressed in mathematical terms that have a strict relationship to the essence of the objects. This is obviously not provable, and in fact could yet be found to be in error, but they believed it because not to would negate any possibility, at least for the physics of their time, of knowing much about the reality of the objects and their behavior.

[See Menas Kafatos and Robert Nadeau, "The Conscious Universe",

New York: Springer-Verlag New York, Inc., 1990 pp101f]

Sometimes their unrecognized assumptions from their culture (that heaven and earth differed), had to give way to unrecognized assumptions from their investigations (that they did not differ).

18. Timothy Ferris, "The Mind's Eye", New York: Bantam Books, 1992, p. 11

Michael Ruse's, "Mystery of Mysteries: Is Evolution a Social Construction?" Cambridge: Harvard University Press, 1998 is an extended discussion of the presence of assumptions among evolutionists and how they colored their science. His conclusion is that there is less of this today, but that remains to be seen.

George Greenstein, "Portraits of Discovery", New York: Knopf, 1999 has many stories of personalities and assumptions getting in the way of "objective" science.

19. Karen Armstrong makes the point that in the history of religion up to now,

> "To expect to have faith before embarking on the disciplines of the spiritual life is like putting the cart before the horse. In all the great traditions, prophets sages, and mystics spend very little time telling their disciples what they ought to *believe*. Indeed, it is only since the Enlightenment that faith has been defined as intellectual submission to a creed. Hitherto, faith had been seen as a virtue rather than a prerequisite...Faith was thus a carefully cultivated conviction that, despite all the tragic and dispiriting evidence to the contrary, our lives did have some ultimate meaning and value. You could not possibly arrive at faith in this sense before you had lived a religious life. Faith was thus the fruit of spirituality, not something that you had to have at the start of your quest."
>
> [Karen Armstrong, in the foreword to "Every Eye Beholds You", Ed. by Thomas J. Craughwell, New York: Book of the Month Club, 1998, p. xiii]

20. Neil de Grasse Tyson, "Things People Say", *Natural History*, Vol. 107, No. 6, July/August 1998, p. 78

21. The evidence suggests that there was a belief in god long before the advent of any of the modern religious traditions. Therefore, the presumption of a god would not be considered the result of revelation as such except for a "natural revelation" which Paul (or Luke), for one, claims is present and convicts atheists of willful ignorance in ignoring. [Acts 17:30]

22. A tribesman converted to Christianity and ordained a Catholic priest reports that,

> "In African traditional settings, events of life need explanation...For our rural folks, witchcraft becomes a medium of explanation—it is our attempt to answer the question "WHY?...An important answer is witchcraft. A witch is someone who uses his or her psychic power to harm others."
> [Vincent Boi-Nai, SVD, "Witchcraft Mentality", *Divine Word Missionaries*, Summer 1998, p.12]

In the novel by Jewell Parker Rhodes, "Voodoo Dreams", New York: St. Martin's Press, 1993, p. 212 another explanation of the belief in spirit acting with human motivation is given.

> "With gods who acted human, people felt less alienated. With gods who were also aspects of nature, followers could explain some of the randomness, the fickleness of their fate. It was understandable that Ezili, with her many lovers, would be jealous of younger women and could cause a pretty girl's death or rape. Just as it was understandable that Agwe' could have an upset stomach and cause towering waves to drown sailors...Sometimes you could save yourself, sometimes not. Saving yourself meant having *loas*, or ancestors, who favored you because of your continuing prayers and gifts. Or else you gained special blessings through sacrifice to the supreme god Damballah."

Though it is a novel, it is based on an actual life and the motivation for engaging in voodoo is possibly authentic.

> [See Joe Nickell, "Voodoo in New Orleans", *Skeptical Inquirer*, Vol. 26, No. 1, January/February 2002, p. 15]

Anne Rice writes novels about vampires and other esoteric creatures, including a young girl with magic abilities. She writes that,

> "God is magic...and so are the saints. Angels are magic. And ghosts, if they be truly the apparitions of souls who once lived on earth, are magic as well...You understand...I don't say that all these

magical elements are equal. What I am saying is that what they have in common is that they are divorced from materiality, divorced from the earth, and from the flesh. But they partake of the realm of pure spirituality where other laws—laws unlike our physical earthy laws—might exist."

[Anne Rice, "Merrick", New York: Alfred AKnopf, 2000, p. 68]

Again, it is a novel but people read them with sometimes a belief that this might be true. Certainly the popularity of television programs and motion pictures with the same theme points out people's willingness to believe, or at least to suspend judgment.

See also "Heartbeats of Vodou" in *Discover*, Vol. 107, No. 10, December 1998-January 1999, p. 30f which points out that, "Traditional Vodou priests reject black magic, believing it entails making a pact with the devil or with evil spirits, which in time will drag down and destroy the practitioner." They do not deny, of course, that these spirits can affect humans.

In other societies, especially European and colonial Puritans, the witchcraft was ascribed to a demon working through a human being.

23. There are many sources for these tales. One I read years ago which gives a "retelling" of the Greek myths is, Roger Lancelyn Green, "The Tale of Troy", Harmondsworth: Puffin Books, 1958. It is obvious that the gods are in charge, and that they have a plan. Proteus, the old man of the sea, tells Menelaus, "it's all arranged up in Olympus." (p. 168)

24. Graham Greene, "Monsignor Quixote", New York: Simon and Schuster, 1982, p. 22

25. According to Webster, teleology is "a doctrine explaining phenomena by final causes" or "being directed toward an end or shaped by a purpose." See "Philosophy in the Flesh" p. 378f for Aristotle's definition, which was largely followed for centuries.

26. Luke 17:11-19

27. The Greeks and Romans of the 4th century, for example,

"...used lead sheets to write binding spells of magic power. The spells, over lovers, business rivals, or personal enemies, were to be activated by spirits of water or the underworld, which is why [they are found] in springs, wells, or even in the graves of the dead. They were also hidden near where they were supposed to work: for example, curses on a rival chariot team could be buried in a hippodrome."

[Barbara Burrell, "Curse Tablets from Caesarea", *Near Eastern Archaeology*, Vol. 61, No. 2, p. 128]

28. Paul Davies quotes David Bohm as saying,

"In my opinion, progress in science is usually made by *dropping* assumptions...So often, major progress in science comes when the orthodox paradigm clashes with a new set of ideas or some new piece of experimental evidence that won't fit into the prevailing theories. [Paul Davies, "Abouttime", New York: Simon & Schuster, 1995, p.199]

Paul Davies himself points out that,

"The philosopher Thomas Kuhn believes that scientists adopt certain distinct paradigms that are tenaciously retained and are abandoned only in the face of glaring absurdities...Experimental scientists pride themselves on their objectivity, yet time and again they unwittingly massage their data to fit in with preconceived ideas."

[Paul Davies and John Gribbin, "The Matter Myth", New York: Simon & Schuster, 1992, p. 23]

The introduction of Thomas Kuhn's *The Structure of Scientific Revolutions,* is included in "The World Treasury of Physics, Astronomy, and Mathematics" p. 787, or of course you can read Kuhn's book itself.

See also Steve Fuller, "Thomas Kuhn: A Philosophical History for Our Times", Chicago, University of Chicago Press, 2000 for a critical treatment of Kuhn.]

See Ian Barbour, "Religion in an Age of Science", New York: HarperSanFrancisco, 1990, p. 51f for an extended discussion of paradigms in science and religion.

The 17th Century investigator Ernst Stahl said that burning articles emitted a substance called Phlogiston and since this seemed natural it was accepted until there was not just good but overwhelming evidence to reject it. Another substance, the "ether", filled every part of the universe according to yet another theory, and never mind that no one could find a trace of it. Since it fit the theories of the time it was not until it could not fit new facts that it was discarded. Lamarck's theory of inherited characteristics leading to the development of new species is still popular, though not among scientists. These changes are, according to Kuhn, more than just a shift in the understanding of details, they are often a shift in "paradigms" or models of reality.

Everyone just presumed that the planet Mars had canals since the astronomer Giovanni Schiaparelli named features he saw on the planet "canali" which just meant channels in Italian but was immediately understood as constructed canals, and even in the face of many subsequent observations that found no waterworks, you could still scare a nation with tales of the Martians' arrival.

> [Stephen Jay Gould takes off from this point in one of his famous examples of the history of scientific thought, both good and bad, in "War of the Worldviews", *Natural History*, Vol. 105, No. 12, December 1996/January 1997, p. 22.]

In non-scientific assumptions, many a Christian has been surprised to discover that Jesus did not say that "an eye for an eye" was a good way to seek justice, and many continue to presume that they have their facts straight and that Jesus himself condones their more barbarous feelings. It was long just presumed that Christians were forbidden to be involved in the lending of money with interest and everyone just "knew" that such a practice was the condemned "usury". A study of Church history can find any number of examples of Conciliar and Papal documents that state as true something that the authorities of an older time had listed as "anathema", or accursed, finally realizing that their assumptions had to be reexamined and their conclusions changed. Unfortunately, the ecclesiastical documents are not quite as straightforward as Einstein was when he forthrightly labeled his theory of a cosmological constant as the biggest mistake of his life.

> [Ironically, it may not have been a mistake. The indications that the universe is actually accelerating make a cosmological constant of some kind possible or even necessary.
> See Karen Wright, "Very dark energy", *Discover*, Vol. 22, No. 3, March 2001, p. 71.]
> However, I am using Einstein's willingness to shift paradigms as the example, not whether or not he was correct.

Too often the Roman publications preface their once heretical pronouncements with the statement, "as our illustrious predecessor said", followed by something completely antithetical to what the illustrious one had indeed taught. There has often been little or no understanding of the underlying process that had taken place, a "paradigm shift" of viewpoints that were not necessarily contradictory but were often perceived as being so. Hans Kung sees that practice that Kuhn has described in science as also occurring in theology and, "since we are dealing in this work with paradigms, basic assumptions that have been long in ripening, are deeply rooted, profoundly influential, often conscious and often unconscious," the disputes that can arise are often "hard, and seemingly so irreconcilable." [Hans Kung, "Theology for the Third Millennium", Trans. by Peter Heinegg, New York: Doubleday, 1988, p. 126]

> [Both Kuhn and Kung see a paradigm not just as a change of a model for understanding but "an entire constellation of beliefs, values, techniques, and so on shared by the members of a given community."
> See the reference in Hans Kung, "Theology for the Third Millennium", p. 132]

29. Arthur Peacocke, "God and the New Biology", San Francisco: Harper & Row, 1986, p73

30. "If methodical investigation within every branch of learning is carried out in a genuinely scientific manner and in accord with moral norms, it never truly conflicts with faith. For earthly matters and the concerns of faith derive from the same God. Indeed, whoever labors to penetrate the secrets of reality with a humble and steady mind is, if even unawares, being led by the hand of God, who holds all things in existence and gives them their identity." ["Gaudium et Spes", *36]*

This seems to me to say, or at least has been interpreted to say, that if a conflict does arise the scientist must not be methodical enough, or has not followed moral norms. That god "holds all things in existence" is what we are investigating.

31. Hans Kung, "Theology for the Third Millennium", p. 143. The book treats extensively the paradigm changes required.

Fritjof Capra and David Steindl-Rast, "Belonging to the Universe", New York: HarperSanFrancisco, 1991 have a section on paradigm shifting in science and theology and another section proposing a new paradigm.

32. What will the religion that takes this article seriously look like? How will it define reality? It will presume that everything can be explained, that there is a cause for everything that exists and everything that happens, and there is some reason why that cause has operated to bring this about. This may include the assumption that there is some logical explanation behind everything and it did not just happen by chance, or it may presume that chance has a large role to play but that "something" would occur is a logical necessity. Science presumes this as well and approaches every phenomenon with the idea that it can be explained. Do galaxies exist? Then there must be some reason why they do. Do most birds build nests and incubate their eggs? Then there must be an evolutionary explanation for both of these behaviors, nesting and incubating. In fact, "the exception proves the rule" and when one finds a bird that does not nest it is presumed that there is some reason for this that makes sense. There is a cause for the formation of galaxies and for the behavior of animals and it can be investigated and understood, at least up to a point, by the human mind. And if the mind cannot visualize or readily understand it, it is not because there is no cause but that the mind is not constructed along lines that can grasp the cause. Quantum physics is a good example of this. There are many non-intuitive conclusions from quantum physics that do not make sense to the human mind, just as the theory of relativity's use of more than three dimensions is hard to visualize, and our brain and senses are not constructed to be able to grasp these things. But many can be described mathematically even if a picture cannot be drawn of them, and it is presumed that all the unknown factors in the universe will eventually be known.

The ancient peoples just factored in a god or some such being whenever they neared an area of human ignorance as to causality. The Greeks were perhaps the first to recognize that the causative agent may not be a conscious, living being. The "prime mover" or "prime matter" or "ur-matter" of Plato, Aristotle, Lucretius and other philosophers was not seen as necessarily an intelligent being, but the ground of all other being from whom or which others derived their ability to move, and indeed to exist. They reached this conclusion by noticing that no human artifact, nothing constructed by people, would exist if a human did not in fact make it. The same is true of our own being. We did not bring ourselves into existence and we certainly do not will ourselves out of existence. There must be an ultimate "something" at the beginning of the line of causation which began everything and whose own existence is necessary.

[See "Aristotle's Physics", Trans. by Richard Hope, Lincoln: University of Nebraska Press, 1961. In the whole discussion of a prime mover, there is the conclusion that it must be eternal, immutable and immovable, but no requirement that it be rational or even intelligent.]

This "something" does not have to be alive and knowing. We might understand this better with an example of the Great Plains of the United States. Why is so much of the land west of the Mississippi so flat? A brief and too simple, but accurate, answer is that it lies east of the Rocky Mountains and is the result of erosion from those lofty heights into a long-gone sea. Other things must be factored in, of course, including the action of glaciers, but we are going for a simple example. One cause of the prairies is the Rocky Mountains. But the cause of the mountains is found in plate tectonics that tells of the movement of the mantle under the Pacific Ocean colliding with and moving under the plate on which the continental United States is found. This subducting raised the Rockies to great heights in the same process that is still forming the Himalayas, and millions of years of weathering has transformed the heights of the Rockies to the dirt of the plains where it has been moved by rain and flood. But there is no such being as "plate tectonics", it is a mental construct to explain a phenomenon. This is because the movement of the plates itself demands some explanation and is caused by some other process. But this cannot go on infinitely. There must be some ultimate cause of the motion and some ultimate "ground of being" of the plates themselves and the earth of which they are a part. Call this ultimate explanation a "prime mover", but it need not be any more alive than "plate tectonics", though unlike the latter it must be more than a mental construct. In practice, it is usually presumed, there is some sentient being behind every occurrence. It may be a human being or an animal, or it may be a divine being. The being may have the best interest of the human in mind or it may not. It may be a rational being or irrational, conscious or unconscious, sane or mad. Why does the dung beetle push the ball of manure around? The human may find it strange, but the beetle has some purpose. It is not aware of the purpose itself, but there is one. Why does the king tax the farmers so mercilessly? Because he has a war to finance, or a palace to build or some other venture in need of capital. Even if he is mad, he has some purpose that makes sense to him. Why does it rain? Not because, as we now know, the sun warms the atmosphere and moves the water vapor around until it condenses, but because the gods are smiling on that overtaxed farmer, or on his king, or maybe the sun itself is a god, or the clouds, or the rain. The gods, too, may be insane or mean or greedy or just uncaring, or even capricious, but there is some reason for their actions, whether the reasons make sense to anyone but themselves or not. In the Christian tradition, to jump ahead for a moment, the purpose god has in mind is variously described. "The world was made for the glory of God." says one formulation, ["Catechism of the Catholic Church" #293 citing "Dei Filius", can 5:DS3025] which might seem self-centered, but actually refers to god's communicating to the creatures out of love.

> "The glory of God consists in the realization of this manifestation and communication of his goodness, for which the world was created. God made us `to be his sons through Jesus Christ, according to the purpose of his will, to the praise of his glorious grace'."
>
> ["Catechism of the Catholic Church" #294 citing St. Irenaeus.]
>
> "The ultimate purpose of creation is that God `who is the creator of all things may at last become `all in all,' thus simultaneously assuring his own glory and our beatitude'." ["Catechism of the Catholic Church" #294 citing "Ad Gentes" of Vatican Council II and referring to 1 Corinthians 15:28.]

All these quotes are very much in line with the presumption that there is a purpose in god's actions. And even though the myths and stories are full of instances of strange or even deranged behavior by the gods, in the long run there must be consistency, otherwise the world could not go on. The universe can tolerate a fair amount of chaos, even from the gods, but there must also be the assurance that most of the time things will go as planned. Otherwise I will not plant the crops, for example, unless I have some assurance that most years I will have a harvest. The gods can be crazy or capricious, but there are limits.

33. "Quantum Questions", Ed. by Ken Wilber, Shambhala: New Science Library, 1985, p. 110

34. M. Basilea Schlink, "Realities", Grand Rapids: Zondervan Publishing House, 1966, p. 29

35. Don Basham, "Deliver Us From Evil", Old Tappan: Chosen Books, 1972, p.175

36. Howard Jones, "Mutiny on the *Amistad*", New York: Oxford University Press, 1987, p. 149

According to Webster, teleology is "a doctrine explaining phenomena by final causes" or "being directed toward an end or shaped by a purpose."

See "Philosophy in the Flesh" p. 378f for Aristotle's definition, which was largely followed for centuries.

S.C. Gwynne in "An Act of God?", reported in *Time*, Vol. 154, No. 25, December 20, 1999, p. 58 on a couple who lost a daughter in the shootings at Columbine who say, "God is using this tragedy to wake up not only America but also the world." As stated this would not fit the pattern I am describing here, since it would mean that God acted after the event to help them bring some good from it. However, the father "believes Rachel's death was meant to be." From her diary and letters she evidently believed that God had it planned that she would die young for some purpose, a belief that her family shares.

37. "Wrinkles in Time", p. 71

38. Edward J. Melvin, C.M. "A Nation Built on God", Huntington: Our Sunday Visitor, Inc., 1975, p. 103

39. Catherine Drinker Bowen, "Miracle at Philadelphia", New York: Book of the Month Club, Inc. 1986, p. 216

40. "A Nation Built On God", p. 105

41. John North, "The Norton History of Astronomy and Cosmology", Ed. by Roy Porter, New York: W.W. Norton & Co., 1995, p. 523

42. The rate is linearly proportional to distance, so a galaxy compared to ours is receding at a certain rate, but one twice as far away is receding at twice that rate.

There are reports that the universe's expansion may be accelerating instead of being linear, so stay tuned.

43. Robert Gilmore, "Alice in Quantumland", New York: Springer-Verlag, 1995, p. vi.

44. See Jeffrey Winters, "Let There Be Matter" in *Discover*, Vol. 18, No. 12, December 1997, p. 40. One usually thinks of transforming the energy of atoms into nuclear energy or starlight. "But like any equation, E=mc2 works in both direction...That is, it should be possible to convert energy into matter. Now a team of physicists has accomplished just that: they have transmuted light into matter."

For a very readable explanation and history of Einstein's famous equation, see David Bodanis, "E=mc2", New York: Walker & company, 2000

45. The "Black Body" referred to is an idealized object that radiates all the energy it has absorbed, or that it contains. It is called "black" because anything of another color reflects away energy coming at it from outside and does not absorb it and re-radiate it back later. The universe, of course, does not absorb radiation from outside itself, but it radiates its own energy as if it had received it from another source. The radiation consists of the movement of the component "parts", mainly photons, of the universe, just as an iron poker heated up will radiate the heat back caused by the agitated motion of its component atoms. The poker will radiate at different temperatures as this motion decreases, and the temperature goes down from white hot to warm and eventually to almost absolute zero. It can never hit absolute zero, of course, as long as the universe around it continues to heat it. The universe radiated very high energy (temperature) in the past. Today that radiated background energy is much lower.

46. See Edward Speyer, "Six Roads from Newton", New York: John Wiley & Sons, Inc., 1994, p. 138f for a history of the discoveries leading to and from this statement.

47. For a simple presentation of the situation see, Etienne Klein, "Conversations with the Sphinx", Trans. by David LeVay, London: Souvenir Press, 1996, p. 109ff

48. Again for a readable explanation of the paradox, see "Conversations with the Sphinx", p. 145ff.

Put simply, Schrodinger imagined a cat in a situation where it is both dead and alive, just as a photon can be a particle and a wave. It only becomes one or the other when an observation is made. Much discussion has followed his thought experiment and there is no agreed upon answer.

See David Lindley, "Quantum Mechanics Gets Real", *Science News*, Vol. 151, No. 9, March 1, 1997, p. S18 for one explanation called "decoherence" which I will not attempt to explain. As far as the cat goes, however, he makes some sense when he says, "A cat...remains dead or alive long enough for that state to be recorded; a superposed dead-and-alive cat, however, can never exist long enough to be noticed" since a cat is such a complex quantum system.

See R. Lipkin, "Schrodinger's cat: Two atoms in one?", *Science News*, Vol. 149, No. 21, May 25, 1996 for a report on one atom being in a "superposition of two `coherent-state wave packets.'." In other words, one object existed in two places or conditions at the same time.

49. See "Eric A. Cornell and Carl E. Wieman, "The Bose-Einstein Condensate" in *Scientific American*, March 1998, p. 40.

> "If two or more atoms are in a single quantum-mechanical state, as they are in a condensate, it is fundamentally impossible to distinguish them by any measurement. The two atoms occupy the same volume of space, move at the identical speed, scatter light of the same color, and so on. Nothing in our experience, based as it is on familiarity with matter at normal temperatures, helps us comprehend this paradox." But when brought to a low enough temperature, the atoms do not operate in a manner that can be described by classical physics which would treat the atoms as discrete particles in the manner of ping pong balls. The result, "differs from [the] classical conception because of quantum effects that can be summed up in three words: Heisenberg's uncertainty principle." [p. 44]

50. Paul Davies, "The Cosmic Blueprint", New York: Simon and Schuster, 1988, p. 4

See "Time's Arrow and Archimedes' Point", for a wide ranging discussion of the symmetry of time and its consequences for understanding quantum theory and its effects on causality.

51. "Reading the Mind of God", p. 210

For a historical investigation of the notion of a vacuum and its various definitions, see Henning Genz, "nothingness".

52. K.C. Cole, "A Hole In the Universe", New York: Harcourt, Inc., A very readable explanation for the lay person, especially of string theory.

53. Frank J. Tipler, "The Physics of Immortality", New York: Doubleday 1994, p. 236

54. See Philip Yam, "Exploiting Zero-Point Energy" in *Scientific American*, Vol. 277, No. 6, December 1997, p. 82

See Lawrence Krauss, "Beyond Star Trek", New York: Basic-Books, 1997 for a discussion of particle-antiparticle interaction being measurable and containing energy.

55. "Reading the Mind of God" p. 212

See also John Gribbin, "Schrodinger's Kittens and the Search for Reality", Boston: Little, Brown and Company, 1995, p. 120f for background on this phenomena of quantum fluctuations and experimental evidence of its existence.

56. Another explanation is that there really is antimatter, but at such a remote distance from the universe we can see and interact with that there is no danger of mutual annihilation.

See James Trefil, "Greetings from the Antiworld", *Smithsonian*, Vol. 29, No. 3, June 1998, p. 61f for a popular treatment of antimatter.

See Gregory Tarle and Simon P. Swordy, "Cosmic Antimatter", *Scientific American*, Vol. 278, No. 4, April 1998, p. 36f

See Helen R. Quinn and Michael S. Witherell, "The Asymmetry between Matter and Antimatter", *Scientific American*, Vol. 279, No. 4, October 1998, p.76 for a good explanation of problems with the Standard Model and efforts to solve them with new (in 1999) accelerators.

57. "nothingness", p. 261 & the related discussion.

58. Nothing has become very fashionable lately. We have already met Henning Genz's book, "nothingness". Then there is K.C. Cole's, "The Hole in the Universe", that does an excellent job, especially in explaining string theory and the appearance of the universe from the vacuum. In reading it, I kept waiting for her answer to the ultimate question, "why is there a vacuum with these qualities rather than nothing at all", but it turns out that there is no answer as yet. However, this is another scientist who feels no need for anything extrinsic, such as god, to be involved.

59. Trinh Xuan Thuan, "The Secret Melody", New York: Oxford University Press, 1995, p. 241

60. See Timothy Ferris, "The Whole Shebang", New York: Simon & Schuster, 1997, p. 306.

61. Webster defines "contingency" mostly as the quality of not being expected, or occurring by chance. One definition says "dependent on or conditioned by something else." This latter is much closer to what philosophers usually mean by the word. A contingent being is one whose existence, properties, or actions are not due to their own nature. Any "creature" or anything created by god is therefore contingent on god.

62. Charles A. Hart, "Thomistic Metaphysics", p. 222 citing Thomas in I Sentences D 19, Q.2, a. I. "Charley Hart" as the book was affectionately known to many seminarians, was the ultimate explanation of the cosmos according to their professors.

63. "Thomistic Metaphysics" p. 224

64. "The Matter Myth" pp 64-65

65. "Wrinkles In Time" p. 116

66. John D. Barrow, "The Origin of the Universe", New York: BasicBooks, 1994, p. 1

67. For a popular treatment of the concepts of relativity and the behavior of light, time and space, see Tony Hey and Patrick Walters, "Einstein's Mirror", New York: Cambridge University Press, 1997

See also Clifford M. Will, "Was Einstein Right?" New York: Basic Books, Inc., Publishers, 1986

For an explanation of exceeding the speed of light by quantum tunneling, see David H. Freeman, "Faster Than a Speeding Photon", *Discover*, Vol. 19, No. 8, August 1998, p. 76f

See also, Raymond Y. Chiao, Paul G. Kwiat and Aephraim M. Steinberg, "Faster than Light?" in *Scientific American: The Mechanics of Sight*, 1998, p. 98, originally printed in 1993.

See Paul Davies, "That Mysterious Flow", *Scientific American,* Vol. 287, No. 3, September 2002, p. 40, for another brief explanation of the question of time. The whole issue treats aspects of time.

68. "The Matter Myth" pp. 76-77

Richard P. Feynman, "Six Not So Easy Pieces", Reading: Helix Books, 1997 chapter 3 is on the Special Theory of Relativity. Chapter 4 is on Space-Time, which we are about to consider.

69. It is a "concept" because we three-dimensional beings cannot imagine four dimensions, just as the inhabitants of "Pointland" cannot imagine "Flatland" with its lines and figures, who in turn cannot imagine anything with "height". See Edwin Abbott's classic "Flatland", New York: Dover Publications, 1952, though the revised edition came out in 1884 when the identity of the 4th dimension was unknown.

70. Stephen Hawking, "Black Holes and Baby Universes", New York: Bantam Books, 1993, pp. 45-6. In confirmation of Hawking's theories on black holes, results gained from observation by the Rossi X-Ray Timing Explorer, a NASA satellite, indicate that space around a black hole or neutron star is being twisted by the gravity, causing a blinking of material being pulled in to the center.

Hawking's discussion of time is an echo of his book, "A Brief History of Time" where he reaches the same conclusion about the lack of necessity for a creator. Thomas P. Sheahen, "Incorrect Reasoning in *A Brief History of Time*", "*The ITEST Bulletin*" Ed. By Robert Brungs, S.J., Vol. 33, No. 2, p. 13 claims that Hawking makes an error in translating a mathematical concept of time to the common usage of the word.

71. William Faulkner's narrator in "The Wild Palms" was not speaking of relativity in this sense when he muses that he is,

> "...supported by [time] in space...just on it, non-conductive, like the sparrow insulated by its own hard non-conductive dead feet from the high tension line, the current of time that runs through remembering, that exists only in relation to what little of reality...we know, else there is no such thing as time."
>
> [William Faulkner, "The Wild Palms" New York: Vintage Books, 1939, p. 137]

Faulkner may well have been ahead of his "time".

There is also an opinion either that time does not really exist, as explained by Julian Barbour, "The End of Time", New York: Oxford University Press, 1999, or that time as we experience it may not be reality, as Huw Price suggests in "Time's Arrow and Archimedes' Point", op. cit. At least I think that is what Price is saying. I have read both of these works but I must admit to not understanding them well. However, if it is true that one or both (I think both) do away with the need for time as a measure of cause and effect, I believe my conclusions below will remain the same.

72. See "The Ontological Argument", Ed. by Alvin Plantinga, Garden City: Doubleday & Company, Inc. 1965

See Simon Blackburn, "Think", Oxford, Oxford University Press, 1999, p. 152ff.

See van Beeck, "God Encountered", p. 53ff for a sympathetic treatment of Anselm's argument, which however, reaches the same conclusion about its value.

73. Daniel C. Dennett, "Darwin's Dangerous Idea", New York: Simon & Schuster, 1995", p. 24

74. Cyrus H. Gordon has long been pointing out the points of contact between the Hebrew civilization and those of other civilizations of the time, including the Greeks. Howard Marble Stone in *Biblical Archeologist* of March, 1996 cites the following incident.

> "Professor Gordon had been delivering a popular lecture on `The Common Background of Greek and Hebrew Civilizations,' particularly about the Heroic Age both in Late Bronze Age Greece and in pre-monarchic Israel. Citing the Iliad of Homer and the biblical book Samuel, he pointed out that the heroes David and Achilles performed essentially the same warlike exploits in search of imperishable glory, the only bulwark against oblivion. The milieux of Achilles and of David were therefore closer to one another than was that of Achilles to Classical Greece or that of David to the Age of the Prophets in Israel. Following the lecture, a little old lady, wide-eyed with astonishment and admiration, made her way up to the distinguished lecturer and asked, `Does that mean, Professor Gordon, that Achilles was Jewish?"'

Despite the connections, it remains true that Greek and Hebrew traditions took separate paths, especially in regard to their interest in empirical data.

75. "Philosophers and Religious Truth", Edited by Ninian Smart, New York: The Macmillan Company, 1969, p 96

76. Menas Kafatos and Robert Nadeau, "The Conscious Universe", p. 75

77. John D. Barrow and Frank J. Tipler, "The Anthropic Cosmological Principle", New York: Oxford University Press, 1986, p. 128

78. "Philosophers and Religious Truth", p. 97

79. Cited in Stephen Jay Gould, "An Awful, Terrible Dinosaurian Irony", in a collection of his essays in *Natural History* entitled, "The Lying Stones of Marrakech", New York: Harmony Books, 2000, p. 195

80. W. Paley, "Natural Theology—or Evidences of the Existence and Attributes of the Deity Collected from the Appearances of nature", Oxford: Clarendon Press, 1938

81. Richard Dawkins, "The Blind Watchmaker", London: Longmans, 1986, p. 37

See Kenneth Miller, "Finding Darwin's God", New York: Cliff Street Books, 1999, p. 137f for a response to Dawkins's views from a believer in god as a designer.

82. From 1959 we read the following:

> "When we look upon the universe about us, the stars of heaven and all that is on the earth, we must marvel at the wonderful order which is in them all. All things, every creature, has its place in the great world. There must be someone who has thought out and brought this order into being; there must be an all-wise and all-powerful creator to have made all this. Each living thing is a work of art. Plants and animals are so made that they can live and grow and multiply. the human body too is formed marvelously, every part fitted to its purpose. These living beings cannot have given themselves such wonderful arrangements; they were the gift of an all-wise and all-powerful God."
>
> ["The Living Faith" New York: Herder and Herder, 1959 pp. 12-13]

A newer 1985 catechism wisely just pleads ignorance.

> "Creation is the way God gave life and the world to man. It seems that creation was a gradual evolutionary development that took place a very long time ago. We do not know how it happened. The important thing for faith is that we owe all that we have to God. The Bible does not try to explain creation...Millions of years after the events they could only imagine how creation took place".
>
> ["St. Joseph New American Catechism #3", p.39.]

There is here no mention of the argument from design as such.

In 1986, Cardinal Joseph Ratzinger was not so hesitant to preach, "The more we know of the universe the more profoundly we are struck by a Reason whose ways we can only contemplate with astonishment."

[Joseph Ratzinger, "In The Beginning..." Huntington: Our Sunday Visitor, 1990, p. 36]

83. "Catechism of Catholic Church", #306

84. "Catechism of the Catholic Church", # 303

85. "Catechism of the Catholic Church" # 314

Joan Acker, "Creationism and the Catechism", *America*, Vol 183, No. 20, December 16, 2000, p. 6 is not as generous to the authors of the Catechism as I have been saying that,

> "Its failure to engage today's religious/scientific culture adequately is an embarrassment both rationally and spiritually. It plays into the hands of `scientific creationists' and has dangerous ramifications for the future of both science and religion."

Acker mentions paragraph 390 of the Catechism and faults it for being vague since it, "uses figurative language, but affirms a primeval event..." I took this as caution. Acker is probably right in seeing something more in it, namely an unreadiness or inability to deal with the issue. She also points out that the new glossary and index fail to include an entry for Evolution.

Oscar Lukefahr C.M. who is conversant with science, ["We Believe"] commenting on the new Catechism, also does not mention the argument from design, though he does use the argument that says "I cannot imagine how it could all have come from nothing or be the result of chance". Presumably, Lukefahr would agree with Paley in saying that surely there must be an intelligent designer because he cannot imagine any other way it could have happened. In a column for the Knights of Columbus, *Mariner*, Vol. 70, No. 4, June 2001, Lukefahr mentions a book by Dr. Richard Swenson titled, "More Than Meets the Eye" and cites Swenson who, "says such facts [as the complexity of the human body] ought to give us confidence in God's power and attention to every detail in our life." I have not read the book mentioned, but evidently it, like Lukefahr, presumes a Creator-God.

It is a very popular conviction and we often hear not the argument as such, but the presumption, expressed by the faithful. On returning from a canoe trip down a beautiful river, the believer is very apt to ask how anyone could see such sights and not believe in god, thus expressing the belief not only that god engineered these things, but that it was done for the delectation of the humans who would later see it. There is puzzlement, of course, over why god would design ticks to chew on them, but this question is better left unexplored.

The Psalms speak of the wonders of god and how wonderfully made we are and there is no doubt about who designed these things. These Psalms are very popular in Christian prayer.

A 1995 Encyclical letter of the current Pope, John Paul II, "Evangelium Vitae", arguing for respect for all life, finds a cause of disrespect for life in the scientific rejection of the design argument when it says,

> "Once all reference to God has been removed, it is not surprising that the meaning of everything else becomes profoundly distorted...This is the direction in which a certain technical and scientific way of thinking prevalent in present-day culture appears to be leading when it rejects the very idea that there is a truth of creation which must be acknowledged or a plan of God for life which must be respected."
>
> [*Origins,* Vol. 24: No. 42, April 6, 1995, p.698]

86. "The Blind Watchmaker" p. 37

87. For an explanation of Ptolemy's views, see "Astronomy and Cosmology", p. 104ff. For an extended treatment of Galileo and his discoveries and difficulties, see p. 332f.

88. Richard Dawkins, "River Out Of Eden", New York: BasicBooks, 1995, pp. 97-98.

Daniel Dennett adds, "One of Darwin's most fundamental contributions is showing us a new way to make sense of `why' questions." ["Darwin's Dangerous Idea" p. 25]

Another cosmologist, Trinh Xuan Thuan says that Paley's argument is very convincing, but,

> "...it does not quite agree with modern science. The latter tells us that extremely complex systems may be the result of perfectly natural evolutionary processes that follow well-understood physical or biological laws, and that there is no need to invoke God as watchmaker. Complexity does not necessarily presuppose a creator and a grand design." ["The Secret Melody" p. 243]

Thuan adds,

> "Complexity and organization may arise spontaneously in a universe that is expanding and contains stars. The hand of a Creator no longer seems necessary."
> ["The Secret Melody" p. 244]

This is an echo of Stephen Hawking's famous question, "What place, then, for a creator?"

> [Stephen Hawking, "A Brief History of Time", New York: Bantam Books, 1988, p. 141]

89. "The Blind Watchmaker" p. 37

90. Matt Cartmill in "Oppressed by Evolution", Discover, March 1998, p. 78f treats of both the Christian right and the liberal left arguing with such a conclusion or its implications. Specifically he cites Richard Dawkins and Stephen Jay Gould whom I have quoted with approval as going too far in their statements and points out that, "the National Association of Biology Teachers deleted the words *unsupervised* and *impersonal* from its description of the evolutionary process." His own conclusion is that, "all that we scientists can do is admit to our ignorance and keep looking. Our ignorance doesn't prove anything one way or the other about divine plans or purposes behind the flow of history. Anybody who says it does is pushing a religious doctrine."

Notice that I have said the "need" for a creator, or the "necessity" of a designer. Neither I nor most scientists feel that science has proven that there is none, nor is such a proof likely.

91. For Copernicus, see "Astronomy and Cosmology", p. 279f
For Galileo and contemporaries see pp. 299f.

92. "A Brief History of Time" p47

93. "Reading the Mind of God" p. 106

94. "A Brief History of Time" p.116.
See Angela Tilby, "Soul: God, Self and the New Cosmology", New York: Doubleday, 1992, p. 120 for a version better disposed to the papacy.

95. See "Astro News", *Astronomy*, April 1997. The "Hubble Constant" is a measure of this ratio. Studies in 1996 result in a Constant of 64 in one case and 65 in another study. "A Hubble constant of 65 implies a 14-billion-year-old universe, give or take a couple billion years."

96. "The Secret Melody" p. 57

97. Neil de Grasse Tyson, "In Defense of the Big Bang", *Natural History*, Vol. 105, No. 12, December 1996/January 1997, p. 76 lays to rest any ideas amateurs like myself may have that the Big Bang idea is in trouble when he says,

> "Regardless of what you may have read or heard to the contrary, the big bang is supported by a preponderance of evidence and has become the most successful theory every put forth to explain the origin and evolution of the universe."

He explores briefly the limitations of the theory which have been discussed in such a way that it might seem that the theory itself is in disrepute and explains why it is not

.

98. See "Onward to the Edge", *Natural History* July, 1996 p. 60f.
See also "Science News" Vol 149, #3, January 20, 1996, p36

99. J. Patrick Henry, Ulrich G. Briel and Hans Bohringer, "The Evolution of Galaxy Clusters" *Scientific American*, Vol. 279, No. 6, December 1998, p. 52

100. P.J.E. Peebles, "Principles of Physical Cosmology" Princeton: Princeton University Press, 1993 pp. 608ff makes it clear how much we do not know about galaxy formation. If you have the math, look up this treatment.

For a more popular treatment see, Alan Lightman, "Time for the Stars", New York: Viking, 1992, p. 49f

101. John Gribbin, "Companion to the Cosmos", Boston: Little, Brown & Co. 1996, p. 157

See also Michael Hawkins, "Hunting Down the Universe: The Missing Mass, Primordial Black Holes, and Other Dark Matters" London: Addison-Wesley, 1997.

See the report on the so-called "Higgsino" as a possible contributor to the dark matter of the universe. Philip Yam, "Mirror, Mirror", *Scientific American*, Vol 275, No.1, July 1996, p. 21
See "The Whole Shebang", Chapter 5

102. Neil de Grasse Tyson in "Darkness Visible" in *Natural History*, Vol. 106, No. 1, February, 1997, p. 76f gives an understandable account of the curving of space being used to measure "dark matter", the invisible component of the universe.

Evidence from the satellite "Rossi X-Ray Timing Explorer" has been interpreted at MIT to be evidence of warping of space by neutron stars and black holes. See *Discover*, Vol. 19, No. 2, February, 1998, p.16

103. There is evidence of the existence of galaxies, and not just quasars, from just 2 billion years after the Big Bang.

See Ron Cowen, "Found: Primeval Galaxies", *Science News*, Vol. 149, No. 8, February 24, 1996.

104. John Boslough, "Stephen Hawking's Universe", New York: Quill/William Morrow, 1985, p. 96

105. "Stephen Hawking's Universe" p. 94

106. Paul Davies, "The Last Three Minutes", New York: BasicBooks, 1994, p. 25

107. "The Last Three Minutes" p. 25

108. "The Cosmic Blueprint" p 122

109. Carl Sagan gives the following explanation in "Billions and Billions", New York: Random House, 1997, p. 9

BIG NUMBERS

Name would take to (U. S.) number from	Number (written out)	Number scientific notation) per second,	How long it count to this 0 (one count night and day)
One	1	1^0	1 second
Thousand	1,000	1^3	17 minutes
Million	1,000,000	10^6	12 days
Billion	1,000,000,000	10^9	32 years
Trillion	1,000,000,000,000	10^{12}	32,000 years (longer than there has been civilization on Earth)
Quadrillion	1,000,000,000,000,000	10^{15}	32 million years (longer than there have been humans on earth)
Quintillion	1,000,000,000,000,000,000	10^{18}	32 billion years (more than the age of the Universe)

Larger numbers are called a sextillion (10^{21}), septillion (10^{24}), octillion (10^{27}), nonillion (10^{30}), and decillion (10^{33}).
The Earth has a mass of 6 octillion grams.

Scientific or exponential notation also is described by words.
Thus, an electron is a femtometer (10^{-15} m)
yellow light has a wavelength of half a micrometer ($10^{-.5}$ μm)
the human eye can barely see a bug a tenth of a millimeter (10^{-4} m)
the Earth has a radius of 6,300 kilometers (6,300 km = 6.3 Mm)
and a mountain might weigh 100 petagrams (100 pg = 10^{17} g).

A complete list of prefixes goes as follows:

atto-	a	10^{-18}	deka-	--	10^{1}
femto-	f	10^{-15}	hecto-	--	10^{2}
pico-	p	10^{-15}	ilo-	k	10^{3}
nano-	n	10^{-9}	mega-	M	10^{6}
micro-	μ	10^{-6}	giga-	G	10^{9}
milli-	m	10^{-3}	tera-	T	10^{12}
centi-	c	10^{-2}	peta-	P	10^{15}
deci-	d	10^{-1}	exa-	E	10^{18}

110. "Stephen Hawking's Universe" p. 98

111. See Tony M. Liss and Paul L. Tipton, "The Discovery of the Top Quark", *Scientific American*, Vol. 277, No. 3, September 1997, p. 54

112. "The Cosmic Blueprint", p. 124
113. "Stephen Hawking's Universe", p. 91

114. "Wrinkles in Time" p. 35

115. If you want to know more about it, here is a synopsis.

Beginning with the singularity there is a destruction of a sort of a presumed primordial single force and it provides another way of tracing out the chronology of the Big Bang event. Today we recognize four fundamental forces in cosmology; gravity, the weak force, the strong force, and the electromagnetic force. Once this last was considered to be two forces, there was electricity and there was magnetism, but these were found to be one by James Clerk Maxwell in the 19th Century.

The weak force, which operates between leptons, was also thought to be an independent force but is now combined through quantum electrodynamics with the electromagnetic force in the electroweak theory. The strong nuclear force binds quarks together into hadrons that include the protons and neutrons of the atom. It has not yet been successfully integrated as one force with the electroweak theory force but it seems likely that this will be done. This is often called a "Grand Unified Theory" or a GUT.
It is also suspected that gravity can also be joined to the other forces and arise from just one unified force which could be described by a "Theory of Everything" or TOE of what makes the universe tick, a force that was only one force at the very beginning of the universe.

As we work backwards in time, we can see the separation of the forces except gravity at various time levels after the Big Bang. The electrical force and the magnetic force can be separated in your kitchen, although some researchers (Macke e.g.) says electrical force and the magnetic force are inseparable. Perceived separation is the result of the observer's frame.
By about 10^{-11} seconds after the Big Bang the weak force became distinct from the electromagnetic force and restricted to short ranges. At 10^{-36} "...Something analogous to a sudden crystallization of ultra-cool water to ice occurred. Heat was liberated...producing a momentary temperature rise. This broke the symmetry...and the so-called strong nuclear force differentiated..."

[Eugene F. Mallove, "The Quickening Universe", New York: St. Martin's Press, 1987, p. 31]

The wall of Planck time comes at 10^{-43} second. At this point there is a fundamental breakdown in the ability of physicists to describe space, time, or matter. It is presumed that gravity has just broken its bond with the single unified force that existed at the instant of the Big Bang. But nobody knows for sure because there is no quantum treatment of gravity.

["Stephen Hawking's Universe", p. 100]

116. What follows in the text is, I believe, a fairly good layman's explanation of the theory of inflation. I thought I had a good handle on it until I read, "Inflation in a Low-Density Universe" by Martin A. Bucher and David N. Spergel in *Scientific American*, January 1999, p. 63f. The article speaks of inflation before the Big Bang. I do not believe that this has any effect on my argument that the universe can come from nothing, but it may affect the chronology.

117. "The Quickening Universe", p. 27
118. The inflationary hypothesis is not without its problems. Huw Price mentions some in "Time's Arrow and Archimedes' Point", op. cit.

119. See "Wrinkles In Time", p.174 for this quote and the book for the COBE discoveries.

See Henning Genz, "nothingness", p. 288f for another readable treatment of inflation.

120. "The Conscious Universe" p. 156

121. "The Last Three Minutes" pp 33-34

122. "The Last Three Minutes" p. 31

123. "The Conscious Universe" p. 156

124. "The Quickening Universe" p. 31

125. "The Last Three Minutes" p. 35

Tim Folger, "The Magnificent Mission", *Discover*, Vol. 21, No. 5, May 2000, p. 48 explains that,

> "Vacuum-spawned particles are constantly flickering in and out of existence around us, arising from and sinking back into the void. During inflation, this process, like everything else in the universe, was magnified tremendously. The rapidly expanding early universe imparted enough energy to these particle wannabes that instead of quickly subsiding into the vacuum, they remained in the real world. The sudden influx of countless particles from the vacuum was like a stone thrown into the dense particle pond of the early universe, sending out ripples—pressure waves. And pressure waves through a gas are nothing more than sound waves. The entire universe rang like a bell."

The MAP project is designed to find the remnants of this sound in the variations in the cosmic microwave background explored by COBE.

126. "The Quickening Universe" pp. 29-30

127. "Reading the Mind of God" p. 212

128. "Black Holes and Baby Universes", p. 92

See also Murray Gell-Mann, "The Quark and the Jaguar", New York: W.H. Freeman and Company, 1994, chapter 14

See Michael J. Duff, "The Theory Formerly Known as Strings", *Scientific American*, Vol 278, No. 2, February 1998, p. 64 for an overview of the state of the theory for the layperson.

See Brian Greene, "The Elegant Universe", New York: W.W.Norton & Company, 1999 which is a book entirely about the superstring theory, its strengths and weaknesses.

See "Before the Big Bang", p. 223f

129. Michio Kaku & Jennifer Thompson, "Beyond Einstein", New York: Anchor Books, 1987, p. 141f

130. One of the best explanations of the state of string theory can be found in Brian Greene, "The Elegant universe". See a review of this book by Chris Quigg, "Aesthetic Science", *Scientific American*, Vol. 280, No. 4, April 1999, p. 125

K.C. Cole, "A Hole In the Universe" is also excellent.

Stephen Hawking, "The Universe In A New York: Bantam Books, 2001, also considers string theory.

131. "Black Holes and Baby Universes", p. 93

132. "The Quark and the Jaguar", p. 99-100 and see the chapter in which the quote is contained.

133. However, George Smoot in "Wrinkles in Time", p. 188 probably states the opinion of most physicists and cosmologists when he says, "It is here that Guth has pointed to a harmonious marriage between astrophysics and particle physics-a match so beautiful it has to be right."

134. D. Russell Humphreys has an imaginative solution using Relativity as an explanation. According to this nuclear weapons engineer, the earth is close to a black hole and so billions of years can pass in other parts of the universe, thus giving the results of astrophysics, but only a few thousand years pass on earth. Since there is no evidence that this is the case and since if it were life could not exist on earth because of resulting gamma radiation, there seems to be no reason to posit such a scenario unless you begin with the idea that the Bible must be correct.

135. The question of what came before the Big Bang is sometimes said to be meaningless since time itself came into existence with the singularity. Some inflationary scenarios see the quantum fluctuation that begins it all happening in some pre-existing space-time and a new universe "budding" from an old one. The "baby universe" idea could mean a process that is eternal, with no beginning and no end, in which case there is still no need for a creator to get it all started. Or it could mean that the initial beginning of the series began *ex nihilo* as we are suggesting here for our universe. Either way the argument stands.

136. "The Cosmic Blueprint" p. 125

137. “Genesis and the Big Bang” p. 44

138. “Stephen Hawking’s Universe”, p. 91 Note that the author is not speaking of biblical creation.

139. I found an interesting indication of the truth of the argument against the need for a cause in Lawrence Sklar, “Philosophy of Physics”, Boulder: Westview Press Inc., 1992. Though he is treating the absence of a causal link between two members of a quantum system, for example measuring the spin on one entity having an effect on the other at a great distance away, and he is not talking about the emergence of the universe from a vacuum, he does say that,

> “Perhaps the most interesting impact of quantum theory…is the claim of many that for the first time we have a theory of the world that allows us to deny, for a given event, that *any* past event could ever be found that was causally adequate to explain why the given event, rather than some specifiable alternatives to it, occurred. [In quantum physics]…reasons exist for denying the existence of the needed cause, not just reasons for thinking that such a cause had merely eluded our grasp.” (p. 204)

I warn you that on p. 205 he says, “we need to follow out a long argument” and he certainly does. But he arrives at p. 223 to say that the best evidence (for Bohr’s understanding and contra Einstein, Podolsky and Rosen) is that, “the features attributed to a system when it was measured and found to have certain values of an observable quantity were `brought into being’ by the measurement and not already present in the system.”

This is not a claim that a scientific measurement brings anything into existence *ex nihilo*. But it does show a lack of need for the kind of causality that the philosophers often insist on.

140. Patrick Glynn, “God: The Evidence”, Rocklin: Prima PublishingFORUM, 1999, is a former atheist turned believer and it is the anthropic principle that turned the tide for him. He is a source for the many factors that must be “just so” for the universe to exist as it does. He is rather big on near-death experiences and the evidence for god from faith’s effect on human lives and health. He even brings in Pascal’s wager that ends with the odds in favor of believing in god.

141. "The Secret Melody" p. 232

See John Leslie, "Universes", New York: Routledge, 1989, for a defense of the strong
Anthropic Principle and for philosophical discussion of the meaning from a Neoplatonist viewpoint.

142. "Darwin's Dangerous Idea" p. 166

143. See Nima Arkani-Hamed, Savas Dimopoulos and GeorgiDvali, "The Universe's Unseen Dimensions", *Scientific American*, Vol. 283, No. 2, August 2000, p. 62 which theorizes that,

> "...our entire three-dimensional universe [could] be just a thin membrane in the full space of dimensions. If we consider slices across the extra dimensions, our universe would occupy a single infinitesimal point in each slice, surrounded by a void."

There could be other universes as close as a centimeter away in another dimension.

Martin Rees, "Just Six Numbers", New York: Basic Books, 2000 describes the basic numbers that it is sometimes claimed must be "just so" for our universe to exist as it does, with stars, galaxies and intelligent life. He draws no conclusions but leans toward multi-universes.

See also, "Darwin's Dangerous Idea", p.176f

I am speaking here of the possibility of other universes coming into existence just as ours did. There would be no question of whether they come into existence "before" or "after" ours since time has no meaning except inside a universe. There is another meaning to the concept of "many worlds", however, which I am not considering. This is the possibility of an almost infinite number of universes arising because of quantum effects in "our" universe. This is also known as an "alternate histories theory."
As Paul Davies and John Gribbin explain in "The Matter Myth" [p. 228f],

> "Taking the many-worlds theory to its logical conclusion, we are led to suppose that countless times every second each human being is split into duplicate copies, each copy inhabiting a slightly different universe. Necessarily, each copy will only perceive *one* universe, and be aware of only *one* self."

See also, "Abouttime", p. 231f.
See also, "The Conscious Universe", p. 112f.
See also, "Lonely Hearts of the Cosmos", p. 123f.

See also, Henry P. Stapp, "Mind, Matter and Quantum Mechanics", New York: Springer-Verlag New York, 1994, p. 112f.
See also, "The Physics of Immortality", p. 203
See also, "The Quickening Universe", p. 58f

144. See "Thomistic Metaphysics", pp. 241ff if you really want to know how a Thomist explains contingency and the need for a final cause. However, I don't recommend it.

145. John Gribbin's "Companion to the Cosmos" is a good source for a quick idea of how galaxies form.

146. The recent discussion of the possibility of life on Mars and the existence of planets around other stars has certainly made this possibility more imaginable.

147. "Darwin's Dangerous Idea" p. 36

148. Philip Kitcher, "Abusing Science", Cambridge: The MIT Press, 1982, p. 7

149. Robert Wright, "The Moral Animal" New York: Pantheon Books, 1994, p. 23

150. See Jeffrey H. Schwartz, "Sudden Origins", New York: John Wiley and Sons, Inc, 1999, p. 7 and the index for Fisher's work and influence.

151. How all this works on the level of the cell and embryo in species conservation and differentiation is way out of my level. You might refer to John Gerhart and Marc Kirschner, "Cells, Embryos and Evolution", Malden: Blackwell Science, 1997 for a complete treatment.

See Mark Ridley, "Evolution", 2nd Edition, Cambridge: Blackwell Science, Inc., 1996 who gives the whole story in great detail.

152. "Abouttime" p. 34

153. E. Richard Moxon and Christopher Wills, "DNA Microsatellites: Agents of Evolution?" "*Scientific American*", Vol 280, No. 1, January 1999, p. 94

154. "Darwin's Dangerous Idea" p.50

155. Robert Wright, "Three Scientists and Their Gods", New York: Times Books, 1988, pp. 205-206

156. For an account of the Miller-Urey experiment to create life by sparking such a soup, see Paul Davies, "The Fifth Miracle", New York: Simon and Schuster, 1999, p. 86.

This same Stanley Miller reported in the April 2000 issue of "Proceedings of the National Academy of Sciences" that sparking a soup of methane, ammonia, nitrogen and water created parts of PNA, peptide nucleic acid, which had been suggested as a more likely candidate for the first genetic material.

R. Lipkin, "Early life: In the soup or on the rocks?", *Science News*, Vol. 149, No. 18, May 4, 1996, p. 278, reports on an experiment that points to warm rocks onto which amino acids and nucleotides were washed and allowed to dry as the medium for the assembly of molecular chains long enough to be precursors of life.

The same scenario is investigated in Robert M. Hazen, "Life's Rocky Start", *Scientific American*, Vol. 284, No. 4, April 2001, p. 77 where minerals provide a scaffold for the assembly of molecules.

The start of life may even have been from outside our world. Re-creating the conditions at the beginning of our solar system, scientists,

> "...zapped a mix of water, methanol, ammonia, and carbon monoxide ices with the kind of energetic rays emitted by hot young stars. In the end, about 2 percent of the frozen gastransformed into oily organic molecules. When dipped in water, the compound spontaneously forms multiwalled chambers similar in size to living cells. Some of these bubbles glow under fluorescent illumination, meaning they transform ultraviolet into visible light, In short, the proto-cells provide two services essential to life. They have membranes that protect the chemistry going on inside, and they convert destructive high-energy radiation into useful, less intense energy."
>
> [Kathy A. Svitil, "Interstellar Seeds of Life", *Discover*, Vol. 22, No. 7, July 2001, p. 13

Similar are the findings of those studying organic compounds that might have originated in interstellar clouds and comets. David Blake and Peter

Jenniskens, "The Ice of Life", *Scientific American*, Vol. 265, No. 2, August 2001, p. 45. theorize that,

> "These balls of ice and rock could then have carried the organic compounds on a collision course with the young Earth. After reaching this planet, the organics could have participated in the chemical reactions from which the first living organisms arose."

157. Stuart Kauffman, "At Home In The Universe", New York: Oxford University Press, 1995, p. 31

Alexander Oparin, "was impressed by the fact that oily substances and water don't mix, and sometimes produce a suspension known as a coacervate, in which the oil retreats into tiny droplets. The oily blobs superficially resemble biological cells." For a discussion of his ideas, see "The Fifth Miracle", p. 85.

158. See "At Home In The Universe" chapter 3 for the "catalytic closure" explanation.
See "The Blind Watchmaker" chapter 6 for a discussion of how much luck we are allowed in the process. See pp 148ff for the "inorganic mineral" theory.
See "The Blind Watchmaker", p. 37f and 77ff for the example of the evolution of the eye and a discussion of the topic being considered here.
See Richard Dawkins, "The Selfish Gene", New York: The Oxford University Press, 1990, for the "primeval soup" model.

See Richard Dawkins, "Climbing Mount Improbable", New York: W.W. Norton & Company, 1966 for the eye and other examples of the process.

See "The Matter Myth" pp.288f for the Miller-Urey experiment involving an electrical spark through a chemical mixture.

See Richard Fortey, "Life" New York: Alfred A. Knopf, 1998 for an up to the minute (in 1998) survey of the development of life on earth.

See John Maynard Smith and Eors Szathmary, "The Origins of Life" New York: Oxford University Press, 1999, p. 153f for the eye as well an a treatment of the topic under discussion.

See Christian de Duve, "The Birth of Complex Cells", *Scientific American*, April 1996, p.56 for an explanation of the appearance of eukaryotic cells from prokaryotic cells.

159. See Karen Wright, "When Life was Odd" in *Discover*, Vol. 18, No. 3, March, 1997, p. 52. Her article gives information about the Ediacarans (pronounced ee-dee-ACK-a-rans), single celled life forms who may have been the precursors of many Cambrian species.

160. See "Climbing Mount Improbable", chapter 5 for an extended discussion of the evolution of the eye.

See Chet Raymo, "Skeptics and True Believers", New York: Walker and Company, 1998, p. 148f.

See Stephen Jay Gould, "Creating the Creators", *Discover*, Vol. 17, No. 10, October 1996, p. 43f. who talks about the same question which has been called,

> "'The problem of the incipient stages of useful structures.' I prefer a catchier label based on a primary example: 'The 5-percent-of-a-wing problem.'...for 5 percent of a wing confers no benefit whatsoever in flight."

His explanation is almost the same as mine in the case of the eye except that feathers have other uses which can be put to the service of flight later, whereas sensitivity to light is just intensified.

See Robert Jastrow, "The Enchanted Loom", p. 95

161. See "Abusing Science", p. 102f for a discussion of the Creationists' argument that there has not been enough time for evolution to occur.

162. "The Blind Watchmaker", p. 78
See Lubert Stryer, "The Molecules of Visual Excitation", *Scientific American: Science's vision: The Mechanics of Sight* 1998, for a detailed explanation of the cascade of molecular reactions that transforms light on a retina into a nerve impulse. Significant is the finding that,

> "protein synthesis is one of the most fundamental metabolic activities of any cell. therefore it seems likely that elongation factor *Tu*, which takes part in that work [and is vital to the visual process] originated earlier in evolution...One of the many fascinating things about evolution is that mechanisms evolved for a particular function may later be modified and applied in different functions. That, I think, is what happened to the mechanism of *Tu*. After evolving to mediate protein synthesis, it was retained for billions of years and ultimately put to work in the transduction of hormonal and sensory stimuli." (p. 30)

163. See Robert R. Jackson, "The Eyes Have It", *Natural History*, Vol. 107, No. 3, April 1998, p. 30 for a short treatment of jumping spider eyes. In fact, these spiders have a pair of principal eyes and six secondary eyes, built on different principles.

Time of November 15, 1999 reports on a parasite of paper wasps with an eye with one hundred complete eyes in the shape of the normal insect compound eye. Also mentioned is that a similar arrangement appeared in trilobites from 230 million years ago. This eye thing just keeps going and going.

164. "River Out of Eden" p. 83.

165. One big change was from invertebrate to vertebrate. One researcher,

> "suggests that a random mutation gave an animal, which probably resembled amphioxus, a double set of chromosomes and hence a second copy of all genes. Such a mix-up is not implausible...because chromosome duplicates occur frequently in nature...After the initial gene doubling that occurred more than 500 million years ago, the first set of genes went on performing its original role, while the duplicate set was co-opted to perform new functions...Those extra genes—particularly the additional developmental genes—allowed the hypothetical vertebrate ancestor to evolve entirely new body structures."
> [Richard Monastersky, "Jump-Start for the Vertebrates", *Science News*, Vol. 149, No. 5, February 3, 1996, p. 74]

166. Meredith F. Small, "Floral Arrangements", *Natural History*, Vol. 108, No. 4, May 1999, p. 46f

167. Jared Diamond, "Evolving Backward", *Discover*, Vol. 19, No. 9, September 1998, p. 64
It has long been thought that snakes "evolved backward" from burrowing lizards, but there is evidence that at least some may have descended from giant marine lizards called mosasaurs.

> [See P. Smaglik, "Retelling the tale of a two-legged snake", *Science News*, Vol. 151, No. 16, April 19, 1997, p. 238.]

168. Stephen Jay Gould, 'This View of Life', *Natural History*, Vol. 105 #7, July 1996, p.72

See also his extended treatment in "Full House" New York: Harmony Books, 1996. As Gould states in the introduction, the purpose of his work is to give a, "general argument for denying that progress defines the history of life or even exists as a general trend at all. Within such a view of life-as-a-whole, humans can occupy no preferred status as a pinnacle or culmination." [p. 4]

169. Annie Dillard, "Pilgrim At Tinker Creek" in the Book-of-the-Month Club edition "Annie Dillard", Harper & Row, 1990, p.135

170. Douglas D. Richman, "How Drug Resistance Arises", *Scientific American*, Vol. 279, No. 1, July 1998, p. 88

Randolph M. Nesse and George C. Williams, "Evolution and the Origins of Disease", *Scientific American*, Vol. 279, No. 5, November 1998, p. 86, points out that, "natural selection is unable to provide us with perfect protection against all pathogens, because they tend to evolve much faster than humans do." The same article is very interesting in explaining why we have diseases and brings up something called "Darwinian medicine" which takes into account the reason why we are affected by diseases and physical degeneration. This is another good piece of evidence for the validity of the theory of evolution.

171. Stephen J. Gould, "The Power of This View of Life", *Natural History*, Vol 103 #6, p. 6

172. "Pilgrim at Tinker's Creek" p. 176

173. See David M. Raup, "The Role of Extinction in Evolution" in "Tempo and Mode in Evolution", Ed. by Walter M. Fitch and Francisco J. Ayala, Washington, D.C., National Academy Press, 1995. Raup explains that, "It is conventional to divide extinctions into two distinct kinds: background and mass extinction. The term `mass extinction' is most commonly reserved for the so-called `Big Five' events: short intervals in which 75-95% of existing species were eliminated. (see table)

Comparison of species extinction levels
for the Big Five mass extinctions

Extinction Episode	Age before present Myr	Percent extinction

Cretaceous (K-T)	65	76
Triassic	208	76
Permian	245	96
Devonian	367	82
Ordovician	439	85

(Extinction data are from Jablonski (1991).

Although the Big Five were important events, their combined species kill amounted to only about 4% of all extinctions in the past 600 Myr.

See also Richard Monastersky, "Life's Closest Call" on the Permian extinction which some ascribe to a build up of carbon dioxide in the seas, others to volcanic introduction of carbon dioxide into the atmosphere. In *Science News*, Vol. 151, No. 5, February 1, 1997, p. 74f

See also Douglas H. Erwin, "The Great Paleozoic Crisis: Life and Death in the Permian", New York: Columbia University Press, 1993

See also Peter W. Ward, "Greenhouse Extinction" on the Permian extinction in *Discover*, Vol. 19, No. 8, August 1998, p. 54f

See also Douglas H. Erwin, "The Mother of Mass Extinctions", *Scientific American,* Vol. 275, No. 1, July 1996, p. 72. A bit of a meteorite which reputedly caused the extinction of the dinosaurs, plus 70% of other animal and plant species, has reportedly been found in the northern Pacific.

174. "Darwin's Dangerous Idea" p. 133.

175. "Darwin's Dangerous Idea", p. 59

176. Not who the winner will be, of course, but that there will be a winner. Tic-tac-toe is also an algorithm, but it does not require that there be a winner. There is the old sucker bet that wagers there is someone in the room who can win ten coin tosses in a row. Once betting the farm on it, merely pair off enough people (1024 to be precise) and have them go at it. The winner of one contest takes on the winner of another. One person will emerge who has won ten times in a row. Guaranteed.

177. An article in *Technology Review,* Vol;. 100, No. 7, October 1997, p. 22f entitled, "A Study in Complexity" by Robert Lee Hotz, speaks of Stephen Wolfram,

> "...and his colleagues believe the complexity of the universe belies an underlying simplicity in which a few basic rules give rise to complicated and unpredictable behavior. Indeed, if one conceives of God as a clever programmer, then one can imagine our vast, expanding universe as the elaborate consequence of an algorithm that set the conditions of the cataclysm known as the Big Bang. Everything that has followed—from black holes and organic chemistry to the rise of human consciousness and the spontaneous melody of a jazz improvisation—is an inevitable result."

Whether or not this research into an all-encompassing algorithm is successful remains to be seen, of course, but even then, many would quibble if this would necessitate a god. It would also be a far cry from the kind of god previously envisioned.

178. A typical reaction is found in a letter to the editor in the *St. Louis Post-Dispatch* newspaper for Saturday, February 24, 2001; though similar examples can be found any time there is an article or editorial about evolution. The writer claims that,

> "Darwin himself indicated that thousands of 'links' would be found to prove his theory. Decades later, still no links and no proof. The Bible, on the other hand has stood the test of time. The evidence for evolution is found in a few bones, teeth and the imagination of those who would rather believe anything than the idea of a God that they might have some obligation to."

Fortunately, the paper printed another letter from a Professor of Geology that did not address the numerous examples of "links" but did point out that, "many scientists, including myself, have strong religious beliefs but have no difficulty reconciling these beliefs with modern scientific theory."

I doubt, however, that the first writer would be open.

179. See "Thomistic Metaphysics", p. 86 which defines "essence" as a principle of limitation.

> "Considered statically as a principle of limitation, or no-more-being, saying that which makes the being what it is and without which it would not be that being, the name *essence* is used."

However, Hart goes on to point out that "In no way does it oppose the position on the entirely distinct problem concerning their origin, which may readily suppose a conceivable process of evolution involving substantial change."

Thomas Aquinas, of course, had little idea, if any, of an evolutionary process, nor did many teachers of Thomism right through quite recently.

Indeed, if a species evolved into another species, it would in effect change its essence. The essence of a wolf would not change into the essence of a dog if the individuals changed. Since there was no experiential knowledge of a species changing into another until Darwinism, there was no reason to suppose that these changes actually occurred.

180. "Darwin's Dangerous Idea", p.38

181. "Humani Generis" August 12, 1950

182. The Best of Fulton J. Sheen, "From the Angel's Blackboard", Ligouri: Triumph Books, 1995, p. 29

183. R. Monastersky, "Winging It: An unusual approach to flight" in "*Science News*", Vol. 151, No. 10, p. 143

184. See "Climbing Mount Improbable" chapter 4 for an extended discussion of the evolution of flight.

185. "The Blind Watchmaker" pp.90-91

186. Michael D. Lemonick, "Dinosaurs of a Feather", *Time*, Vol. 151, No. 26, July 6 1998, p. 82

See Josh Fischman, "Feathers Don't Make the Bird", *Discover*, Vol. 20, No. 1, January 1999, p. 48

See Kevin Padian and Luis M. Chiappe, "The Origin of Birds and Their Flight", *Scientific American*, Vol. 278, No. 2, February 1998, p. 38

See R. Monastersky, "China yields a flock of downy dinosaurs", *Science News*, Vol. 156, No. 12, September 18 1999, p. 183.

However, there seems to be some controversy over the Chinese feathered dinosaurs.See Michael D. Lemonick, "Down-Covered Dinosaur", *Time*, Vol. 157, No. 18, May 7 2001, p. 56 reports that new finds make it a certainty that flightless dinosaurs had feathers in a relative of T. rex dating from 124 to 147 million years ago.

187. The missing-link argument is still with us. It was used in an action by the Kansas School Board, now thankfully reversed, to insist that other explanations than Darwin's be taught in school. In Minnesota a high school teacher went to court for the right to teach even though he had publicly made statements such as, "Look at how complex this system is. It's hard for

me to believe this all came about by chance mutations over billions of years." However, this was not his only objection. He also cited, "the amazing lack of transitional forms in the fossil record. There has never been a creature discovered that could be considered a logical intermediate of any two major classes of animals or plants." Eugenie Scott, executive director of the National Center for Science Education countered, "there are transitional fossils out the ying-yang."
Eloquently stated. [*Time*, Vol. 156, No. 2, July 10, 2000, p. 60]

Anole lizards of the Greater Antilles are another example, though not of intermediate fossils. Rather, a species of lizard has changed (evolved) to fill many niches in the environment. An article by Jonathan B. Losos in *Scientific American*, Vol. 284, No. 3, March 2000, "Evolution: A Lizard's Tale", p. 64 is really interested in the fact that the same morphological changes occurred on several islands because this body type is best suited to that particular niche, but it serves for our purposes here as well. Evolution in action, again.

188. Stephen Jay Gould, "Hooking Leviathan by Its Past", in *Natural History*, Vol. 103 #5, May 1994, p.10. See the entire article for the full story of the discovery of the "missing link".

See Carl Zimmer, "The Equation of a Whale", *Discover*, Vol. 19, No. 4, April 1998, p. 78f for another history of the controversy

See Richard Monastersky, "The Whale's Tale", *Science News*, Vol. 156, No. 19, November 6, 1999, p. 296 linking the whale to hippopotamuses.

See Philip D. Gingerich, "The Whales of Tethys", *Natural History*, Vol. 103, No. 4, April 1994, p. 86

189. Gould adds to his own article in *Natural History* in his book "Dinosaur In A Haystack" p. 373 about the discovery of another whale, *Rodhocetus kasrani*, which also shows the intermediate steps to the modern whale.

See also Kate Wong, "Cetacean Creation", in *Scientific American,* January 1999, p. 26

190. Jules Verne, "20,000 Leagues Under the Sea" Trans. by Anthony Bonner, New York: Bantam Books, 1962, p. 279

191. Another is the pufferfish which blows itself up into a sphere of tough flesh and often spines. In studying its evolution, it is found that there is not

much change in the mechanism by which the final result is achieved. Over the eons of evolution there is a seemingly great distance between the pufferfish and the sunfish from which is developed,

> "...but the sequence of nerve impulses that control these traits has barely changed."

Indeed, many examples have been found,

> "...in which evolution holds on to an old neuromuscular pattern while changing part of the body it controls, thus creating a new behavior. Birds, for example, fire some of their shoulder muscles in much the same pattern as reptiles."
> [Carl Zimmer, "How the Pufferfish Got Its Puff", *Discover*, Vol. 18, No. 9, September 1997, p. 30]

192. See the refutation of this objection, especially as articulated by Henry Morris in "Abusing Science", pp.100f

193. Wolfhart Pannenberg, "Toward A Theology of Nature", Louisville: Westminster/John Knox Press, 1993, p. 23

194. "At Home in the Universe" p. 33

195. James Collins, "A History of Modern European Philosophy", Milwaukee: The Bruce Publishing Company, 1954, p.831

196. James A. Michener, "The Covenant", New York: Random House, 1980, p. 403

197. See "The Blind Watchmaker" chapter 3 on `Accumulating small change.'

Are mechanical robots alive? See Paul Tachtman, "Redefining Robots, "*Smithsonian*", Vol. 30, No. 11, February 2000, p. 97. Mark Tilden of the Los Alamos National Laboratory in New Mexico manufactures analog (not digital in being controlled by a computer) that can solve problems of locomotion and reach simple objectives. A comparison of the robots to insects is inescapable. Their manner of operation is perhaps comparable to our own on some level. As Tilden says,

> "...many studies of consciousness are based on the concept that we are rational animals. No! We are a solid core of pure chaos bounded by linear systems keeping us regulated toward some level of

cohesiveness with our world. We are chaotic creatures who are made rational by our environment." [p. 101]

If we began with robotic creatures sent from another solar system, as we may well someday do ourselves, which then evolved into humans, at what point would the transition occur?

198. W. Ford Doolittle, "Uprooting the Tree of Life", *Scientific American*, Vol. 282, No. 2, February 2000, p. 90

199. Richard Leakey, "The Origin of Humankind", New York: BasicBooks, 1994, p. 104

200. "Humani generis", 1950. Cited in the series, "Great Religions of Modern Man", in "Catholicism" ed. by George Brantl, New York: George Braziller, 1962, p. 46

See a discussion of this action as a capitulation to science in "A Glorious Accident", p. 238

See Rainer Koltermann, "Evolution, creation, and church documents", *Theology Today,* Vol. 48, No. 2, Summere 2001, p. 124f for this and other papal statements in context.

201. See "God and the New Biology" p.78f

202. "Catechism of the Catholic Church" #364

In what to me is a stunning contradiction to this teaching, a Vatican official objected to human cloning saying that, in the words of the report in the *St. Louis Review*, January 16, 1998, "a human clone would be created in the image and likeness of man, not God." If the bishop cited is consistent with Catholic doctrine he would have to hold for the divine creation of the soul of the person no matter how the body was formed. This would make the person an image of God. Otherwise, the clone would also fall outside of the rights and privileges of all other humans and would, presumably, be an animal that could be used by humans for their purposes.

The same official claimed that the cloning of a human from the genes of one individual would be, "an affront to Almighty God and to the laws on the transmission of human life." How he knows that the standard practice of human intercourse is the only permissible means of conception I presume is based on Church teaching about *in vitro* fertilization deriving from Tradition since there is certainly nothing in Scripture dealing with the matter. Presumably, "natural law" is invoked so see the section on natural law.

203. "Genesis and the Big Bang", p. 23. Indeed, in order to keep the time frame of Genesis in synch with the geological time frames, Prof. Schroeder claims that Neanderthals and Cro-Magnons, the latter firmly in the *Homo sapiens* camp, did not receive this additional spirit, and hence are not really human.

204. See any translation of Plato's "Phaedo" such as "Five Great Dialogues: Plato", Trans. by B. Jowett, Ed. by Louise Ropes Loomis, New York: Walter J. Black, 1942, or perhaps something more modern. In it, Plato has Socrates justify his willingness to accept death since it will separate his soul from his body, the very thing he has been trying to do in the pursuit of philosophy.

205. "God And The New Biology" p. 88

206. "Catechism of the Catholic Church" #365

207. "The Catechism of the Catholic Church" #366

208. William Faulkner expresses this hope very well in his "Nobel Prize Acceptance Speech" when he says,

> "I decline to accept the end of man. It is easy enough to say that man is immortal simply because he will endure; that when the last ding-dong of doom has clanged and faded from the last worthless rock hanging tideless in the last red and dying evening, that even then there will still be one more sound: that of his puny inexhaustible voice, still talking. I refuse to accept this. I believe that man will not merely endure: he will prevail. He is immortal, not because he along among creatures has an inexhaustible voice, but because he has a soul, a spirit capable of compassion and sacrifice and endurance."
>
> [Cited in "Essays, Old and New", Edited by Robert U. Jameson, New York: Harcourt, Brace and Company, 1955, p. 335, but available from many sources.]

209. See *Discover*, Vol. 19, No. 11, November 1998, p. 28

210. David Darling, "Soul Search", New York: Villard Books, 1995. p. xv This burial with flowers is also referred to by Richard Leakey in "The Origin of Humankind", p. 155, who puts it in context of looking for evidence of consciousness.
However, *Natural History*, March 2000, p. 51 reports that the, "flowers may well have been deposited at Shanidar by the Persian jird, *Meriones persicus*" a local rodent, and cites Jeffrey D. Sommer's conclusion that, "the flower pollen recovered near Shanidar IV is more likely to have resulted from the activities of rodents than Neanderthals."

See B. Bower, "Child's bones found in Neandertal burial", *Science News*, Vol. 148, No. 17, October 21, 1995, p. 261.

211. Charles Pellegrino, "Return to Sodom and Gomorrah", New York: Random House, 1994, p. 19

212. "Darwin's Dangerous Idea" p. 63

213. We are going to consider this question at some length. For more information read George Lakoff and Mark Johnson, "Philosophy in the Flesh", who begin with the position of science that,
"The mind is inherently embodied. Thought is mostly unconscious. Abstract concepts are largely metaphorical." They then ask,

> "What would happen if we started with these empirical discoveries about the nature of mind and constructed philosophy anew? The answer is that an empirically responsible philosophy would require our culture to abandon some of its deepest philosophical assumptions." [p. 3]

I am asking the same question of the Faith. What happens when we consider it beginning with empirical discoveries instead of assumptions?

214. Georges Van Riet, Ph.D, "Thomistic Epistemology" Vol. 2, Trans. by Donald G. McCarthy, St. Louis: B. Herder Book Co., 1965, p. 89

215. There are excerpts from Aquinas, Descartes, Locke, Berkeley, Hume, Kant and others, many treating or touching on the need for a spiritual quality to thought and volition in William L. Kelly, S.J. & Andrew Tallon, "Readings in the Philosophy of Man", New York: McGraw-Hill Book Company, 1967 all gathered in one place.

216. "The Matter Myth" p.308

For a discussion of Western philosophy's rejection of dualism versus the Buddhist ideas on the soul or psyche, see Jean-Francois Revel and Matthieu Ricard, "The Monk and the Philosopher", Trans. by John Canti, New York: Schocken Books, 1998

217. "Mind, Matter and Quantum Mechanics", p. 41

218. Benedict de Spinoza, "Ethics, Part V. Prop. XXIII", Translated by R.H.M. Elwes, in "The Rationalists" New York: Doubleday & Co. In. 1960, p. 395.

219. F.J. Sheed, "God and the Human Condition", Volume One, New York: Sheed and Ward, 1966, p. 274.

220. See Edwin Arthur Burtt, "The Metaphysical Foundations of Modern Science", Garden City: Doubleday & Company, 1954, p. 127f for Hobbes' objections, as well as Gilbert, Boyle, Newton and others, and of course Gilbert Ryle of "ghost in the machine" fame.

221. John M. Oesterreicher, "Five in Search of Wisdom", Notre Dame: University of Notre Dame Press, 1952, p. 2

222. "Demon Haunted World" p. 325

223. "Darwin's Dangerous Idea" p. 263

224. *Time* March 25, 1996, p. 50

225. Rudolph Chelminski in "Your Opponent Must Be Destroyed" in *Smithsonian,* Vol. 28, No. 10, January 1998, p. 45, records the antics of chess masters through the centuries. It does seem unlikely that a computer would ever be such a character.

226. Margaret Wetheim, "The Pearly Gates of Cyberspace", New York: W.W. Norton, 1999.

227. "Darwin's Dangerous Idea" p. 44

228. John F. Ross, "A few miles of land arose from the sea—and the world changed", *Smithsonian*, Vol. 27, No. 9, December 1996, p. 117

229. See a number of books and articles by Stephen J. Gould, including his regular column in *Natural History*. "Lucy on the Earth in Stasis" in Vol. 103, No. 9, September 1994 is a good
example based on the history of our own ancestors.

See a treatment of punctuated equilibrium in Daniel Dennett's, "Darwin's Dangerous Idea" pp. 282-299.

See also Gary Cziko "Without Miracles" and his conclusion "So actually nothing in the theory of punctuated equilibrium is in any way fundamentally inconsistent with Darwin's conception of evolution." p. 316

See also "The Quark and the Jaguar" p. 238f

See Ian Tattersall, "Becoming Human: Evolution and Human Uniqueness", New York: Harcourt Brace & Co., 1998

See "Showdown on the Burgess Shale", *Natural History*, Vol. 107, No. 10, December 1998-January 1999, p. 48f. Simon Conway Morris and Stephen Jay Gould square off on the meaning of the fossils found in the Burgess Shale and elsewhere which Gould claims is an example of a radical leap in the Cambrian age.

See Jeffrey H. Schwartz, "Sudden Origins", on the discovery of "homeobox genes" which can explain the sudden appearance of new species. Natural selection continues to play a role in evolution but there is a new actor on stage.

Lynn Margulis and Dorion Sagan, "The Beast with Five Genomes", *Natural History*, Vol. 110, No. 5, June 2001, p. 38 bring yet another vehicle of evolution to our attention, namely "symbiogenesis" in which one life form takes up residence in another, or one corrals the other and uses it to its own ends. One example used is a termite which contains within its gut a number of organism, one of which itself contains five distinct creatures, each with their own genome. Chloroplasts in plants and mitochondria in most animals, including us, are other examples of captured beings some of which are still capable of living an independent life. The authors conclude, "eventually we may well realize that natural selection operates not so much by acting on random mutations, which are often harmful, but on new kinds of individuals that evolve by symbiogenesis."

However, Carl Zimmer, "An Explosion Defused?", *Discover*, Vol. 17, No. 12, December 1996, p. 52 says there is evidence that there was no explosion but that, "the genes suggest that a leisurely paced and comfortably spaced course of evolution was well under way half a billion years before the putative Cambrian explosion."

230. Fritjof Capra, "The Web of Life", New York: Anchor Books, 1996, p. 228.

See also Stuart Kauffman, and Ilya Prigogine as well as Humberto Maturana and Francisco Vaela on "Autopoiesis".

231. "The Conscious Universe" p. 94

232. Jack Werren, "Genetic Invasion of the Insect Body Snatchers" in *Natural History*, June 1994 p. 36f. "Humans have hundred of thousands of copies of a transposonlike element called Alu that makes up more than 5 percent of our DNA. This parasite is relatively benign, although every once in a while it causes a harmful mutation by inserting itself in the wrong place."

For an extended discussion of transposons, see Allan Campbell, "Transposons and Their Evolutionary Significance" in "Evolution of Genes and Proteins", Ed. by Masatoshi Nei and Richard K. Koehn, Sunderland: Sinauer Associates Inc., 1983, p. 258f

See Ayala Ochert, "Transposons", *Discover*, Vol. 20, No. 12, December 1999, p. 59 on the role of transposons in the development of chromosomes in multicellular organisms and again in the appearance of vertebrates.

233. "*Science News*", Vol 148, #21, November 18, 1995, p.330

234. Mark Ridley, "The Problems of Evolution". Oxford: Oxford University Press, 1985, p. 9.

235. "The Secret Melody", p. 34

236. "The Origin of Humankind" p. 46

See B. Bower, "Ancient human ancestor emerges in China", *Science News*, Vol. 148, No. 21, November 18, 1995, p. 327 for an account of a discovery in China of a find resembling *H. habilis* and *H. ergaste*r but prior to *H. erectus*.

H. habilis itself comes from a long line of ancestors, including *Australopithecus* (*aethiopicus, africanus, afranensis* {Lucy being the most famous}, *anamensis*), *Ardipithecus ramidus ramidus*; and *Ardipithecus ramidus kadabba*, recently discovered (2000-1) and more than 5 million years old which walked erect. The relationships of these ancestors is still being worked out, but it is obvious to most scientists that modern humans are descended from a long line of ancestors, just as is every other animal alive today. It is also obvious that we are the result, not of linear evolution but are a branch of the family, just as our ancestors were.

See Stephen Jay Gould, "Unusual Unity", *Natural History*, Vol. 106, No. 3, April 1997, p. 20 who champions the "bushiness" of all evolved species.

237. There is an interesting article by Noel T. Boaz and Russell L. Ciochon, "The Scavenging of `Peking Man'", *Natural History*, Vol. 110, No. 2, p. 46. It is not about the history of *homo erectus* but about the discovery of the largest cache of fossils of this species in China.

238. "The Origin of Humankind", p. 83

239. For another scenario, see Robert Kunzic, "The Face of an Ancestral Child" in *Discover* Vol. 18, No. 12, December 1997, p. 88. The following gives the sense of the argument.

> "Certainly the scenario [presented]—a complement to Stringer's "Out of Africa theory"—is tidy enough. The ancestor of all humanity, in this scenario, was *Homo ergaster*, living in East Africa between 1.5 and 2 million years ago. Sometime during that period *ergaster* migrated to Asia, becoming *erectus* on the way. Dates that are still controversial put *erectus*in Java as early as 1.8 million years ago; even more controversial ones have it in China 2 million years ago. The dates are a mess, but the concept is simple: *Homo erectus* was essentially an Asian species. Contrary to long-held belief, we are not descended from it. When those who would become *erectus* left Africa, the rest of the *ergaster* population stayed behind. After another 500,000 years or so of evolution, they gave rise to *Homo antecessor*. Sometime around a million years ago, perhaps a bit earlier, *antecessor* also decamped, following in the footsteps of *erectus*. But in the Near East it turned left into Europe, finally reaching Spain after many millennia...The European population of *Homo antecessor* gave rise to *heidelbergensis*, which spread all over Europe and even to Boxgrove, England...Perhaps 100,000 years later, *heidelbergensis* had evolved, in complete isolation from its African forebears, into Neanderthals...Meanwhile, in Africa, the populations of *antecessor* that had never left were embarked on a very different evolutionary path—one that eventually led to modern *Homo sapiens*. After that the story is straight Out of Africa.

Modern humans left that continent a little before 100,000 years ago and colonized Europe and Asia, gradually replacing Neanderthals and *Homo erectus.*"

The theory is undergoing new discussion with the discovery of a child who seems to be the result of a fusion of modern humans and Neanderthals. This would cast some suspicion on the orderly progression and replacement of one group by the next group from Africa and give some weight to the theory of Multiregional evolution.

See Kate Wong, "Is Out of Africa Going Out the Door?", *Scientific American*, Vol. 281, No. 2, August 1999.

See Michael D. Lemonick and Andrea Dorfman, "Up From the Apes", *Time*, August 23, 1999, p. 50 for a popular treatment of human emergence

See Robert Kunzig, "Learning to Love Neanderthals", *Discover*, Vol. 20, No. 8, August 1999, p. 68

See Ian Tattersall, "Once We Were Not Alone", *Scientific American*, Vol. 282, No. 1, January 2000, p. 56 for a discussion of different *sapiens* ancestors and contemporaries living at the same time.

See Kate Wong," Who Were the Neantertals?", *Scientific American*, Vol. 282, No. 4, April 2000, p. 98

See Joao Zilhao, "Fate of the Neandertals", *Archaeology*, Vol. 53, No. 4, July/August 2000, p. 24f

See "From Our Faithful Correspondent...", *Discover*, Vol. 23, No.1, January 2002, p. 66 for a summary of the problems with "out of Africa."

See John J. Shea, "The Middle Paleolithic: Early Modern Humans and Neandertals in the Levant", *Near Eastern Archaeology,* Vol. 64, No. 1-2, 2001, p. 38.

240. See Dinosaur In a Haystack", p. 101ff

See also "Return to Sodom and Gomorrah", p. 70ff

See Ian Tattersall, "A Hundred Years of Missing Links", *Natural History*, Vol. 109, No. 10, December 2000, p. 63 for a discussion of the many kinds of hominids which existed at the same time, even though only one survives.

See Carl Zimmer, "After You, Eve", *Natural History*, Vol. 110, No. 2, p. 32

241. Sally Carrighar, "Wild Heritage" New York: Ballantine Books, 1965, p. 28

See the review of a new edition of Darwin's "The Expression of the Emotions in Man and Animals", New York: Oxford University Press, 1998 by Mark Ridley in *Scientific American*, Vol. 278, No. 5, May 1998, p. 104.

242. James Trefil, "Are We Unique?", New York: John Wiley & Sons, Inc., 1997, p. 3.

243. For a good popular explanation of the development of the brain through the "smell brain" and the "vision brain", from reptiles and mammal-like reptiles through computers, see Robert Jastrow, "The Enchanted Loom: Mind in the Universe", New York: Simon and Schuster, 1981.

See Richard Leakey, "The Origin of Humankind", p. 139 treats of the origin of the mind. He finds a continuity of brains, of course.

244. This is using the term loosely. See Karen Wright, "The Tarzan Syndrome" in *Discover*, Vol. 17, No. 11, November 1996, p. 89ff where the author shows that,

> "Only apes, it seems, alone among all the animals, can truly distinguish themselves from the world around them. But only the naked apes, apparently, can conceive of not just `self' but `other'."
> "Chimpanzees may inhabit a cognitive realm that includes a subjective notion of "me" but not "you"." (p. 94)

In human infants, on the other hand, "these conceptions of self and other as conscious, mental agents seem to develop in tandem." (p. 92)

There is no question, however, that there is a continuum in brain capability, not a need for another mechanism to explain the differences.

See Meredith F. Small, "These Animals Think, Therefore...", *Natural History*, Vol. 105, No. 8, August 1996, p. 26

245. Richard Dawkins, "The Evolved Imagination" in *Natural History*, Vol. 104, #9, Sep- tember 1995, p.24

246. Ravens can seemingly visualize the process of pulling food on a string to within reach even before they try it. Crows cannot. There is a difference in the emergent quality of their brains suggestive of our own imagining.

See Bernd Heinrich, "A Birdbrain Nevermore", *Natural History*, Vol. 102, No. 10, October 1993, p. 50

247. "The Cosmic Blueprint", p. 186

248. "The Cosmic Blueprint", p. 187

249. Robert K.J. Killheffer, "The Consciousness Wars" in *Omni,* Vol. 16, #1, October 1993, p. 56

250. "The Quickening Universe" p. 217

251. Francis Crick, "The Astonishing Hypothesis", New York: Charles Scribner's Sons, 1994, p. 11 offers this definition,

> "of the separate parts, its behavior can, at least in principle, be *understood* from the nature and behavior of its parts *plus* the knowledge of how all these parts interact."

See also Marc Hauser, "Wild Minds: What Animals Really Think", New York: Henry Holt, 2000

252. Michael D. Lemonick, "Glimpses of the Mind" in *Time*, Vol 146, #3, July 17, 1995, p. 47

253. Bil Gilbert, "New Ideas In the Air at the National Zoo" in *Smithsonian* Vol 27, #3, June, 1996, pp.34f

254. See a review of Eugene Linden, "The Parrot's Lament" in *Time*, Vol. 154, No. 10, September 6, 1999, p. 56 and of course see the book.

255. See Marc D. Hauser, "Games Primates Play", *Discover*, Vol. 19, No. 9, September 1998, p. 48. Human beings are capable of lying to one another and knowing precisely what they are doing. At this time it seems that, "nonhuman primates do not make the connection between what others see and what they know. They are con artists, knowing how to deceive without knowing that they are deceiving."

However, some tests indicate that some primates may be on their way to having this kind of conscious knowledge. What is interesting is that very young human children seem to exhibit the same patterns as these primates until about the age of four.

> "During the course of evolution, some organisms acquired an understanding that they are deceiving. This event represented a renaissance in thinking, an awakening of mind. It allowed not only for true Machiavellian deception but also for self-reflection, an

> understanding of mortality, and an appreciation for how and why belief systems diverge and converge. We humans are unquestionably part of this renaissance. But we have yet to determine when or how it started—or why."

Once again, we seem to be seeing a continuum, not a break.

256. John R. Searle has developed all the philosophical questions I am attempting to treat in this section in his masterful work, "Mind, Language, and Society", New York: Basic Books, 1998 and I believe reconciles the personalness of our consciousness with a biological basis.

257. Stephen Jay Gould, "On Embryos and Ancestors", in *Natural History* July/August 1998, Vol 107, No. 6, p. 22. Also contained in "The Lying Stones of Marrakech".

258. Francis Crick, "What Mad Pursuit", New York: Basic Books, Inc. Publishers, 1988, p. 138

259. "Soul Search" p. xiv

260. For a puzzling commentary on the confusion of the personalities of an aging father and his son, see Robert Sapolsky, "Ego Boundaries, or "The Fit of My Father's Shirt" in *Discover*, Vol. 16 #11, November 1995, p.62f.

261. Antonio R. Damasio, "Descartes' Error", New York: G.P. Putnam's Sons, 1994. Chapter 1 tells the story of Gage and the book contains numerous examples of brain damage leading to a change in personality. Gage's story is also told in "Are We Unique", p. 79f

A similar occurrence is treated by Frank Vertosick, Jr., "A Bullet to the Mind", *Discover*, Vol. 17, No. 10, October 1996, p. 38

262. Oliver Sacks, "The Man Who Mistook His Wife for a Hat", New York: Summit Books

263. "Abouttime" p. 269 speaking about Daniel Dennett

264. "The Conscious Universe", p. 141

265. "The Conscious Universe" p. 139

266. From "Alice in Quantumland",

"What the Heisenberg uncertainty relations are telling us is that we are looking at things in the wrong way. We have a preconception that we *ought* to be able to measure the position and momentum of a particle at the same time, but we find that we cannot. It is not in the nature of particles for us to be able to make such a measurement on them, and the theory tells us that we are asking the wrong questions, questions for which there is no viable answer. Neils Bohr used the word *complementarity* to express the fact that there may be concepts which cannot be precisely defined at the same time: such pairs of concepts as justice and legality, emotion and rationality. There is, apparently, something fundamentally wrong with our belief that we *should* be able to talk about the position and momentum of a particle, or of its exact energy at a given time. It is not clear why it should be meaningful to talk simultaneously of two such different qualities, but it appears that it is not."

267. "Soul Search", p. 114

268. Antonio Damasio, "The Feeling of What Happens", New York: Harcourt Brace, 1999.

In an article, "Remembrance of Things False", Bruce Bower treats how memories are distorted by the brain so that we can "remember" things that never occurred. [*Science News*, Vol. 150, No. 8, p. 126, August 24, 1996]. If the brain were merely reporting to a non-physical entity we would not expect such mistakes.

269. Gerald M. Edelman and Giulio Tononi, "A Universe of Consciousness", New York: Basic Books, 2000, p. 200.
To be honest, I was a little disappointed since the dust jacket description of the book claims that the authors, "present an empirically supported full-scale theory of consciousness." They do not manage quite that.

270. "A Universe of Consciousness", p. 219

271. Candace B. Pert, Ph.D., "Molecules of Emotion", New York: Scribner, 1997, p. 141

272. "Molecules of Emotion" p. 187

273. Sarah Blaffer Hrdy and C. Sue Carter "Hormonal Cocktails for Two" in *Natural History*, Vol 104 #12, p. 34

274. Jeffrey Kluger, "Following Our Noses", *Time*, March 23, 1998, p. 72

See "The Best Ways to Sell Sex", *Discover*, Vol. 17, No. 12, December 1996, p. 78f. One of Jared Diamond's examples of how we have continued our ancestor's ways of advertising to the opposite sex is that,

> "...the pubic and armpit hair grown by both men and women in adolescence is a reliable...signal that we have attained reproductive maturity. Pubic and underarm hair are well placed to trap any pheromones we might secrete into our sweat and urine, which may signal our sex and (if we're women) the phase of our ovulatory cycle to those with discriminating noses."

His article has many other examples of parallels between ourselves and other animals.

Steven Pinker, "How the Mind Works", New York: W.W. Norton & Company, 1997 has more than 550 pages of examples of why we react to things as we do, how to explain free will (we are determined scientifically, but free morally), why we fall in love and stay together even when a more desirable partner becomes available, why parents take care of their children and much more.

275. "The Moral Animal", pp. 53-54

There are other evolutionarily designed stratagems in the war between the sexes which are not under our control. There is a male "imprinted gene" which tends to make the placenta grow larger than might be good for the female but which benefits his offspring. The woman's body meanwhile has developed defenses against this. See Robert Sapolsky, "The War Between Men and Woman", *Discover* Vol. 20, No. 5, May 1999, p. 56.

See also Matt Ridley, "Genome", New York: HarperCollins, 1999, chapter 15 for this and other wars of the sexes.

See also Jared Diamond, "Why Is Sex Fun?" New York: Basic Books, 1997 for a number of other examples.

276. Marilyn vos Savant, "Ask Marilyn" *Parade Magazine*, February 8, 1998

277. Jim Robbins, "Wired for Sadness", *Discover*, Vol. 21, No. 4, April 2000, p. 77

278. "Three Scientists and Their Gods" p. 277

279. "The Demon-Haunted World" p. 315

280. *Natural History* June 95 p. 77

See "The Origins of Life", chapter 13 for an extended discussion of the evolution of language in humans.

See Martin A. Nowak, "Homo Grammaticus", *Natural History*, Vol 109, No. 10, December 2000, p. 36 for a discussion of the animal precedents of human language.

281. "Three Scientists and Their Gods" p. 289

282. See Jeffrey Moussaieff Masson and Susan McCarthy, "When Elephants Weep", New York: Delacorte Press, 1995, for a thorough discussion of emotions in animals. Much of the evidence is anecdotal, but so is the evidence for human emotion.

283. "The Demon-Haunted World" p. 268
I realize that I am passing over here the question of whether animals have emotions like humans. Behaviorists such as B.F. Skinner and others question whether this is true, and many question whether our own emotions are really part of a conscious decision-making process or if we are really just as programmed as the "survival machines" that animals are taken to be. The Skinnerian approach is falling out of favor today, replaced by a willingness to admit emotions in animals and conversely to admit the influence of emotions and chemical responses in human consciousness and decision making.

See Frans de Waal, "Good Natured", Cambridge: Harvard University Press, 1996, and his article in *Discover*, July 1997, p. 50.

On the matter of whether animals are aware of death and its meaning, Gary Kowalski, "The Souls of Animals", Wallpole: Stillpoint, 1991, p. 9f tells of Koko the gorilla who seemingly knew not only that his pet cat had died but that gorillas also die. He also finds an awareness of death in elephants. Kowalski's parallels of human behavior and emotions with those of animals are sometimes a bit of a stretch, but the spirit in which he writes is gaining more acceptance and not being dismissed as unwarranted anthropomorphism quite as readily.

284. See Frans de Waal, "The Ape and the Sushi Master: Cultural Reflections of a Primatologist", New York: Basic Books, 2001

285. Tom Clancy has a novel, "The Bear and the Dragon", New York: G.P. Putnam's Sons, 2000, showing how a difference in the perception of reality between the United States and the Peoples' Republic of China could have disastrous consequences. He makes it very believable.

286. "The Quickening Universe" p. 129

287. "Darwin's Dangerous Idea", p. 516

288. Kenneth E. Boulding, "Beyond Economics", p. 186

289. Andrew M. Greeley, "What a Modern Catholic Believes About God", Chicago: The Thomas More Press, 1971, p. 19

290. Susan Blackmore, "The Meme Machine", Oxford: University Press, 1999. Chapter 15 is on Religions as memeplexes, but the whole treatment is helpful, as is the foreword by Richard Dawkins.

See also Malcolm Gladwell, "The Tipping Point", New York: Little, Brown, 2000 for many examples of how ideas and trends are passed on and have effect.

291. "Darwin's Dangerous Idea", p. 365

292. For a discussion of whether chimpanzees have "culture" in any sense comparable to human, see Meredith F. Small, "Aping Culture", *Discover*, Vol. 21, No. 5, May 2000, p. 53

Recent research (2003) has shown that Oran Utans have culture since on one side of a river they do things one way and on the other side they do not. Adults teach the young the new way.

293. "The Astonishing Hypothesis", p. 6
Elizabeth Kadetsky, "Truth Building", p. 30 disagrees with Crick and says that,

> "scientific advancements in understanding the brain do not threaten the idea of the divine so much as challenge modern believers to define the soul in new ways...'as a unique pattern of relationships embedded in the brain' [according to William Newsome]"

This quest does not seem so helpful to me, just reactionary.

294. "The Physics of Immortality", p. xi

295. "Darwin's Dangerous Idea", p. 144

296. "The Mind's Eye", p. 75

297. Edward O. Wilson, "Naturalist", Washington: Island Press/Shearwater Books, 1994, p. 332.

For a comprehensive treatment on the development of the field of study of the effect of genes on behavior, of which Seymour Benzer is a guiding light and pioneer as is Wilson, see also Jonathan Weiner, "Time, Love, Memory", New York: Alfred A. Knopf, 1999. Among many other topics, Weiner defends sociobiology and says that, "Biologists who hate sociobiology maintain that we can learn very little about human nature by studying the instincts of other animals because the powers conferred by interdigitating neurons like these lift us too far beyond the other animals for their instincts to illuminate ours." (p. 183). However, he shows parallels between ourselves and fruit flies (of all things) and concludes that, "although our brains are fancier, they are made of the same stuff."

Another excellent explanation of what sociobiology is and what it is not will be found in John Alcock, "The Triumph of Sociobiology", New York: Oxford University Press, 2001. What it is is, "the systematic study of the biological basis of all social behavior". [p. 5] He has many examples of good scientific explanations of the social behavior of animals, including but not mainly human animals. He also does a good job with answering the critics of this approach.

Sociobilogy and Wilson et. al. has had detractors. See Ursula Segerstrale, "Defenders of the Truth", Oxford: Oxford University Press, 2000 for an account of the battle, much of it having little to do with good scientific method.

298. "The Quickening Universe" p. 203

299. For a list of experiments illustrating the influence of the unconscious see for example Bruce Bower, "The Mental Butler Did It", *Science News*, Vol. 156, No. 18, October 30, 1999, p. 280

Agen

1. **Statements**

   A. Number of years you have known: Primary Proposed Insured _

   B. I have ordered/obtained the following requirements: ☐ APS ☐ MD Exam ☐ Oral Fluids *(as state permits)* ☒ Parame

   C. If requirements are scheduled, please provide name of examiner

   ____

   D. Did you personally see the proposed insured(s) on the date of th the answers yourself? *(If no, please provide details in the Remar*

   E. Does any proposed insured have any existing or pending annuit If yes, do you have any information that indicates that any propo of any existing or pending life insurance policy or annuity with ar *(If yes, please provide details in the Remarks section below and completion of replacement-related forms even when other life ins being applied for.)*

   F. Are you aware of any information that would adversely affect an *(If yes, please provide details in the Remarks section below, anc*

   G. Did you provide the client with a Limited Temporary Life Insuranc

   H. Have any of the proposed insureds or the owner submitted an a American International Group, Inc. within the last 30 days?

   I. If primary proposed insured is a child or age 18 and over and nc amount of insurance is in force on the father? $ __________

   J. Are you related by blood or marriage to any proposed insured? *(If yes, relationship)* ____________________

R

2. **Details and Explanations** *(Please include information on any split*

Bruce Bower also treats what we can learn about our ability to make choices and interpret the world around us from those dealing with split-brain individuals whose hemispheres do not communicate. In one telling example he reports,

> "One man had a picture of a chicken claw flashed to his left hemisphere and a picture of a snow scene presented to his right hemisphere. From the ensuing selection of pictures, he correctly chose a shovel with his left hand (controlled by the right hemisphere) and a chicken with his right hand (controlled by the left hemisphere). When asked to explain his choices, he responded, `Oh, that's simple. The chicken claw goes with the chicken, and you need a shovel to clean out the chicken shed.' [The researcher] concluded that the left brain observed the left hand's choice of a shovel—which stemmed from the right brain's nonverbal, inaccessible knowledge—and proffered an explanation based on its own fowl information."
>
> [Whole-Brain Interpreter", *Science News*, Vol. 49, No. 8, February 24, 1996, p. 124]

If all the information is being handed on to a "little man", why the confusion?

300. See "Descartes's Error", chapter 8 for a discussion of our decision making process.

301. Ask William R. Clark & Michael Grunstein, "Are We Hardwired?" Oxford: Oxford University Press, 2000,

> "Can we imagine that the connection between genetic and behavioral variability we see in animals is suddenly lost in evolution with the emergence of human beings? Are we, as a species, fundamentally different from other species in the mechanisms governing our behavior?" [p. 260]

His conclusion is, of course, that we are not removed from the same mechanisms, but as we will see below, we have additional mechanisms.

302. "The Physics of Immortality", p. 199f

Edward Speyer, "Six Roads From Newton", mentions something similar in is attempt to allow room for something of free will in a Newtonian deterministic universe. He has three possible ways to accomplish this. The first is not very satisfactory, merely pointing out that we are all a part of "the

Great Clockwork" and cannot absent ourselves from it, but as he says, this is true, "whether we have free will or not."
The second made no sense to me at all, I confess.
The third also proves nothing about free will that I can see, but it does bring in the indeterminacy inherent in the universe.

> "...recognition of Mr. In Between, of halfway houses lying between determinism and `pure' free will. Instead of seeing every event as either Caused or Uncaused, as Aristotelian logic would have it, we see events as slightly out of focus, a little fuzzy. Everything that happens is caused, but there are accidental happenings, degrees of potentiality and possibility. Recognition of unpredictability need not open the door to the supernatural, but merely to uncertainty. The limits on causality are mainly due, not to spooks, but to noise and to the limited accuracy of our scientific concepts and definitions, and to our mistaken presumption that the physical universe is as tight as our mathematical descriptions of it...Kurt Godel...proved that a mathematically tight set of axioms and postulates *must* contain an indeterminate case which cannot be handled. The standard example of Godel's theorem is: In a town with only one barber, the barber shaves all those, and only those, who do not shave themselves. So, does the barber shave himself? If he does, he doesn't. And if he doesn't, he does."
>
> [Edward Speyer, "Six Roads from Newton" p. 161]

I am not sure what Speyer thinks he has said about free will, but his indetermination fits in with the work of others on free will and the brain.

303. See "Mind, Matter and Quantum Mechanics" p. 91 for the rejection of "chance" and the rest of his book for a quantum mechanical explanation of consciousness and decision making.
304. "Are We Hardwired" points to a chaotic unpredictability, "embedded in crucial organs such as the brain" [p. 268] and shows the desirability of such an arrangement. Still, this is not free will as we think of it and may be very deterministic. However,

> "Chaos may indeed force us to experience things not scripted in either genes or experience, but we have an extraordinary power to learn. We can see and understand fully how our behavior affects our own lives, and the lives of those around us. Perhaps therein lies the definition of moral choice, and thus the nature of free will; the ability to choose among personal and social possibilities dictated by neither genes nor experience." [p.269]

The authors do not claim this is the final word but a direction in which to seek. However, the whole book leads to free will as an extension of abilities found in other species, not to something external to ourselves.

Matt Ridley, "Genome", chapter 22 treats the issue from a similar viewpoint.

305. Marian Stamp Dawkins "Through Our Eyes Only?" Oxford: W.H. Freeman & Co. Ltd., 1993, pp. 54f

306. As long ago as 1966, the theologian Raymond J. Nogar, O.P., recognized that there was a real problem in insisting that there is a soul in his book, "The Lord of the Absurd", Notre Dame: University of Notre Dame Press, 1966. He wanted to keep the soul, asking,

> "...how...do you express the fact that the spirit of man emerged simultaneously with the evolution of the human body, but not *out of* cosmic matter and energy? That God *specially* created the human spirit without upsetting the natural unfolding of the material creation?" (p. 47)

He really wanted to just say this is so, but was challenged and had to admit,

> "It is a fact which must be acknowledged, that most philosophers and theologians are brought up as apprentices to a religious point of view...They have an axe to grind...From my own experience I can number on one hand the philosophers and theologians I know who are really concerned about the factual evidence for evolution." (p. 52f)

His honesty and openness is admirable. What we are trying to do thirty years later is still take evolution seriously.

307. "Are We Unique" p. 7

308. "Are We Unique" p. 111f & p. 225

309. "The Enchanted Loom", p. 162f

310. "Portraits of Creation", article, "The Scientific Investigation of Cosmic History", p. 123-4

311. "Evolutionary Humanism and the Faith" Raymond J. Nogar, O.P. in "Is God Dead", "Concilium", Vol. 16, New York: Paulist Press, 1966, p. 55

312. "Portraits of Creation", p. 123

313. Otto Eissfeldt, "The Old Testament", Trans. by Peter R. Ackroyd, New York: Harper and Row, Publishers, 1965, p. 35 says,

> "The numerous descriptions of the theophany...are splendidly majestic and of sublime grandeur, and they show that Israel was quite capable of producing a colourful mythology. But real myths are not to be found in the Old Testament, at least none which originated in Israel, but only some borrowed from elsewhere."

This is explained because Israel had but one God, at least officially and it takes two contending to make a myth.

> "What the Old Testament offers in the way of myths or allusions to myths has quite clearly come into Israel from outside and has been in large measure at least deprived of its really mythical character."

See also "The Anchor Bible: Genesis" Trans. by E.A. Speiser, Garden City: Doubleday & Company, Inc. 1987, p. 10f on ancient myths in the creation story, p. 18f on Eden, p. 25f on the fall, p. 46 on the nephilim and p. 54f on the flood.

The discovery of texts from Ugarit has given a better understanding of the milieu from which the Israelite myths took their form. See Andre Caquot, "At the Origins of the Bible", *Near Eastern Archaeology*, Vol. 63, No. 4, December 2000, p. 225

314. Joseph Campbell, "The Power of Myth", New York: Doubleday, 1988, p 54

315. "Demon Haunted World", p. 316

316. Letter 138, 1

317. "Catechism of the Catholic Church", #301

318. "Reformed Confessions of the 16th Century" Ed. by Arthur C. Cochrane, Philadelphia: The Westminster Press 1966, p. 197. See also p. 131 of the same book for the Geneva Confession and p. 166 for the Scots Confession.

See also the Westminster Confession of Faith in "The Book Of Confessions", Philadelphia: Office of the General Assembly of the United

Presbyterian Church in the United States of America, #6.024 and numerous other confessions.

319. "At Home In the Universe", p. 6

320. "God and the New Biology", p. 95

321 [For an example of how this happens architecturally, see Ann Irvine Steinsapir, "The Sanctuary Dedicated to Holy, Heavenly Zeus Baetocaece", "*Near Eastern Archaeology*" Vol. 62, No. 3, September 1999, p. 182f]
But, having no anthropologists around to explain the cultural shifts that have occurred from worship of the feminine to the masculine principle, one theorizes a story of the male god Marduk conquering the goddess and using her body to form the earth. As Joseph Campbell explains it,

> "The Semitic people were invading the world of the Mother Goddess systems, and so the male-oriented mythologies became dominant, and the Mother Goddess becomes, well - sort of Grandmother Goddess, way, way back.It was in the time of the rise of the city of Babylon. And each of these early cities had its own protective god or goddess. The characteristic of an imperialistic people is to try to have its own local god dubbed big boy of the whole universe, you see. No other divinity counts. And the way to bring this about is by annihilating the god or goddess who was there before. Well, the one that was here before the Babylonian god Marduk was the All Mother Goddess. So the story begins with a great council of the male gods up in the sky, each god a star, and they have heard that the Grandma is coming, old Tiamat, the Abyss, the inexhaustible Source. She arrives in the form of a great fish or dragon-and what god will have the courage to go against Grandma and do her in? And the one who has the courage is, of course, the god of our present great city. He's the big one.So when Tiamat opens her mouth, the young god Marduk of Babylon sends winds into her throat and belly that blow her to pieces, and he then dismembers her and fashions the earth and heavens out of the parts of her body. This motif of dismembering a primordial being and turning its body into the universe appears in many mythologies in many forms. In India it comes up with the figure of Purusha, the reflection of whose body is the universe.Now, the mother goddess in old mother-goddess mythologies was herself already the universe, so the great creative deed of Marduk was a supererogatory act.

There was no need for him to cut her up and make the universe out of her, because she was already the universe. But the male-oriented myth takes over, and he becomes-apparently-the creator.]

[Joseph Campbell, "The Power of Myth", p. 170]

These people were not telling the story so much to explain the origin of the sky but to explain why one could safely ignore the goddess and worship the son-conqueror. It also explained that the titular head of the city, Marduk, was more powerful than Tiamat, the titular of the enemy. If it also fit the way the world appears, with a dome for a sky as the arched body of a goddess, well, that is what a good story does, but it was not the point of the story in the first place. But myths take on a life of their own. If the story explains the question of how the world came to be, then this becomes the dominant point of the story and the reason it is told and retold. Once it becomes an explanation of the origin of the world, then it can be manipulated to make points about that question rather than its original purpose. The reworking of the Sumerian creation myths by the Hebrews was not done to uphold an article of faith.

Claus Westermann pointed out some time ago that,

"For the man of the Old Testament it was not possible that the world could have originated in any other way. Creation was not an article of faith because there was simply no alternative. In other words, the Old Testament had a different understanding of reality from ours, inasmuch as there was no other reality than that established by God. They had no need expressly to *believe* that the world was created by God because that was a presupposition of their thinking."

[Claus Westermann, "Creation" Trans. by John J. Scullion, S.J., Philadelphia: Fortress Press, 1974, p. 5.

Interestingly, Bill Moyers's book, "Genesis: A Living Conversation", New York: Doubleday, 1996, brings together a number of commentators on the book of Genesis. Granted the forum, it is still amazing how much the participants just presume the same things the original authors presumed; god the creator, humans in the image of god, etc.

See the review of Moyers in "*Time*", October 28, 1996.]

Most likely the ancient pagan myths were used by the Hebrews and made to assert the superiority of the god Yahweh over the god Marduk and his ilk. The first chapter of Genesis, relying on much older myths, considers that the material out of which the universe was created was already present, but without form or order. Matter and void were all mixed together where they formed nothing useful. The work of creation is an ordering of the matter and a separation of it from the void, and the emphasis was on who had the power to make such a separation, rather than on how it was done. However, the question of why there was matter in the first place, or why there were gods, was not asked.

Lucretius, following on Epicurus, argues that nothing can come from nothing. He holds, "Let this then be for us our cornerstone—that ne'er, by power divine, doth aught rise out of nothing" for Lucretius has little place for intervention in the material world by the gods.

[Lucretius, "On The Order Of Things", Book I, Translated by Charles E. Bennett, New York: Walter J. Black, 1946, p.10.]

This actual lessening of the place of a deity would seem a late development in the history of religious thought and one that came as a reaction to the prevailing ideas. Lucretius was not widely accepted in his time or later and his "views attracted little notice; throughout the Middle Ages until the Renaissance, he was scorned as the devil's disciple"

["Classics of Western Thought: The Ancient World" Ed. by Stebelton H. Nulle, New York: Harcourt Brace Jovanovich, Inc., 1968, p. 228.]

The editor continues, "However, with the scientific revolution that began in the seventeenth century, his doctrine of atoms...was taken up as never before, and his great poem gained recognition..." The translation given here of that cited above is, "...the inner law of nature whose first rule shall take its start for us from this, that nothing is ever begotten of nothing by divine will."

The whole idea of relating to the gods was to influence their intervention in the world, not to contemplate their majesty in creating. But once the assumption was made that god was involved in a creative work, then the nature of this creative work could also be investigated. The responsible god becomes not just the orderer of chaos but the author of the original material as well. The religion based on these assumptions, and nurtured through the centuries by other assumptions based on the myths of creation, did indeed view god not just as the sustainer but the principle, and probably only, explanation of how things came to be in the first place. God has become a Creator. And not only the creator, but the creator "ex-nihilo" or out of nothing, a teaching that arose in the Church out of theological motives rather than philosophical. [See Jacques Fantino, "Whence the teaching creation *ex nihilo*?" "Theology Digest", Vol. 46, No. 2, Summer 1999, p. 133f.]

Out of nothingness the universe came to be at the command of god. This has been officially part of Catholic doctrine since 1215 when it was proclaimed in Council. This creating need not have been done all at once, nor in six days. William of Ockham and his followers felt that there was a continuous creation from nothing. In the Jewish tradition, Maimonides insisted on creation from nothing against the eternal existence of the universe. Of course the understanding of what this doctrine means about the necessity of a creator has undergone much development, especially in this century.

In earlier times everyone could accept the assumption that there was a creator responsible for the existence of the universe, since no other mechanism would be conceived, as Lucretius made clear. When the literal truth of the biblical creation story was called into question, Christians had to respond in some manner. Some responded by denying the findings of science, but the Catholic church tried to be more honest, at least after the first knee-jerk reactions against this new information, and investigate some of the assumptions that had led to the Church's prior position. If the Bible could no longer be read in the same literal way that had sufficed for centuries, and if it was the word of god, how did one go about understanding it? You used the same scholarly methods as with any other literature. The meaning of the Hebrew word "bara'" is a good example. It is the word used in the book of Genesis for the creative activity of god and is normally translated, "created". The use of the word in the first verse of Genesis has been presumed to mean that god made the universe from nothing, and much theological speculation has been based on that presumption. But that is probably not the case. There are many treatments of the use and meaning of the word and I cite only two here, the first of which says,

> "In biblical usage it [bara'] obviously expressed the idea `to bring something into being,' to bring into being some specific reality that had not existed before. However, it does not *as such* signify giving existence to something that has never before existed *in kind*...Nor is *bara'* used only of physical entities. In the language of the Old Testament, God `creates' historical events...; conditions...; praise on the lips of mourners...much as the signature on a painting announces that it was conceived, willed, and executed by the one whose signature it bears."
>
> [John H. Steck, "What Says the Scripture" in "Portraits of Creation" Grand Rapids: William B. Eerdmans Publishing Co. 1990, pp. 207-208]

After considerable discussion of the use of the word in the Bible, John Steck, just quoted, concludes that,

> "contrary to claims often made, while the verbs employed affirm in richly varied ways divine effectuation—the idea that God with sovereign power and will has made and ordered all things—they do

not specify *ex nihilo* or instantaneous creation or the absence of process or mediating agencies."

["What Says the Scriptures", p. 221]

Another commentator on the Scriptures points out that, "Scripture scholars hold that the idea of `creation from nothing' was not an important consideration for the biblical authors." Rather, the picture presented in Genesis is of a "formless wasteland".

"This primordial mess is generally described as `chaos.'...The Priestly author [of one biblical creation story]...did not mean that God first created this mess, then put order and beauty into it."

[Leonard Foley, O.F.M. "From Eden to Nazareth", St. Anthony Messenger Press, 1983, p. 3]

This finding did not mean that the Church abandoned the position that god created the world out of nothing, but it did mean that that creation need not be in the marvelous, almost instantaneous manner which we had thought, nor that the Church had to oppose the ideas from science. In the 1950's, even the Pope would embrace the Big Bang theory since it was seemingly in accord with the spontaneous creation of the universe from nothing pre-existing. The physicist-theologian John Polkinghorne feels that,

"it is...correct to see this doctrine of creation [*ex nihilo*] as implicit in the Biblical doctrine that God is the creator of heaven and earth, that he can do all things, that nothing is beyond his power."

[John Polkinghorne, "The Faith of a Physicist", Minneapolis: Fortress Press, 1996, p. 74]

Polkinghorne says this immediately after giving an account of the beginning of the universe in the Big Bang.

[In an article entitled "The Word of God in Creation: A Response to Dr. J. Cyril Hanisko", Dr. Rudolf Brun reaches many of the conclusions that I do about the impact of science on the content of the faith and our concept of god. But even now he feels bound to say that,

"*Creatio ex nihilo*" is central to the Christian understanding that God is free. It is also central for the understanding of the relationship between the Creator and creation." ["Institute for Theological Encounter with Science and Technology Bulletin", Ed. by RobertBrungs, S.J., St. Louis, Vol. 32, No. 3, Summer 2001, p. 7]

He then goes on to show that nature is creative as well. In my opinion this first is a gratuitous assumption.]

God was identified with the Greek notion of a "Prime Mover", a being who or which had existence as part of its definition and which was the cause of all that is contingent. The prime mover need not be a personal being, and need not be rational or conscious, but certainly in Christian tradition it is these things and more. Wrapped up in the figure of god is the prime mover, the creator, the sustainer and the orderer. Contained in this concept are a number of "obvious" axioms in philosophy. One we have seen is that "*ex nihilo, nihil fit*", or that "nothing comes from nothing." Therefore any motion or movement must ultimately come from this Prime Mover, **which itself disproves the axiom by moving with causation.** The philosopher Jacques Maritain brings in a corollary axiom in a lecture on the principle of causality which he claims states that, "every contingent being has a cause" which he says may also be expressed,

> "more philosophically in terms of potency and act. We shall affirm: That every being compounded of potency and act inasmuch as it is potential does not pass of itself to act, does not *reduce* itself to act. It passes to act by the operation of another being in act which causes the change. *Nihil reducit se de potentia in actum.*"
> [Jacques Maritain, "A Preface to Metaphysics" New York: Mentor Omega Books, 1962, p. 129
> Literally the phrase means that, "nothing reduces itself from potency into act." *Reducit* should really be translated as "moves itself from the state of potency into actuality."]

The phrase basically means that nothing changes itself from what it could be but is not to actually being that. Nothing which can become more intelligent, for example, actually becomes smarter all by itself. It must have the impetus of another. For a Christian philosopher like Maritain, that other is ultimately god, and he says that,

> "nothing therefore, absolutely nothing, could be produced here below, not the slightest stir of a leaf, not the tiniest ripple on the surface of water, not the lightest movement of the emotions, not the most insignificant act of thought or will, if the created universe were not *open* to the action, virtually transitive, of the pure *act*, which, as Aristotle said, touches it without being touched by it, if a continual current of causal efficacy were not being poured without ceasing into creatures from the bosom of subsistent Intellect and subsistent Love." [Jacques Maritain, "A Preface to Metaphysics", New York: New American Library, 1962, p. 129f]

322 In fact a teleology may be constructed in which one can learn what that purpose is by studying the steps to accomplish it, precisely because the gods take those steps in order to accomplish that end. The end may well be for the gods' own purposes and not in the best interest of the humans, but if we are lucky they will at least not hurt us too much in the process. One of Stephen King's characters feels that "God pisses down the back of your neck every day but only drowns you once." [Stephen King, "The Drawing of the Three", New York: New American Library, 1987, p. 116]

If we are exceedingly fortunate, the will of the gods is the benefit of their beloved humans, though this is not considered the case in most religions. The best that one can do is find out what the gods intend and find a way to profit from its fulfillment yourself. And so Zeus or Hermes have their own gain in mind according to their own long range plan, usually demanding the subjection of the other gods, and most of what they cause to happen on earth is to further that end. The trick is to make yourself their vassal so that you profit when they do, and catch the crumbs from their table. For Christians, however, we are exceedingly fortunate. God has a purpose in mind and that purpose is not selfish but is to reveal godself to us as a loving god who wants us to live in god's presence forever.

SOMEONE IS IN CONTROL OF THIS PLAN

Everything is directed to this end of bringing us to live with god. This was certainly the faith of the Scriptures and of the Church throughout the centuries, although the message was corrupted in innumerable ways on the journey, turning god into one who tests and probes and is all too willing to punish, though also to reward those who are righteous. But through it all the assumption remained that things happen and things exist in order to carry out god's purposes. This was the common understanding of Christians and remains so today. John Barrow and Frank Tipler explain that,

> "Nineteenth century biologists also saw teleology at work not only in the adaptation of living things, but also in the over-all relationship of living beings to each other...Pre-evolutionary biology regarded the living world as organized into a Great Chain of Being with single-celled organisms at the bottom of the chain, mankind somewhere in the middle, the angels above him, and God at the top. This picture of living creatures was static; the species were created to fit into this ordering at the beginning of time and were ordained to remain so ordered for all time. God's purpose never changed since he was unchanging. A species could never become extinct."
>
> ["The Anthropic Cosmological Principle" p. 128]

Even though the angels were above us as beings, we were above them in the love of god bestowed upon us. The Son never became an angel, he became human. The role of the angels is to be messengers to and guardians of humans, and their whole purpose of existence is to carry out god's loving designs for us.

> [Thomas Aquinas has quite a bit to say about angels. "Angels mean messengers and ministers. Their function is to execute the plan of divine providence, even in earthly things." Also, "God's universal providence works through secondary causes...The world of pure spirits stretches between the Divine Nature and the world of human beings; because Divine Wisdom has ordained that the higher should look after the lower, angels execute the divine plan for human salvation: they are our guardians, who free us when hindered to help to bring us home."]

Indeed, all of creation was made for us. The sun shines for us and the moon lights our way at night. And all this because god designed it that way to bring about god's purpose and plan. But suddenly Darwin's, "central theory of natural selection held that no direct causes worked at such comprehensive levels and that all general, and apparently purposive, patterns arose as a side consequence of struggle among individual organisms for reproductive success."

> ["*Natural History*", Vol 102, No. 9, Sept 1993, p. 12]

Applying this new view to other fields, it became clear that not only organisms but the sun and moon themselves were the result of natural processes that could well have gone some other way. The assumption that god has a purpose and plan for creation was now questioned and it took very little time before the assumption of god as a creator and finally the assumption of the very existence of god was also challenged. It took a while for Christians to respond to the challenge. Sir Charles Bell responded with a vengeance in his 1833 *Bridgewater Treatise* when, thinking that he had destroyed the evolutionists' arguments once and for all, he concluded that,

> "It is, above all, surprising with what perverse ingenuity men seek to obscure the conception of a Divine Author, an intelligent, designing, and benevolent Being—rather clinging to the greatest absurdities, or interposing the cold and inanimate influence of the mere `elements' in a manner to extinguish all feelings of dependence in our minds, and all emotions of gratitude."
>
> [Stephen Jay Gould. "The Gift of New Questions", "Natural History", Vol 102, No. 8, August 1993, p. 11. In another article mentioning the *Bridgewater Treatises* Gould points out that the Darwinians called them the "bilgewater treatises".]

The First Vatican Council in 1870 maintained this high tone when it declared anathema, or excommunicated, anyone who would claim that the ultimate purpose of creation was not to share in the goodness of god and thereby to give glory to god. [Robert Butterworth S.J., "Theology of Creation", p. 16f] Science was said to be able to explain the "how" of things, but not the "why", and on one level this is true. However, the apologists did not seem to realize that science had questioned that there need be a "why" and they continued to assume that there must be some purpose to it all. But to its credit, the Church slowly came to realize that it had to answer the objections and not merely proscribe their adherents. By our own day we have seen many responses to the questions raised by Evolutionism. Some continue to insist on a division of labors between science and theology and claim that science can make no proclamations about whether or not there is a purpose behind the existence of the universe. And in their better, or weaker, moments, scientists agree. Edward Fredkin admits that it is hard to believe that everything is an accident. Perhaps, he theorizes, the purpose is to compute answers to someone's questions, but who or what that someone might be he has no idea. ["Three Scientists and Their Gods", p.67f]

Kenneth Boulding, an economist, promoter of science, and author of religious poetry, says that evolution can account for everything that exists, but still there is something weird out there.

["Three Scientists and Their Gods" p.171f]

So if science cannot make apodictic proclamations about the purpose of the universe, maybe theology can.

Others take up the challenge by recognizing the findings of evolutionary theory and using them to the advantage of the faith. God, they say, is still the creator and has a purpose in mind for creation, but this more modern design argument, in the words of George Ellis,

> "...Would see God designing the Laws of Nature and the Boundary Conditions of the Universe, in such a way that life (and eventually humanity) would then come into existence through the operation of those laws...through the process of evolution"
>
> [George F.R. Ellis, "Before the Beginning: Cosmology Explained", New York: Boyars/Bowerdean, 1993, p. 106]

And not only the laws of nature, but the makeup of humans itself is under the control and follows the purposes of god, but through the instrumentality of the evolutionary process. Arthur Peacocke says that,

> "...one who believes in, is committed to, God as Creator can affirm that it must have been God's intention that human society should

have attained its present economic and technological complexity since he created man, through evolution, with just those abilities which made such complexities inevitable and he, at least, must have known that it would be so."

[Arthur Peacocke, "God and the New Biology", p. 107]

But of course to one who is not so committed there is still a legitimate question.

Another response is to claim that scientists do not recognize the basic contingency of the universe. The theologian John Haught wants to put science and religion, especially the traditional understanding of god as creator, on the same page. To do this, he brings back into prominence a time honored understanding of god as a lover pouring the self out in service to the beloved, but first noting that,

"...most of us who are intellectually most at home with scientific skepticism find nothing whatsoever in the new scientific emphasis on chaos and complexity that would lead us to religion. In fact, the utter spontaneity of matter's self-organization,...Would seem to render more superfluous than ever the idea of an ordering deity."

[John F. Haught, "Science and Religion", New York: Paulist Press, 1995, p. 149]

But having made this admission, one that theologians not long ago would never have done, he brings back again the lover image.

"The God of whom we are speaking here is one who is concerned that the universe through and through...possess a self-coherence and capacity for self-creation that makes it appear, at least at first glance, as though it does not need god at all...Theology...thinks of God as infinite, self-emptying love. Such a God is said to be `kenotic'...A `self emptying' God. In this theological perspective, it is only because of God's self-emptying love that a self-organizing universe can come into being" ["Science and Religion" p. 160.

See also John Polkinghorne, "The Faith of a Physicist"]

The universe in this view is every bit as contingent as the universe of the Scholastics, though it does not look like it. However, it is free and independent in whether or not it responds to its creator and the means it uses to develop its own potential. This same contingency is found in Peacocke. Even genetics and the seemingly free, even chaotic, history of our genetic development is dependent on the creator god's intentions.

"...The genetic constraints upon our nature and action are, from a theistic viewpoint, what God has determined shall provide the matrix within which freedom can operate."

["God and the New Biology" p. 111]

This acceptance of the findings of science and the use of these findings to re-interpret the traditional faith has been beneficial to scientists who have the "gift of faith", putting their work in the perspective of their faith. It has been even more fruitful for theology, helping to put aside many assumptions which are now found wanting, and to strengthen the understanding of many more aspects of the faith that were previously ill-understood.

Some others have responded from a different perspective. Angela Tilby sees the work of creation not as a chore but as a process, very much involved with quantum indeterminacy.

"In the quantum picture creation is a game in which chance plays a major part. It is as if God creates, not with any one end in view, but out of the sheer joy and delight of seeing what will happen."

[Angela Tilby, "Soul: God, Self and the New Cosmology" p. 175]

The universe is no less contingent and dependent on god, but the process is much more free. There is no hidden agenda or a subtle using of the laws of evolution to arrive at a foregone conclusion in the mind of god. As she phrases it, there is no blueprint being followed. The universe is free to follow the dictates of chance, and certainly every conscious and self-determining being within it is free to follow it's own lights. John Haught reaches somewhat the same conclusion when he argues that god does influence the universe, but in persuasive, not coercive, ways. ["Science and Religion" p. 47f]

The faith that is based on the assumption of a creator-god has thus had to re-evaluate its understanding and, except for some fundamentalists, has been honest in facing the challenge.

[I have been accused of being unfair to fundamentalists, and perhaps I am, coming from my own experience. See Karen Armstrong, "The Battle for God", New York: Alfred A. Knopf, 2000. Part Two is almost 200 pages on Fundamentalism and gives it fair treatment.

But as just one example of what I consider the willful blindness of many creationists and fundamentalists, I ran across an article by Stephen Caesar, "The Recent, Catastrophic Formation of the Grand Canyon", in "Pulpit Helps", Vol 26, No. 7 put out by AMG Publishers in Chattanoogo, TN., p. 6 in which he concludes that science is "moving toward a more biblically-oriented scenario." However, he has just said that the younger dates being proposed by some, "don't mesh perfectly with biblical chronology since the geologists place the catastrophe at about 5 million years ago. Further, they do not hypothesize a world-wide flood, but an extremely large local flood." Personally, I don't know how you talk to someone who can justify Noah's flood from that kind of evidence.

For a brief explanation of the formation of the Grand Canyon, see W.K. Hamblin & Laura hamblin, "Fire and Water", "*Natural History*", Vol. 106, No. 8, September 1997, p. 35. Their explanation includes not only riverine erosion but dams of lava forming large lakes which when subsequently breached caused huge rushes of water and debris.]

The faith has been enriched by putting aside the more crass forms of creationism and introducing into god's governance much more freedom and autonomy for creation.

[Ditto for creationists, but see Robert T. Pennock, "Tower of Babel: The Evidence against the New Creationism", Cambridge: The MIT Press, 1999 who gives a detailed rebuttal of the latest forms of creationism.]

[Niles Eldredge, "The Triumph of Evolution dna eht eruliaF fo msiniotaerC", New York: W.H. Freeman and Company, 2000 is a lighter but still comprehensive treatment of the creationist (fundamentalist) approach.]

Still, the assumption remains that god is the creator. In the Faith's spirituality, its eschatology, its moral teaching, apologia, catechesis and evangelization and indeed in almost every aspect of the faith and the way it is lived, the role of god as creator is introduced. The motive for responding to god in worship is frequently gratitude for god's creative activity. The prayer for the 2nd Sunday in ordinary time put's it succinctly, "Almighty and ever-present Father, your watchful care reaches from end to end and orders all things in such power that even the tensions and the tragedies of sin cannot frustrate your loving plans." God is seen as the giver of our life, the maker of beautiful scenery, the author of moral laws through the constraints of our nature. God is the ground for hope since built into the history of the universe and its inhabitants is the plan of god to bring about a new creation. All that we see is only temporary and when it has served its purpose it will be replaced by a less "chancey" universe. Morality is based on the natural law which is the will of god expressed through the reality of creation. Does it work to have everyone stealing from their neighbors? No? That is because god frowns on theft and built the prohibition into the laws of society from the very moment of creation. In Christian apologetics the role of god as creator is taken as a self-evident truth which no one can deny and therefore a good starting point for justifying faith. How can anyone look at the proverbial sunset, the Christian asks, and not know that there is a god? The assumption that god is a creator god is not just an adjunct to the faith but its bedrock.

323. Related to this understanding of god as in charge of creation, guiding it to a predetermined end, no matter how loosely that control is exercised, is the conviction that god knows the end result and more, god knows the future. In the earliest understanding of god there was no problem. God was in total control of everything and had already worked out what would happen in minute detail. When Tevye in "Fiddler On The Roof" asks if it would spoil "some vast eternal plan if I were a wealthy man," the presumption is that it would. He has his role to play and it is not his place to question god.

As the universe was understood as a clockwork mechanism, inexorably following the physical laws, it was easy to say that god, and god alone, knows the future since it is so inflexible. We humans can know it a little, in the crassest form, if we can follow out the precise actions and reactions of an event. If I set up a row of dominoes set at precise angles to one another, I can predict how they will fall when set in motion. But there is a large margin for error since my knowledge is so incomplete. With god's complete knowledge of everything, surely god can follow out the chess moves of the universe in an infinite progression and know what will happen at any moment. God need not be responsible for setting the initial conditions which will inevitably lead to a result, but if not god, who?

The discovery of the role of chance and the role of chaos in physics has made this ability of god questionable, but even with these factors thrown in it is perfectly possible to reach the conclusion that "only God has the wisdom to understand the final law, the throws of the quantum dice. Only God can foretell the future." ["At Home In The Universe", p. 29]
Some physicists may use the word "god" facetiously in this context, but the believer might take it literally and feel justified in doing so.

Ordinarily the plan of god is carried out in the "normal" course of events, though on another level there is no normal course. Everything is being done by some agent with their own agenda, even the rising of the sun, which in some religious belief systems must be accomplished each day. There is nothing guaranteed about it, just as the return to spring must be accomplished by the actions of the gods, mimicked and aided by the religious rituals of the humans. If these are omitted or if the gods do not act, there is no regularity. If the maiden is not sacrificed to the gods in the Andes of Chile the problems that the people have been experiencing may continue. Aztecs in Mexico reacted to an eclipse with, "a tumult and disorder, all were disquieted, unnerved, frightened...People with light complexion were slain. All offered their blood."

> [E. C. Krupp, "Echoes of the Ancient Skies" cited by Steve Mirsky, "When the Sun Disappears and Dolphins Do Back Flips", *Technology Review*, Vol. 100, No. 3, April 1997, p. 34]

In Heather Pringle, "Temples of Doom" *Discover,* March 1999, p. 78, "For the Aztecs, human sacrifices were ironically the stuff of life. Only human blood, they later told Spanish priests, could give the sun strength for its daily climb from the underworld."

The eclipse was an act of the gods, and an especially frightening one, but then bringing the sun up each morning and keeping it out throughout the day was also an act of the gods, so on one level everything is a "miracle". Yet the gods are mostly regular and so one can speak of a miracle as something outside of that regularity, and contrary to it. Ordinarily the river floods in the spring, or ordinarily it does not. If it does what it does not usually do, that is a work of god or the gods and a miracle.

Emily Harwell tells of a Dayak tribesman of Borneo explaining that,

> "When my grandfather was a boy, one drought lasted seven years, a period we call *Kore Ogok.* Ogok was a woman who married her own son, which is forbidden. Her offense caused the rains to stop and brought unbearable heat for seven years. Only when the two of them were killed, as is dictated by our law, was there rain."

Harwell also explains that, "the drought, fires, and economic crises...have left many villagers wondering what human misdeeds have brought on this latest adversity. For the Dayaks, there is no such thing as a `natural disaster'."

> [Emily Harwell in "Unnatural Disaster", *Natural History*, Vol. 108, No. 6, July-August 1999, p. 36]

There is no mention of gods in her report and it is possible that some peoples believe more in impersonal forces that get out of balance because of human behavior rather than a personal deity who takes offense. However, it is a short step from an unnatural disaster to a miracle.

Normally a band of thirty men cannot rout an army of ten thousand. [Judges 6, 11f] If they do, it must be a miracle. As a general rule a cancer runs its course and kills the victim. If there is a spontaneous remission, especially if accompanied by prayers and incantations, it must be a miracle. But then, we will get to that momentarily.

324. This is yet another tenet of the deist faith (and the Christian faith) which is based on the assumptions listed in the previous chapter. Is it a doctrine of the faith that miracles can occur? Thirty years ago and more it was recognized that there is a problem justifying faith in god as a miracle worker. In 1965 it was certainly noted in an avant-garde series of theological reflections on Vatican II that,

> "we are aware also of the reserve that greets the account of the miracles. The development of science and skill, the understanding of different religions, historical criticism—these have largely diffused skepticism where they are concerned. It is a fact that, for many Christians, the miraculous, far from constituting a motive for faith, actually creates a difficulty."
>
> [Henri Bouillard, S.J. "Human Experience as the Starting Point of Fundamental Theology" in "Concilium: The Church And the World", Volume 6. New York: Paulist Press. 1965. p. 86]

Believing in Jesus as a teacher and prophet was much easier than believing in a wonder worker, especially when so many other wonder workers had been debunked. The famous "Death of God" movement of the 60's made it clear that,

> "Our world has come of age in the sense that now we do have fully adequate patterns of explanation for all the events of our natural and social lives without appealing to any forces or entities outside the circle of empirical events. We now think in terms of a self-sustaining world...No appeal to any external power is necessary or helpful in elucidating the pattern of events."
>
> [Emerson W. Shindeler, "Taking the Death of God Seriously" in "The Meaning of the Death of God" ed. by Bernard Murchland. New York: Vintage Books. 1967. p. 111]

This was nothing new, of course. David Hume had questioned the possibility of miracles long before our own century. But in our time it has become very clear that many things we have taken as miracles are not to be so explained. Is it still possible to put any credence in the likelihood that god will pull one out for us? A 1995 Gallup poll said that 79% of Americans believe that miracles are possible.

Karl Rahner would like to think so, too, but in terms that may make a fundamentalist queasy. He would rather talk of the miracle of god's love or the miracle of the resurrection of Jesus and the promise it holds out to our faith. With this faith,

> "There is no need to talk of the laws of nature being suspended if one admits the fact of miraculous cures, because at a propitious moment the levels of physical life which have been disordered may so fit back together that organic laws yield to spiritual ones."
> [Karl Rahner, "Do You Believe in God" Trans. by Richard Strachan, New York: Newman Press, pp. 96-7]

The action of god's love brings about a physical cure by bringing the person into line with physical wellness, but the wellness is attained by purely physical means. If there is any meaning to "faith healing", this is probably it.

> [Of course it is often hard to separate healing through faith from the placebo effect of medicines. See Walter A. Brown, "The Placebo Effect", *Scientific American*, Vol. 278, N. 1, January 1998, p. 90 for a description of the effect of mere treatment on illness, never mind the sugar pills.]

Arthur Peacocke also finds a place for the traditional assumption of the possibility of miracles, but tempered by science, especially the discovery of the role of chaos. He would like to see the two principles of the chaos present in the world and the control exercised by god combined and says,

> "In human life we must accept, for the stability of our own mental health and of our faith, that reality has a dimension of chance interwoven with a dimension of causality-and that through such interweaving we came to be here and new forms of existence can arise. This acceptance of chance as part of the mode of God's creativity is more consistent with the fundamental creativity of reality than the belief- stemming from a Newtonian, mechanistic, determinist view of the universe with a wholly transcendent God as the great Lawgiver - that God intervenes in the natural nexus for the good or ill of individuals and societies. We must learn to accept these conditions of creation and of creativity in the world..."

[Arthur Peacocke, "God and the New Biology" p. 99-100]

Richard Swinburne, another theologian-philosopher, concludes that

> "I am, however, inclined to think that we do have enough historical evidence of events occurring contrary to natural laws of a kind which God would have reason to bring out to show that probably some of them (we do not know which) are genuine miracles...We are rational to believe, while allowing the possibility that evidence might turn up later to show that we are mistaken."
> [Richard Swinburne, "Is There A God?", New York: Oxford University Press, 1996, pp. 120-121]

James Brady, an author of popular novels, is on firm ground for believers when a character in "Holy Wars", the priest-hero is skeptical of the ability of a Sicilian boy to effect a cure and on being questioned says that, "I believe there *have* been miracles. That there are things we cannot explain. That God moves in strange ways..." [James Brady, "Holy Wars", New York: Simon and Schuster, 1983, p. 286]

But of course when the cure is effected, it is the skeptic who looks foolish. Miracles are expected.

325. "Soul", p. 125

326. "Soul", p.125

327. "The Quickening Universe", pp. 139-140

328. "The Blind Watchmaker", p. 139

329. Psalm 114:4

330. "The Blind Watchmaker", p.161

331. "The Demon Haunted World", p. 230

332. The "Catechism of the Catholic Church" tries to have it both ways. #374 says that,

> "the first man was not only created good, but was also established in friendship with his Creator and in harmony with himself and with the creation around him."

#376 claims that,

"As long as he remained in the divine intimacy, man would not have to suffer or die...and finally the harmony between the first couple and all creation, comprised the state called `original justice'."

Seemingly, then, the Catechism is on the side of this original condition being a reality. However, #375 says that it is, "...interpreting the symbolism of biblical language in an authentic way." But there does not seem to be any symbolism present. It takes this as historical. I think it had better listen to its own scripture scholars and theologicans who find the Genesis picture symbolic.

333. "Philosophers & Religious Truth", p. 160

334. "Philosophers & Religious Truth", p. 160-161

335. "God and the New Biology", p. 93

"The only state of primeval `innocence' that...a scientific account of the emergence of man might allow is of a kind that can be attributed to all non-human, non-self-conscious mammalian organisms, namely, one innocent of any sense of responsible choice and of the relationship of power to moral choice."

336. "God & New Biology", p. 93

In his preface to the book, "Is Original Sin in Scripture?", by Herbert Haag, Trans. by Dorothy, New York: Sheed and Ward, 1969, Bruce Vawter, C.M. a noted scripture scholar at the time, pointed out that,

"...only with considerable foot dragging has [the Church] now begun to come to terms with the Darwinian revolution that lies in the background of this book."

Judging by the "Catechism of the Catholic Church", #'s 388ff, the foot dragging continues.

337. Charles Colson, "Loving God", Grand Rapids: Zondervan Publishing House, 1983, p. 103

338. "Catechism of the Catholic Church" #386

339. "Catechism of the Catholic Church" #390.

#385 speaks of the Fall and mentions St. Augustine's role, citing the "Confessions" where Augustine says, "I sought whence evil comes and there was no solution." E.L. Doctorow in his own "City of God", blames

Augustine for the concept of Original Sin as it is often understood, calling it, "a nifty little act of deconstruction – passing it on to the children, like HIV."

The preacher in John Steinbeck's, "The Grapes of Wrath", New York: Bantam Books, 1939, p. 88 accepts the story of Adam's sin and sees it as something to which all humans are liable. He explains that,

> "I got to thinkin' how we was holy when we was one thing, an' mankin' was holy when it was one thing. An' it on'y got unholy when one mis'able fella got the bit in his teeth an' run off his own way, kickin' and draggin' and fightin'. Fella like that bust the holiness. But when they're all workin' together, not one fella for another fella, but one fella kinda harnessed to the whole shebang – that's right, that's holy."

340. "The Moral Animal" p. 13.

See also, Robert Wright, "Science and Original Sin", *Time*. Vol. 148, No. 20, October 28, 1996, p. 76. Wright points out the evolutionary origin of our darker side. He also mentions our innate tendency to justify our bad actions and make them good, a tendency which is also close to the scriptural and theological view of sin.

341. Lyall Watson, "Dark Nature" New York: HarperCollins Publishers, 1995 p. 87

342. The Moral Animal p. 12

See also Edward O. Wilson's talk in "Religion, Science, and the Search for Wisdom", Washington: United States Catholic Conference, 1987, p. 82f

Also Marc D. Hauser, "Morals, Apes, and Us", *Discover*, Vol. 21, No. 2, February 2000, p. 51

Christopher D. Ringwald, "Encoding Altruism", *Science & Spirit*, Vol. 12, Issue 5, September October 2001, p. 18f is speaking about the possibility that civilizations on other planets would have altruism, but in speaking of it he seems hard-pressed to find a basis for altruism in anything but evolution.

343. Richard Conniff, "The Enemy Within" in *Smithsonian,* Vol. 29, No. 7, October 1998, p. 90. The author points out that,

> "the same reproductive quirk does not occur in termites; genetically, their offspring are as good an investment as their siblings. And yet most termites also give up the chance to reproduce, and termite

> soldiers routinely defend the colony with their lives. In one spectacular display of self-sacrifice, the soldiers of a termite species swell up their abdomens till they explode, spattering their guts all over any adversary that threatens the colony."

This may have been even better reproductive strategy in the past, as it is for ants now. Or the termites may just be stuck in an evolutionary deal which does not lead anywhere for the individual, but is at least as good as the alternative. Since termites have no choice in the matter on the individual level, the practice persists.

344. Susan Milius, "Who's Dying for Sex?", *Science News*, Vol. 156, No. 20, November 13, 1999, p. 312. It is not that the male nourishes his eggs by feeding the mother. He is often not big enough to make a meal. But he can transfer sperm as long as the female is otherwise engaged, as in eating him. Thus he fertilizes more eggs.

345. "Darwin's Dangerous Idea", p. 470

346. Steven Pinker, "Against Nature" in *Discover*, Vol. 18, No. 10, October 1997, p. 92f.

The author points out that it is not the genes that are making decisions,

> "The metaphor of the selfish gene must be taken seriously: people don't selfishly spread their genes; genes selfishly spread themselves. They do it by the way they build our brains. By making us enjoy life, health, sex, friends, and children, the genes buy a lottery ticket for representation in the next generation, with odds that were favorable in the environment in which we evolved...Our goals are sub-goals of the ultimate goal of the genes, replicating themselves. But the two are different. Resist the temptation to think of the goals of our genes as our deepest, truest, most hidden motives. Genes are a play within a play, not the interior monologue of the players. As far as we are concerned, our goals, conscious or unconscious, are not about genes at all but about health and lovers and children and friends."

347. "The Origins of Life", p. 132 makes this point and says,

> "We do not doubt that inclusive fitness theory [genetic] will explain a lot about human behaviour, but we are equally convinced that culturally acquired beliefs are important. For biologists, the question that must be answered is why it should be that humans can

so readily be influenced by myth and ritual to do things that do not increase their inclusive fitness..."

The book's chapter 12 gives a good explanation of the idea of memes.

348. For an extended discussion of the passing of the old understanding of the basis of morality and of its modern explanation see "Philosophy in the Flesh", chapter 14

349. "Catechism of the Catholic Church", #1706

Francis S. Collins, head of the Human Genome Project, is a Christian scientist for whom the presence of a conscience is an indication of the work of a personal god. *Scientific American*, February 1998, p. 29 says of him,

> "Humans have an innate sense of right and wrong that `doesn't arise particularly well' from evolutionary theory, he argues."

The reader will have to decide if my argument here is convincing, and/or will have to look up the sources and see for themselves.

350. "Catechism of the Catholic Church", #1776 citing "Gaudium et Spes" #16

351. Daniel Jonah Goldhagen, "Hitler's Willing Executioners" New York: Alfred A. Knopf, 1996.

See Saul Friedlander, "Nazi Germany and the Jews" for a different slant, finding a less unified attitude toward the Jews in German society.

352. Mark Twain, "A Connecticut Yankee in King Arthur's Court", New York: Airmont Books, p. 99. The citation continues,

> "All that is original in us, and therefore fairly creditable or discreditable to us, can be covered up and hidden by the point of a cambric needle, all the rest being atoms contributed by, and inherited from, a procession of ancestors that stretches back a billion years to the Adam-clam or grasshopper or monkey from whom our race has been so tediously and ostentatiously and unprofitably developed."

353. The "Theological Declaration of Barmen" declared in 1934 that

> "...as members of Lutheran, Reformed and United Churches we may and must speak with one voice in this matter today...we may

not keep silent, since we believe that we have been given a common message to utter in a time of common need and temptation."
["The Constitution of the United Presbyterian Church in the United States of America, Part I Book of Confessions" Philadelphia, 1966 # 8.08]

But the issue was not the treatment of Jews but division in the Church caused by government action. They never got around to speaking with one voice on anti-Semitism.

354. "Toward a Theology of Nature" p. 2

355. "The Moral Animal", p. 148

356. Gallup reported in 1995 that almost 75% of Americans believed in angels. Religious goods catalogues are full of gizmos and doodads featuring angels. Usually cute little cherubs, or beautiful Caucasians with wings. The Archdiocese of St. Louis has installed an "Angel of Harmony" next to the Cathedral, but at least this one has African-American features.

What does science have to say about other beings in the universe besides the plants and animals of earth? Nothing, of course, without evidence. And that evidence must be something that impacts our senses. The search for extraterrestrial intelligence, or SETI, listens with powerful antennas for a communication from elsewhere. A rock from Mars could contain physical traces of life on another world. If there is no evidence, the presumption has previously been that we are alone in the universe when it comes to consciousness, and that our planet may be alone in possessing life. This was especially true when the earth was considered the center of the universe and it all existed to support or amuse us. Today the presumption among many is that if life arose on our planet it probably arose elsewhere as well since the mathematical chances of there being other planets suited for life are so great. But no one can say without hard evidence, which should not merely be, "I saw it on `X Files' the other night." What evidence is there for the existence of alien life forms operating on our planet? Despite claims about an incident in Roswell, New Mexico, and similar claims about UFO sightings, there is little or nothing which can claim a level of scientific evidence that "little green men" or any other alien life forms have ever visited our planet, unless it is nucleic acids in comets and meteorites or evidence of some simple life form in rocks from Mars. There is no end of theories and scenarios, of course. Some scientists object that it would be passing strange to travel for thousands of light years and then not make your presence known. This is countered by UFOlogists by reference to some "prime directive" ála Star Trek not to intervene in the evolution of life on some other planet (ours in this case). But then if the aliens are successful in keeping under cover, how does one prove they are present? Others would say that the aliens have not kept under wraps, they are in fact walking about the surface of the planet in broad daylight. We earthlings are, they propose, the result of a "life bomb" sent from some other planet, somewhat like Superman from Krypton, except that we arrived in the form of DNA which either inoculates the planet and begins from one-celled organisms or springs forth from the pod ready-made, or as clones. In the rather confusing recent Stephen King novel, "The Presence", this is the plot which all happens to come together in Hawaii. Once again we have to confess, "who can say them `nay'," but without evidence we are back to believing in leprechauns—or angels and devils.

> [This same connection of belief in angels and in UFO's, etc. is found in "Skeptics and True Believers", p. 97].

Anyone can ascribe any cause they want for unexplained events, they are by definition unexplained. Anyone from the Federal government to visitors from Planet X to angels or devils can be given the blame for something when we don't know what really caused it. But the reasonable person, looking at all the unusual phenomena for which some perfectly natural explanation is later found, normally concludes that the same kinds of answers will eventually be found for the small percentage that remain a mystery. This is true of alien life forms and it ought to be true for the reasonable Christian in the question of angels and devils.

> [Thomas Aquinas did not discuss how many angels could dance on the head of a pin. This was from those mocking the Scholastics. Lewis Mumford must not get the joke because he tries to bring angels into the world of science when he says, "The question of how many angels could dance on the point of a pin no longer is absurd in molecular physics, with its discovery of how broad that point actually is and what a part invisible `messengers' play in the dance of life." Angels as RNA or enzymes is really pushing it.]

There is far more reason to accept evidence from the spontaneous remission rate for a certain form of cancer, which can be documented and quantified, even if not yet explained, than to ring in some supernatural creature whose job is to roam the earth working wonders or wreaking havoc. The chemical changes in the body brought about by the physical action of the brain which takes a positive attitude towards a cure is a far more likely explanation for "faith healing" than the interference of a god who wishes to reward a person who has done god the favor of believing in god's power or existence, whether god works directly or by sending an angel with an unlikely Irish accent as in the TV show "Touched By An Angel".

The devil gets blamed for many things that we bring on ourselves or are natural occurrences. The evidence, however, is the same as for angels. "But angels and demons are in the Scripture," the believer objects. Of course they are, but not until the (human) authors of Scripture are exposed to them through contact with the Zoroastrians and Magi of Persia. They do make convenient literary foils and stand-ins for god, especially for those authors who are a little squeamish about having god sit down under a tree for a leisurely lunch with Abraham. But there is no evidence that they are any more real than other characters in myths or novels. In societies that just accept the existence of supernatural beings, no evidence is necessary, of course.

The same kind of material turns up in time of the Dead Sea Scrolls

> "Magical Incantations against evil spirit were certainly a very significant part of popular religion in Graeco-Roman antiquity. Everyone believed in such spirits, and it was thought prudent to take steps to ensure one's protection...The local magician or scribe would write out a spell or two, often on a bowl...The inscribed object would then be buried under one's house."

It seems that these Judaism of the time was not immune to these ideas.

> [Robert H. Eisenman and Michael Wise, "The Dead Sea Scrolls Uncovered" New York: Penguin Books, 1992, p. 257]

.In the Middle Ages, Christopher Marlowe's Doctor Faustus made a compact with the devil.

> "This conveyed much more horror to an Elizabethan audience than it does to us, for Elizabethans by and large believed implicitly in such bargains...A legend developed that during a performance [of the play], in one scene in which Faustus called up devils, the actors counted one more devil than the scene called for and realized that Satan himself was in their midst. In terror, they stopped the play; the audience bolted from the playing place."
>
> [In foreword to: Christopher Marlowe, "The Tragedy of Doctor Faustus", Edit. by Louis B. Wright, New York: Washington Square Press, Inc. 1959, p. xx]

The reader may detect an "ad hominem" type of argument here, and certainly I personally do find it amazing that people so easily believe in something for which there is so little evidence. Carl Sagan in "The Demon Haunted World", New York: Random House, 1995, points out the seriousness of our gullibility when he reports on police officers willingness to believe in Satanism as an explanation for crimes.

> "Here are some excerpts from FBI expert Lanning's analysis of `Satanic, Occult and Ritualistic Crime,' based on bitter experience, and published in the October 1989 issue of the professional journal, *The Police Chief.* `Almost any discussion of Satanism and witchcraft is interpreted in the light of the religious beliefs of those in the audience. Faith, not logic and reason, governs the religious beliefs of most people. As a result, some normally skeptical law enforcement officers accept the information disseminated at these conferences without critically evaluating it or questioning the sources...For some people Satanism is any religious belief system other than their own.' Lanning then offers a long list of belief systems he has personally heard described as Satanism at such conferences. It includes Roman Catholicism, the Orthodox Churches, Islam, Buddhism, Hinduism, Mormonism, rock and roll music, channeling, astrology and New Age beliefs in general. Is there not a hint here about how witch hunts and pogroms get started?"

See also the article "Yankee Doodle Druid" in *Discover*, Vol. 19, No. 2, February 1998, p. 84 about the willingness to believe in "America's Stonehenge" in New Hampshire.

When others find out that I do not see any reason to believe in angels, the response is either the expression of a generalized belief in especially angels which it is obvious is more an "I hope it's so" kind of attitude, or the person relates some incident which proves the matter for them. This is usually along the lines of "I don't have any other explanation, so I will give the credit to an angel (or to god)." Movies like "Angels in the Outfield", which have angels helping a ball team, are popular and of course everyone who sees them will swear that they don't take it seriously. But it all fits with a general belief. Less noticed is that these movie angels work their wonders because a little boy prays, but the prayer often begins, "God. If there is a God." A god who proves him/herself by sending angels is what we would really like to have.

Among Roman Catholics today there is not so much enthusiasm for the devil. He seems to have lost some credibility, though why evil should seem less "real" than goodness in today's world is a good question. Robert Wernick, says,

> "Today, a few scant hundred years later [than 999], he has dropped so far out of sight that some believe he is gone for good. It may be true that 48 percent of Americans tell the pollsters they believe in the existence of the Devil and another 20 percent find his existence probable. But...they do very little talking about him out loud. Reported physical appearances of the Devil are far rarer than sightings of UFOs. In practical terms, people have banished him from public life."
> [Robert Wernick, "Who The Devil is the Devil?", *Smithsonian*, Vol. 30, No. 7, October 1999, p. 113]

His is a good article on the history of belief in the Devil.

The devil retains his popularity among the fundamentalist Catholics as well as other fundamentalist Christians, so don't worry for him. The Redemptorists have been known to throw a little fear of hell around as well, and who do you suppose lives there? Still, as with so many of the Christian beliefs and assumptions, one cannot say that they are wrong, but one can definitely say that there is no evidence demanding credence.

357. The idea of a final judgment on each person's life is common to many religions. It is sometimes combined with a belief that the final, final judgment will come at the end of the world. The Catholic catechism says,

> "On judgment day at the end of the world, Christ will come in glory to achieve the definitive triumph of good over evil which, like the wheat and the tares, have grown up together in the course of history." ["Catechism of the Catholic Church", #681]

We have to say a "final, final", because a judgment is thought of as occurring at the person's death, since they somehow continue to exist in either a state of blessedness or of damnation. But then there is the image of everyone who ever lived being gathered like sheep and goats, or wheat and weeds, at one time. So there must be two judgments, which seems a little redundant.

The fate of the universe is thought of differently in various belief systems.

"The unfolding of a world-order may...be cyclical in its pattern, a pattern in which decay or destruction may occupy as much space-time as creation. Sometimes the sequence is final; the whole creation may be doomed to destruction, as in the Scandinavian *kagnarok* (where the gods themselves appear to be included in the holocaust); or else creation having served its purpose, may be revoked, as in the Christian Last Judgment."
[David Maclagan, "creation myths", London: Thames and Hudson, 1977, p. 22]

But never mind this. Does science have anything to contribute to the discussion? Obviously not, when we are talking about judgment after one's death. The concept of a soul that could be judged is problematic so just what it is that is rewarded or condemned is a good question, of course, and so if the human being is what the existentialists say and "what you see is what you get", then what you see is obviously in the grave and not undergoing any judgment. But though science can remove any reason for positing a soul in the first place, based on the functioning of the human body or mind, it cannot definitively say that there is not a "spiritual" component of a human that could survive physical death. This is not to back off from my denial of a soul. There is no reason to think that there is any such entity. But it is impossible to prove that something does not exist, so we must allow that it could be, however improbably. Some readers have landed on the previous sentence with both feet, saying in essence and sometimes saying literally, "Aha! You can't prove the soul doesn't exist!" Well, no, I must admit, but if an opponent wishes to occupy such shaky ground, welcome to it. The faith will hardly be served by doing so, however.

358. If the universe is closed, then eventually it must cease expanding and begin contracting. Just what the end of that process would be is unknown. Would it just contract back into a singularity and continue as the granddaughter of all black holes? Would it wink out of existence since its energy debt is repaid? Would it "bounce" back into another Big Bang, just a little diminished from the one that preceded it? On the other hand, if it is an open universe, it will continue to expand forever. Neither of these scenarios looks for a sudden "end of the world". In the open universe there is no end and in the closed there may be but it is the result of a natural process. The religious "end of the world" view sees the universe as an arena whose function is to be a background for the human drama, or a divine drama. This view has been quite popular because it puts humans at the center of the

stage, but it is apparent that we are barely on stage at all in the scale of the universe and quite unneeded for its operation. Why a judgment on us would be of any significance to the universe, much less mark it's end, is hard to fathom. Could god step in at some time in the natural history of the universe and change it or put an end to it prematurely? Certainly we cannot deny the possibility. The god assumed to exist has the power to do anything, and there are plenty of Christians around waiting for just such an intervention, whether in the form of a "rapture" or a final judgment. There is also the question of why god would start the universe out on some course which looks today like an "open universe", and then cut it off before it runs its course? Those who hold for a final judgment answer this in various ways, but most seem to still have an idea of a static universe that is merely the stage upon which the main characters perform. When the stars of the show, us obviously, not solar masses, have reached some predetermined state, or when god just gets tired of fooling with the whole bunch, then god can dispense with the scenery. Cosmologists do not see the universe in this way and to try to talk to them about the faith while maintaining this view of a god-induced finale would be futile.

359. "The Astonishing Hypothesis" p. 3-4

360. "Augustine Day by Day" p. 65 (sermon 16A, 13)

See Peter Brown, "Augustine of Hippo", Berkeley: University of California Press, 2000, p. 285 for some background and consequences of this position of being citizens of heaven but also being involved in the affairs of the world.

Thomas A Kempis in "The Imitation of Christ", Notre Dame: Ave Maria Press, 1989, which has been used for centuries to form Christians, has this same outlook. As he says,

> "This world is not your permanent home; wherever you may be you are a stranger, a pilgrim passing through. You will never find peace unless you are united with Christ in the very depths of your heart. Why do you look around here to find peace when you do not really belong here? Your place is in heaven, and you should see everything else in terms of heaven." p. 65

This is not exactly being "At Home in the Universe".

361. Rev. Bernice A. King at a program in St. Louis, June 15, 1999

362. Henri de Lubac, "The Mystery of the Supernatural", New York: Herder & Herder, 1967, p. 109 argues, however, that these supernatural aspects of our nature are the results of sanctifying grace and are not innate in our natures.

> "In short, for Christians created nature is no kind of divine seed...It is indeed our `capacity' for it—to take a word used by Origen, St. Bernard, St. Thomas, and many others—but that does not make it a participation in it, even initially or distantly, `which needs but to be developed and enriched'."

363. "Before the Beginning", p. 115

364. "The Demon-Haunted World", p. 204

365. "The Demon-Haunted World", p. 204

366. "The Demon-Haunted World", p. 269

367. "Without Miracles", p. 160

368. "Three Scientists and Their Gods", p. 172

369. Claudia Wallis, "Faith and Healing", *Time*, Vol. 147, No. 26, June 24 1996.

> "Studies show that the relaxation response is controlled by the amygdala, a small, almond-shaped structure in the brain that together with the hippocampus and hypothalamus makes up the limbic system. The limbic system, which is found in all primates, plays a key role in emotions, sexual pleasure, deep-felt memories and, it seems, spirituality. When either the amygdala or the hippocampus is electrically stimulated during surgery, some patients have visions of angels and devils. Patients whose limbic systems are chronically stimulated by drug abuse or a tumor often become religious fanatics. `The ability to have religious experiences has a neuro-anatomical basis,' concludes Rhawn Joseph, a neuroscientist at the Palo Alto VA Medical Center in California."

For another, sympathetic, take on the whole question of religious experiences, especially in near-death visions, see Melvin Morse, M.D.,

"Parting Visions", New York: Villard Books, 1994. On p. 123, on the subject of the brain, Morse says,

> "One of the many things proven by death-related visions is that our brain is equipped to have spiritual visions. In fact, death-related visions shatter forever the notion that human beings do not have a spiritual side. They are proof that paranormal abilities are not the creation of con men and hucksters but are real events that happen to real people."

In my view, it is not a question of these experiences being staged by dishonest people, but that they arise from the limbic system or some other physical process and are not proof of anything spiritual.

370. "Darwin's Dangerous Idea", p. 130

371. Discussion of life after death and the hope of immortality naturally suggests the topic of Eschatology. Indeed, it should be the other way around. "In the course of history this proclamation [of the Christian hope for the future] itself has been strangely concentrated on the individual..."

> ["Sacramentum Mundi" Vol. 2 Ed. by Karl Rahner S.J. New York: Herder and Herder, 1968 p. 242]

Eschatology is properly the theology of the planned fulfillment of *all* creation. As it is explained by theologians,

> "The actual universal propositions contained in eschatology which precede its various material propositions would likewise form part of a properly elaborated eschatology and would include: the intrinsically limited character of time and its historical configuration from genuine beginning to a genuine irreplaceable end; the uniqueness of each part of sacred history; death, and the "change" effected as an event by God, as a necessary mode of genuine fulfillment of time (since the Fall); the fact that the end has already come with the incarnation, death and resurrection of the Logos made flesh; the presence of this end as constituting the fact of the victorious mercy and self-communication of God (in contradistinction to a double outcome, on an equal footing which would be specified by man's freedom alone); the special character of the time now still unfolding "after" Christ; the persistent character of this period as a conflict (with Antichrist), which necessarily becomes more intense as the end approaches; the question of the convergence of the natural and supernatural finality

of man and cosmos (the factors of a "natural" eschatology which involves more than merely the "immortality of the soul"), etc."

["Sacramentum Mundi" Vol 2 p. 245]

As with all of these topics, science can contribute nothing to the discussion about whether or not there is a plan that is working itself out or whether there will be an intervention by god at some future date. If one believes that either or both of these is true, it will have an impact on one's present-day behavior and that is why Eschatology should not be limited to future events or the Last Things. It has an impact on today, especially since, for the believing Christian, the "kingdom" has already been initiated in the Christ event and these are already and simultaneously the last days and the first days. Science presumes that the universe is going along at its own pace but it has nowhere where it must go (except a "big crunch" or a slow burn out depending on whether it is open or closed. The most recent opinion is against the universe closing back on itself and in favor of expansion into virtual nothingness.

See Michael D. Lemonick, "The End", *Time*, Vol. 157, No. 25, June 25 2001, p. 48)

Christianity presumes that the universe is going along at god's pace and that there is a destination. These are not mutually exclusive views and one is free to accept them both without conflict. This is especially evident when Eschatology addresses the fate, not of the whole of the universe but of one of its parts, namely an individual human being. The universe can and will go its merry way with or without me, but I am interested in what lies in store, if anything, at the time of my death. If the Eschaton is a "new creation", then it can occur without impact on the universe and can exist at the same "time" as the universe, recognizing that time can be defined differently for different realities. Still, science cannot be an ally of faith in this, and,

"...the Christian...theory of history as the perfecting of the world by God cannot be confirmed within history but can only be accepted in faith."

[Ingo Hermann, "Total Humanism" in "Is God Dead", Concilium Vol. 16 New York: Paulist Press, 1966 p. 167]

Yet, why can it not be confirmed within history?

"According to Christian doctrine, such a manifestation of the eternal within temporal reality will occur in its fullness in the eschaton (last times), but by anticipation it occurred in the midst of the ongoing sequence of events in the resurrection of Jesus. This event persuaded the Christian community that the eternal Logos was incarnate in Jesus."

[Pannenberg, "Toward a Theology of Nature" p. 25]

Why is there not more impact on the history of the universe, or at least on our small planet, if indeed the very purpose of the universe has been attained, at least in embryo? For believers, of course, there has been an impact that is measurable, at least in human evolution, and perhaps, almost like Chardin, prior to us. Pannenberg suggests that,

> "There...seems to be a tendency toward increasing participation in the divine spirit and Logos in the course of the evolution of creatures, approximating the eschatological presence of the eternal in the temporal. The human mind is distinguished by a unique degree of openness to the presence of the eternal which is expressed in the experience of an amplified presence that overlaps, though in a limited way, past and future events. The participation of the human mind in the eternal Logos through the ecstatic power of the spirit may account also for the possibility and specific character of human knowledge of the created world. In a trinitarian perspective the work of the Logos and that of the spirit in the creation of the world belong closely together. Can this be expressed in a language that takes account of modern science? If von Weizsicker's suggestion is followed, namely, that the ancient philosophical Logos doctrine can be reformulated in terms of modern information theory, then it does not seem completely inconceivable that a field theory of information can do justice to the cooperation of Logos and spirit in the creation of the world." ["Toward A Theology of Nature" p. 26]

This may make sense to the believer, but to the scientist, and especially the evolutionist, it is hardly a compelling argument. There is no evidence that any force outside the universe has had any effect on the development of life or consciousness. But then again there is no evidence that there was not such influence, how could there be?

> [It is interesting that Tipler, "Physics of Immortality" p. 13f says, "Wolfhart Pannenberg has suggested that there may exist a previously undiscovered universal physical field (analogous to Teilhard's "radial energy") which can be regarded as the source of all life, and which can be identified with the Holy Spirit. There are no undiscovered energy fields of significance to biology; conservation of energy and the size of the energy levels in biology preclude it."

> But Tipler then tries to use the same idea to back up his argument and says, "However, I shall argue in Chapter VI that the universal wave function (provided it satisfies an `Omega Point' Boundary Condition) is a universal field with the essential features of Pannenberg's proposed new `energy' field."]

Pannenberg admits that there is little the two views can say to one another, but he looks hopefully for religion to take account of the findings of science, and science to be open to the effort.

> "The...question...of eschatology...needs to be raised in its own right because it points to one of the most obvious conflicts between a worldview based on modern science and the Christian faith: Is the Christian affirmation of an imminent end of this world that in some way invades the present somehow reconcilable with scientific extrapolations of the continuing existence of the universe for at least several billions of years ahead? To this question there are no easy solutions. Scientific predictions that in some comfortably distant future the conditions for life will no longer continue on our planet are hardly comparable to biblical eschatology. On the other hand, some people are always quick to expurgate the religious traditions from elements that seem to make no sense to one period in the development of scientific insight. Perhaps one should, rather, accept a conflict in such an important issue, accept it as a challenge to the human mind to penetrate deeper still into the complexities of human experience and awareness. It does not seem unreasonable to expect that a detailed exploration of the issues involved in the question concerning time and eternity may lead one day to more satisfactory ways of including biblical eschatology in an interpretation of the natural world that should take appropriate account of modern science."
>
> ["Toward A Theology of Nature" p. 26-27]

Pannenberg recognizes the problem and is consistent with himself since he had earlier written saying,

> "Jesus Christ, the end of history, is not available to us as the principle of a `Christologically' grounded total view of world history. Christ's Resurrection, the daybreak of the eschaton, is for our understanding a light which blinds as Paul was blinded on the Damascus road...our participation in this event, the hope of our own resurrection, is still hidden under the experience of the cross. No one can make the eschaton into a key to calculate the course of

> history, because it is present to us in such a mysterious, overpowering, incomprehensible way."
>
> [Wolfhart Pannenberg, "Redemptive Event and History", in "Essays on Old Testament Hermeneutics", Ed. by Claus Westermann, Atlanta: John Knox Press, 1963, p. 314f]

Whether this faith position can be reconciled with science, as Pannenberg hoped in 1993 is problematic. He has evidently worked on it for 30 years but the best he can do is to counsel that we accept the tension and believe that we will figure it out someday.

Though Eschatology properly starts with the promise of renewal of all creation, what grabs most people's attention is their own fate. Let the universe take care of itself. But if we are to be raised from death somehow and judged on the conduct of our lives, to what kind of existence? The traditional choices are two, heaven and hell, and in fact these are seen as the goals of our existence. Certainly heaven is described as such in many a spiritual tract, as Augustine depicted it in his spirituality. Some even believe that some people, presumably others, have been intended or predestined for hell, and some bikers and others sport tattoos proclaiming not only their belief that this is their fate, but their willingness to go there. With this as a presumption, we can easily miss the question of why we cannot live forever and skip the whole eschatology thing. In fact, it might be possible to live, if not forever, at least a very long time, but we have not inherited the genes to do this.

> "Doubtless a eugenicist could breed a race of superlatively long-lived humans. You would choose for breeding those individuals who put most of their resources into their own bodies at the expense of their children: individuals, for example, whose bones are massively reinforced and hard to break but who have little calcium left over to make milk. It is easy enough to live a bit longer, if you are cosseted at the expense of the next generation. The eugenicist could do the cosseting and exploit the trade-offs in the desired direction of longevity. Nature will not cosset in this way, because genes for scrimping the next generation will not penetrate the future."
>
> ["River Out of Eden" p. 127]

The world just did not have the kind of resources to do both, and it was just as effective to have the genes live in a series of individuals as in one particularly long-lived one. It is theoretically possible, of course to be practically immortal and still reproductively successful, but nature has not gone in that direction. So we, like everything else, have a limited life span. Not as short as a mayfly, fortunately, but not as long as we would wish. Unlike everything else, we are left to ponder our fate and ask not just about the possibility of life continuing but where and what kind of life it might be. Science is going to be about as much help here as it was in the question of the possibility of life after death. It has changed our speculations about *where* such places as heaven and hell may be. Certainly heaven is not above the crystalline sphere of the heavens and hell is not under the earth, despite the testimony of Russian tabloid journalists who claim that voices and moans have been heard from deep wells. Consequently those who speculate about such things today are more likely to speak of heaven and hell as a *state of being* rather than a place. Science has pointed out that there is a perfectly natural reason why we die which has nothing to do with only being here in exile anyway. We are a part of the universe and in the universe that has evolved everything dies, including the universe, and most assuredly including us. Whether an external agent will do anything about this state of affairs is another matter.

372. So that we touch all the bases, let us consider now another subject, dear to the heart of every Catholic Thomist (and there are few non-Catholic Thomists), namely, *transubstantiation.* The term is only used in one case, that of the presence of Christ in the Eucharist.

> "Transubstantiation is described by the Council of Trent as the change of the whole substance of bread (*totius substantiae panis*) into the body of Christ, and of the whole substance of wine (*totius substantiae vini*) into the blood of Christ, the species of bread and wine remaining unchanged (Session XIII, can. 2; D884)."
>
> ["Sacramentum Mundi" Vol. 6 p 292]

Of course this theological position is conditioned by the understanding of the times in which it was developed, which were decidedly philosophical and not scientific. Objects were described in metaphysical terms such as "matter and form" instead of as unions of atoms and molecules. To change, the bread would, with our new understanding of the make-up of "stuff" mean that billions and billions of atoms would have to be transubstantiated, not to mention quarks and other compositional particles. This is clearly unsatisfactory and new explanations were sought in our times, including "transfinalization" and "transignification".

It had always been obvious that there were problems with the Aristotelian concepts. There is obviously such an object as "Alton Lake" north of St. Louis. One can go there, sail on it, swim in it. But is there really such an object as "Alton Lake" in the same way that there is that lamp post or that squirrel? Even the lamp post is a bit of a fiction. If I hang a flag from it, did it change from being "really" a lamp post to being "really" a flag pole? Or did I just change the function of this conglomeration of atoms and put a different name on it? With living things the philosophical constructs of form and matter make a little more sense. It is obvious that this living squirrel is not the same being as a pair of squirrel gloves, so I can get by with saying that the "form" of the material involved has changed. And certainly when I die my body is no longer "me", but I fervently hope that there is a "me" that continues to exist, the form or soul. But the old Thomistic explanation of how the bread and wine can be the body and blood of Christ through transubstantiation, the form changing while the matter remains the same, is no longer explanatory, though the American bishops seem to think it is and use it in a 2001 pastoral statement, "The Real Presence of Jesus Christ in the Sacrament of the Eucharist."

Because of the findings of science, new explanations have had to be sought for the conviction that Christ is "really" present in the Eucharistic food. And this is not to mention the different kind of presence, a real but not physically perceptible presence that is postulated for all the sacraments. The church has never really said that the water of Baptism is not really water but the cleansing presence of Christ, and that there has been a substantial change in the materials of the sacrament, but that it has not is perhaps an accident of history. The faith of the Church is clearly that Christ is present in the oil, the water, the bodies of the sacramentally married, the imposed hands of the bishop, but the concept of transubstantiation was never applied to these since it did not seem to fit. It makes me wonder if since it was never applied to the other sacraments, it should not have been applied to the Eucharist, either.

Thanks to the scientific explanation of the reality of things in place of the older, philosophical "matter and form" idea, the Church has been forced to reconsider what it means to say that the bread and wine have undergone some transformation, and the discussion has been good for the exegesis of the faith. On the other hand, if there is no such thing as the essence or substance of bread or wine, and a light post or squirrel or any other "essence" is merely a mental construct which is no longer useful, it appears obvious that there is no such thing as the essence of a human being either, since we are composed of the same atoms and molecules as any other object in the universe. If the existence of the building blocks of matter is an ephemeral existence and an accident of evolution, so is our existence. If their being has no meaning or purpose, apart from what we give to the structures they form, then the universe itself, and we within it, have no meaning or purpose other than what we construct, or some other being constructs for us. We must obviously return to this thought in our concluding chapter. But it seems to be well for the Church that it has rethought the concept of transubstantiation, since it has so little grounding in reality. The bottom line seems to be that the reality of anything at all in the sense that it is an independent structure deserving of a name and having a purpose to other independent structures within the universe must come from its meaning *relative to* other structures. Otherwise it is just another random agglutination of quarks, or superstrings, or whatever is the bottom line of the universe. The concept and reality of "squirrel" has a meaning to us because they exist in the same plane of existence as we, and it has a meaning to everything else on our level, and evidently to levels above ours, but it has no "meaning" in the universe, so there is no underlying "form" which could be changed into another form, as in the Eucharist.

However, there are other ways to view the situation. WE are part of the universe, not something separate, and therefore if WE recognize the reality of something, then the universe recognizes it in and through us. If a squirrel is something real and independent to me, not only from dogs and stones and galaxies, but even from other squirrels, then is there not a reality there in the universe since I am the universe as much as any other? Isn't everything connected?

In the construct of those holding the strong Anthropic Principle, and among those who hold that there is someone who has put purpose into the universe, or into us, as well as the followers of Chardin and company, the kernel of all that will be is already contained in what is. There is a final cause operating backwards that demands that "squirrel", for example, be already present *in potentia* or else it could never arise. Things therefore have an identity in themselves and may not be just random groupings of molecules, since they serve some purpose or carry out some prearranged agenda. There need not be a conscious planner behind this, it may just be that for the universe to go anywhere at all it must begin with the right conditions to develop all that it will ever develop, though this is not a final cause. Or, for the theist-scientist, there may have been some "fine tuning" of the initial conditions of the universe so that it will go where it desired to go and evolve what the tuner is interested in, in which case it is a final cause again. And under the premise that nothing can come from nothing, except perhaps the universe itself, there must already be in the beginning the seeds of what will later be. This kind of position takes on added strength when one recounts all the variables that must be just so in order to arrive where we are, and that suits theists just fine since it lends credence to the explanation that someone had it all figured out in the first place. True, they may allow that the identical physical configurations at the beginning may turn out differently if one could rerun the "program", and that there is no necessity that god would know that god's conscious creations would end up with two legs, two arms, and the potential for male-pattern baldness, leaving these details to the chaotic machinations of evolution. But that there would be *some* conscious beings eventually who could respond to god's overtures of love was built into the system from the beginning.

Can this be true? It may seem so from scientists' own findings about evolution and chaos. A system, such as the universe, is called "linear" if cause and effect are proportional and in general the whole is the sum of the parts. Conversely, a non-linear system is one in which the total is more than the sum of the parts, among other things. Either may be quite simple or quite complex. Paul Davies points out that,

> "Generally speaking, a nonlinear system must be understood in its totality-which in practice means taking into account a variety of constraints, boundary conditions and initial conditions. These supplementary aspects of the problem must be included in the study of linear systems, too; but there they enter in a rather trivial and incidental way. In the case of nonlinear systems, they are absolutely fundamental in determining what is going on."
>
> ["The Matter Myth" p. 46]

If you stack up coins every time a clock strikes the hour, and add one coin every time, the stack will grow at a constant rate, and if you graph the situation you will have a straight line. But if you add one for every swing of the bell, one coin at one o'clock, three coins at three o'clock, etc., there will be a jagged line on your graph, though this is still a linear situation since cause and effect can be understood quite easily. However, if you add one coin every time a coin toss on the hour comes up heads and 10 coins every time it is tails, a random element has been introduced and you cannot predict where you will be on the chart after a given amount of time. This is a non-linear system and you don't know, nor can you predict, where you will end up at any chosen point. Still, there are laws in operation here since it is obvious that in the long run heads will come up as often as tails (except for the problem of the head side being slightly heavier than the tail side), but we cannot tell from initial conditions or from the rules of the game precisely how things will turn out. Many scientists, however, hope that they can discover previously unknown laws that do explain the situation. Even in chaotic systems, says Stuart Kauffman,

> "If...core phenomena of the deepest importance do not depend on all the details, then we can hope to find beautiful and deep theories,"
> ["At Home In The Universe" p. 18]

but he certainly does not see any Designer involved. Still, he says some things that allow the theist to place god in control of the initial conditions which then seemingly inevitably result in the universe somewhat as we know it now, such as the statement,

> "...One hopes to explain, understand, and even predict the occurrence of these generic emergent properties; however, one gives up the dream of predicting the details."
> ["At Home in the Universe" p. 24]

The question then becomes,

> "...how would such laws of emergent order, if they should someday be found, be reconciled with the random mutations and opportunistic selections of Darwinism? How can life be contingent, unpredictable, and accidental while obeying general laws?"
> ["At Home in the Universe", p. 24]

His answer has nothing to do with a Designer, however. He concludes that

> "Ultimately, it must be a natural expression of a universe that is not in equilibrium, where instead of the featureless homogeneity of a vessel of gas molecules, there are differences, potentials, that drive the formation of complexity. The flash of the Big Bang 15 billion years ago has yielded a universe said to be expanding, perhaps never to fall together in the Big Crunch. It is a nonequilibrium universe...If we are...natural expressions of matter and energy coupled together in nonequilibrium systems, if life in its abundance were bound to arise, not as an incalculably improbable accident, but as an expected fulfillment of the natural order, then we truly are at home in the universe." ["At Home in the Universe", pp. 19-20]

His idea that we are "natural expressions" gladdens the hearts of the theists and the Anthropists who can say, either, "see, someone intended that we should be here," or "see, if we were not here the universe would not be here either, for we create it, even though it created us." I am fairly sure that Kauffman and others who are pursuing this line of thought have exactly the opposite in mind, but the result is the same. It allows that we are already contained in the universe from the very beginning or else we could never have arrived. This is not that far from Plato's "ideal forms" which have some idealized existence apart from the universe. Remember Charles Darwin on Plato and monkeys.

However, there seems to be another explanation altogether. The universe is recognized as a nonlinear system. This has not always been so. It was presumed that if one knew enough, or perhaps only knew a few basic things, all the rest would follow. Many of the savants of the Enlightenment compared the universe to a clockwork that could be completely understood once you got the drift of its gears. You could know the whole by the sum of its parts, and in a real sense everything that was to be was contained in what is now. But the universe is not like that, we have discovered, and does not always reach the result that would ordinarily be expected. This is not because of the intervention of some "miracle" or suspension of the laws of physics, but because the laws of physics themselves allow for the result to be more than the sum of the parts.

> "It is no longer possible to proceed by analysis, because the whole is now greater than the sum of its parts. Nonlinear systems can display a rich and complex repertoire of behavior, and do unexpected things." ["The Matter Myth" pp. 45-46]

This is not to say that one cannot begin with a complex structure, be it a building or a squirrel, and work backwards through a series of simple steps to its origin and explanation. Certainly you can, and evolution has constructed everything that is through just such a series of "simple-minded" changes, additions and subtractions. But the linearity does not extend in both directions. One cannot begin with a set of initial conditions, no matter how precisely you know them, and predict what will result. This is not really a very popular idea among theists nor reductionist evolutionists. The theists would like to see the emergence of something new and unpredicted (as well as unpredictable) as the result of the action of god. The reductionists would like to be able to explain everything as purely natural (and, not so coincidentally, not owing to such divine intervention), and they certainly have the better case, even though a chaotic system can in principle be predicted if you have an infinite computer—or an infinite divine mind.

Even so, the most die-hard reductionist evolutionists, including Steven Weinberg, can say that,

> "As we look at nature at levels of greater and greater complexity, we see phenomena emerging that have no counterpart at the simpler levels, least of all at the level of the elementary particles." [Steven Weinberg, "Dreams of a Final Theory", New York: Pantheon Books, 1992, p.39]

They certainly do not intend to say that the higher level phenomena are not explainable in terms of lower-level functions being put to new uses or that something new and miraculous has occurred, but they will admit that one cannot predict from simple structures what is the effect of combinations that may later occur. Nor are we saying that the new entities and their unpredictable powers are not the result of the same evolutionary process that produced the simple structures on which they are built. But we do contend that, though the reductionists are correct in saying that everything can be reduced to fundamentals and that these fundamentals are the source and basis of all that is, one cannot begin at the other end, the fundamentals, and explain how they will develop.

To summarize, while from one point of view there is nothing "real" above the level of quarks or their constituents, (as "real" as they are, which is "not very"), everything being just a shifting around of the basic building blocks, from another point of view, new "things" do come into existence which have a reality of another order. And as these recombine, newer things come into existence. Laws that describe the functioning and interactions of the quarks are inadequate to describe the new realities and new laws must be discovered. Higher organization brings about new "realities" and new beings. In a reductionist universe, one can know all the possibilities, it is true, but we now realize that one cannot know which of the possibilities will become a "reality". And there is a difference between those possibilities that do evolve and those that merely could evolve. There is a difference between actuality and potentiality-without-actuality. Therefore, for example, while we said that the concept of transubstantiation is no longer an illustrative argument for what occurs in the Eucharist, one cannot say that *something* does not happen in the bread and wine that cannot be inferred from the interactions of the constituent parts.

This has been a long introduction to where I believe we can go in understanding what the Church intends to say about the Eucharist. I believe that the Church is saying that something "real" has become something else just as "real," and I believe that science can allow for this.

373. "Abouttime" p. 277

374. "Companion to the Cosmos", p. 81

375. Jack Cohen & Ian Stewart, "The Collapse of Chaos", New York: Viking, 1994, p. 20.
See the extended treatment of chaos on p. 189f.

376. "At Home In The Universe", p. 29.
See the chapter and later chapters for a discussion of this theme.
See Stephen Hawking, "The Universe in a Nutshell", p. 107, where he says,

> "We now realize that the wave function is *all* that can be well defined. We cannot even suppose that the particle has a position and velocity that are known to God but are hidden from us. Such "hidden variable" theories predict results that are not in agreement with observation. Even god is bound by the uncertainty principle and cannot know the position and velocity; He can only know the wave function."

377. "The Conscious Universe", p. 145

378. "At Home In The Universe" p. 26

379. "Demon-Haunted World" p. 330

380. "Is God Dead" p. 56
381. Matt Cartmill, "Oppressed by Evolution" in *Discover* Vol. 19, No. 3, p. 83

382. Stephen Jay Gould, "Nonoverlapping Magisteria", *Natural History*, Vol. 106, No. 2, p. 61 puts the context of this encyclical well when he points out that Pius XII had reluctantly given evolution the status of a hypothesis. With John Paul II's pronouncement, Gould says,

> "...sincere Christians must now accept evolution not merely as a plausible possibility but also as an effectively proven fact. In other words, official Catholic opinion on evolution has moved from `say it ain't so, but we can deal with it if we have to' (Pius' grudging view of 1950) to John Paul's entirely welcoming `it has been proven

> true; we always celebrate nature's factuality, and we look forward to interesting discussion of theological implications.'"

Even in the periodical *Theology Today*, I found an article favorable to the faith but critical of the papal position on evolution, saying that,

> "...it does not go beyond *Humani Generis*, provicded one does not take into account the statements that evolution is 'a hypothesis to be taken seriously' (Pius XII) and is 'more than a hypothesis' (John Paul II). There yet remains, however, the dualistic statement: the body has its origin in evolution; the spiritual soul has its origin through God's creative hand."

Koltermann finds fault with the papal position that the soul is a contingent essence which must be created immediately *(creatio continua)* by god, pointing out that in the Church's position the body, too, is a contingent entity. [Koltermann, Vol. 48, No. 2 op cit.]

383. Even those who are calling for a renewal of faith in Jesus' healing of the physical condition of those prayed for and anointed usually recognize that

> "...deep spiritual healing is by far the most beautiful fruit of the healing ministry. I often say to the people who come to a healing service: 'The Lord is not interested in just healing your symptoms or certain parts of your body, but wants to heal your whole person from the inside out."
>
> [Matthew Swizdor, O.F.M. Conv. "Lay Hands on the Sick", Granby: Franciscan Friars, St. Hyacinth College, 1983, p. 50]

384. Naturally, the first assumption to be questioned must be that everything has a reason and can be explained. There are a host of agents affecting any situation, each working on their own or in tandem, and identifying proximate causes is all a part of explaining "why" something happens. Some of these agents may be conscious and rational but for the most part they will be unconscious, just objects following the impersonal workings of the universe and its laws. Even when the agent itself is rational, there may be much of the unconscious about the motives. This is the great understanding of both psychiatry and sociobiology. In any case, they will form a chain of causation. For example, when it rains on my parade I can explain the atmospheric mechanism which produces rain clouds, the effect of the energy of the sun on our planet, the reasons for the chemical reactions that produce such energy, and so on. In the case of a conscious agent, I can explain why a person has chosen to do a certain action, including both

conscious motivations and unconscious. And along the chain we can ask questions such as, "why is it this way and not another?" and, as humans do, we will make value judgments, placing either blame or praise, on many. Susan's question about why god is harming her by giving her an abusive husband could also begin to be answered from the other direction, by asking why he is abusive. Here we are in the realm of human personality and motivation and the going is particularly sticky, unlike the chain of causality of a rain cloud. He may have chosen to be abusive through a clear and precise process of reasoning, but this is hardly ever the case. Most likely he is as he is because of his own experiences leading to a view of what is real, which makes it logical to act in this manner. He may have been, and statistically speaking probably was, himself abused, or experienced the abuse of women by other men. He learned that this is how the world works and it makes sense to him to continue in this pattern. He may have just watched too many John Wayne movies and gathered that to be a man means to physically intimidate others, though John Wayne rarely used his strength against a woman. Rooster Cogburn won an argument with Katherine Hepburn by pointing out the differences in their sizes, but he was not called upon to make good on his threat. Who knows what the boy watching on cable TV is going to make from this lesson on being a man? The chain of this kind of human influence on one another can continue through generations in one form or another. Many of the problems affecting African-Americans can be traced to the period of slavery, both in their own attitudes toward themselves and one another and their relationship to White society, and in the attitudes of the descendants of the slave-holders, or those who could have been slave-holders in the past, just as the problems of many European Americans can be traced to the prejudices they learned here.

In the novel, "Voodoo Dreams" by Jewell Parker Rhodes, one common opinion is expressed by a Catholic priest that,

> "The colored races are the children of Ham, cursed long ago...Ham, Noah's youngest son, looked upon his father's nakedness...When Noah awoke, he knew his son's sin. Noah cursed Ham, the father of Canaan, and all descendants from his blood. He said, `cursed be Canaan; a servant of servants shall he be unto his brethren.' Black people share the burden of Ham's curse." P. 110

Belief in this curse was not restricted to the United States. South Africans believed that the non-white races would always be "hewers of wood and drawers of water" for the more "advanced" race. To the extent that those on either side believe this, there will be an effect on the culture and behavior of the society. This is the kind of influence that can be passed on for generations.

> ["To Make Our World Anew", Ed. by Robin D. G. Kelley and Earl Lewis, New York: Oxford University Press, 2000 is a good history of African Americans by two Black professors. It shows in many places the effects of slavery on both the Black and White population.]

Having conjectured about why Susan's husband is like he is from a psychological point of view, we can ask why any human beings are like this. Why are the males at once so aggressive (and, the females ask, insecure?) Why are the females often too docile long past the point where many a male would rebel?

> [I am following one line of thinking here which some say is prejudiced in favor of males. For another viewpoint on women's roles in evolutionary history, see Natalie Angier's "Woman: An Intimate Geography" published by Houghton Mifflin, or Dianne Hales, "Just Like a Woman: How Gender Science is Redefining What Makes Us Female" published by Bantam.
>
> In Eleanor E. Maccoby, "The Two Sexes: Growing up Apart, Coming Together", Cambridge, Belknap Press, 1998 there is a strong antagonism to saying that any gender traits are hard-wired. All are learned behavior. I believe the truth is, as usual, in the middle somewhere, and if this is true, then my argument is unaffected.
>
> Barbara Ehrenreich, "The Real Truth About the Female", *Time*, Vol. 153, No. 9, March 8, 1999, p. 57f]

Here we leave the area of human psychology and enter that of human evolution. And here it makes perfect evolutionary sense for the male to be aggressive, coming off millions of years of efforts to gain reproductive ascendancy over other males as well as defending hearth and home from invaders.

> See Richard Wranggham and Dale Peterson, "Demonic Males", Boston: Houghton Mifflin Company, 1996.

The Achuar Indians have an extremely violent society and studies of them show why aggression pays off. A researcher points out,

> "Remember that the [aggressive warrior] is really only a vehicle for the trait...Let's say we have 10 individuals with the psychology to be [aggressive]. Nine get wiped out; they push it too far. One of

them succeeds and gets 20 wives. Overall, it's still a good reproductive payout."

Mary Roach, "Why Men Kill" in *Discover*, Vol. 19, No. 12, December 1998, p. 106.

There has been some objections to the reports on the Achuar, to the effect that they were putting on a show for the researchers.

It also makes evolutionary sense why the female would tend to suppress feelings of aggression since in the past her physical resistance would often gain little, but hiding the children and staying with them to keep them quiet would be a good reproductive strategy.

[See "Demonic Males" for information on species where the female is the aggressive sex.]

In fact, let the aggressive husband draw all the attention to himself with his roaring and bragging. So much the easier to slip the little ones into the bush while the attacker's attention is focused elsewhere. Husbands are expendable. Of course, the male's aggression needs to be mostly directed outward from the family, but if a little is misguided inward, there is not too much problem. However, there is definitely a problem when the male lives in a society that both frowns on aggression towards the neighbors and tolerates assault on the family. The male is practically encouraged to direct his hostility toward the wife. This is particularly true of the "macho" male who has learned and/or been taught that the wife is not his equal but his servant, can always be replaced, and serves little use other than for sex, reproduction and housekeeping. Some Fundamentalists teach men to be the head of the wife as a scriptural injunction, leaving just what this means up to each man. Billy Graham is reported to have told his prospective fiancée who wanted to be a missionary that, "Woman was created to be a wife and a mother." When she demurred, he told here, "If that's the case...we will call a halt, during which time you should search the Scriptures until you find out just what is God's place for women. And when you are willing to accept god's place, you can let me know."

[This relationship seems to have worked out but I would question the basis of his demand. I suspect there are many women who would have kissed him goodby after such a statement.

See Patricia Cornwell, "Ruth, A Portrait" condensed in *Reader's Digest*, Vol. 151, No. 908, December 1997, p. 97]

If the male does not value her as a friend, confidant, partner and an asset in promoting his own growth as a person (and few genuine male chauvinists are into personal growth), then there is even less reason to curb his aggression. "...Primates, including humans, suppress aggression or make peace not for peace's sake but in order to preserve something valuable."

["Good Natured", p. 181]

See also Frans de Waal, "Are We in Anthropodenial?", *Discover*, Vol. 18, No. 7, July 1997, p. 50

If Susan's husband is violent and abusive towards her, to him it makes sense, though probably on some unconscious level. If it did not, he would not do it. He has inherited the tendencies that lead to his behavior and under some circumstances his aggressive actions may be appropriate, though probably not in the society in which he lives today, and not directed towards his wife. Changing the example to living with alcoholism, another frequent problem bringing up the "why is god doing this to me?" question, we would reach the same results, even though the causes of the inherited problems are different.

Why then were our progenitors like they were? Why does the world operate as it does? Why is there any aggression at all? (Bonobo chimpanzees are not aggressive towards each other, and their replacement social interaction, sex, is an inspired choice.)

["Demonic Males" tells the story of the bonobo, including the female dominance, and makes some parallels with humans.]

All of these questions themselves have some answers from the realm of science,

[See books on the subject including "The Moral Animal" "Good Natured" "Dark Nature", "When Elephants Weep" and their bibliographies.]

but it is obvious that we will arrive at some point where there is no known explanation, and some point even farther back in the chain where there is no possible scientific explanation. After all, the most primitive living forms probably got their nutrients from the sea around them and did not have to resort to eating their neighbors. It is certainly possible that evolution could have followed a different path and suppressed aggressive tendencies in favor of live-and-let-live. Yet aggression did appear, and probably quite early. Is this the point at which we bring in god? Did god begin the whole chain of evolution along certain lines of physical laws knowing what the outcome would be? Here, we must remember, we are at the beginning of a chain of causation that will lead to Susan and her question, "Why is god doing this to me?" and it will be a valid question since she presumes god put the aggression into animals. Or did god just begin the existence of matter and allow it to evolve as it will, absolving god from blame in some way? Or perhaps the "free will argument" applies here and god was not able to give both freedom from evil and freedom of choice at the same time. However, if god was not able to impose laws on evolution that would forbid or limit aggression, for example, then god is absolved of blame, but seems to be a rather impotent god and not one that people of faith would like to claim. If god was able but did not give direction, as the deists would hold, then god is to blame by default in the same way that Dr. Frankenstein was to blame for his monster, and the splicer of genes would be responsible for the tomato that eats Cincinnati. Don't start what you do not wish to or cannot finish.

Mark Twain in "The Mysterious Stranger (and other Stories)" New York: New American Library, 1962, p. 252f, (Twain's copyright 1916) concludes his story with this thought:

> "...a God who could make good children as easily as bad, yet preferred bad ones; who could have made every one of them happy, yet never made a single happy one; who made them prize their bitter life, yet stingily cut it short; who gave his angels eternal happiness unearned, yet required his other children to earn it; who gave his angels painless lives, yet cursed his other children with biting miseries and maladies of mind and body; who mouths justice and invented hell - mouths mercy and invented hell - mouths Golden Rules and forgiveness multiplied by seventy times seven, and invented hell; who mouths morals to other people and has none himself; who frowns upon crimes yet commits them all; who created man without invitation, then tries to shuffle the responsibility for man's acts upon man, instead of honorably placing it where it belongs, upon himself; and finally with altogether divine obtuseness, invites this poor abused slave to worship him."

And Twain's conclusion? He will not live with such a god, and so he finds,

> "...there is no God, no universe, no human race, no earthly life, no heaven, no hell. It is all a dream, - a grotesque and foolish dream."

I do not agree with Twain's conclusion because I do not believe that we can lay such things at god's doorstep.

It is possible, of course, that this is just the way god is, namely with a mean streak, uncaring, and so that is the way the universe is, but there is no necessity, as we have seen, that there be a god to begin it at all. And indeed it would be better for god's reputation if god were not involved in the chain of causality in any way, just as science supposes that god is not. If god is not at the start of everything, what, then, stands at the beginning of the chain of causality? None other than our old friend, chance, which by definition is not "reasonable."

John Leslie, "Universes" is a Neoplatonist and says that,

> "The notion that an ethical requirement or set of requirements *could itself be creatively effective* may supply a philosophically tidy answer to the question of `why there is something rather than nothing', i.e. why any person or thing is ever more than merely possible. And this question has no other answers which are in the least plausible, I think. Either something or other - a divine person creatively responsible for all things outside himself, or a universe or universes - just happens to exist, or else ethical requiredness is responsible for the matter. There are no further options." [p. 167]

I am sure that if I met Dr. Leslie he would eat me alive philosophically and I must admit that his book was a real struggle. He treats many of the questions I raise and I recommend it, though it is a tough go for a non-philosopher. However, his neoplatonic "ethical requirements" strike me as a replay of Anselm's Ontological Argument with a smattering of, "I cannot imagine it could be otherwise." He makes much of the strong Anthropic Principle that I do not believe should be given the weight. I must chose his option that both the universe, and perhaps many others, do "just happen to exist" and I believe that god also "just happens to exist." This gives the favored place to chance.

It could have happened otherwise than it did. It is certainly conceivable that the universe could have evolved in a manner different from what we experience. We are not too concerned if one star eats another or even if the core of the earth eats some of the mantle, as long as it does not work to our disadvantage. We are concerned that plants compete with each other for sunlight and for nutrients, that animals eat plants, that life preys on life, for this leads to Susan's husband preying on her. And it could be otherwise. Most plant cells exist on sunlight, not on predation. Cells can exist on the inanimate nutrients around them; they need not become consumers of one another in order to exist. The cells of our own bodies do not slaughter one another. The brain cells do not have to chase the skin cells through the veldt and break their little necks in order to get something to eat. The whole pattern of life on earth could be different than it is and not have the ingredients of aggression and predation. Nor are these needed in order to evolve intelligent life, though it may have taken a much longer time period. So why is the manner of making a living for one creature so often at the expense of another? Chance. No need for a god who likes violence, nor even for one who will just permit it when it could be stopped. The god who is said to be good but who puzzles us with a tolerance for "evil" need not be the god we believe in. We can believe in a good god (or of course we can believe in no god at all) and still account for all the causes without invoking a god. Our faith in an all-controlling god who is ultimately the cause of all that is and all that happens must give way to a faith in some other kind of god (or no god at all). We did say that this god seems rather impotent, but we will examine this question in the last chapter. The first item of faith that we encounter which must be open to change is our conception of god as the reason behind all that is, the controller of all processes, and the ultimate explanation for why things happen.

385. First and foremost, the faith says that there is a God. This "G"od is a personal being and not just some inanimate force in the world. No one goes to church to worship the First Cause or the Prime Mover, they go because they think they have, or wish to have, experienced a being with whom they can relate. The philosophers may be content with a non-personal god, but then even Thomas of Aquin wrote poetry and music celebrating the love of a parental god for the people. This god always existed, though Susan may well have some understanding that using "always" in regard to god is not accurate since god exists outside of time. Even to speak this way is hardly accurate since there is no "place" outside of time. Though some physical cosmologists may interpret this as a sign that there can be no god, the Christian merely accepts it as an attribute of god who is the creator of time. The attribute is known as Transcendence and places god outside of everything that is not god, meaning the entire universe for one thing. This does not mean that god is unrelating to other realities, but that god owes existence to nothing but godself. In fact, all the so-called "attributes of God" are recognized, in our more honest moments, as only analogies to things and qualities that we can recognize, but these concepts are wholly inadequate to explaining god.

Brian Swimme says something of the same about the universe,

> "I call the universe a green dragon to remind us that we will never be able to capture the universe with language...because the universe is a singularity! To speak, you need to compare things...but there is only *one* universe. We cannot compare the universe with anything." [Brian Swimme, "The Universe is a Green Dragon", Santa Fe: Bear & Company, 1984, p. 25] (I am not sure what he does with multiple

universes, but then we don't know about them and so cannot make comparisons.)

He does the same in "The Hidden Heart of the Cosmos", changing "quantum vacuum" to "all nourishing abyss" to avoid connotations from the normal use of vacuum.

[Brian Swimme, "The Hidden Heart of the Cosmos", New York: Orbis Books, 1996]

The value of this explaining by analogy,

> "...has been limited and negative, namely, to have enabled Christian thought to preserve a *faith* in God's transcendence while *reasoning* about him as if he were not transcendent...It has enabled Scholastic philosophy merely to acknowledge the inadequacy of its hellenic conception of God - or rather, to admit it without consciously acknowledging it." ["The Future of Belief", p. 178]

If one is going to say that god is transcendent, then one must also say that god is *not* transcendent, as well. If one is going to say that god exists, one must also say that god does not exist as we exist since existence is a quality of the being we know and only by analogy a quality of god. However, speaking in this manner is often the best we can do. There is also a tension between seeing god as transcendent and at the same time immanent in the world, but being present by choice is far different than owing one's existence to another. It has always been the faith that god is transcendent but out of love chooses to be enmeshed in god's creation and not an interested or uninterested bystander.

There is only one god, not many. Monotheism is the norm in the Western world, though it may not be the same god who is considered the only one. This was not a conclusion easily reached. There seem to be many powers in the world and it is easy to think of each one as distinct from the others. Each one may be divine and a force to be reckoned with. The storm is not the flood and neither is responsible for fertility of field or wife, a role often assigned to the moon.

> [See the treatment on the moon as a goddess of fertility in Diane Burton, "Many Moons", New York: Prentice Hall Press, 1991]

Of course, the god who looks over my city-state is not the same as yours, in fact mine can lick yours any day. During the period of Solomon's Temple, it is problematic whether Israel was a real monotheism, or just a practical one, with the worship of only one god being observed more in the breach than in reality. "The monotheism of Amos," for example, "was not a philosophical theory of the universe; Amos did not declare that there is and can only be one God. It was a practical monotheism reached apparently in consequence of the prophet's personal experience of the righteousness and power of Yahweh."

> [George A. Barton, "The Religion of Ancient Israel", New York: A.S. Barnes & Company, Inc., 1961, p. 95]

Certainly the worship of a god who was not an idol but was an imageless god was not restricted to them. Other peoples also knew that the object before them was not god and the god was often localized not by representation as a statue or stone pillar but the god's presence in an empty seat

> "...is indicated by the provision of a resting place for the god, such as a throne, pedestal or animal mount. The god does not reside in the object itself but, rather, exists only in close proximity to the object."
>
> [Review of the book "No Graven Image? Israelite Aniconism in its Ancient Near Eastern Context", reviewed by Victor Hurowitz in "Picturing Imageless Deities", *Biblical Archaeology Review*, Vol. 23, #3, p. 46]

This does not preclude belief in and worship of other gods. There is good evidence that even Yahweh may have had a consort, Asherah.

> [Larry G. Herr, "The Iron Age II Period: Emerging Nations" in "Biblical Archaeologist", Vol. 60, No. 3, September 1997, p. 114f. On page 128 we read,
>
> "By far the most famous artistic/religious item from the tenth century is the cult stand from a cultic room at Taanach. It includes four registers of religious symbolism, probably in alternating tiers of Yahweh and Asherah symbols. Asherah is represented by a nude figure and a tree of life nibbled by goats. Cherubim and a solar disk allude to Yahweh's presence. This stand most likely represents an Israelite theology that saw Yahweh (an El deity) possessing, like El, a consort named Asherah. She is by far the most frequently mentioned goddess in the Bible, at times invading the sanctity of the Jerusalem temple (2 Kgs 21:7). Although the popularity of this Yahweh/Asherah worship has been suppressed by later `orthodox' biblical writers and editors, archaeology indicates it was the most prominent form of Israelite worship."
>
>> [See also J. Glen Taylor, "Was Yahweh Worshiped as the Sun?", *Biblical Archaeology Review*, Vol. 20, No. 3, May/June 1994, p. 53]

Pamela Gaber with William G. Dever, "The Birth of Adonis?", *Archaeology Odyssey,* Vol. 1, No. 2, p. 49. points out that the Cypriot goddess, the "Wanasa" was known as Aphrodite to the Greeks and Venus to the Romans.

"For the Cypriots, the Wanasa was more than a fertility goddess: She was the supreme deity, mistress of all, creator of life and death, heaven and earth. She was all-powerful and all-knowing. And her worshipers were forbidden to create images of her; she could only be symbolized by a standing stone, called a beryl in the Cypriot tongue.

Both of these practices-aniconism (the prohibition against graven images) and the use of a standing stone (massabah in Hebrew)-are characteristic of Israelite cultic practices. So we are led to ask, Is there a connection between worship in ancient Cyprus and worship in ancient Israel? Again, the link is the Phoenicians. Popular religion among the Israelites as opposed to the 'official' religion promoted in the Hebrew Bible, especially the Book of Deuteronomy-was similar to Phoenician religion. The Bible presents a purified, elite monotheism devoted exclusively to the worship of Yahweh. The orthodox, nationalistic parties that produced the Hebrew Bible proscribed the worship of Ba'al and suppressed all but the faintest traces of a theology that included a consort of Yahweh. But both Ba'al and this female goddess continued to live on in Israelite popular religious practices as well as in Phoenician (formerly Canaanite) practice.

At Kition the Canaanite and Phoenician god Ba'al is paired with the goddess Astarte. In Israel of the Iron Age (12th-6th centuries B.C.), Ba'al is coupled with the Canaanite goddess Asherah rather than with Astarte. However, as we know from 14th/13th-century B.C. cuneiform tablets from the Canaanite city of Ugarit, on the Mediterranean coast of modem Syria, Astarte and Asherah (along with the goddess Anat) had very similar natures and roles; they were manifestations of the great Mother Goddess. These three fertility goddesses were frequently fused into a single goddess, as was apparently the case in ancient Israel. But this composite goddess was always paired with a male god.

The male-female pair was proscribed in official Israelite religion, but it lived on in popular Israelite religion, as we now know from recently discovered Hebrew inscriptions. In 1968 at Khirbet el-Qom 10 to 12 miles west of Hebron (probably biblical Makkadab),

American archaeologist William G. Dever found an eighth-century B.C. tomb inscription invoking the blessing `by Yahweh...and his Asherah.' In 1978 Israeli archaeologists excavating at another eighth-century B.C. site in the eastern Sinai desert - a combination fort-shrine at Kuntillet `Ajrud-found several Hebrew inscriptions mentioning Ba`al and El (the chief god of the Canaanite pantheon and, in the form `Elohim,' a name sometimes used to refer to God in the Hebrew Bible) along with the blessing formula `by Yahweh...and his Asherah.' Several scholars have argued that the term "Asherah' here cannot denote merely the tree - like symbol of the Mother Goddess, but must refer to the deity herself. In the Phoenician cult, paired deities were the rule. Ba`al's consort was Asherah. And Asherah was the consort of Yahweh in Israelite popular religion.
The word `asherah' occurs (disapprovingly, of course) about 40 times in the Hebrew Bible. Only a few of these references, however, designate the goddess herself. The verbs used in passages referring to the word `asherah' make it clear that what is being referred to is the goddess's *symbol* - either a living tree or a treelike wooden pole. An asherah was something that could be `made,' `erected,' `planted,' `cut down' or `burnt.' In Deuteronomy 16:21, for example, the Israelites are admonished, `You shall not plant any tree as an asherah beside the altar of the Lord.' In Judges 6:25, Gideon is told to `cut down the asherah that is beside' the altar of Ba`al."

See also the extended discussion of the worship of other gods in Israel and the existence of Asherah in Ephraim Stern, "Pagan Yahwism: The Folk religion of Ancient Israel", *Biblical Archaeology Review*, Vol. 27, No. 3, May/June 2001, p. 20]

However, by or during the time of the Babylonian Exile the Hebrews had been pretty much forced to conclude the existence of only one god. It was either this or admit that theirs was a rather weak divinity who had not been able to save them from defeat. Not willing to admit this, they denied the existence of the other gods and ascribed their condition of exile to the one god teaching them a lesson and purifying them by punishment. With the development of Christianity there was again a problem since some would suggest the existence of two gods, one being the Father and the other the Son, Jesus, and perhaps the Spirit as a third. These heresies were worked out in favor of a single divine being. In the Orient there is a tolerance for many gods and goddesses, and even in the West it is hard to see an identity between the god of the fundamentalists and the god of the liberals, but the Catholic position is certainly that there is one god who is responsible for everything.

There is only one God, but there are three persons in that God, the Father who is the creator of the universe, the Son or Logos or Word who would become Jesus and is the Redeemer, and the Holy Spirit, formerly known in English as the Holy Ghost, who has been a rather nebulous presence of God to most Catholics. Some prefer to enumerate the three by different names such as Parent-Friend-Lover, and this is perfectly legitimate since Father-Son-Spirit is itself an analogy and maybe not the best one. This arrangement is known as the Trinity of Persons. These three live in such perfect love and union with each other that they are in fact one being, which of course is a complete mystery to human beings. Susan may have been given examples of a three leafed clover which though three is one, and a triangle with its three sides to one figure, but it was always recognized that these were just analogies and did not give a real explanation. Since god loves so deeply that it can be said by John that "God is love" [1 John 4:16], we can by analogy apply to god what we see in a human who loves. One attribute we see in human love is the desire to communicate one's love to the other. One who loves from afar and never tells the other of their feelings is looked upon as tragically flawed. The lover hopes that the other will love in return, but this is not necessary to their own love, but the other must by all means know that they are loved. Love is seen then as desiring to communicate itself. The more one loves, the greater the number of recipients of that love. In god this love is the foundation of being. The love of Father and Son is the Spirit.

> [There is a long history of agreement and controversy in the manner in which the Persons of the Trinity are related to one another. For a recent treatment of the long-standing "*filioque*" argument see Albert Patfoort, "The *filioque* before the Council of Ephesus." *Theology Digest*, Vol. 46, No. 1, Spring 1999, p. 3 but read it for the flavor of the discussion unless you are a devotee of Trinitarian theology.]

God's love, being a perfect love, could not be only a love of oneself, even if there is a distinction of persons in god. God's love desired to communicate itself to other being as well. And so god set out to create other beings god could love, and who ideally would love god in return. Too often the return would be understood as a prerequisite for god's love being offered, but this would be an aberration of the true Judeo-Christian understanding. From the very first, and even before the very first, god loves us and prepares the universe for us.

386. We will take up in the final chapter what can be said of god as a personal being. What needs to be said here is that historically speaking we arrived at the idea of a human "person" by reference to each one being an image of a "person" in god. Historically the divine person was defined philosophically. Then the concept was extended to the human person. However, if are not created in the image of god, then our explanation of the value of a human "person," and our explanation of the nature of humans which insists upon a rational soul in the image of god will have to be reevaluated both because of the non-existence of an entity such as "soul" and because we are the image of the universe rather than god.

Henri de Lubac in his classic work, "The Mystery of the Supernatural" speaks of a human openness to the divine which is present because of our created status as rational, spiritual beings. In this we are separated from the rest of the universe by an immense gulf. See p. 133f where he says,

> "For there is nature and nature. If, in contrast with the supernatural order, the being of angels and men as resulting simply from their being created must be called natural, we must allow that their situation, in relation to other natures, is `singular and paradoxical'; for it is the situation `of a spirit which is to become subject and agent of an act of knowledge for which it has no natural equipment, and which is thus to be fulfilled by getting beyond itself'. If, then, there is a human nature and an angelic nature, we cannot use the terms wholly in the sense in which we speak of animal nature, for instance, or cosmic nature. If every created spirit, before being a thinking spirit, is itself `nature', if, before even being thinking, `it is spiritual nature', then it must also be recognized that, in another sense, spirit is in contrast with `nature'. Even in the terminology of the scholastics, and of St Thomas especially, *natura rationalis* or *creatura rationalis* is not a *res naturalis*. Spiritual beings cannot be confounded with beings known simply as `natural beings': `naturalia', `entia naturalia, `res naturales', `causae naturales', `creaturae naturales', `corpora naturalia', `formae naturales'. These `natural beings' are in effect those `which their nature condemns to be no more than what they are...Such beings, so to say, `are existed'

> (*sont existes*) more than they exist'. The spiritual being, on the other hand, `is not completely defined by its natural form; it transcends it in a certain way...Its nature is not simply to be nature in the sense of finite, determinate, particularized nature...' For it is endowed with what St. Augustine calls *mens*, or *principale mentis*, or in St. Paul's phrase, *spiritus mentis*. [Cites Augustine's *De Trinitate*.]"

De Lubac presumes the presence of a rational, spiritual element setting humans apart from the rest of the universe. He cites another theologian, called only "Berulle", as saying,

> "...There are two sorts of creatures, one spiritual and rational, the other natural...The natural have not so much a likeness as a trace of the Creator imprinted upon them; the spiritual and rational have that likeness, and beyond their natural being...they can approach and be united with God."

De Lubac continues,

> "Hence every spirit, whether in a body or not, enjoys certain privileges which, making him `in the image of' the Creator, make him at the same time superior to the whole order of the universe. This is what another spiritual writer, the Capuchin Constantine of Barbanson, careful of scholastic exactitude, called `the separation of spirit from the completed nature.'
>
> There is no point in continuing to stress a distinction as simple and obvious as this, and one which is verified on the most immediately empirical level...It was universal in Christian antiquity. [Cites Clement of Alexandria saying, `The domain of nature is plants, seeds, trees and stones'.] It is based upon scripture: Pere Festugiere has shown this for the idea of the `pneuma', and Karl Barth could write in his *Church Dogmatics*: `...in the OT and NT man is addressed as a natural being also, but still as this distinct natural being, a natural being distinguished by spirit'." This is the truth Cassiodorus was summing up when he gave his two definitions of the soul, as *species naturalis* and *spiritalis propriaque substantia*. We find it everywhere, in varying forms, in the works of the great scholastics, especially in St Bonaventure who distinguishes, among the `naturalia', the `bruta' from the `rationalia', and St Thomas who contrasts `res naturales' with `res humanae', and `naturalis' or `creaturae naturales' with `natura humana'. [Cites Summa contra Gentiles]"

Needless to say, all this thinking, distinguishing and theologizing will have to be redone, if anyone cares anymore.

387. To say that this god "always existed" may in fact make more sense because of our understanding of time as an attribute of the universe rather than a free-standing entity. We have learned that to ask the question, "what happened before the Big Bang?" is a nonsense question. There can be no "before" when nothing of the universe exists. Just how a being can exist in which there is no component of time is unknowable to us who are so immersed in it, but the possibility is certainly there. If there are many universes, each completely independent of one another, is there perhaps some "time outside of time" by which the duration between the coming-to-be of one universe could be related to the coming-to-be of another? The idea of "always" seems to presume that there is, and that god knows how many divisions of time (seconds, hours, kiloparsecs, etc) occurred between the "time" that our universe came into existence and when another did. This is not what the faith has meant to say about god "always existing". It has meant to say that time is not a factor for god. Nothing in our universe can exist without a relationship to others that can be measured in time, or more precisely space-time, but this says nothing about whether "something" can or cannot exist without such a relationship.

388. This is a "transcendence" which is absolute, but which does not preclude involvement of some kind in the existence of other entities that do exist in time. These concepts of transcendence and immanence need not be excluded because of any of the findings of science, and a faith which attributes them to its god will probably not have to change its thinking much here, except that the reason for god's immanence is understood as entirely by choice and not required by our purported contingency on god's sustaining us in existence.

389. The first avenue of revelation has traditionally been found in creation itself. The "proofs" of god's existence have been elucidated in this way for centuries and perhaps millennia. Just as Augustine could invite his reader to consider the beauty of the world around them and realize the greater beauty of him who made them, [Augustine, "*Sermo* 241" 2:PL 38, 1134] so Thomas Aquinas could find other pointers to the creator in the order of the world: its constant change; the inability of an object to put itself into motion, yet to be in motion; and other natural phenomena. One might say that if you begin with the assumption that there is a creator who wishes to be known,

then you may well be struck by these seemingly self-evident proofs. For Aquinas,

> "the notion of creation determines and characterizes the interior structure of nearly all the basic concepts in [his] philosophy of Being...[but] it is scarcely ever put forward explicitly; it belongs to the unexpressed in St. Thomas's doctrine of being."
>
> [Josef Pieper, "The Silence of St. Thomas", Trans. by John Murray and Daniel O'Connor, Chicago: Henry Regnery Co., 1957, p. 48]
>
> [One can read Aquinas, of course, but it is notoriously tough going. Reading about his thought in something like Frederick Copleston, S.J., "A History of Philosophy", Vol. II, "Mediaeval Philosophy", Part II, Garden City: Image Books, 1950 gives a good overview. At the same time it gives an idea of the thinking of philosophers in the Catholic Church since Copleston was used in many seminaries for decades and is still a recognized source.]

Thomas assumes that being a creature is basic to the essence of all that is except the creator. Indeed, "Things are good - *all* things. The most compelling proof of their goodness in the very act of being lies in their createdness...What is, is good, because it was created by God."

> [Josef Pieper, "Guide to Thomas Aquinas" Trans. by Richard & Clara Winston, New York: New American Library A Mentor-Omega Book, 1962, p. 116]

Certainly many a Christian finds it hard to believe that an atheist could look at a fly's eye or the Grand Canyon and not see the hand of the creator, but they *begin* with this assumption. Indeed there may well be something within the human being that is led to this line of thinking. It certainly seems to be the basis for much of primitive religious experience and surely someone somewhere in the distant past must have approached the world and its beauty without that presumption having been taught them by others and yet reached this conclusion. Richard Dawkins attributes this openness to this kind of "proof" to the presence of "memes", a cultural equivalent of genes that confer some survival value on the group that contains it. In our readiness to embrace a religious faith this value would perhaps be in preventing suicide and encouraging altruism among the clan. This would preserve not only the "meme" of the kinship group but the commonly shared genes as well.

> [Richard Dawkins, "The Selfish Gene". See the last chapter of his book.]

This seems to be the same presumption preserved in the culture. Edward Wilson in "On Human Nature" attributes religious feelings to an even more biological root, the genes themselves, and also sees them as adaptive for the transmission of the genes to future generations since there is purpose and meaning in life provided by the purpose of god which one is predisposed to believe in.

> [Edward O. Wilson, "On Human Nature", Cambridge, Mass: Harvard University Press, 1978.]

The Church, too, recognizes that there is something innate in humans predisposing them to belief, though it does not mention anything genetic and really prefers to explain it as a gift from god, who wills to "give (a person) the grace of being able to welcome this revelation in faith." ["Catechism of the Catholic Church", #35] It thereby recognizes something inherent in human makeup that disposes us to interpret the world around us as a product of the action of some superior being, and even calls those who would reject this conclusion culpable in some degree. Yet it also knows how little proof there really is here and calls upon god to gift us with the ability to "welcome" these hints from creation. The argument from order seems to be the most popular of the "proofs" among people of faith and science claiming that god can be known through creation. The argument by William Paley is that if you have something so ordered that it could not be accidental, then it must have a maker. The limitations of this manner of knowing god from creation have always been recognized. Even though Paul is not so willing to excuse ignorance of god because of the paltry evidence of creation, saying of those who would refuse to see the truth revealed in the world around them, "that is why such people are without excuse…they made nonsense out of logic and their empty minds were darkened.", [Romans 1:20-21] the author of Acts takes a more sanguine view and has Paul say, "God overlooked that sort of thing when people were ignorant", [Acts 17:30] recognizing that while god can be known in this manner, god cannot be known very well. The New Catechism upholds the belief that creation is a source of knowledge of god, though it recognizes the difficulty of the enterprise when it states

> "Our human understanding, which shares in the light of the divine intellect, can understand what God tells us by means of his creation, though not without great effort and only in a spirit of humility and respect before the Creator and his work."
>
> > ["Catechism of the Catholic Church", #299. See also #286f]

Our own "being" or "human nature" has also been recognized as an avenue of knowledge of god. It would be, the argument maintains, a cruel joke of nature to place within us a longing for meaning in life, for life without end, for some ultimate ground of being in which we could be rooted and then not have any way of fulfilling these wishes. If we have them, then surely there is some way of attaining them, which of course begs the question of whether or not these feelings are from the creator or from evolution. So, the argument goes, if they are beyond our own capability, there must be some being who can grant them to us. Besides this, these feelings and emotions seem so immaterial. Certainly, they are not shared with the rocks that are content to lie about completely insensate and without any fulfillment. These feelings are shared only minimally by the beasts, if at all. There must be something within us that transcends the material world around us. How much more does this exist in the author of love, of purpose, of meaning? God then can be known, at least by analogy, by knowing ourselves. Just as the rest of creation reflects the attributes of god, so our own composition partakes of that of god. We are made "in the image of god" and though the Scripture scholars tell us that this can be variously interpreted, the most common understanding is that we have some immaterial component of our being, usually called the "soul", which is not the product of physical evolution but is something akin to the very substance of god, though the Church is very careful to say this is only by analogy and that we do not really have a share in the divine essence. We are not god, nor are we a piece of god. We have a human essence, not the essence of god. Certainly we humans do feel that we are different from the rest of the universe around us. In speaking of cloning, a young person was quoted as worrying whether a clone would even have a soul. "How do you know, like, *what's* walking down the street? It's like out of a Stephen King book...You can't recreate that." [Parade Magazine, July 27, 1997 p. 17] If we are different, surely god must be different in an analogous way, and so by seeing how we are made we can get some idea of the attributes of god. This knowledge will still be only fragmentary and open to much misinterpretation. It is a limited knowledge at best.

And yet, the faith maintains, we have not been left to these avenues of knowing the one who loves us. God has chosen to reveal godself much more directly.

> "Man's faculties make him capable of coming to a knowledge of the existence of a personal God. But for man to be able to enter into

real intimacy with him, God willed both to reveal himself to man and to give him the grace of being able to welcome the revelation in faith." ["Catechism of the Catholic Church", #35]

God did this by entering into a covenant with humanity, beginning in the garden of Eden and continued after the great flood, neither of which must be taken literally, but often are. The theme of Covenant was presumably written back into the oral history of the people of Israel and is found in the covenant with Abraham and his progeny. Its presence in the story of the Exodus is problematic as well. Would Abraham or any of his descendants leaving Egypt centuries later have understood that god had made a covenant with them? Would they have even known the term? Perhaps. Perhaps not. The jury is still out on this.

[See the article "God as Divine Kinsman: What Covenant Meant in Ancient Israel", "*Biblical Archaeology Review*", Vol. 25, No. 4, July/August 1999, p. 32 and the book to which it refers, Frank Moore Cross, "From Epic to Canon", Baltimore: Johns Hopkins Univ. Press, 1998.]

Certainly the members of the dynasty of David, if there was such an individual, would have believed in a covenant made with their house and would have known the word "Covenant", and those who went before them did feel that this god was their very own god who had cast in god's lot with them, whether they used covenant terminology or not.

[There is a controversy among the so called "minimalists" and "maximalists" about the historicity of the Bible.

Philip R. Davies promoted the claim that, for instance, there was no historical David, or Solomon either, but that David is like King Arthur in English legends. William G. Dever responds to the revisionist position in *Near Eastern Archaeology*, March 1998, Vol 61, No. 1 in his article, "Archaeology, Ideology, and the Quest for an `Ancient' or `Biblical' Israel".

For another example, see two articles in *Biblical Archaeology Review,* July/August 1998, Vol. 24, No. 4, one titled "It's Not There" which claims that the archeology of Jerusalem shows that there was not a city for David to conquer. Another, titled "It's There" which claims that the evidence is for the existence of a city on the site at the time of David.

P. Kyle McCarter in response to Richard Friedman, "A Long-Hidden Book in the Bible?", *Bible Review*, Vol. XV, No. 2, April 1999, p. 39 certainly argues for a historical David.

Hershel Shanks, "Has David Been Found in Egypt", *Biblical Archaeology Review*, Vol. 25, No. 1, January/February 1999, p. 34.
Gabriel Barkay, "What's an Egyptian Temple Doing in Jerusalem", *Biblical Archaeology Review* Vol. 26, No. 3, p. 48 presents evidence for the settlement of Jerusalem prior to David.
Biblical Archaeological Review, Vol. 23, No. 4, July/August 1997, p. 26 contains an article, "Face to Face: Biblical Minimalists meet their challengers"
The discussion is ongoing. Refer to these magazines or the books of the scholars involved.]

In this covenant relationship, it is believed, god revealed godself to them. Erich Fromm points out the pivotal nature of the concept of covenant.

> "The idea of the covenant constitutes, indeed, one of the most decisive steps in the religious development of Judaism, a step which prepares the way to the concept of the complete freedom of man, even freedom from God. With the conclusion of the covenant, God ceases to be the absolute ruler. He and man have become partners in a treaty."
>
> [Erich Fromm, "You Shall Be As Gods" Greenwich: Fawcett Publications, Inc. 1966, p. 23

Beginning on a primitive level of one god among many, and that god a rather arbitrary one, they would eventually progress to believe only in their god, a god who entered into a relationship with them and with all of humanity, indeed with all life. And they would reach this conclusion not because of human logic but, they felt, because god was constantly revealing godself through actions recorded in their history. "Remember the wonders the Lord has done" would be a constant refrain. And if god has done these things in the past, surely god will do so in the present, and in the future as well. This god then is much more than a creator, and even more than one who sustains everything in existence by an act of will. This god is a god who cares, and more than this, is willing to get involved in working things out for god's people. The Hebrew god, and the Christian god, is a god who is creative throughout history, not just once and for all.

> "The faith of Israel was radically different from the religious beliefs of the other peoples of the ancient Near East precisely in that it was concerned with events of *history*."
>
> [James Plastaras, C.M., "The God of Exodus", Milwaukee: The Bruce Publishing Company, 1966, p. 2]

God creates a people at Sinai, and even before in the time of the patriarchs. God creates them anew in returning them from exile in Babylon, (though some feel that the latter is actually the model of creative activity that is read backwards into their history to the Exodus). God makes all things work to their advantage, even those events that at first seemed destructive of their identity and their being. The slavery in Egypt, far from reducing them or ultimately destroying them, makes them grow in numbers and leads to their constitution as a coherent political unity, first as tribes and then as a kingdom. The total defeat by foreign powers of the dynasty in Samaria and then the house of David in Jerusalem, rather than removing them from the world stage, recreates them by god's action working through Cyrus who allows them to return home, renewed in their faith, now a faith in a monotheistic god. This god is willing not only to make the normal course of physical events work to their advantage, but even to suspend this sequence; willing to work miracles for them. While the plagues of Egypt may well have natural explanations, the whole point is that they are works of god apart from the usual workings of nature. The stopping of the Jordan River so that they could cross dry-shod is done by god as a deliberate act of creation, not a sustaining of normal events. Surely they knew then, as biblical scholars know now, that landslides can occur on the Jordan which stop or even reverse the flow until the blockage is overtopped. Nevertheless, they did not marvel at god's timing, they marveled at the wonder. What god was willing and able to do for the nation, god was able to do for the individuals in the nation as well. All this manipulation and interference in nature and history must be rethought.

390. Richard Conniff, "The Natural History of Art", *Discover*, Vol. 20, No. 11, November 1999, p. 94f, says

> "...biologists believe that ancient history - the rise and fall of different body types, survival strategies, and instincts for habitat - has a way of shaping the DNA, where it becomes a sort of ghostly puppet master."

One way for evolution to move the individuals to good survival strategy is to make the object attractive. Why do we find something beautiful? Partly, and possibly mostly, because it serves an evolutionary end. The landscapes we find beautiful are precisely those which contain what we need to survive, adequate food, water, shelter, a view of approaching enemies. "Most animals appear to be biologically prepared to recognize highly specific indicators of good habitat," says Conniff.

Why do we insist on lawns, even in the desert, when they are so expensive and time and water consuming? Because we find them beautiful. And why would this be? Maybe because they resemble the savannah of our African ancestors?

391 God, then, sets out to create human beings. Of course, today we understand that the process of evolution is involved in this creative action and that human beings could look very different than they do. God did not necessarily have a particular form in mind, and there even may be creatures on other planets in other solar systems who look like a squid. What was required for god's purposes was that the being be rational and conscious and able to understand god's message and profession of love. The universe was created as a home for these beings, a support system for god's beloved. Yes, god has some feelings for animals and as one scientist pointed out, must have a special fondness for beetles since god created so many of them.

> [Stephen J. Gould, "Dinosaur in a Haystack", New York: Harmony Books, 1995, p. 377f, goes into the origin of this possibly apocryphal story.]

It was all done for the sake of those beings capable of receiving and perhaps responding to god's love. Of course in its crassest form this understanding has led repeatedly to exploitation and misuse of the rest of creation, but the ecology movement has worked long and hard to build a case for respect for all of god's creation. The belief that we are the masters and the rest the servants has not made this idea easy to convey.

While the purpose of creation was said to be to provide other beings who could know god, culminating in humans, this also leads to the conclusion that one can know god through the creation. Cyril of Alexandria long ago (died 444 a.d.) took his opponents to task for not recognizing god in god's works,

> "...they who ought to have been struck with amazement on beholding the vaultings of the heavens; they, who ought to have worshipped him who reared the sky as a dome, who out of the fluid nature of the waters formed the stable substance of the heavens...and whatever thou canst thyself discover or conceive, `from the greatness and beauty of the creatures mayest proportionably see the maker of them'."

[Saint Cyril, "On the Maker of Heaven and Earth" in "Orations from Homer to William McKinley" Volume III, Ed. by Mayo W. Hazeltine, A.M., New York: P.F. Collier and Son, 1902, p. 1199f]

The universe is created through the word of god and thus, like any word, it communicates something of the speaker. The Second Vatican Council says, "God, who through the word creates all things and keeps them in existence, gives men an enduring witness to himself in created realities." ["*Dogmatic Constitution on Divine Revelation*" chapter 1] Paul claims that the Gentiles are liable to judgment because they have not seen the evidences of god that are present in creation. There are, according to Aquinas, any number of things that we can postulate, albeit analogously, to god because that quality shows up in nature. True, some dispute that this process produces any knowledge of god. Karl Barth,

> "...rejected wholeheartedly any theological `analogy of being' which would have allowed human beings to reason from nature to God in the way Aquinas did." [Angela Tilby, "Soul", p. 212.]

Some of Barth's reason for rejecting the possibility of knowledge of god from nature came from his insistence on god's transcendence, and some came from opposition to the anti-Semitism of his times in Germany, believing, "that such a `natural theology' was the root of the religious syncretism and anti-Semitism of the `German Christians' - those who supported Hitler's national socialism."

> [Mark Galli, "Karl Barth", in "*Christian History*", Vol. XIX, No. 1, Issue 65, p. 23]

Karl Rahner admits that the attempt is suspect in the light of modern science, especially since most of us "...do not face a nature that God has made but one that we ourselves have made."

> [Rahner, "Do You Believe In God", p. 46]

Yet for all our understanding that no god is needed to keep the moon in its orbit, and for all our appropriation of god's powers in our ability to clone life and perhaps one day to create it in a test tube, it is the rare Christian and Catholic that does not feel that the universe is not only evidence of the goodness of the creator and a source of information about who god is, but a proof for god's existence.

Susan may well not put any credence into the literal interpretation of the Creation stories of Genesis, though there is many a fundamentalist who does. Whether it took seven days or millions of years, it is believed deep down that it was all done for the purpose of providing a home for rational beings. Humans are considered the goal of all god's creative activity. All of creation is pointed at producing this result, though why god messed around with dinosaurs for so many millions of years is a bit of a mystery to Susan. Maybe to give paleontologists something to do or to amuse the children, or perhaps just to show off god's power and imagination. Certainly today there would be an understanding that species have evolved and that they did not drop ready-made from the hand of god just as they are. Yet, there would also be enough exposure to the biblical stories of creation and Noah's ark to put some doubt in the evolutionary scenario, or at least to say that they are describing the same phenomena in different terms.

> See Howard J. Van Till, "The Fourth Day", Grand Rapids: William B. Eerdmans Publishing Company, 1986
> And see Shroeder, "Genesis and the Big Bang", for attempts to reconcile the biblical and scientific picture.
> For a description of efforts to find the ark and reports that it had been found, see Christopher P. Tourney, "Who's Seen Noah's Ark?", *Natural History*, Vol. 106, No. 9, October 1997, p. 14]

There is still a feeling that things have always been pretty much as they are. A recent study shows that 45% of Americans think that god created humans very much as we are and very recently. This in contrast to 7% of Britons. Only 10% of Americans accept the Darwinian version of human evolution over millions of years. Carl Sagan reports that,

> "I meet many people offended by evolution, who passionately prefer to be the personal handicraft of god than to arise by blind physical and chemical forces over aeons from slime. They also tend to be less than assiduous in exposing themselves to the evidence"
> ["Demon Haunted World, p. 325]

When pressed many Catholics today will say that if the scientists say that there have been billions of years of development, that need not be a hindrance to one's faith. It was not so long ago, however, that the Church rather officially believed that species were stable, as did the rest of society.

> "Their individual members came and went, but the species itself remained unchanged and unchangeable...Essences were unchanging, and a thing couldn't change its essence, and new essences couldn't be born-except of course by God's command in episodes of special creation. Reptiles could no more *turn into* birds than copper could turn into gold."
> [Daniel C. Dennett, "Darwin's Dangerous Idea", New York: Simon & Schuster, 1995, p.37-38

For an extended study of the idea of unchanging essences, see "Philosophy in the Flesh", chapter 16f]

There are, of course, people today who will insist on the biblical picture. The same is true for the appearance of humanity. The Catechism continues to insist that, "God created everything for man..." ["Catechism of the Catholic Church", #358]

While Susan, and the catechism, admits to the evolution of *Homo sapiens* from primate ancestors, she will do it grudgingly and would much prefer to believe in the special creation of human beings apart from the animals. And when we turn to the "higher powers" of mind and will, it is certainly the understanding of the Church that there is some other principle than the material at work to endow human beings with these capabilities.

> "By virtue of his soul and his spiritual powers of intellect and will, man is endowed with freedom, an `outstanding manifestation of the divine image'."
> ["Catechism of the Catholic Church", #1704]

This soul is the direct work of the creator in a special action and not subject to the processes of evolution.

> "The Church teaches that every spiritual soul is created immediately by God-it is not `produced' by the parents..." ["Catechism of the Catholic Church", #365] so is not subject to the process of evolution.

The Catholic Answer, Vol. 12, No. 4, September/October 1998, p. 19, responds to a question about the Church's position on evolution by pointing out that a literal interpretation of the creation accounts is not required by the Faith. However, it claims that there are some things that are required and it lists these:

> "(a) The entire creative process began with Almighty God. (b) At a certain point in evolution, God took what He considered to be `apt' matter and `breathed life' (metaphorical language) into that being, thus endowing it with an immortal soul. (c) God sustains all creation until the end of time. (d) God is directly, personally involved in the creation of every single human being since our first parents (by giving each and every one an immortal soul). If any scientific theory casts doubt on any of these teachings, such a theory cannot be held by a Catholic."

Does not the Scripture of the Christian faith present god as a creator god, and therefore responsible for what is created and how it works? Do the findings of modern science leave any room for an acceptance of this role for god? If not, can we still claim that the Scriptures are true in any sense? Of course, the fundamentalist will claim, a priori, that the Scripture is true and accurate and anything that presents a different view must be wrong. Janet Raloff, "When Science and Beliefs Collide", "*Science News*", Vol. 149, No., June 8, 1996 reports that,

> "...in [an] ongoing analyses of nearly 40 local religious congregations, roughly half of them spiritual home to Christian fundamentalists. For these fundamentalists...the way to interpret the world 'is to quote the appropriate chapter and verse in the Bible,' rather than to form hypotheses and test them." [p. 360]

The study also shows that believers in witchcraft (Wiccans) and what is being called "Postmodernism" also reject much of science because of their manner of viewing the world.

It is very hard, probably impossible, and certainly counter-productive to try to argue against this position so I will not even try. Those who hold it are not even consistent with their own position. If they were they would observe Saturday over Sunday as the Sabbath and would be missing a lot of hands and eyes. (Matthew 5:29; Matthew 18:9; Mark 9:47) It would seem that the extreme of this fundamentalist position would die out since it is so patently in error, even prior to factoring in the findings of archeology and the other sciences, but it has the advantage of being clear cut and very often consoling, since the faithful of this sort are among the one hundred forty-four thousand who will be saved and are, not coincidentally, spared a lot of bothersome decision-making. So the David Koresh's and Jim Jones's will probably always have some following.

It is a mystery to me how some groups can so consistently be wrong in their predictions of the "end of the world" or the "coming of the rapture" or the "seventh day." In the last century the Millerites (or Muellerites) in the northeast planted no crops or at least did not harvest their crop in anticipation of the end time and were crestfallen when it did not occur on the expected date. And then did it all again a year later. Contrary to what we would expect, such groups often become even more earnest in their faith after such a failure. My own conjecture is that not to do so would prove them a fool instead of just mistaken in their calculations.

Most Christians are not extremists and try only to preserve some place for some kind of inerrancy in Scripture and are willing to re-interpret their understanding of the Bible in the light of the findings of science. How much room can be left for them to view god in traditional terms? It might seem "not much" will have to be the answer since the Bible is based so strongly on god as a god of history and a god intimately involved in the physical world, especially in the manipulation of human history. "Hielsgeschicte", a German word for the concept of god working in the universe to such an extent that all history can be viewed as salvation history, and the special intervention in history known as the "incarnation", are two fundamental concepts in Christian faith. Can we leave any room for god to act in this manner? Indeed the Bible's first words are about god's role in bringing about the universe and it makes some strong statements that science seems to contradict. Certainly the time frames, the presumed creation of each species unique to itself, the special creation of humans, the watery chaos that somehow precedes it all and threatens to engulf the ordered creation, the existence of a privileged human state in the garden, Noah and the Deluge, and many other incidents in the story can no longer be taken as historical events just as written.

Reported in *Biblical Archaeology Review*, Vol. 23, No. 2, March/April 1997, p. 10,

"...7,500 years ago sea levels rose dramatically and salt water cascaded through the narrow Bosporos strait into the Black Sea. [There]..water levels rose about 500 feet, perhaps in only a matter of months...Scholars have long noticed similarities between the Biblical flood story and an earlier flood story in the Gilgamesh epic (probably written down in the early second millennium B.C.)...The similarities may suggest either that Noah's story was modeled after Gilgamesh or that both stories derive from something else; they may even preserve memories of an actual event. If the Black Sea deluge was that event, the memories stretched back some 3,500 years."

See also Philip and Phylis Morrison, "Noah's Flood?" in *Scientific American*, February 1999, p. 105 reporting on the same phenomenon.

See William Ryan & Walter Pitman, "Noah's Flood", Simon and Schuster and a review of heir book by James Trefil, "Evidence for a Flood", *Smithsonian*, Vol. 31, No. 1, April 2000, p. 18

The question of where Cain would find a wife (unless he married a much younger sister); who were the people that Cain was afraid of (maybe his much younger sister?); and other inconsistencies in the received story have always bothered believers and been the cause of derision among non-believers. Many, of course, simply abandon the Bible because of these contradictions and put it down as so much myth. Others investigate the meaning of "myth" and find ways to show the Bible's relevance if only it is understood for what it is and not as a science book. Ian Barbour notes that,

> "In the twentieth century, the Roman Catholic church and most of the mainline Protestant denominations have held that scripture is the human witness to the primary revelation, which occurred in the lives of the prophets and the life and person of Christ." Even "many traditionalists and evangelicals insist on the centrality of Christ without insisting on the infallibility of a literal interpretation of the Bible." [Ian Barbour, "Religion in an Age of Science", p. 9]

Some qualified scientists, and many unqualified, have tried to give the Bible room by commendably taking science seriously yet showing that there are no contradictions. Gerald L. Schroeder is one of these respected scientists and he claims he can show,

> "...that the duration and events of the billions of years that, according to cosmologists, have followed the Big Bang and those events of the first six days of Genesis are in fact one and the same. *They are identical realities that have been described in vastly different terms.* (his italics)" ["Genesis and the Big Bang", p. 26]

Even the six days of creation are reconciled with fifteen billion years of evolution, but of course, he says, one must look at it differently. Not that one of these "days" equals billions years as some have claimed. No, his argument goes to the scientific: "Einstein demonstrated that when a single event is viewed from two frames of reference, a thousand or even a billion years in one can indeed pass for days in the other."

["Genesis and the Big Bang", p. 34.]

In other words, he uses relativity to uphold the literal interpretation of the Genesis story. This is all very clever but not acceptable to most scientists, or even to most scripture scholars, who would prefer to say that the words of Genesis were never intended, even by god, to give a scientific account of creation. Trying to reconcile them is a favorite pastime of fundamentalist scientists and of course, no one can say their conclusions are false. God *could* have put the story in some kind of code, for example, to be discovered later when we became scientifically literate, just as god *could* have encoded the future in the Bible, spelling it out in every eighteenth letter, but this is a pretty tricky god and not the straightforward being encountered by Moses and Jeremiah.

[Michael Drosnin in "The Bible Code" claims that this is just what god did. Most commentators and reviewers are not convinced.
See David van Biema, "Deciphering God's Plan", *Time*, June 9, 1997.
See Shlomo Sternberg, "Snake Oil for Sale", *Bible Review*, Vol. XIII, No. 4, August 1997, p. 24 who concludes that the, "...method can produce any desired `hidden message' in any sufficiently long text."
See Ronald S. Hendel, "The Bible Code: Cracked and Crumbling", in the same issue with Sternberg.]

[Another phenomenon that is sometimes taken as a sign of divine authorship of the Scripture on a letter by letter basis is to choose a word in the first verse of Genesis 1 in the King James version. Count the letters. Move forward that many words. Count the letters in that word. Go forward that many words and continue doing this until you get into the third verse. Your count will always end on the word, "God", no matter which word in the first verse you begin with. Of course, the fact that there is a convergence in any sufficiently long text and that this is known as a "Kruskal Count" will not matter if you consider this a proof of divine activity.]

Obviously the most honest response will be either to show where the Bible is correct and how some finding of science is wrong or, more likely, to take another look at just what the Bible is in the light of our new knowledge. This can be and usually is an honest quest for truth and we have obviously misinterpreted the Scripture on many occasions, something that must be seriously faced and corrected. However, it can also be self-serving as when John Henry Newman says that if something of the faith seems to be at odds with science then one of three things must be true: the science is mistaken; the science does not really contradict the faith; or what science seems to be denying really never was revealed as true in the first place, we just thought it was. In this way the Faith can never be in error, but it is not a very honest position.

[John Henry Newman, "The Idea of a University", New York: American Press, 1941, p. 445.
See van Beeck, "God Encountered", p. 100f for a good treatment of Newman in context with other Catholic apologists]

This process is not new and one can refer to any number of treatises on the interpretation of Scripture in the light of form criticism, archaeology, paleontology, and many other disciplines. Our knowledge of what we are reading in the Bible has been greatly enhanced by re-examination in the light of modern findings and there is no need to insist on any hidden revelation of scientific truth in order to profitably read the Bible, including the account in Genesis.

392. Who can dispute that the universe reveals not only the existence of god, but the attributes of god as well? Certainly every early natural philosopher prior to and including Isaac Newton, and quite a few following him, including perhaps Darwin himself, at least enough to please his wife, have studied the universe not only for its own sake but also to learn about its creator. Indeed, many studied it only for this reason.

> "Making the natural world a fit object for religious contemplation was the primary goal of scholarship [in the Middle Ages], and making the theories fit observations was only tangentially related to that mission." ["Reading the Mind of God" p. 34]

Newton was deeply entrenched in alchemy, at least partly in response to Descartes' picture of the universe as a machine, perhaps made by god but now left to its own devices. For alchemists, "the alchemical world was more like a plant or animal than a machine" and so it left more room for divinity.

> [See Jennifer Lee Carrell, "Newton's Vice", *Smithsonian*, Vol. 31, No. 9, December 2000, p.131]

The 16th Century Belgic Confessions makes clear the common thought that god could be known through two "Books", the book of the Bible and the book of Nature. [Belgic Confession (1561), Art. 2] Gradually over the centuries, abetted by an increasing laicization of scholarship and secularization of society, this motive of understanding god from god's creation lessened. Scientists began insisting that their field concerned only things that could be sensed and probed, and of course the tenets of faith often could not be so treated.

> [See the article on "Simon Newcomb: Astronomer with an Attitude", *Scientific American,* Vol. 279, No. 4, October 1998, p. 88f which notes that,
>
> "In his 1878 address as retiring president of the AAAS, he discussed the relation between natural science and the Christian religion,

> calling for the separation of scientific reasoning and theological arguments...Newcomb used the scientific method as a criterion by which to differentiate scientific knowledge from religious belief, portraying the two as distinct but complementary...He placed the scientific use of language at the center of his methodological argument, insisting that true science restricts itself to terms that `have exact literal meanings, and refer only to things which admit of being perceived by the senses, or, at least, of being conceived as thus perceptible.' Traditional natural theology failed to meet this methodological test and therefore should be divorced from modern science, he declared."

The idea of understanding god from nature was also diminished when it became apparent that not much of god's motives could be discerned from nature. True, it seemed, there is much beauty in the world, so the creator must be beautiful; there is order so the creator must be orderly and regular, a regular mathematician in fact; and life being so ubiquitous in the part of the universe we inhabit, the creator must have a special fondness for living beings. St. Augustine phrased the thought prayerfully, "You, Lord, created heaven and earth. They are beautiful because You are beauty. They are good because You are goodness. They exist because You are existence".

[Confessions 11,4]

However, there is also so much destruction and randomness and what does that say about the creator? A modern editor comments,

> "Its easy to see why some...early physicists who first penetrated the reality of the nonintersecting, coplanar orbits of the planets thought that the hand of a creator was discernible. They were unable to conceive of any alternative hypothesis that could account for such magnificent precision and order...But in the light of modern understanding, there is no sign of divine guidance here, or at least nothing beyond physics and chemistry. Instead we see evidence of a time of remorseless and sustained violence, when vastly more worlds were destroyed than preserved." [*Omni*, Vol 15 #1 October 19 92, p. 48.]

Nevertheless, even in the time of the 1st Vatican Council in the last part of the 19th Century, the assembled bishops,

> "emphatically stated the conviction of the Roman Catholic tradition that there is a form of natural revelation mediated through the created cosmos...It is possible to see the work of the sciences as a

way of learning to read the book of the cosmos which...is at least a potential revelation of the divine."

[Zachary Hayes, *New Theological Review,* August 1995, p. 7]

This view was not restricted to the Church but influenced the popularization of science at the time because, "educators hoped that boning up on birds [and] bugs...Would not only teach the kiddies some natural science but also instill in them an appreciation of nature's beauties, a respect for her creatures, and a reverence for her author." [*Natural History* February 1991, p. 64]

Is this still a tenable position? It is certainly still held, and not only by those impressed with the beauty of a sunset or the intricacy of a preying mantis. We have seen the teachings of the Catholic Church, and they are much more nuanced than those of some other Christian denominations, but the conviction remains that the universe reveals god. It was present from the very beginning of the faith. Paul's two main statements about the pagans' knowledge of god (Acts 17, 16-30; Rom. 1, 18-22) attest to his conviction that god is immanent in god's creation and is to be found there and not in temples built by humans. This is not surprising for a person of Paul's time, nor is it restricted to the Christian faith. Hans Urs von Balthassar explains,

> "In the early, `mythical' period,...being was surrounded with such a luminosity - in its highest moments at least existence was so much a gift and a kindness - that being could not be explained otherwise than by means of the presence of `a God'. Subsequently, in the `philosophical' period, there was the overwhelming force of the order of the world,..."
>
> [Hans Urs von Balthasar, "Meeting God in today's World" in "Concilium: The Church and the World", New York: Paulist Press, 1965, p. 25]

This obviously spoke of the existence of an ordering god. In modern times, the theologian von Balthassar feels that with the revelation of god through the Incarnation, the Church has allowed the revelation of god through creation to become superfluous. Depending on revelation as found in the Hebrew scriptures especially, coupled with Greek philosophical thought, it did not need the lessons from nature. He submits that,

> "From the 3rd century down to the 19th, Christianity formed an alliance with antiquity; and now this ally has had its weapons taken away. The Godward orientation of the thought of antiquity is now rejected as `mythological' and primitive by the enlightened understanding. Also, the theology that is tied up with this kind of

> philosophy is instructed to `de-mythologize' itself if it would be up to the standard of our age. [Ibid. p. 30]

He believes that god is still knowable through god's created universe, and that because of god's immanence in the world,

> "...in the present moment in the history of the Church the responsibility for the effectiveness of the revelation of God in the universe is placed in the hands of Christians. Where the sign of salvation, love of man for man, shines brightly, the hieroglyphics of the book of the external world can, to some extent, be deciphered and God discovered. The more unmistakably the whole of Christianity orients itself toward the cross, the more it appropriates the wise foolishness of the cross and lets its presence shine forth - in deeds and not merely in words - the more theophanous will it make the world again." [Ibid. pp 36-37]

In other words, since humans have placed themselves at the pinnacle of the universe, and for our own understanding that is just where we are, the universe reveals god when humanity reveals god-like life and behavior. As long as humans were somehow outside the universe and the universe was made for our service, then its order and beauty and magnificence revealed god to us, just as our own beauty and consciousness did. When science fixed us permanently in the realm of the universe and denied us any god-like qualities which were not attributable to the normal working of the forces of the universe, we were forced to reconsider our use of ancient philosophies or Thomistic theologies. Our acceptance of the universe as a window on the mind of god had to be rethought. For von Balthassar, in contrast, since we are very much a part of the universe, the universe still reveals god, especially through us.

Also in response to the challenge of science, Karl Barth

> "...believed that faith was endangered if it tangled with philosophy. He rejected wholeheartedly any theological `analogy of being' which would have allowed human beings to reason from nature to god in the way Aquinas did. The doctrine of creation is for him...an article of faith. The doctrine of creation is...`Neither native to (man) nor accessible by way of observation and logical thinking'." ["Soul" p. 212]

Teilhard de Chardin still held a place for the universe revealing god, but the revelation was only tentatively to be found in the universe as it is. The final revelation of god through the universe would only be found in the future.

> "Teilhard's vision of the universe...deliberately started from the scientific story and found in it not hopeless enigma, but evidence of emergent design. He did not argue that evolution is directed in all its details. He thought that it was shaped in such a way that it would eventually converge at a still-to-be-realized point of communion with god." ["Soul" p. 213

For another treatment of evolution and faith from the Teilardian perspective, see Juan Luis Segundo, "An Evolutionary Approach to Jesus of Nazareth", Vol. 5, Ed. & Trans. by John Drury, New York: Orbis Books, 1988.
For the flavor of the work, consider this citation,

> "The fact is, you see, that in the hypothesis in question the differentiation or non-differentiation of species is allegedly due to genetic chance working from inside and the external environment working from outside. But is it possible that there is another *internal* factor that is not chance?" [p. 59]

Jesus as the primordial "Word" expressing itself in creation from the first, then in the Incarnation and finally in eschatology is the conclusion of the work.

These ideas are not nearly as universal as they once were and the usual position today is that,

> "there is...nothing in the vision of physical reality provided by modern physical theory that requires one to endorse any metaphysical or ontological positions...Such a belief requires, as Kierkegaard pointed out, a `leap of faith' that may have little or nothing to do with the dictates of reason."
> ["Conscious Universe" p. 186]

So von Balthassar, Kierkegaard, Barth, Chardin and others look to god's revelation within humanity, and humanity as that part of the universe which reveals god. Virgil Elizondo says,

> "God began to reveal himself to man from within the heart of man. Since man comes from the stuff of the world, we might go a step further and say that God has been continually revealing himself through the marvels of his evolving creation."
> ["The Human Quest", Huntington: Our Sunday Visitor, Inc., 1978, p. 42]

Karl Rahner puts it in this way,

> "If mystery no longer confronts him so plainly and directly in the nature which surrounds [humanity], it is now welling up out of his own nature. We call this mystery God."
> ["Do You Believe in God?" p. 53]

Officially, the Catholic Church still sees nature external to humanity as revelatory of god. The First Vatican Council accepted this, as noted above. The Second Vatican Council repeats the pronouncement of the First, and asserts that,

> "all creatures bear a certain resemblance to God, most especially man, created in the image and likeness of God. The manifold perfections of creatures...all reflect the infinite perfection of God. Consequently we can name God by taking his creatures' perfections as our starting point..." ["Catechism of the Catholic Church" #41]

Vatican II also admits that, "In the historical conditions in which he finds himself, however, man experiences many difficulties in coming to know God by the light of reason alone", ["Catechism of the Catholic Church" #37] and calls for the enlightenment of revelation. There would be none of this hedging if not for the findings of science about the workings of the universe. If there were not now an at least tacit understanding that the need for a creative, sustaining god is not required for a sufficient understanding of the existence and workings of the universe, we would still be in the position of the Middle Ages or the world of Newton who had recourse to a helping hand from the divine from time to time to explain certain mysteries of the universe and felt justified in doing so. Today the professionals know that some rethinking is needed, but little of this has filtered into popular piety and apologetics. We have already taken note in our treatment of the presence of design and purpose in the universe of the many catechisms and individuals who feel they have learned something about god by watching god's creation. There is a whole "creation spirituality" movement which certainly finds god in nature, though usually only in the beauty and serenity of nature, not in the violence and disregard of life. Many people claim they can find god better in the woods than in church on Sunday morning, but you can bet they are not out turning over leaves to watch the mayhem occurring there. If they are in the woods at all on a Sunday morning instead of reading the funnies, they are probably sopping up the serenity which comes from everything else in the woods being afraid of them and hiding out, and blissfully unaware of the life and death struggles going on around them. In fact, if the study of nature were invoked to reveal anything about the qualities of god it would have to be that their god is far worse than the most uncaring and insensitive god ever invented in the human mind, given the scope of the violence in god's creation. Nature would reveal this too, as well as god's "inordinate fondness for beetles", as Huxley put it, since god created so many of them. Not to mention that parasites busily munching their way through others' bodies or taking over their behavior that outnumber other species four to one.

[Carl Zimmer, "Do Parasites Rule the World?", *Discover*, Vol. 21, No. 8, August 2000, p. 80]

393. One of the most important doctrines to be examined, beyond god as creator, is the Incarnation which we will consider in due course. If god can use the matter of the universe to express godself in human form in the man Jesus, then god can use the matter of the universe in other ways as well: to manipulate history, for example, or to work miracles. If we find little or no evidence that god has, in fact, changed or guided history, possibly using the process of evolution; or that god has had to step in at crucial points to navigate through some shoals which evolution could not handle, the appearance of life or consciousness being the most often put forward; then not only god's willingness to enter into the process is questioned, but god's *ability to do so is questioned as well.* This is certainly one of those areas where neither side is going to be able to offer any proof. The evolution from non-living to living may have been the result of god's "fiat", but then again if god did not issue the orders the universe was perfectly capable of bringing it about anyway. The following is certainly an "*ad hominem*" argument, but the history of living beings, especially conscious beings, attests that if someone can do something, they will. Not the individual, perhaps, but the species. If mountain goats are capable of walking up sheer cliffs to get to grass, they will. (Things do not work the other way around. The goats do not desire to get to the mountaintop and so they evolve suction cups on their feet. First the ability and then the use of it.) We worry about human cloning because, as is often said, despite laws forbidding the practice everyone knows that if humans can do something they will do it. It just seems logical that if god could affect the universe in a physical way, god would. Our prayers, the worthy ones at least, would be answered in a less equivocal way. The evil would be punished and the good rewarded. God's loved ones would be cared for. Yes, I am aware of all the standard answers to this objection asserting that god does indeed do these things, and Susan would hear them all, but they are not satisfactory, as we have already said and will say again. No, it just makes more sense to say that if god does not manipulate the physical universe to do god's will in any evident manner, it

is not because god has chosen not to, or does so in ways that are not evident to us, it is because god *cannot.* This is certainly the most controversial of the conclusions that this book will reach and we will discuss it at greater length later. The reason I mention it now is to forewarn the reader that it is coming and get us thinking about it as a possible result of our re-examination of the faith. It is also a timely place to tip our hand since it shows that another area that may have to be reinterpreted is the Incarnation. If god cannot do with the universe what god wishes, then god cannot use it to form a body for godself. If the universe does not reveal the mind of god, since it could all have come about by other means; if we cannot discover the attributes of god from investigation of god's creation, since the attributes or "accidents" of the universe are not contingent on anything in god; if we cannot discover our own nature by looking at ourselves as the image of god but must see ourselves as the image of the universe, all of which seems to be true; then we cannot do these things because god is not to be found in the universe nor its operations. God is truly "transcendent" in a far more radical manner than we ever suspected, and if god is that transcendent, then god cannot incarnate godself in a human body any more than in the laws of physics or the initial conditions of the universe.

There has certainly been the idea of the "Cosmic Christ", the "Logos" as the "Word" incarnated in the universe. Dermot Lane, "The Reality of Jesus" gives a good short treatment of this concept and maintains that,

> "there is an underlying unity between creation and Incarnation. Creation, or better, the underlying presence of God in the world is that reality which is ordained towards the mystery of the Incarnation. The goal of creation is Incarnation so that in a real sense creation only reaches its point of completion in and through the reality of Incarnation."
>
> [Dermot Lane, "The Reality of Jesus", New York: Paulist Press, 1975, p. 134]

However, if we find something of god in Jesus then we must find some other way to explain what and how since the explanation of divine and human forms inhabiting one human body, or one assemblage of the stuff of the universe, no longer suffices, even if there were such a thing as a human form or soul. The meaning of the message and person of Jesus will have to be re-examined.

394. We are challenged as to the truth not only of the *process* of creation as described in the Scripture, but also whether or not god can be honestly

described as a creator at all. So much of our spirituality and theology and the motivation for many of what we would call "Christian actions" depends on god being the creator of the universe. As such, god has the consequent right to invite and expect care for it and for its inhabitants, and participation in bringing it to the fulfillment of its purpose. We are invited to study its character in order to discover "the mind of god" and god's attributes. Add to this a healthy dose of fear and awe at god's power and authority, and to deny god is its creator is not an insignificant challenge to the faith. Adrian van Kaam is a noted author on spirituality and his book *On Being Yourself* is a classic. At the very beginning he says,

> "I find in my deepest self the mystery of my own Origin, which is the Origin of all that is. In these depths I feel at-one with God. I feel also at-one with every person and thing that emerges from the same Divine Ground."
>
> [Adrian van Kaam, C.S.Sp., "On Being Yourself" Denville: Dimension Books, Inc., 72, p. 25]

If god is not the "divine ground" of all being, if the mystery of van Kaam's origin which he shares with every other person and thing is the mystery of the uncaring, unfeeling, impersonal universe, this entire spirituality is extinct. An allegory entitled *Hinds' Feet on High Places* speaks of a spiritual journey toward god in terms of gazelles on the mountains. It is a beautiful story and speaks much about being open to god's teaching and guidance. It begins with a preface saying,

> "As Christians we know, in theory at least, that in the life of a child of God there are no second causes, that even the most unjust and cruel things, as well as all seemingly pointless and undeserved sufferings, have been permitted by God as a glorious opportunity for us to re-act to them in such a way that our Lord and Saviour is able to produce in us, little by little, His own lovely character."
>
> [Hannah Hurnard, "Hinds' Feet on High Places", Old Tappan: Fleming H. Revell Company, 1973, p. 6]

If god has no power to "permit" any such thing but must be restricted to the role of observer and, perhaps, reactor, this spirituality must be reconsidered.

In its explanation of Baptism, the World Council of Churches says that though the world is wounded, "God its creator has entered it in person in Jesus Christ...accomplishing in his own humanity the true vocation of man...In him, therefore, man is restored to his created status."

["One Lord One Baptism" World Council of Churches Commission on Faith and Order, Minneapolis, Augsburg Publishing House, 1960, p. 38]

If "man" has no created status and never did, we cannot be "restored" to it. The whole basis for the WCC's understanding of Baptism, and the same will be found in Catholic sacramental spirituality as well, is removed. A spirituality can and must be based on both what we have received from god and what we share with one another as members of the universe. In fact, our being-at-home in the universe is what unites us to every person and thing, and that can be very deep and satisfying in itself as we will see when investigating the faith of the agnostic evolutionists. What god has done for us in calling us into relationship with godself and through god with one another is another element in a spirituality. This is certainly different than what van Kaam and the World Council of Churches had in mind.

395. See "The Whole Shebang", p. 305

396. The mainstream position taken goes something like this. God exists in and of "himself". Being is a necessary attribute of god. In fact the ontological argument of St. Anselm through a sophisticated sophistry tries to argue that one can know that god exists since existence is a logical and necessary part of the definition of god. This argument has been roundly (pardon the pun of the circular argument) discredited by philosophers.

[See a discussion of this argument in Edward Harrison, "Masks of Universe", New York: Macmillan Publishing Company, 1985, p.265f. See also "The Ontological Argument" op. cit.]

Still it is also widely assumed that there just "must" be a god. God created everything that exists, it is presumed. It was created out of nothing, "*ex nihilo*", and god had a purpose in creating it. This purpose resides in god since nothing else has any meaning or existence apart from god. And that purpose is, "not for increasing his own beatitude, nor for attaining his perfection, but in order to manifest this perfection through the benefits which he bestows on creatures, with absolute freedom of counsel." as the Second Vatican Council phrases it.

"*Dei Filius*", 1:DS3002 as cited in "Catechism of the Catholic Church", #293]

This is not a selfish god who creates for any benefit for self, like the gods of the Egyptians who create in order to have a work force. This is a loving god who, like any lover wants to communicate godself to others, to share what god has with others. A human person can share themselves with other humans who already exist. God, when the only being in existence, must first create others with whom to share. The human parent knows how much more sharing and loving can be given to one's own child, one's own creation who was brought into existence and nourished in that life by the parent. This Christian god is much more than a creator who merely sits back and watches the mechanism unwind, but is a loving parent who wants nothing but the best for the offspring. "The glory of God is man fully alive."

[Irenaeus, "Adversus Haeres", 4, 20, 7:PG 7/1, 1037]

After all, the human can love a pet and feel an attachment to an inanimate object such as the land or a prized possession, but we feel that they are lacking something if their affection is not given to another human person who can respond to it and understand it. So god's affection is directed to those conscious beings of his creation far more than to the unconscious. Most believers, indeed most humans, believe that somehow we are the goal of the universe. The book of Genesis is hardly alone when it places humankind at the apex of the creative activity. This approach is still with us as when Gerald Schroeder informs us that,

> "...mankind is formed from the previously existing substance, in fact, the same substance used to form fowl and land animals (Gen 2:19). However, a special ingredient not mentioned before is summoned at this juncture. God breathes a *neshamah*, a `soul of life,' into this creature and man became a living being..."
>
> [Gerald L. Schroeder, Ph.D. "Genesis and the Big Bang", p. 149]

Teilhard de Chardin tried to link this conviction of the centrality of humanity with the findings of science. Leslie Dewart claims that Chardin's,

> "thought was not an apologetic endeavor to `harmonize' science and the traditional
>
> conceptualization of the Christian faith. It was a creative attempt to follow through to its ultimate consequences a scientific and fully contemporary (scientific) experience in the light of a Christian faith which, on the one hand, functioned to make that scientific experience religiously meaningful, but which, on the other, required re-interpretation and re-conceptualization in the categories of contemporary experience for the very sake of illuminating that everyday scientific experience."
>
> ["The Future of Belief", New York: Herder & Herder, 1966 p. 43]

Chardin found that the phenomenon of human intelligence was the climax of eons of preparation.

> "To culminate in man at the stage of reflection, life must have been preparing a whole group of factors for a long time and simultaneously-though nothing at first sight could have given grounds for supposing that they would be linked together 'providentially'."
>
> [Pierre Teilhard De Chardin "The Phenomenon of Man" New York: Harper & Row. 1959 p. 170]

His "providentially" may be in quotation marks, but it is no secret that Chardin considered this whole movement of the universe up through consciousness to the Omega Point to be intended by god.

> "God, the Centre of centres. In that final vision the Christian dogma culminates. And so exactly, so perfectly does this coincide with the Omega Point that doubtless I should never have ventured to envisage the latter or formulate the hypothesis rationally if, in my consciousness as a believer, I had not found not only its speculative model but also its living reality."
>
> ["The Phenomenon of Man" p. 294]

The Catechism states that we occupy "...a unique place in creation..." [Catechism of the Catholic church #91] which would not seem so chauvinistic if it had not previously stated that, "Man is the summit of the Creator's work..." [Catechism of the Catholic Church #343] This article of faith says little about who god is, but says much about how we can approach god in our relationship. The faith therefore states that god is the creator, *ex nihilo*, of everything that is and that this creation was done with the express purpose of leading to the creation of humanity so that the creatures could respond to the creator, not to serve god as menials, as in many of the pagan myths, but to love god in return.

397. For some discussion on this see Capra and, "Belonging to the Universe", p. 86f. See also p. 167f for a discussion of eschatology and the environment.

398. "The Church in the Modern World", no. 69

399. Bill McKibben, author of "The End of Nature" and numerous other books on religion and the environment is interviewed in "Alive Now", Nashville: The Upper Room, 2000. The issue is devoted to "Creation" and

has many examples of the approach I am saying is no longer valid. In the course of the interview, McKibben cites the Evangelical Environment Network of Madison, Wisconsin's advertisements,

> "...that showed pictures of plants and animals and said, `God made this. Who are we to wipe it out?' That's a lot better argument than saying that this plant may cure you of some disease."

Unfortunately I do not believe it is a valid argument.

403 That god is the creator is also not a true concept on which the base the worth of a human being. Science is not going to give us a reason for respecting each other except the built-in tendency to take care of others, at least those closest to us, as a way of caring for our gene line. Faith has taught that others should be respected because they are made in the image and likeness of god and that therefore care for the other is care for god. We are fond of saying that we should treasure other people because they reveal god or because they are all made by god's loving hand and should not be abused. A text on Justice is typical when it says that,

> "All people not only share basic needs, but also possess God-given worth and dignity. Thus people have a right to a sense of self-esteem and personal power."
>
> [Joseph Stoutzenberger, "The Christian Call to Justice and Peace" Winona: Saint Mary's Press, 1987, p. 19]

What if they reveal nothing of god; are, like all peoples, mere accidents of the universe; have no divine purpose in living and are completely expendable, at least to the universe? If we have more right to the beings of this universe than god does, having evolved as part of it and being at home in it while god is but an "interloper" and perhaps only an observer, then much of our theology of creation, as well as that based on the Incarnation making all things holy goes out the window. Once again the idea of god incarnating godself in human flesh comes into play since god is present in the other, especially the baptized other, by virtue of their union with the resurrected Christ. Note that this presumes that everyone is in the image of god by virtue of creation, the original creation of the human species as recorded in Genesis and the special creation of the rational soul.

In "Living the Gospel of Life: A Challenge to American Catholics", adopted by the full body of American Bishops at their November 1998 General Meeting, the bishops point out that,

"Virtual reality and genetic science may give us the illusion of power, but we are not gods. We are not our own, or anyone else's, creator. Nor, for our own safety, should we ever seek to be."

Their rationale for granting protection for basic human rights is based on the premise that, "No one but the Creator is the sovereign of basic human rights..." They do recognize this statement as the assumption that it is, even using the word by saying,

"Without the assumption that a Creator exists who has ordained certain irrevocable truths about the human person, no rights are `unalienable,' and nothing about human dignity is axiomatic." (pp. 11-12)

Unfortunately, they are correct.

Sr. M. Timothy Prokes, Professor of Theology and Spirituality at Notre Dame Graduate School of Christendom College, Alexandria, Virginia, has an article titled, "Convergence of Trinitarian Mutuality and Technological Truth" in "*The Future of the Family: Proceedings of the ITEST Workshop October 1998*", St. Louis: ITEST Faith/Science Press, 1999, p. 47f. At the beginning of her article, Sr. M. Timothy says,

"It seems helpful at the outset to state basic assumptions and to note how several terms are understood here. First, it is assumed that theology is `organic,' in each moment of history a renewed search into the mystery of divine-human relationships. Closely allied with this is the assumption that theology necessarily has profound implications for daily life and the development of creation. The word *convergence* in the title indicates an assumption that human technology is not intrinsically at odds with God-life, but rather is one expression of the human capacity to collaborate with divine creativity."

The article is an excellent example of what one can do with traditional theology to speak to a technological society. However, the really "basic assumption" which is not recognized is that there is a creator god and that moral positions can be taken based on this assumption.

Statements about health decisions are often made based on the assumption of god's right as creator over a human life. Speaking about the question of nutrition and hydration of someone in a "persistent vegetative state", Archbishop Justin Rigali of St. Louis begins by saying,

"As Catholics, we believe that God is the author of life...We believe that euthanasia and/or assisted suicide is a usurpation of

> God's authority over all human life and is, therefore, morally wrong." [Statement issued Friday, June 30, 2000.]

If god is not the source and author of life in the universe and if we ourselves do not participate in the life of god but rather are part and parcel of the universe alone, and actually nothing but an assemblage of parts which will dissolve back into the chaos from which we came, both individually and collectively, then why should I care about others?

If god did not make the universe with us in mind, and we are not the ultimate goal of creation, if we have no created soul, then human nature itself is not sacrosanct. Genetic engineering may well change us into something else and there can be no theological or biblical injunction against doing so. Ted Howard and Jeremy Rifkin, "Who Should Play God?", say,

> "As for the argument that biological and social revolution can coexist, nothing could be more impossible. The assumption underlying social activism is that biologically we are sacred and inviolate. The assumption underlying genetic engineering is exactly the opposite. If genetic engineering were allowed to continue at its current pace, the Homo sapiens species would experience no more than five or six more generations before being irreversibly replaced by a new, artificially engineered organism. Though this new species would include some of our characteristics, it would in many ways be as different from us as we are from our closest relatives, the primates." [New York: Dell Publishing Co. Inc., 1977, p. 224]

It is interesting that their book mentions God only in the title.

In addition, if god is not the giver of life and the creative source of our existence, then what reason do we have to believe that god can be the *re*creative source of our continued existence. The argument of the mother of the seven Jewish martyrs of Maccabees was,

> "I do not know how you came into existence in my womb; it was not I who gave you the breath of life, nor was it I who set in order the elements of which each of you is composed. Therefore, since it is the Creator of the universe who shapes each man's beginning, as he brings about the origin of everything, he, in his mercy, will give you back both breath and life, because you now disregard yourselves for the sake of his law."
>
> [2 Maccabees 7:21f. This reading is used on Wednesday of the 33rd week of the year.]

If this argument no longer makes sense since it is the process of evolution and the controlling agency of the DNA that formed them in the womb and the breath of life is only a metaphor, what reason would there be for the seven brothers to undergo martyrdom for the sake of the Law? We must have different answers to these questions as well if we are to be able to speak to our world.
Indeed, it is not only theistic religion that is affected by this different understanding of the basis for human dignity. Jefferson's Declaration of Independence proclaims that, "We hold these truths to be self-evident, that all men are created equal, that they are endowed by their Creator with certain unalienable Rights, that among these are Life, Liberty and the pursuit of Happiness." Without the presumed creator of human rights, can we find another basis for human dignity and fair treatment of all?

The question of individual dignity is also posed by the question of the Incarnation of Jesus as god's physical presence assuming a human nature. Cardinal Francis George points out that, "presupposing that there is one human race, because there is a human nature, there can be one incarnate Lord who is the Savior of the whole world. That is the basic theological datum that stands behind a lot of the moral discussion."

> [Francis Cardinal George, OMI, "Biotechnology and Some Thought on the Body", Proceedings of *The Genome, Plant, Animal, Human,* Ed. by Robert Brungs, SJ & Marianne Postiglione, RSM, St. Louis: ITEST Faith/Science Press, 2000, p. 190

Our moral discussions will have to change if the nature of this incarnate Lord is rethought. Also for our purposes here we can turn this around and, presuming that there is such a thing as "human nature", ask how one person possessing it can indeed redeem everyone else through of a solidarity and unity with both the human and divine if in fact such a unity is metaphorical rather than ontological.
401. Much has been written on prayer other than the prayer of petition in which we implore god to change the operation of the universe, or some corner of it. We can pray to open ourselves to the wisdom of god. We can pray for the courage and strength to put this wisdom into operation. Of course, this is *not* usually the content of much of our prayer. When we think of prayer we begin with prayer asking god to do something, petitionary prayer. This is based on certain assumptions. A barren woman normally remains barren, and certainly after the age of childbearing is past. However, god can suspend this "law of nature" for those who petition it or when it suits god's purposes. Samuel is both sought by his mother and petitioned

for in prayer at the temple [1 Samuel 1:11f] and he is needed for god's purpose in establishing the Davidic dynasty, and so god creates a special set of events to bring about his birth. The Psalms are filled with petitions for and thanksgiving for god's listening to requests. Psalm 4, for example, is a prayer for rain and Psalm 6 a prayer for healing from a wasting disease.

> [For a scholarly commentary on this and other psalms see "The Anchor Bible: Psalms I", Trans. & Notes by Mitchelle Dahood, S.J., New York: Doubleday, 1966 and the other two books on the Psalms in the series.]

Why not, for god does not really have to suspend some laws truly inherent in the physical world, for they are only laws from the human viewpoint. God need only do things in a different way than god usually does them. The miracle is not that god has the power over nature to keep it from acting as it normally would by its own constitution. The miracle is that god would change god's mind about what to do. For the laws of nature are arbitrary and exist only "in the mind of God". Nothing happens without god willing it, and god may merely will something else. When there were gods who were natural phenomena, when there was a god who was the sea or the sky, when there was an Apsu and a Tiamat, then the god Marduk had to struggle against them and perhaps change their "laws", their idea of how things ought to unfold. He could do it or not depending on how much power the god possessed.

> [The Greeks and Romans developed a temple at Pergamum dedicated to Asklepios, the Roman god of medicine. *Biblical Archaeology Review*, Vol. 24, No. 5, September/October 1998, p. 76 reports that, "An inscription over the door of this sacred medical center, called an Asklepieon, read, 'In the name of the gods, Death may not enter here.'" This control over nature by the gods was ubiquitous in the ancient world.]

The god of the Hebrews, at least after some development of their theology, had no competition, nothing to gainsay god's will, and certainly not some physical laws encoded into the being of matter. There is only one ground of being, god, and god's will is the law.

The official public prayers of the church are not so much directed to convincing god to act contrary to the natural course of events, but to have god act contrary to god's own preconceived plan (presuming that god is not preconceiving the act of saying the prayer itself). We do not pray that god would change the weather patterns of el Nino, but just to make it rain. True, we are more sophisticated now and claim that there are natural laws governing such things as drought and flood which god normally uses, perhaps as a shortcut to having to be involved in every little decision, but the underlying assumption is that god is really in control and does make all the nitty-gritty decisions which affect our lives. Or else why would we sometimes preface our prayer for relief from some "natural disaster" by saying "we have sinned, O Lord", or "we have learned our lesson, O God"? God has brought about this problem for god's purposes. Susan does not really feel that her marriage has been a source of grief for her just out of chance. She feels that god has been involved in everything leading to her situation and she wants to know why god has chosen this path rather than another for her.

Abraham Lincoln saw god playing both sides when he mused that,

> "...if we shall suppose that American slavery is one of those offenses which, in the providence of God, must needs come, but which, having continued through His appointed time, He now wills to remove, and that He gives to both North and South this terrible war, as the woe due to those by whom the offense came, shall we discern therein any departure from those divine attributes which the believers in a living God have always ascribed to him?"
>
> [in Bruce Catton, "This Hallowed Ground" New York: Pocket Books, Inc. 1956, p. 488]

In the same vein Lincoln wrote,

> "...the nation's condition is not what either party, or any man, devised or expected. God alone can claim it."
>
> ["The Living Words of Abraham Lincoln" Hallmark Editions, 1967, p. 43]

This fits well with his upbringing.

> "His parents were Separate Baptists, a small denomination that taught God's absolute control over each and every human choice, down to the smallest events, so that no one really exercised free will in choosing."
>
> [Allen C. Guelzo, "Lincoln and the Abolitionists", *The Wilson Quarterly*, Vol. XXIV, No. 4, Autumn 2000. p. 63]

Lincoln, in fact, may not have prayed at all in petition, believing more in a natural order ruled by "Providence" and not amenable to change. His listeners, however, would probably have heard with their own ears and added a petition for change.

There are other reasons, of course, to pray and other forms of prayer that have nothing to do with changing god or the course of history.

> "Prayer...deals with four things: adoration, thanksgiving, forgiveness of sin, procuring of spiritual and temporal benefits...The prayer of offering has no real place in any of the four categories put forward, and yet it is at the heart of all spirituality..."
> [Maurice Nedoncelle, "God's Encounter With Man", New York: Sheed and Ward, 1964, p. 101]

The forms of prayer that have nothing to do with "changing god's mind" are often the content of courses on prayer to deepen the spirituality and the relationship of the person to god. Much of liturgical prayer is not really petitionary of favors except in the relational, spiritual, and personal development realm, though they sometimes are phrased in a pleading manner. These forms of prayer are most done in groups seeking to praise god or recognize god's activity, but in individual prayer, the appeal to god's power to change things often predominates.

> [Of course, praise may not be an unselfish motive to pray. In speaking of the role of drums in West African culture, one author notes that, "by praising something, extolling its attributes over and over, you gain influence with it. It can be a god or a king or your grandfather. In West Africa, everyone has praise poems associated with them, but in Cuba the praises are reserved for the gods...The belief [is] that the goddess would become more engaged in the lives of those who praise her."
> [Michael Kernan, "The Talking Drums", *Smithsonian,* Vol. 31, No. 3, June 2000, p. 28]

A small book on prayer intended for teaching young people says that,

> "Prayers of supplication are prayers of asking. Most of us practice this kind of prayer more than any other. We take to heart the Lord's teaching, 'If you ask for anything in my name, I will do it' (Jn 14:14)."
> [Michael Pennock, "Prayer & You" Notre Dame: Ave Maria Press, 1985, p. 33]

In spite of this, the book cited has far more material about conversing with god, being open to god, meditation, and centering than it does about petitioning.

A classic book on prayer by Hans Urs Von Balthasar scarcely mentions petition but speaks of conversation and contemplation at great length.

[Hans Urs von Balthasar, "Prayer" Trans. by A. V. Littledale, New York: Sheed & Ward, 1961]

George Custer claimed that he did not pray "as others do", but nevertheless he admits that "on the eve of every battle in which I have been engaged, I have never omitted to pray…to commend myself to God's keeping, asking Him to forgive my sins, and to watch over me while in danger…This belief…makes me brave and fearless as I am."

[Evan S. Connell, "Son of the Morning Star", New York: Harper & Row, 1984 p. 112.]

Presumably, he prayed in this fashion on the hill overlooking Sitting Bull's camp.

For the warriors of the Second World War, we have the song, "*I Had a Little Talk With the Lord*" where a soldier "in a foxhole in Bougainville" prays for rain (I am not sure why, perhaps to delay the battle) and he, "found the strength to fight for my land" when the rains come down. I suppose there are truly no atheists in foxholes, or on the hills around the Little Big Horn.

[Copyright 1944 by Santly Joy Inc., New York]

Indeed, it has been the constant work of spiritual directors throughout the ages to expand beyond petition the faithful's understanding of prayer and include the loftier motives and forms of prayer. There is justification for some scientists' and philosophers' objections to prayer as a way to "get results". Ken Wilbur feels that there need not be a conflict between science and religion but that a conflict does arise because,

> "…during the youthful period of mankind's spiritual evolution human fantasy created gods in man's own image, who, by the operations of their will were supposed to determine or, at any rate, to influence the phenomenal world. Man sought to alter the disposition of these gods in his own favor by means of magic and prayer. The idea of god in the religions taught at present is a sublimation of that old concept of the gods. Its anthropomorphic character is shown, for instance, by the fact that men appeal to the Divine Being in prayers and plead for the fulfillment of their wishes."

[Ken Wilbur, "Quantum Questions", Shambhala: New Science Library, p. 109]

Wilburn in "Quantum Questions" tries to show that agnostic physicists are often closet mystics, following in the footsteps of spiritual mystics who also do not stress petition but rather reflection as a proper stance before god.

Carl Sagan brings up the same objection when he writes,

> "...there's a category of prayer in which God is begged to intervene in human history or just to right some real or imagined injustice of natural calamity - for example, when a bishop from the American West prays for God to intervene and end a devastating dry spell...What is implied here about the limitations of a supposedly omnipotent and omniscient deity?" ["Demon Haunted World" p. 276]

On the other end of the spectrum, Larry Dossey M.D. tries to find scientific evidence that petitionary prayer is effective but he must continually put in qualifiers or apologies, a state of affairs that Sagan is quick to pounce on when Sagan gloats,

> "Even a recent book advocating the efficacy of prayer in treating disease (Larry Dossey, "Healing Words") is troubled by the fact that some diseases are more easily cured or mitigated than others. If prayer works, why can't God cure cancer or grow back a severed limb?..." Dossey quotes Stanley Krippner, M.D.: "'[T]he research data on distant, prayer-based healing are promising, but too sparse to allow any firm conclusion to be drawn.'" And Sagan feels compelled to add, "This after many trillions of prayers over the millennia." ["Demon Haunted World" p. 234]
>
> [The best experiment on the effect of prayer on humans yields this conclusion by Dossey,
>
> "I feel the Byrd experiment is suggestive but inconclusive and inherently ambiguous. It simply contains too many problems that prevent us from drawing firm conclusions about the possible power of prayer. In fact *all* the human prayer studies we have examined so far fall into this category."

He continues,

> "In view of these problems, many researchers feel it is easier to study the effects of prayer in simple, *nonhuman* living systems. Prayer experiments in simpler life forms are much less ambiguous..."
>
> [Dossey, M.D., "Healing Words", New York: HarperSanFrancisco, 1993, p. 186.]

Whether it is god's intervention or the effect of some sort of energy or non-divine mechanism gets rather lost in the book.

[For another response to Dossey and to Herbert Benson whose thinking is similar to Dossey, see Chet Raymo, "Skeptics and True Believers", p. 93]

In August 2001, on "Downtown" on ABC television, a piece was run about the healing power of prayer. Dr. Gary P. Posner, who was interviewed as a skeptic about the results of experiments wrote a letter to "The Skeptical Inquirer" complaining that when his interview was used it was cut down to 20 seconds, which however was at least enough to cause the real star, Deepak Chopra, to admit that the studies were indeed tenuous. However, the viewer poll was nine to one in favor of the healing power of distant prayer. This is hardly surprising given all of prayer's promoters.

[*Skeptical Inquirer*, Vol. 25, No. 6, November/December 2001, p. 9. Posner's prayer-related articles are at http://members.aol.com/garypos/prayer.html]

It would be nice to refute these criticisms by saying that Christians have long ago given up praying as if they need to inform god of something or sway god's intentions or grab god's attention, but the prayer of many if not most believers is the same as that of the first woman who buried a seed in the ground and had then to acknowledge that she was powerless to continue the action farther and so she prayed that some being who could would take over from here. We want god, if god exists, to do what is beyond our capability and so we pray the prayer of petition. And this is usually the direction of petitionary prayer. Most Christians do not pray that god would change things that they are perfectly capable of changing themselves. To do so is recognized as a form of hypocrisy, as when one prays that god will find shelter for this homeless person but is unwilling to make a phone call to try to locate a place for them or get in their car to drive the person there. Susan is not praying that god will do something that she can handle on her own, she is praying because she cannot effect a change. She believes that god could do it all but expects her to do her part. Nevertheless, she also believes that god can "pick up the slack" when she herself is powerless. Even the apologists for prayer, though they dismiss the crasser forms of begging the powers-that-be as unworthy and shortsighted, still feel they must maintain that god *could* do what the pray-er is asking of god, but does not do so in favor of something better. One authority explains that god does us a favor by not answering our prayer, thereby leading us outside of ourselves to the wider reality. But as he states it, "...God's answer to our prayers is a way of

making us understand the meaning of his negations so that we may thereby transcend ourselves." ["God's Encounter With Man" p. 85]
This sounds very farsighted on god's part and surely I need to "transcend" myself much more than I need to win the lottery. Still, the author felt compelled to grant god the power to "negate". One can only answer "no I will not" if one can, in fact, do the thing. I cannot legitimately say that I will lay the stars at your feet when in fact I cannot, never mind the song writers' protestations that a lover can in fact accomplish this task for the beloved. Aquinas grants god even more power when he claims that our prayers do not change god's mind, after all god is changeless, but they obtain for us what god had previously decreed would be ours *through our prayer!* God can teach us a lesson on relying on someone other than ourselves and remain totally unchanged and unmoved at the same time. However, all of this, sophisticated or crass, makes sense of prayer, and especially of petitionary prayer, only if god can have an effecton those things that we would wish changed, and this is precisely what is questioned and must be rethought.

402. The claim that there must be direction, or even direct intervention, in order for structure, or life, or consciousness to arise, or even for the universe itself to arise, has about as much validity today as the man who claimed that his prized hound is responsible for keeping alligators out of Alaska, as borne witness to by the fact that there are none resident. There are many other reasons for the absence of alligators, and most of them far more credible than that given by the dog's master, even if he himself firmly believes it to be true. One cannot *dis*prove his claim, of course, short of removing the dog and seeing if alligators then move in, but one certainly cannot prove it. The same is true for claims that god has a plan. It cannot be proved nor disproved, but since there are other explanations for how the universe got to be what it is, and these do not require a divine plan, nor divine creation, nor divine guidance or sustaining activity, all claims to such things are suspect and the presumption that such a plan is in operationis certainly challenged. Even the retrenched position that god is working only in human history and pretty much lets the rest of the universe go its own way is difficult or impossible to prove, or disprove. The theist can make the claim, as the later prophets and the authors of the Wisdom literature do, that it was god who brought the people of Israel back from the Exile, even using the Persian Cyrus as an instrument for god's purposes, (2 Chronicles 36:22; Ezra 1:1; Isaiah 44:28; Isaiah 45:1) but any good historian or sociologist can explain the same events from a different perspective. God's physical intervention in the larger matters of human history, and certainly god's manipulation of the

personal history of individuals must be examined. This is not a trivial question for theology. For example, Professor R.A.F. Mackenzie speaks for many biblical theologians when he first admits that,

> "...we must recognize the fact that any theology rests ultimately on a basis of personal commitment. Its *raison d' etre* is an act of faith made by the theologian in the validity of the datum which he proceeds to discuss and analyze." But he then goes on to claim that the Old Testament, "taken as a whole is not a work of theology: it is a work of history...It contains a story of divine activity, and yet it is not mainly, as are comparable narratives of the exploits of other gods, presented in mythological terms. It narrates God's deeds, as performed in the full light of human history and experience."
> [R.A.F. Mackenzie, S.J., "Faith and History in the Old Testament" New York: The Macmillan Company, 1963, p. 13]

He therefore claims a place for theology among the sciences that deal with the interpretation of physical data. That god was involved in *any* of the activities recorded in the Bible is itself an affirmation of faith that is so basic to Christianity that Mackenzie does not even realize the assumption he is making. In the light of our investigation in the previous chapter, we must respond to the objection that the faith interpretation of the biblical material is indeed "mythological." This is made even more evident when we realize that today many faithful scripture scholars themselves question the historicity of much of what has been presumed to be actual events reported by eyewitnesses.

403. The most impressive expression of god's power to effect change is the miracle! We no longer, at least most of us, call it a miracle and say that god or Jesus has prevented the rain from freezing before we got to our destination, even though we might have "prayed" to whoever might be listening, "please don't let it freeze before I get there." And if science makes it harder and harder to maintain that in fact god is involved in these daily coincidences (after all, not everyone is going to make it home before the ice and some of the other drivers were praying and beseeching just as loudly as we, and given the number of people doing so, it is mere dumb luck that I am among the number who make it to safety - this time). On the other hand, when Pope John Paul II visited St. Louis, the "Pink Sisters", a cloistered community of Nuns, were given credit by many for praying for good weather, to the chagrin of other "nuns in veils" doing the same but without the publicity. There were, in fact, two days of beautiful weather

surrounded by days of freezing rain and then miserable cloudiness. Try to tell their fans that this was a coincidence! Many will allow the possibility of chance, but what of those really impressive occurrences that we label "miraculous"?

We must exclude from consideration those events so labeled out of ignorance of the true cause. Certainly we put no credence in the claim for a miracle coming from the cargo-cultist who ascribes divine activity to the ability of the C47 to fly over his village. We should also put none in the claim of the medical doctor with four degrees and fifty citations who labels a recovered child a "miracle baby" merely because she has no idea of the cause of the remission or change in condition or sells short the body's resilience. Still, there certainly are claims made for the miraculous intervention of god in changing the very laws of physics, and the possibility of such an occurrence needs to be considered. As we saw earlier, there is no evidence for miracles that holds up to scientific investigation. We are left then with the profound hope that god has control, but no reason to believe that god does. And if god does not control, or does not choose to control, physical events of which there is little, read "no", "scientific evidence", then, as we have seen, the believers are going to have to revise their understanding of the purpose of prayer and of the meaning of the miracles of Jesus.

Aside from fundamentalist insistence that the miracles of Jesus happened as recorded, the process is well advanced that explains them in other terms. Even a catechism for youth, Thomas Zantig, "Jesus of History, Christ of Faith", Winona: St. Mary's Press, 1982, points out that, "sorting out which of the events included in the gospels occurred just as they are described continues to be a major challenge to biblical scholars." (p. 140) It relates that in Jesus' own time the "wonders" would have just been expected, but in our own time we need to take into account "psychosomatic illnesses" and the "power of suggestion" but most of all what the reported actions tell of the kingdom and of the nature of god.

There are certainly more scholarly treatments of the question of miracles, for example, Harold Remus, "Jesus as Healer", Cambridge: Cambridge University Press, 1997, but the fact that even Zantig's very popular exposition questions the meaning of the miracles is telling.

Chapter 2 of Russell Stannard's "The God Experiment", Mahwah: HiddenSpring, 1999, puts the best face he can on the question but concludes that, "it can be held that miracles are not intrinsically impossible. Which is not to say they necessarily happen - merely that belief in miracles is not irrational." [p. 26] In order to reach even this conclusion he presumes that god is somehow in charge of god's creation and that there is a higher law than the physical. It is this law that is not broken in the miraculous.

Simon Blackburn in, "Think", treats of miracles on p. 176f. His is an interesting conclusion.

> "...a deity that sets the laws of nature into motion and never relents at least has a certain dignity. One that occasionally allows hiccups and intermissions, glorified conjuring tricks, is less impressive. Why just those miracles, just then?...A little miracle or two snuffing out the Hitlers and Stalins would seem far more useful; than one that changes water into wine...It is what one might have expected antecedently, knowing that the world was under the regime of a good God. But the world as we know it does not confirm it." [p.184]

Deepak Chopra is a well known writer on spirituality who has plenty of room for miracles and even tries to explain them in scientific terms. See "How To Know God: The Soul's Journey into the Mystery of Mysteries", New York: Harmony Books, 2000. For example, he speaks of healers in the Philippines who claim to reach into a person's body and snatch the bad growths. Far from exposing them as frauds as many scientists have done, he tries to explain it by saying,

> "As we know from our quantum model, any object can be reduced to packets of energy...The medicine man turns a mental image into physical reality - in fact, this is what all miracle workers do. At the quantum level they `see' a new result, and in that vision a new result emerges." [p. 141]

Much of his thinking depends on god as creator. On his stage 5 of spirituality, he explains god making all things work for the good of the believer. He explains,

> "Things work out because they are meant to. Otherwise the world would be a churning cauldron, and it isn't. Everywhere I look I see patterns and symbols; there is incredible beauty and order." [p. 197]

The world is indeed a churning cauldron of potential and patterns brought about by the processes of evolution, not the well-ordered and patterned universe where god is in control manipulating events. Chopra has any number of "miraculous" events, including finding himself in front of the house of a woman he has been seeking to employ, even though his getting the phone number is four or five times removed. It is sometimes difficult to explain these things, but that is why they are called coincidences. I have to admit that I found it curious that Candace Pert recommends his work, but then I am not well versed in her position.

404. If god seems to have no role in the physical processes of the universe, nor even in the direction of human history, then what would be the consequences in the life of a Christian? We can see that morality cannot be based on any "natural law" set down by the creator in the initial conditions of the universe, or in the initial conditions of the human soul.

Much is made in the Catholic tradition of the role of "nature" in guiding moral actions. Since god is the author of all of creation and it obeys god's bidding and follows god's laws and purposes, one way to know right from wrong is to examine how things actually happen in nature and pattern one's behavior on the "natural law". At its worst this just retains the status quo and things are deemed to be bad because they have always been considered so, or good because society approves of them. At its best the concept of a natural law demands holding up our actions for scrutiny against truths deeper than the superficial appearances. However, those guides to behavior that arise from the "nature of the beast" are deemed not very reliable. Worse, they carry sanctions that are often too subtle, or the consequences of violation are too far in the future to be much of a constraint. Don Juan may end up a bitter, frustrated and lonely old man if he keeps up his lustful ways, since lust is against the natural law, but right now he is having a great time, and he too can justify his behavior by reference to the natural order of things. After all, birds do it, and bees do it. In "The King and I", the king of Siam claims that the honeybee must flit from flower to flower, but the flower must not ever go from bee to bee to bee. It is just not "natural".

Macbeth asks his murderous henchmen, who have been told that their target is a man who has wronged them,

> "...Do you find
> Your patience so predominant in your nature
> That you can let this go? Are you so gospeled
> To pray for this good man and for his issue,
> Whose heavy hand hath bowed you to the grave

> And beggared yours for ever?"
> [Macbeth, Act III, Sc.1, l. 93f]

In other words, it is presumed that natural law is not strong enough, so the Gospel must give further guidance. Even so, they have not been so convinced by the Gospel as to forego revenge and pray for their enemy.

Taking others' property may remove the possibility of meaningful friendships in the life of the thief and it would seem that this makes such theft immoral just by the nature of the action, but the consequences only occur if the theft is revealed. So what is the natural consequence of the action if it is not discovered? Is it moral to take from others if they have no chance of discovering who is responsible? And of course taking from another clan and sharing the spoils with one's own tribe will not cause expulsion from human commerce but rather a hero's status with those who share the treasure.

Surely, it would be a deterrent to masturbation if one really did go blind from such "self-abuse" but no such natural consequence seems to be forthcoming. No, "nature" and just responding to those things that our own conscience says are wrong is not a very reliable guide to moral deportment. Nevertheless, it would be true to say that belief in a "natural law" embedded in our nature by god is firmly entrenched in Catholic moral teaching.

The catechism sees the human conscience as at least partially innate and natural to the human person. This presumably is why it does not function with much reliability. Seven year olds know "right from wrong", but only in a rudimentary manner.

> "Kyle O'Pruett, a professor of clinical psychiatry at the Yale Child Study Center, illustrates this point with a test: `Tell a seven- or eight-year-old, `Johnny broke one teacup throwing it at his sister. Sara broke eight teacups helping Dad load the dishwasher. Which kid did the worse thing?' The average seven-year-old will pick Sara because she broke more. By 11, they have it sorted out that intentionality is part of the moral system. Not when you're seven.'"
> [John Cloud, "For They Know Not What They Do?", *Time*, Vol. 152, No. 8, August 24, 1996, p. 65]

God must implant this ability in stages! Still, the Catechism insists that the conscience is of divine origin.

> "Deep within his conscience man discovers a law which he had not laid upon himself but which he must obey. Its voice, ever calling

him to love and to do what is good and to avoid evil, sounds in his heart at the right moment...For man has in his heart a law inscribed by God." ["Catechism of the Catholic Church" #1776]

Most Catholics believe that this is true and have no empathy for people who have no remorse for their evil deeds, feeling that since they themselves would feel guilty for doing such a thing, the other person felt guilty as well and did it anyway. Reports that many people, and too many of our young people, have no conscience seemingly, and that this is traceable to a lack of guidance and the teaching of a value system are not readily accepted since conscience is seen as something built in.

[See "Conscience grows on temperamental grounds" in *Science News,* Vol. 151, No. 13, March 29, 1997, p. 189]

Granted the conscience has to be formed and one can convince oneself that something wrong is right, but still it is looked upon as reflecting the natural law. At the same time, as the Catechism admits, there is some kind of divine influence that must occur. Thus conscience is midway between a purely natural law morality and a morality imposed by the divine mandate. This is somewhat true to the teaching of Thomas Aquinas that, perhaps surprisingly, "does not lead one to expect pronouncements on morality that would be infallibly true for every time and place without exception." Thomas, "does not support the attempt to do ethics by the deductive use of principles conceived as static derivatives of an immutable nature."

[Daniel C. Maguire, *Moral Absolutes and the Magisterium*, in "Absolutes in Moral Theology?", Edited by Charles E. Curran, Washington: Corpus Books, 1968, p. 76]

In its better moments, the ethical teaching of the Church has recognized that human nature is not static and so there can be no universal truths deducible from it. Since natural law is so unreliable, most of the moral direction provided by the Church is not based on a reasonable analysis of natural law but on revelation from a higher power which can provide more immediate sanctions. Most moral teaching and exhortation then is not done in the name of negative logical consequences of the activity under discussion, but on whether or not it is in harmony with what has been revealed to be good Christian behavior.

[See William Barclay, "Ethics in a Permissive Society", New York: Harper & Row Publishers, 1971, p. 14f, who makes the point that,

"The very first thing to say about the ethic of the Old Testament is that it is an *ethic of revelation.* In this case, if we start out by simply looking at words, then we get off very much on the wrong foot. The word *ethics* comes from the Greek word *ethos*; and *ethos* means a *habit* or *custom.* Are we then to say that ethics simply consists of habits and customs and conventions which have become fixed and stereotyped so that things which were once the usual thing to do have become the obligatory thing to do?

Take another word; take the Greek word for law - *nomos.* If you look up *nomos* in the Greek- dictionary you will find that the first meaning given for it is *an accepted custom.* Are we at the same thing again? Is law something which has become so habitual, so conventional, that it has finished up by becoming an obligation? Is it simply a case that the *done thing* has become *the thing that must be done*? Take still another word; take the Greek word for justice - *dike*; in Greek *dike* means *an accepted standard of conduct*-and obviously this is an entirely variable thing, quite different in one society from another, quite different in Central Africa and in the Midlands of England or the Highlands of Scotland. Are we back at the same thing again? Is justice simply stereotyped custom, habit and convention? When we talk about ethics, law, justice, are we really only talking about habits and customs-or does it go deeper than that?

In the Old Testament it goes far deeper, for, as it has been put, for the Old Testament *ethics is conformity of human activity to the will of God.* Ethics for the Old Testament is not what convention tells me to do, but what God commands me to do.

Second, the ethics of the Old Testament are rooted in history. There is one thing that no Jew will ever forget - that his people were slaves in the land of Egypt and that God redeemed them. To this day that story is told and retold at every Passover time. `You must remember that you were a slave in the land of Egypt and that the Lord your God rescued you' (Deuteronomy 7.18; 8.2; 15.5; 16.12; 24.18,22). That is the very keynote of Old Testament religion.

That saying has two implications; it means that for two reasons God has a right to speak. First, he has the right to speak because he did great things. Second, he has the right to speak because he did these great things for the Jews. The Jew would say: 'God has a right to tell me how to behave, for God has shown that he can act with power-and act with power for me.'

> For the Old Testament the idea of ethics is tied up with the idea of a covenant. A covenant is not in the Old Testament a bargain, an agreement, a treaty between two people, in this case between God and Israel, for any of these words means that the two parties are on the same level. The whole point of the covenant is that in it the whole initiative is with God. The idea is that God out of sheer grace - not because the nation of Israel was specially great or specially good but simply because he wanted to do it - came to Israel and said that they would be his people and he would be their God (Deuteronomy 7.6-8; 9.4,5).

However, that very act of grace brings its obligation. It laid on Israel the obligation for ever to try to be worthy of this choice of God."

Bishop Donald Wuerl in an article in *Columbia*, April 1999, p. 8f unites both and says that,

> "in our search for the 'objective norms of morality' that enables us to inform our conscience and keep it from error, we turn obviously to the word of God. While there is planted within us a natural moral law that inclines us to do good and avoid evil, God, in his infinite mercy, has chosen to speak directly to us. First he gave us the Ten Commandments. Then he gave us the Old Testament prophets. Finally, he sent his Son."

Susan was not taught not to lie to others because it damages the trust so necessary for human community-building, something inherent in the very definition of what it is to be human. No, she was taught not to lie because there is a commandment from god that she should not. Good thing, too, since lying does not seem to be excluded from the kind of behavior built into the human system. The more advanced of our primate cousins are perfectly capable of lying to the other members of the group when it suits their purpose. The basis for morality is seen as the revelation of god. Though the teaching of moralists is based on the "natural law" concept ["Catechism of the Catholic Church", #1954 f] and therefore the proscriptions are not viewed as arbitrary but rather logical deductions based on the nature of reality, for many a normal Christian they are seen as somewhat unfounded in reality or even whimsical or despotic. This is more true of laws that are only tentatively based on the natural law. Their purpose is seen not so much as saving from negative consequences but as a test of obedience. Why should there be a law that says one must attend Mass on Sunday? The explanation that the law is there to back up the value of gathering for the sharing of faith is not the normal one. Rather, it is presumed, god just picked out this day and commanded that everyone should keep it holy by attending Mass. As a result many Catholics respond to the law not by coming together to share their faith by song and gesture and prayer, but by mere physical presence, much like attending the king. It is presumed that going fishing or painting the porch would be more beneficial and enjoyable, and given the amount of enthusiasm with which many approach liturgy it surely would be. However, moral behavior is not based on doing the natural or desirable thing but on doing the commanded thing. This is certainly true in the areas of sexuality. Behavior is not determined by the goodness and rightness of the actions but by their lawlessness. Of course this does not square with the authentic Catholic view of sexuality, but inconsistency is nothing new to human thought and behavior. Too often the approach to sexuality is Calvinistic, requiring justifying reasons and the bond of matrimony for engaging in this (questionable) activity instead of promoting the goodness of the communication of the person to another through the right use of sex. The Catechism takes this positive approach when it says, "Sexuality, in which man's belonging to the bodily and biological world is expressed, becomes personal and truly human when it is integrated into the relationship of one person to another..." ["Catechism of the Catholic Church" #2337] but spends much more time treating the offenses against Christian sexuality than promoting its use in proper relationships and circumstances. Nowhere does it recommend that a dating couple even hold hands as a help in expressing the developing relationship, much less kiss. Still, aside from the puritans, most Catholics have a rather healthy understanding of the benefits and goodness of sex. Even if their understanding of the Church's position on it sometimes causes tinges of guilt after the fact, most find themselves acting, usually appropriately, first and asking questions later.

Not that there is no recognition among some, perhaps many, teachers in the Church that this is not the best criteria for making moral judgments. A popular author of a question-answer column for Catholics even years ago stated that,

> "...theologians today are having second thoughts about whether the Church can require the observance of her own laws [Sunday Mass attendance, for example] under the pain of mortal sin. She can and should make laws regulating worship, the administration of the sacraments, the organization of her institutions, etc. and some of these she can declare more serious than others. Still, it is hard to

> understand how the breaking of such rules could be serious sins in themselves...Recent moral theology...opposes the oversimplification of the recent past which described sanctity as keeping laws and sin as breaking them."
> [Raymond Bosler, "What a Modern Catholic Believes About Moral Problems", Chicago: The Thomas More Press, 1971, p. 12]

Even the morality of behavior which was formerly described as certainly "contrary to the natural law" is recognized today as much more relative with more room for the effects of human psychology. In a popular treatment of conscience meant for RCIA candidates, there is a distinction between the "classical method" based on natural law and the "personal approach" which it says is used by "the majority of theologians today" which takes into account personal circumstances and "the development of an idea or teaching through the years instead of focusing on human nature in the abstract."

> ["Catholic Update", Edited by Fr. Jack Wintz, O.F.M., Cincinatti: St. Anthony Messenger Press, 1982]

There is a greater recognition of the role of the Holy Spirit in enlightening decision making.

Marc Oraison, "Morality for our Time", Garden City: Image Books, 1968 had already reached the conclusion,

> "that for many centuries, at least in the West, the point of departure for moral speculation has not been the word of God but rather a confused jumble of valuable, rational truths and questionable (though unquestioned) traditions; of correct and fundamental intuitions, and ignorances that loom large in the light of modern science; of elaborated reasoning on good and evil and on a multiplicty of taboos." [p. 72]

This is becoming more understood by the non-professionals in the Church, but the basis for morality as a construct of natural law is still very much the norm in official responses to moral questions.

About as close as I get to this concept is just recognition of "reality". In the presence of gravity it is "natural" for a human being to move toward the earth at a constant acceleration. For someone to deny that this is so and to act on that denial will mean a rude awakening if they test it from too much height. When it comes to the physical parameters of the human's existence, there are laws set down seemingly "*ab initio*" which must be obeyed.

However, the morality problem really comes in human behavior towards other beings.

Is it a tenet of some "natural law" that one human should not enslave another? Most abolitionists were Christians who felt that indeed there was such a law and moreover that god was its author. One of their problems was to, "demonstrate that black people had human feelings and emotions," not to show that if they were human god intended them to be free.

[Howard Jones, "Mutiny on the *Amistad*", op. cit., p. 10. The book shows in many places that the existence of a god-given natural law was just presumed.]

Opponents argued that, "The `God of Nature' never meant the black race to live with whites `in any other relation than that of master and slave'." not to show that if they were human god intended them to be free

"Mutiny on the *Amistad*", p. 49 citing the New York "Morning Herald"

For an extended analysis of the attitudes toward slaves see Edward Ball, "Slaves in the Family", New York: Farrar, Straus and Giroux, 1998. Also enlightening is Neil Henry, "Pearl's Secret: A Black Man's Search for His White Family", University of California Press, 2001.]

Ecology is another example. Is it "natural" that humans should take over all the resources of the planet and completely overwhelm all other species? In one sense it is, since every species tries to propagate its own at the expense of others if need be. Still, we are also "naturally" a part of a whole web of life and our interdependence is well documented. We depend on photosynthetic plants for the oxygen we breath and so it would be unnatural to exterminate them. But then it is also in our nature to be technological. If we first invent a machine to turn our waste exhalations into inhalable oxygen, would it be alright then to exterminate plants? We seem to have a natural affinity for other animals. The creation of great parks and refuges attest to our desire to live in a world populated with other species. A selling point for saving the whales is that they are mammals like us. To eradicate all animals would seem to be unnatural. Yet there are many denizens of the city who spend much if not all of their lives without the company of animals of any kind, and those they see scurrying around their kitchen they eliminate. And these people seem to do very well without them. This is both natural and unnatural, depending on how one looks at it, but is one or the other good or bad? It seems as though the most natural thing among individuals within a species is to be in competition with one another, for food and sex to be sure, since this is seen in so many of the forms of life on earth. It is also true that the search for power over others, for security, for status, and for other intangible things that we attribute exclusively to human life and for which some people have felt a need for a non-physical explanation, are found in many other life forms as well and are not unique to us. Nor are the "rules" of the competition unique that find expression in many animal, and even plant, species. These are built into the human being as well, war being the extreme, but "keeping up with the Joneses" is quite popular, too. Even gossip is a way to triumph over another. To try to remove all competition seems to be a doomed enterprise. Frisbee throwing without seeing who can throw the farthest or the fanciest has been tried as recreation, but it soon palls. We compete for money, for fame, for food in desperate circumstances, and certainly for sex.

Take sex as another example. It is "natural" for the male of many species, our own included it seems, to satisfy the reproductive urges built into us by having as many partners as possible. A bull walrus and his harem is a slacker compared to some human males. And speaking of harems, Solomon and some Turkish beys are the envy of most men. So where is the "natural law" proposed by most religious systems that declare that a male should be faithful to one female? Of course it is in society's interest to see that the young are raised properly, and the norm seems to be that the male can only do well with the children of one female. Too many of these can tax his abilities quite enough. So it would seem to be natural, written into the essence of being human, that the nuclear family should be the norm. However, if the male possesses the resources to support his offspring from many partners, where is the natural law that forbids it? And though human females, as well as the females of some other species, seem to fare better by selecting just one mate, the best they can attract, and profit little reproductively from many matings, all humans seem to be programmed to, sometimes at least, use sex in another way than merely to procreate. We use it in social interactions as well, as do many of our primate cousins. While our [Catholic] society leads us to believe that sharing sex should always bring with it a permanent commitment to a family structure, to a marriage relationship, more and more people feel free to use it to express less deep relationships. After all, a handshake is a sexual act in being a physical contact meant to communicate the relationship of the two parties. It expresses little more than a promise not to raise one's hand to the other in this encounter, but it is not essentially different than the kiss or the hand holding expressing a slightly closer relationship, or the intercourse which can be understood by both parties to express a commitment to a relationship which can fall considerably short of that of marriage and family. The "natural law" adherents will say that there is something built into the human essence that determines what these sexual actions will automatically communicate, but it is often difficult to see where something essential to human nature comes into play and where we are merely dealing with the prejudices of a certain culture. Many societies have not met the criteria of Christian sexual morality yet the individuals seem to be quite well adjusted.

An ethnic group in China called the Mosuo,

> "prefer a visting relationship between lovers - an arrangement theyu sometimes refer to in their language as *sisi* (walking back and forth). At about the age of twelve, a Mosuo girl is given a coming-of-age ceremony, and after puberty, she is free to receive male visitors...Children born from such a relationship live with their mother, and the male relatives responsible for helping to look after them are her brothers."
>
> [Lu Yuan and Sam Mitchell, "Land of the Walking Marriage", *Natural History*, Vol. 109, No. 9, November 2000, p. 60]

The article mentions no problems arising from this form of social relationships. Presumably there are the usual, although the Chinese government seems to have a problem with it, much as the Vatican might have.

What does seem to be built into human nature is to use sex as merely recreational with little or no relationship implied or understood by either party. Prostitution is known as the oldest profession and seems to be endemic to every human culture. Granted it often is exploitative of one of the partners, but it need not be under all circumstances. Sacred prostitution is certainly not recreational sex but it has been a component of many religions and theologies with both sides feeling that their actions further their relationship with the divine, but not with the human partner. Homosexuality is not directed to reproduction, nor is prostitution, yet it may well be "natural", at least to individuals, in a very basic way. If it is written in their genes, then it is as natural for a homosexual to relate to others in this manner as it is to a heterosexual. Indeed, what makes the heterosexual want to kiss a person of the other sex on the lips but finds the same action with one of the same sex, especially for males, repulsive? It is obviously the same operative that makes the taste of meat unsavory to a herbivore, but delicious to carnivores and omnivores like ourselves. It is written in our genes and the way in which the brain operates on the input from the nose or other organs.

> [Homosexuality seems to be much more common in non-human species than we have supposed. See Jeffrey Kluger, "The Gay Side of nature", *Time*, Vol. 153, No. 16, April 26, 1999, p. 70 reporting on Bruce Bagemihl's "Biological Exuberance", St. Martin's Press. Some critics deny the relevance of his findings to human behavior. Be that as it may, it is mainly evidence that our behavior at least lies on a continuum with other animal behavior.]

What is written in our genes is the ultimate in what is natural, and seeking the pleasure of sexual union is certainly written in our genes. Jared Diamond has written a whole book whose title tells it all, "Why is Sex Fun?" in which he explains many of the characteristics of human sexuality. Many animals other than ourselves seem to experience the pleasure of sex, but not the interpersonal communication. We seem perfectly capable of keeping the two separate or uniting them in the one action. For centuries the Catholic position on sex claimed that the production of life was the only purpose of sex. Many a wife was told to submit to sexual actions which the husband demanded even when these actions communicated little or no love and relationship. To reject the attentions of a drunken husband on the grounds that there was no communication of love would have been incomprehensible to most confessors and moralists. Only recently has the Church's position returned to its authentic understanding of sexuality as containing also a relational aspect. ["Catechism of the Catholic Church", #2360-63]

Books such as Rollo May, "Love and Will", New York: Dell Publishing Co., Inc., 1969 did much to bring the Church up to date in its thinking about morality and sexuality. After considering the problems of modern society, May shows part of the solution in our understanding that,

> "love is personal. If love were merely a *need*, it would not become personal, and will would not be involved: choices and other aspects of self-conscious freedom would not enter the picture. One would just fulfill the needs. But when sexual love becomes *desire*, will is involved; one chooses the woman, is aware of the act of love, and how it gets its fulfillment is a matter of increasing importance. Love and will are united as a task and an achievement. For human beings, the more powerful need is not for sex per se but for relationship, intimacy, acceptance, and affirmation." [pp. 310-311]

The claim is made that our nature requires that the two aspects of intercourse, fecundity and relationship, be kept together in every sexual act. The encyclical, *Humanae Vitae* (section 10) says that this is the plan of, "God the Creator, a plan made manifest...by the very nature of marriage and its acts," but the demands of our natural genes say otherwise.

[I will not go into the scientific data here. Suffice it to say that there is much evidence that falling in love is the result of a hormonal system designed to bring parents together. This is not a permanent situation. The hormonal effect diminishes after the offspring are more capable of survival, typically about 7 years. There is another hormonal reaction designed to keep the couple together, but the relationship changes. The Church would do well to teach people about this to prepare them for the process and indeed the rewards of the often deeper relationship can far outweigh the initial hormonal sexual attraction which brought the couple together in the first place. However, for the Church to declare that there is some natural law proscribing divorce and that children-producing pairings lasting for a lifetime are natural seems to be more than suspect.

[The September 25th, 2000 issue of *Time* magazine can give a quick report on the status of the question of divorce at that time in the U.S.]

Who we are and what our nature requires of us to do is whatever in fact we do. Not as individuals, it is true. One individual may act against the nature of the human animal by jumping from a high place thinking they can fly. Or one can have many sexual encounters with absolutely no relational content to the actions. Because of the nature of the beast these individuals will likely come to some harm, emotionally or from the other's spouse. However, whatever the whole species does collectively must certainly be considered "natural" even if destructive. And since any species that contains destructive elements in its composition is not going to last very long, it will also be natural if we exterminate ourselves. Not "good" or "evil", but natural. We are going to have to find some other basis for our morality than what is innate in ourselves, either as a species or as individuals, which we might call a "natural law", but we are also going to have to take what is natural into account and not base our moral decisions on some external revealed source.

405. The faith has not always believed in angels and devils. Certainly there was a long period before their exposure to the Persians that our Hebrew ancestors never heard of genii and perii and the like and cherubim were only guardians at the temple of Ishtar in Babylon for example. Of course the Hebrews probably believed in all sorts of lesser gods and goddesses as everyone of their times did, against whom the prophets continually complained, who stood in the place of the angels and devils of today. If Yahweh did not grant them a boon, they may have readily tried the god down the street or on the high places. Nevertheless, this belief in angels and devils did enter the corpus of the faith and the Scripture and remains there today.

Satans did not start out in the Scripture as evil forces. In the Book of *Numbers* a Satan is sent to bar the way of Balaam, a man intent on an evil purpose. The Satan is not a messenger but a barrier. In the Book of Job the Satan is sent to test the man but not to do evil. However, somewhere a shift occurred and the Satans became not a barrier to evil but the personification of evil. Ever since it has become the custom to demonize other people whom we consider to personify evil.

Lucille A. Roussin, "Helios in the Synagogue", *Biblical Archaeology Review*, Vol. 27, No. 2, March/April 2001, p. 53 in treating the appearance of astrological figures in Jewish mosaics, points out,

> "...that the sun, moon and stars were revered as angels is evident from the text of Psalm 148:1-4: `Praise the Lord out of heaven; praise him in the heights. Praise him, all his angels; praise him, all his host. Praise him, sun and moon; praise him, all you shining stars; praise him, heaven of heavens.'" This was evidently a problem in later Judaism and gradually dealt with by a proscription of images. Roussin points out that the early Christians were on the side of the rabbis who objected, but belief in angels remained in Christianity as well.

We should apply the same skepticism about the existence of angels or devils as we should apply to the presence of alien life forms, but we also have to admit that no one can prove anything either way. Our investigation does, however, provide us reason to re-examine our thinking about angels and devils since they have always been explained as messengers from a god who cares to intervene in our daily affairs. Granted that the existence of helper spirits or enemy spirits does not automatically imply the existence of a god or vice versa, but certainly the Christian faith has always insisted on the connection.

Many people do believe in the personal attention of some supernatural being who causes what others may consider coincidences, or may believe in devils who do the same, though with evil intent, and some believe in both, but most probably live their lives without giving either of them a thought unless someone reminds them of their existence. C.S. Lewis, author of "The Screwtape Letters" of correspondence between devils employed in the degradation of humans, professes to believe in their existence but also admits that, "my religion would not be in ruins if this opinion were shown to be false."

> [C.S. Lewis, "The Screwtape Letters", New York: Macmillan Publishing Co., Inc. 1961, p. vii]
>
> However, Arthur Miller in his play "The Crucible" about the Salem witch trials, notes that one of the witch hunter's "...`lines has never yet raised a laugh in any audience that has seen this play'; it is his assurance that `We cannot look to superstition in this. The Devil is precise.'"
>
> [Arthur Miller, "The Crucible", New York: Penguin Books, 1952, p. 33]

We may not want to talk about the devil much, although he is a standard in folk tales.

> [My favorite is about the devil's complaint that there is no justice. He get's god to put a stone in the path and places a bag of money on it. When someone stubs their toe on the stone they say, "What the devil, I stubbed my toe." But on finding the money they look up and say, "Praise God." There ain't no justice.
>
> ["Afro-American Folktales" Ed. by Roger D. Abrahams, New York: Pantheon Books, 1985, p. 78]

Still, the devil's popularity in plays, movies and TV shows that belief in a person who is evil is very much around, at least enough that Hollywood can exploit it. Even today when people view the movie, "The Exorcist", or reading the book, many people are deeply troubled.

> [See William Ryan, "`The Exorcist' at 25", *Catholic Heritage*, Vol. 8, No. 1, November/December 1998, p. 22]

The tendency to wish for Saints, angels or fairy godmothers is understandable in a universe that can often seem, if not against us, at least indifferent. How healthy this faith is depends on whether or not these beings actually exist.

If we must concede the possibility, indeed the probability, that god does not intervene in the workings of the physical world because god cannot do so, we must also conclude that god's agents do not do so either, for if god's agents can interfere, then god can, and that is what is being questioned. There is no room for angels literally throwing us a rope to get us out of a tight spot. We do not look for god to do so nor for some extension of god. If anyone wants to still maintain that there are angels who help us in this life, and for some strange reasons angels are more popular than ever, the burden of proof is on them to show that what they claim is not some hallucination or wishful thinking or fortuitous coincidence. It is impossible for anyone to prove that something does not exist. If the Irish want to say that leprechauns are more than a figment of imagination or folklore, who can say them "nay". And the anecdotal evidence can be quite strongly claimed, just as it is for little green men or other alien beings, or for the Loch Ness monster for that matter. Most people would say that if you choose to believe in these things, that is fine, but don't look surprised if others think you the fool. We definitely need to re-think our commitment to angels and devils being included in the argument for the faith.

406. The standard understanding goes as follows. The plan of god is not limited to living this life on earth. It is directed to a continuing life with god forever. Indeed, the purpose of this life is to prepare us for the next life, which is the real life. Many popular prayers speak of our being "in exile" on this earth, almost as if we had previously existed in heaven and then been cast out with the mission to work our way back again. The "*Memorarae*" prayed after the rosary speaks of this. It is of course a very Platonic notion. Augustine reminds his hearers that, "we are Christians and our homeland is not here. Like good children, let us turn our steps homeward, that our course may be approved and guided to its conclusion."

> ["Augustine Day by Day", Ed. by John E. Rotelle, O.S.A., New York: Catholic book Publishing Co., 1986, p. 65 from Sermon 16A, 13.]

We are created to be citizens of heaven and our life as citizens of the earth and the universe is merely a prelude to what we are called to be, in fact what we really are.

> "In the light of the mystery of Christ and creation man can begin to glimpse his true reality, his real identity, which is in Christ. He can begin to appreciate the value of things created and redeemed in Christ; above all the value and meaning of the human person,

created and saved from their own selfishness to partake in Christ's own eternal self-giving reality before the Father."
["Theology of Creation" p. 89]

It was not the intention of the creator to build a world in which there would be a succession of beings who could respond in some way, but then die to be replaced by others. No, each responding person is to be kept in existence forever. Life in this world is not only a prelude to life in the next, it is a testing time and provides the winnowing-out period for human destinies.

"Each man receives his eternal retribution in his immortal soul at the very moment of his death, in a particular judgment that refers his life to Christ." ["Catechism of the Catholic Church" #1022]

This "particular judgment" is required since the commonly accepted faith is that each person continues to exist as a disembodied soul, though there is a strong movement among many Christians to allow that perhaps a person does just "die" in the sense of cease to exist as a person until god gets around to raising everyone at the same time. This is a particularly inviting concept to those who are experiencing problems with the idea of the reality of a "soul". In any case, when the world has served its purpose, it will be replaced with a new creation that will be initiated by the resurrection of everyone's body, which, if you believe in a soul, means a reunion of the physical body with the spiritual personality that had continued to exist after death. If you do not believe in a soul it means the coming back into existence of the human person after a period of non-existence. Then comes the day of reckoning.

"The Last Judgment will come when Christ returns in glory. Only the Father knows the day and the hour; only He determines the moment of its coming. Then through His Son Jesus Christ He will pronounce the final word on all history. We shall know the ultimate meaning of the whole work of creation and of the entire economy of salvation and understand the marvelous ways by which his Providence led everything towards its final end. The last judgment will reveal that God's justice triumphs over all the injustices committed by his creatures and God's love is stronger than death."
["Catechism of the Catholic Church" #1040]

God will also pronounce the final judgment on each human person, separating the sheep from the goats. A character of Zora Neale Hurston says that the women of the town go to church so that,

"...they'll be sure to rise in Judgment. Dat's de day dat every secret is s'posed to be made known. They wants to be there and hear it *all*."

[Zora Neal Hurston, "Their Eyes Were Watching God", Urbana: University of Illinois Press, 1937, p. 16]

This general judgment is a little anti-climactic if sentence has already been passed at the individual judgment at the time of death when each person received their just deserts, "either entrance into the blessedness of heaven - through a purification or immediately, - or immediate and everlasting damnation." ["Catechism of the Catholic Church", #1022]

The universe, having been created for god's purpose of housing those to be tested, now becomes rid of the evil that had infested it. For some this is done by a total recreation of the universe, necessary because as physical beings we need somewhere to live. Over the past decades, in response to an understanding that we are creatures of this universe, not just temporary residents, the church has come to believe that,

> "the visible universe, then, is itself destined to be transformed, `so that the world itself, restored to its original state, facing no further obstacles, should be at the service of the just,' sharing their glorification in the risen Jesus Christ."
>
> ["Catechism of the Catholic Church" #1047]

Pope John Paul writes in this vein in pointing out that,

> "it can be said that until recently the Church's catechesis and preaching centered upon an *individual eschatology,*...The vision proposed by the Council, however, was that of an *eschatology of the Church and of the world.*"
>
> [John Paul II, "Crossing the Threshold of Hope", New York: Alfred A. Knopf, 1994, p. 181]

407. Not that there have not always been many approaches to understanding the resurrection. E. L. Doctorow in "City of God", New York: Random House, 2000, has a character explain one period in the history of the doctrine very well when he says,

> "Pagels, working from the scrolls discovered at Nag Hammadi in Egypt in 1945, finds that the early Christians were profoundly divided between those who proposed a church according to apostolic succession based on a literal interpretation of Jesus' resurrection and those who rejected resurrection except as a spiritual metaphor for gnosis emotionally, mystically achieved, as knowledge beyond ordinary knowledge, a perception beneath or above the everyday truth...So there was a power struggle. Gnostic and

synoptic contested with competing gospels. The Gnostics...were routed"

Doctorow's book is a novel, but his interpretation is correct. Doctrine by political process is nothing new in the Church.

408. It is the resurrection of Jesus' body that is the inspiration for all Christian dogma. Without it, it could all be put down as wishful thinking, but if Jesus is really experienced in the Church as risen and present in our world then there is hope that the promise of resurrection for the rest of us is on solid ground. Some try to maintain that there are empirical reasons to believe in the resurrection of Jesus. For a fundamentalist, the fact that it is in the Scripture which is the word of god is proof that everyone should believe, ignoring that there is no empirical proof that the Scripture is in fact god's word. However, the Bible is a written document claiming witnesses who experienced Jesus as a flesh and blood presence and this has some credibility. At least as much as those who claim to have evidence of King Arthur of the Round Table. Swinburne claims that

> "The body of Jesus was never produced. Here we have a serious historical claim of a great miracle for which there is a substantial evidence." Though he admits, "Just how strong that historical evidence is is a matter on which innumerable books have been written over two millennia, and readers must follow up these issues for themselves in some of these books." ["Is There A God" p. 126]

Dermot Lane, points out that the story of the empty tomb was well known and could not have been maintained if there were not some historical truth. However, he is at pains to say that,

> "From a theological point of view it should be noted in passing that the object of Christian faith is the resurrection of Jesus and not the empty tomb as such."
> "*The Reality of Jesus*", chapter 4 of "The Resurrection: A Survey of the Evidence," p. 55]

For most Catholics it is just a matter of belief, but a belief buttressed by what they consider experiences of the presence of Jesus as a real figure. Sometimes this is just an unconscious belief instilled in childhood and not questioned; sometimes an experience of the sacramental presence; sometimes a perception of his presence as a force in the Church community; and sometimes a conviction based on a quasi-sociological argument that if Jesus had not been working in the world it would be a much worse place than it is. Whatever the person's reason to believe, it is true to say that when asked most Catholics will admit to a belief in the historicity of the resurrection of Jesus from the tomb on Easter Sunday. Many theologians point out that it is not a "historical" event in the usual meaning of the word because it did not happen in our history or our time but in god's history that has no time.

Hans Kung, says,

> "Since according to New Testament faith the raising is an act of God within God's dimensions, it can *not* be a *historical* event in the strict sense: it is not an event which can be verified by historical science with the aid of historical methods. For the raising of Jesus is not a miracle violating the laws of nature, verifiable within the present world, nor a supernatural intervention which can be located and dated in space and time...What can be historically verified are the death of Jesus and after this the Easter faith and the Easter message of the disciples. But neither the raising itself nor the person raised can be apprehended, objectified, by historical methods."
>
> ["On Being a Christian", Translated by Edward Quinn, Garden City: Doubleday & Company, Inc., 1976, section C, V. especially p. 349]

On page 350 he points out, however, that,

> "...this cannot be a merely fictitious or imaginary but in the most profound sense a *real* event. What happened is not nothing. But what happened burst through and goes beyond the bounds of history." It is a transcendental happening out of human death into the all-embracing dimension of God."

This makes it no less real an event, though not historical in the same sense as the Battle of the Bulge. Others explain it as only one of many ways of saying that Jesus continued to exert an influence after death as he did before death.

> [John Dominic Crossan, "Jesus, a Revolutionary Biography", New York: HarperSanFrancisco, 1994, see chapter 7]

All these explanations are largely lost on most Catholics, however, and they prefer to believe that on a certain day the body of Jesus rose from the tomb and if we had been there with a recording device something would have registered. The Shroud of Turin has been touted as physical proof and that it shows scientific evidence of physical processes which impacted the cloth of the burial shroud and left their mark. The Shroud has been disavowed officially by the Church, but for some it remains a potent proof.

[David Van Biema, "Science and the Shroud", *Time*, Vol. 151, No. 15, April 20 1998, p. 52.]

409. The Assumption of Mary is poorly understood by many Catholics, being akin to the "apotheosis" of one of the Greek heroes, or even of George Washington who is depicted in a famous painting being taken to heaven in a blaze of glory. Nice for Mary, but having little relevance to me. The feast was instituted in the Church as a holy day of obligation not only to honor Mary but to place her as an example, a "down payment" in fact, on what god has in mind for every human person. Fulton Sheen makes a telling point when he notes that,

> "in 1854 the Church spoke of the Soul in the Immaculate Conception. [Because of the temper of the times]...in 1950 its language was about the Body: the Mystical Body, the Eucharist, and the Assumption." [Most Rev. Fulton J. Sheen, Ph.D., D.D., "The World's First Love", New York: McGraw-Hill Book Company, Inc., 1952, p. 119f]

Even though one does hear questions such as "where, then, is her body?" or "where does Mary live?", just as the resurrection of the physical body of Christ raises similar questions, most Catholics rightly do not bother themselves with these matters but with matters of the relevance of the doctrine to their lives.

410. There are many anecdotes of purported signs of the continued presence of the deceased, at least for a while. On the lighter side, families have been known to joke, when returning from the funeral, that they had better not swat a pesky fly because it might be Uncle Harry. But there are more serious feelings that the dead are somehow still able to listen in on our conversations, take pride in our accomplishments, or be ashamed of our failings. There is certainly a resemblance here to the ancestor worship of some societies. Maybe they spring from the same roots. Even an avowed unbeliever as Carl Sagan reports that

> "Sometimes I dream that I'm talking to my parents...When I wake up I go through an abbreviated process of mourning all over again. Plainly, there's something within me that's ready to believe in life after death. And it's not the least bit interested in whether there's any sober evidence for it."

He adds the information that,

> "More than a third of American adults believe that on some level they've made contact with the dead."

The Church doesn't say much about the question and individual pastors are very wary of agreeing that the reported experience was genuine, having seen many examples of delusion. On the other hand, the Church has long taught people to pray to the Saints who hear the prayer and respond to it, thereby showing an effect of the living on the dead and of the dead on the living. Canonization is not the instrument that makes it possible for the person to be contacted, they must show evidence of being real before they can be declared a Saint, so presumably if grandmother is in heaven she can behave in the same way as Saint John Eudes. Being canonized just gives an assurance of the person's accessibility.

411. The statues of saints proclaim the belief that these saints are not dead, at least in the sense of "dead and gone", but are in contact with the living. This is based on the assumption about god as a god who is the author of life. God brought each person into existence, perhaps through the process of evolution, but directly in the creation of a soul, and god knows each person. God also knows the kind of person they are and some are brought to be with god in heaven. In spite of the belief that a human person is both body and soul, and the bodies of the saints are very much dead - and sometimes spread in splinters and pieces through thousands of churches - the saints can still have an influence on the universe. They can, it is believed by some, bring about cures and manipulate objects and are beseeched to do so. Still, their more important function is to intercede with god to bring about these desired results. The prime intercessor, of course, is Jesus and many Christians pray exclusively to him. He is not only an intercessor with god but is himself god and therefore able to make things happen in the universe. This can lead to a bit of confusion in some prayers, but is basically understandable. What becomes a little murkier is the role of the saints. Mary, in particular, is seen as a "*mediatrix*" of all grace, meaning that any action god has on humans is mediated through Mary. Pope John Paul is reported to want to declare her as such and to go farther and say that all salvation comes through her, though the theologians he polled are said to have advised him not to do so since Christ is the only mediator.

> [See David Scott, "Debating new dogma for a new millennium", *Our Sunday Visitor*, August 10, 1997 for an overview at the time. This action of naming Mary as a *co-redemptrix* has not been taken as yet.]

It has always been problematic to posit that Jesus-God is all-knowing and all-loving but if he is not reminded by his mother now and then he tends to forget someone or other.

> [The Knights of Columbus' magazine, *Columbia*, Vol. LXXVII, No. 10, October 1997, reports on p. 17 that a certain Cardinal Turcotte said at an annual meeting, "I pray that through Mary's intercession God will guide your deliberations." Presumably if you follow his reasoning, there is a chance that if Mary did not ask, god would not guide.]

Or perhaps Jesus is just not motivated to do something good for the person that his mother is motivated to do. The situation becomes even more complicated when other saints are brought into the picture. The mother of Mary, known through legend as Ann, is asked to "...recommend me to [Mary]. She will refuse you nothing..." [*Novena to St. Ann.*] Now we have a saint interceding with a saint who presumably intercedes with the high king. This worked well with the members of the court of Louis XV, but it seems a little strange to have to go through all this politicking in order to win a hearing from a god who knows and cares about each individual. And for real impossible cases, forget Ann and Mary and go for St. Jude!

> [For a treatment of novenas in general and for the role of the Saints prayed to in a novena, see Barbara Calamari and Sandra DiPasqua, "Novena", *Catholic Digest*, Vol. 63, No. 9, July 1999, p. 6. The article gives a history of novenas, which are about 1000 years old, and a balanced view of this prayer form, including asking the question, "why do novenas work?", leaning away from superstition or even to the intervention of the saints and toward the effect on the pray-er of taking the time to pray.]

In addition to hearing prayers, Saints are believed to take the initiative for those who bear their name. The custom of naming a child after the Saint whose feast was celebrated on or near their birthday was not just to give the child someone to look up to, but to give them someone who was looking down on them in protection. Gueneviere of "Camelot" fame gets upset with her patroness when she has not protected her from an unwanted marriage, but changes her mind when the groom turns out to be King Arthur. The birth of twins or triplets on the same day has led to some interesting situations with all three boys named Theodoric or two girls, both named Cunnigunda. While this approach is found in some Catholics, most others get along quite nicely without devotion to the saints. . Most Catholics, except again the ultra-right, don't get too agitated over the Saints and no longer have the same outward devotion to them as we used to see, possibly because access to Jesus has been restored through bringing the altar to the people instead of against a wall or hidden by a screen. On the other hand, many a Catholic has been inspired by the example of the life of a Saint and has done well by patterning some of their behavior or qualities on that of the holy one

412. It is difficult to say what most Catholics believe about heaven.

[See David Van Biema, "Does Heaven Exist?" in *Time* Vol. 149, N. 12, March 24, 1997, p. 71f for a comprehensive evaluation of the state of belief in heaven in contemporary society, including among practicing Christians.]

"What the modern Catholic is experiencing is a failure of the contemporary imagination to grapple with the expanding universe. Heaven and hell are theological realities in search of scientific grounding."

[John Shea, "What a Modern Catholic Believes About Heaven and Hell", Chicago: The Thomas More Press, 1972, p. 11. Though written shortly after the Second Vatican Council, the situation is about the same today.]

At one time the streets of gold and the pearly gates and the young dancing girls peeling grapes for one's delectation may have been an appealing image, but today there is more sophistication.

Heaven was, of course, originally the place of a god's residence and humans were not to be found there (think of Mount Olympus of the Greeks), at least until some belief in a possible continuation of life after death came into being, and then only for the ruling class. The pyramids of Egypt and funeral rights of the pharaoh look for him (or her) to enjoy the delights of the gods.

In the Old Testament we have the story of the garden of Eden where humans did live, presumably in god's garden until they were ejected. There is an interesting explanation of royal residences and temples being built as if they were god's dwelling on earth, with the king and retainers enjoying some participation in it.

[See Lawrence E. Stager, "Jerusalem as Eden", *Biblical Archaeology Review*, Vol. 24, No. 3, May/June 2000, p. 37]

Believing in heaven as the fulfillment of all the potential of our being, including the potential to love and be loved, is a more nuanced view and precisely how this will come about most are willing to leave up to god. There are some who have difficulty with the problem of eternity, especially an eternity of strumming a harp singing the praises of god. This seems like a rather boring existence even after a short period, much less for an eternity, but for most the idea of eternity and its' implications are not considered. What is of more pressing concern for most Catholics, and most Christians, is whether or not they will be reunited with their loved ones. Will I know my mother and be able to experience her love once more? Will they still be older than I am? Will my pet turtle or Old Yeller be there? In pursuit of assurances that death is not a final separation from those close to us, most Catholics are willing to await reunion in heaven. Some replace resurrection with re-incarnation, or they replace Purgatory as a time of growing out of our old self and having a second chance at heaven, or perfection, with being placed back into the world for another try. Just how this is supposed to happen since we don't start from where we left off but from scratch, is a bit of a question. In response to it there are efforts to surface memories of a prior life. Om Seti was a famous elderly woman who moved from Europe to Egypt because she claimed she had lived a prior life there. At least she had the decency not to claim that she was Cleopatra or someone famous, unlike some of the movie stars who hardly ever seem to have been the village idiot in another life. There are a good number of Catholics who are willing to listen to these speculations, but most, when challenged, are not very committed to them. After all, every Sunday they attend Mass they say in the Creed that they believe "in the resurrection of the body", and such repetition has an effect.

For most people the questions are still about an individual's fate after death and judgment. For the individual person, most Catholics say they believe that he or she will go to heaven or to hell, though the concept of an everlasting hell is under suspicion for many. Then there are those who feel that the old vindictive god they knew as children before Vatican II is the real god and that they will see the vindication of their beliefs when the many are found in hell and the few in heaven.

> [St. Anthony Mary Claret devised his own version of St. Ignatius of Loyola's Spiritual Exercises. A section on Hell was printed in *The Fatima Crusader*, Issue 59, Spring 1999 and though it says that the purpose of the meditation is to scare the sinner away from sin, it seems to take some delight in explaining that "One who is damned

loses God's devoted, special Providence which cared for him. As long as a man is alive, he is under the care of God...But a soul which has entered the eternal abyss must hope for none of this. God no longer cares about it and regards it as something which no longer belongs to Him." It then goes on to describe the existence of one in such a state, including torments by the Devil who, "...can twist into the form of a serpent, enter the body, and torment him cruelly with his teeth. As a poisonous snake he can enter the mouth and bite and gnaw and destroy lungs, liver, heart and all the bowels. He can make his victim swallow molten metal and feed him poisonous toads."

Presumably they grow back so the devil can do it all over again. After all, once in eternity would not be too bad.]

Many believe in hell, not as a place of punishment but as the natural consequence of choices the person makes in life, expressed in a lack of relationship with god. David Mills, "Let's Hear It for Hell", *Columbia.* Vol. LXXXI, No. 6, June 2001, p. 10 makes a good point about hell when he calls it a compliment from god, respecting our freedom to make choices. Those in hell, he speculates, are "...forever bereft of every good thing, forever in a state of war and hatred with God, with his fellow creatures and with himself...forever mad, forever enraged, forever impotent and forever lonely." This is certainly an improvement over Claret's views and as he quotes Pope John Paul, "the thought of hell...is a necessary and healthy reminder of freedom within the proclamation that the risen Jesus has conquered Satan...", a reminder that may lose some psychological force if we define hell as non-existence. However, non-existence does not require engineering from god whereas a hell of punishment certainly does. A hell of loneliness is in the middle.

Many presume they will be among the damned. For those who go to heaven, they may do so "immediately" as the Catechism states, or "through a purification" which is known as Purgatory. Formerly this was looked upon as a part of hell since it's purpose was to punish and to burn away the sins that the person had committed. Such suffering could be shortened by gaining indulgences. St. Louis Mary de Montfort, defines an indulgence as,

> "a remission or relaxation of temporal punishment due to actual sins, by the application the superabundant satisfactions of Jesus Christ...A Plenary Indulgence is a remission of the whole

punishment due to sin; a partial indulgence of, for instance, one hundred or one thousand years can be explained as the remission of as much punishment as could have been expiated during one hundred or one thousand years, if one had been given a corresponding number of the penances prescribed by the Church's ancient Canons."

["The Secret of the Rosary", Trans. by Mary Barbout, T.O.P., Bay Shore: Montfort Publications, 1964, p. 108f,]

This is actually a good explanation, putting the exacting of punishment or reparation on the Church and not on god. But Montfort goes on to cite an apparition of a deceased young woman who explains that,

"she had been condemned to seven hundred years of Purgatory...She implored him to ease her pains by his prayers...Two weeks later she appeared to him, more radiant than the sun, having been quickly delivered from Purgatory through the prayers."

If the "punishment" is merely penal, then it makes sense to forgive it if someone else "posts bail". If it is for personal growth, it makes little sense to remove it since the growth is for the good of the deceased and is not punishment at all.

Incidentally, along the lines of one person's gaining an indulgence for another, the Church explains that it can dispense and apply with authority its spiritual treasure which contains the infinite merits of Christ as well as the prayers and good works of Mary and the saints, living and dead. All this is explained in the "Enchiridion of Indulgences" of 1968 and the "Catechism of the Catholic Church", #'s 1471-1479

The idea of punishment that is due for our sins and not forgiven by the Sacrament of Reconciliation is still with us. Writing in the January 2000 edition of "*Columbia*", a magazine for the Knights of Columbus, Greg Burke in "No Pain, No Spiritual Gain" says,

"...when we go to confession, the sins are forgiven, but the temporal punishment remains. And that's the price we have to pay for our sins, either in this life or the next. So we either do penance for our sins here on earth, or we have pain inflicted upon us in purgatory."

This pain is evidently demanded by god, for who else has such power?

In our day, many, if not most, see the idea of purgatory as the "antechamber of heaven, not an adjunct of hell" where, or by which process, we can grow into a person worthy of full admittance and awareness of god's love. Jay Copp, points out that,

> "As far back as the 16th century, St. Catherine of Genoa tried to dispel that image [of purgatory as a place of suffering]. In her groundbreaking Treatise on Purgatory, she described it as less a torture chamber than a place of joy where souls anticipate meeting God. 'There is no joy save that in paradise to be compared to the joy of the souls in purgatory,' she said, describing her vision."
> [Jay Copp, "Since Apostolic Times, Christians Have Prayed for the Dead", in *Catholic Heritage*, Vol. 8, No. 1, November/December 1998, p. 13]

A popular non-Catholic evangelist says that,

> "Although Purgatory is primarily a Roman Catholic notion, the psychiatrist in me takes to it with ease. I imagine Purgatory as a very elegant, well-appointed psychiatric hospital with the most modern and highly developed techniques for making learning as gentle and painless as possible under divine supervision."
> [M. Scott Peck, M.D. "Further Along The Road Less Traveled" New York: Simon and Schuster, 1993, p. 168]

413. Andre-Marie Dubarle, O.P. "The Biblical Doctrine of Original Sin", New York: Herder and Herder, 1964. See chapter 7

In 1962 Rachael Carson published her famous book, "Silent Spring", which began by saying that,

> "The most alarming of all man's assaults upon the environment is the contamination of air, earth, rivers, and sea with dangerous and even lethal materials. This pollution is for the most part irrecoverable; the chain of evil it initiates not only in the world that must support life but in living tissues is for the most part irreversible." [Greenwich: Fawcett Publications, Inc., 1962p. 16.]

Carson was not speaking of original sin or of religion at all, but this kind of observation of evil being passed down to others and contaminating all that it touches is part of the biblical and theological notion of original sin.

414. See Jerry D. Korsmeyer, "Evolution & Eden", New York: Paulist Press, 1998 for an up to date interpretation of the idea of original sin and the impact of science on the concept.

415. This interpretation of the Sacrament of confirmation is still in flux. My guess is that most pastors and those in contact with the people in the

parish are in favor of Confirmation as a Sacrament of Christian Maturity. On the other hand, many theologians, followed by the bishops, want to emphasize it as one of the three Sacraments of Initiation and place it at a much earlier age. Both aspects are legitimate, of course, and time will tell which wins out.

416. Carlo Carretto, in "Why O Lord?", Maryknoll, New York: Orbis Books, 1986, would logically have to refuse the power to help innocent victims since in his chapter 9 he calls them "blessed" as martyrs. Evidently he would not like to save them from such blessings but then most of us do not have his compassion.

417. Walter Kasper, "The God of Jesus Christ", Trans. by Matthew J. O'Connell, New York: Crossroad, 1997, p. 159

418. On the question of the presence of evil, the more sophisticated will put god as far as possible from responsibility for evil in the world and will do so through philosophical and theological arguments. Some of the faithful solve the problem by finding demons (or angels) involved in the minutest occurrences. The devil is busy, they say, and can easily be involved in hiding one's car keys to prevent attendance at church. On the other hand, St. Anthony will help you find them again and your guardian angel may well rescue them from the devil and place the keys where you can easily find them, even though you have no memory of having placed them there. One can deride this belief as no better than belief in leprechauns and genies in bottles, but it is merely the carrying to extremes the belief that everything must ultimately have some explanation and even some purpose. Belief in god's control, either directly or through intermediaries, and either proximately or remotely, is certainly in the scripture and is basic to Catholicism and most or all of Christianity yesterday and today. In a report on a deadly tornado in Alabama that took the lives of 32 people, a survivor tells of his faith, "We feel it was by the grace of God that we were saved." The article relates that,

> "...his wife had kept a winged stuffed animal `guardian angel' on the living room mantle. While sifting through rubble after the tornado, the family found it in what had been the boy's room. To get there, it would have had to sail right over their heads as the storm was tearing the house apart...`To me that's just another sign that God was really protecting us." [*Extension,* July 1998, p. 8]

Why god thought more of this family than any other is left unexplained, and unexamined.

Stuffed angels and devils aside, how can god allow the presence of evil in a universe created good? This is the question that Susan has asked and that every Christian eventually asks. Some answers are rather flippant, as a catechism entitled "*We Believe*" tries to turn the tables and insist that if someone wants to deny god by bringing up evil, then "non-believers must answer a far more difficult question, `If there is no God, why is there good in the universe?'" somewhat ignoring the unbelievers possible retort that perhaps there is nothing good or evil in the universe.

> [Oscar Lukefahr, "We Believe", Liguori: Liguori Publications, 1990, p. 7]

Pope John Paul II connects the presence of evil and its relation with god to the "*Scandalum Crucis*," the scandal of the cross. God is in solidarity with humanity in its fight against evil.

> "He is not the Absolute that remains outside of the world, indifferent to human suffering. He is Emmanuel, God-with-us, a God who shares man's lot and participates in his destiny."
>
> ["Crossing the Threshold of Hope", p. 62]

The Pope honestly admits that this does not answer the question. It all comes down, in John Paul's thinking, to the free-will argument, and "...confronted with our human freedom, God decided to make Himself `impotent'." ["Crossing the Threshold of Hope" p. 64]

The price of our freedom to choose to do evil is god's being falsely accused and misunderstood.

Another approach to explaining the presence of evil takes a little introduction. Just as humans feel that the sinner should suffer to repay the suffering they caused, hence the penalty nature of many penal systems, so there has been a feeling that god would want the same. There was an explanation that our human and therefore finite suffering could never repay the suffering we had caused an infinite god, ignoring that the suffering of the prisoner never really did much to restore the damage done to the victim, either, but we wanted it anyway. However, the suffering of an infinite human/God could "atone". A 1950's style approach was that,

> "Christ was given grace...in view of His purpose in coming upon earth. The office He was to perform was the satisfaction which the Head was to make for His guilty members. The abundance of His grace, therefore, disposed His soul to obey the decree of the Trinity

concerning the cross and all its concomitant sufferings whereby strict satisfaction was to be made." [Louis Chardon, O.P., "The Cross of Jesus", Trans. by Richard T. Murphy, O.P., St. Louis: B. Herder Book Co., 1957, p. 27.]

Susan may have heard this explanation at times, but probably in recent years was taught that god is not one to exact revenge or a pound of flesh. However, something must be done about our lack of loving response to god. It is disobedience to god's direction that landed us in our sorry state, and so it is an example of obedience that is needed. It is by the obedience of Jesus to the Father's will that he should be crucified that we can learn our lesson. And so the price of our redemption in the blood of Christ is not to be thought of as an eye for an eye but as a supreme example. We now know what is expected of us and what we are capable of doing. Jesus has shown the way and now we must take up our cross and follow him. The cross is not a payment to an angry god but the conjunction where the ultimate evil makes contact with the ultimate love.

> "Evil brought to its essential self-expression necessarily encounters and succumbs to the love of God...*That* is the sign of God's all-accepting love, *because* it is the at last adequate sign of evil."
> [Sebastian Moore, "The Crucified Jesus Is No Stranger", New York: The Seabury Press, 1981, p. 8]

If we subscribe to this line of thinking, lo and behold, we have an explanation for why god might afflict Susan with an abusive husband, why her son might have a learning disability and why her sister cannot have a child. This is the cross that they must carry in obedience to god's will, but it is a loving will and if they will carry it to their deaths it will lead to a like resurrection. For this is the bedrock of the faith. The man Jesus who listened to god and responded to god's love in obedience to god's will was raised up from the dead, or perhaps raised himself since he was god. The good do get their reward, though most often after this life.

This is certainly the theology espoused by Graham Greene's fugitive Mexican priest in "The Power and the Glory" when he preaches,

> "...that joy always depends on pain. Pain is part of joy...That is why I tell you that heaven is here: this is a part of heaven just as pain is a part of pleasure...Pray that you will suffer more and more and more. Never get tired of suffering. The police watching you, the soldiers gathering taxes, the beating you always get from the jefe because you are too poor to pay, smallpox and fever,

hunger...that is all pat of heaven - the preparation. Perhaps without them - who can tell? - you wouldn't enjoy heaven so much."

["The Power and the Glory", New York: Bantam Books, 1940, p. 64]

The idea that suffering is a necessary part of life and a requirement for eternal life has the merit of giving some explanation to suffering and evil and absolves god of the most negative aspects of creating and/or allowing evil. It is just such explanation that the human mind seeks, however that does not make it true. But I am getting ahead of myself.

The evil will, of course, get theirs as well. Some Christians take the position that evil people just cease to exist after death, though most opt for some form of punishment, either in purgatory, which is a temporary situation, or in hell which lasts forever. These are the people who have not responded to god's love and are cut off from responding now, in some thinking even if they would want to. The good, however, are those who have kept the commandments and gone beyond them in response to the love of god for them and god's direction in their lives. And this, it turns out, was god's will all along: to assemble a people who would know god's love for them and would respond, and respond for all eternity. As early as the third century, Christians were decorating the catacombs, the burial places of their families, with scenes of banqueting in the kingdom of god.

[Robin A. Jensen, "Dining in Heaven", "*Bible Review*" Vol. XIV, No. 5, October 1998, p. 32f.]

This assembly in the kingdom, or heaven, is the purpose of the creation of the universe. Every last detail has indeed made sense and has been directed to this end by a god who knew all along where it was going and how it was going to get there. Before there was a universe, god knew Susan, what she would look like, what her problems would be and quite possibly how she would handle them.

[The Church's main argument is the "free will defense", which are getting to. For a good overview of the question, see Simon Blackburn, "Think", p. 168f]

The troubled Christian, whom we are calling Susan, or her pastor trying to answer the question of why these things are occurring, take the position they do because that is the faith of the Church which views the world as working according to this model. Is it also the case that Susan did not just happen to marry a man who is abusive? That there was someone who brought this about and knew what would happen? Why would they do such a thing? Perhaps in order to punish her for some transgression, or to teach her some kind of lesson, or to discipline her through adversity in order to take her to a higher realm of personal development. It is assumed that the same control and intervention that is part of the creative and sustaining activity of god and sometimes shows itself in a miraculous intervention, however this is defined, extends to the details of individual lives. This intervention is often looked upon as a sign of the love that this agent has for the object of their attention, but one has to wonder, with friends like this, who needs enemies? Of course Susan may not be the real object whose interests are served at all. Maybe she has been placed in this situation so that she will be a more compassionate mother for the real beloved of the gods, her son. Or it may even be that it is all for the amusement of the gods such as in tragedy when the fate of the human being is already sealed by the events around them.

In any case, she feels that her situation has been engineered by someone according to their own plans and purposes. Since our subject is a Christian and sought advice from a Christian pastor, what kind of religious assumptions did her Christian faith lead her to make? First, of course, as we have seen, she assumes that everything has happened for a cause. To try to tell her that it is just blind chance that she is in this situation will not register with her, nor with her pastor. Both assume that there is indeed some purpose and that it is in fact god's purpose.

> [Read Breena Clarke, "River, Cross My Heart", Boston: Little, Brown Company, 1999, p. 48f. An African American preacher, on the occasion of the drowning of a little child puts the faith perspective well by saying,
>
> > "We do not understand this. Our hearts are torn asunder with grief. We do not understand, Lord. But Lord, we do not question your divine wisdom...Lord Jesus, we do not see your plan." He goes on, to a chorus of Amen's to say that the child is in Jesus' bosom. However, his sermon comes to a faulty end because, "he turned away from the assemblage to hide his profound confusion. In reality, he couldn't see the Lord's plan in taking this little one, but he would not show this doubt to the congregation."]

This belief in god's never giving you more than you can bear and always bringing some good out of evil can, it is true, be interpreted in another way. If one ignores why the evil occurred in the first place, you can make god the hero, working with us against the effects of evil. This "Providence of God" is, "the most popular belief or doctrine in the Black worlds of either church or street." This according to Nicholas C. Cooper-Lewter and Henry H. Mitchell, "Soul Theology: The Heart of American Black Culture", San Francisco: Harper & Row Publishers, 1986. Mitchell and Cooper-Lewter examine the "core beliefs" of African Americans (and the African antecedents from which they sprang), but they see no problem with saying for each of these beliefs such as omnipotence, omniscience and providence that they are based on an assumption.

Susan and her pastor may consult a book such as Daniel Harrington's "why do we suffer?" or Carlo Carretto's "Why O Lord?" and investigate the question of suffering in the Scripture. Harrington minces no words about the incomplete answers to be found there but reaches the same conclusion as Susan's pastor and most others, that suffering is somehow part of god's plan and that thinking of suffering in the life of Jesus,

> "...reminds us that even the one whom we confess to be the Son of God had to struggle to accept God's will made manifest in the grim realities of the cross."
> [Daniel J. Harrington, S.J., "why do we suffer?" Franklin: Sheed & Ward, 2000, p.144.]

Harrington honestly explains both the positive and negative aspects of the various biblical approaches to the question of suffering from the Lament Psalms, the "law of retribution" approach which claims that we are only getting our just deserts when we suffer (Wisdom & the Prophets), the Book of Job and its overwhelming god, the idea of willingly accepting suffering as a sacrifice, the Apocalypse and New Testament. A good resource, though it has few conclusions about an answer to why we suffer.

Carretto, too, is blunt in concluding that,

> "It's God himself who has crippled me." Fortunately for Susan this is the Christian God and so the purpose is not for the benefit of god, but for her good. There is a good purpose why, as Carretto says, "God permits it. God wounds me...But precisely in wounding me he draws out the best in me"
> [Carlo Carretto, "Why O Lord?", p. 78-79.

Carretto also has the answer to why the innocent suffer.

> "The child who dies in a state of innocence is actually the lucky one-the one who pays the price for the unjust and the wicked; he is on the same road as Jesus. He who gives his blood for children heads the great procession of mankind towards the heavenly Jerusalem...At sight of you, love is unleashed in those touched by

the Spirit. At the sound of your scream the good set out for the goal." [P. 71-2]

As Harrington points out, "the memory of Jesus also places before us the hope that God can and does bring life out of suffering and death, and reminds us that suffering and death never have the last word in the Bible." ["why do we suffer", ibid.]

Of course, Susan may have to wait until after death to see the whole picture, but there is a picture to see someday. The teleology that has guided god to cause these events is in her favor. The goal of god is the welfare of god's people, and since this god is all-powerful and never fails in intent, she will be benefited by enduring whatever god has in store for her. "All things work together for the good for those who love God" [Romans 8:28] and "I believe that I shall see the good things of the Lord in the land of the living." [Psalm 27:13 in another translation says, "In the Victor do I trust, to behold the beauty of Yahweh in the land of life eternal; and see Job 19:27.]

These Christians have assumed just what all people of faith have assumed before them, though in altered form today. The question most often asked by them is not whether these assumptions are correct, but, presuming that they are, what is god's purpose in bringing these things about? They have assumed that there is a purpose, that there is some logical explanation of the situation, and that it all somehow fits into a grand scheme leading to the fulfillment of a divine purpose. And what can Susan do to change the situation? Changing the situation is, of course, the purpose behind organized religion.

The death of a star occasions no questions among the other stars, nor even with us unless it is our star. The death of all but a small percentage of animals goes unnoticed by anyone who would lament them and even when it occurs to those individuals or even species known to us, we scarcely complain. "It was just a dog on the road"; "it's only a bird with a brain the size of a pea"; "who cares about a cold-blooded reptile or snail darter"; take your pick of reactions. Looking back on the millions of species that went extinct before we got a crack at destroying them, we are content to allow god to kill them off with impunity, though we sometimes wonder if god could not have created a natural order without all the mayhem. However, when death impinges on us, and certainly in our person, then god must explain godself. And there have always been a good number of apologists for god.

One explanation from ancient times was that there were really two gods, one good and one evil. One cannot blame the good god because it is the other that is responsible. This idea entered Christianity as well and,

> "the form of Gnosticism derived from this [ancient Iranian cosmogony] by Mani (which gave rise to the Manichean heresy) also entailed the belief that the two antagonists, Light Good and Darkness Evil, had existed *ab initio*, but held that the outcome of the struggle between them was uncertain."
>
> ["Creation Myths" p. 19
>
> The Qumran sectarians, or the Essenes if they are the same, also believed in the separation of the world into Light and Darkness. Evidently it was in the air.]

This was rejected in the west and,

> "Manicheism underwent a persistent decline from the 6th century on. It was still being opposed in the 9th century in the eastern part of the Empire...The influence of Manichean thought can be traced in the...Albigenses and Catharists..."
>
> ["Sacramentum Mundi" Vol. 3, p. 375]

so it was a very popular way to exonerate god from blame for evil. It is not gone yet.

> [Charles Bouchard, O.P. "From the President: The Trouble With Angels", *Aquinas News,* Vol 15 #1 Fall 1996, gives a current example of the persistence of Gnosticism.
>
> "Because it has been on best seller lists for more than two years, I recently picked up a copy of The Celestine Prophecy. Billed as a work of fiction, but the copy on the jacket promised that it would `give me chills as I began to perceive its predictions unfolding all around me in intimate relationships and international affairs.' Reading it gave me the chills all right, but not exactly for the reasons the author had hoped.
>
> The Celestine Prophecy describes the search for an ancient lost manuscript in the Peruvian mountains. Despite collusion of the Peruvian government and the Church, who attempt to suppress the manuscript because it threatens their authority, it is discovered gradually, with each new discovery revealing one of nine `insights.'
>
> The Celestine Prophecy is part of a larger category of books which I call `apocalyptic lite.' These include books on angels and those which describe near-death and after-death experiences. I call them

`apocalyptic' because they focus on the goal of human life and the advent of a new period in history which will be marked by dramatic change. They are `lite' because there is nothing frightening or demanding about them as there is in earlier apocalyptic literature such as the Book of Revelation. These new books provide a soft, comforting view of a world filled with insights, vibrations and energy fields, with a smattering of Freud, Jung, and family systems theory thrown in for good measure. Change will not come about violently, but quietly, peacefully, and from within. Let me quote one passage as an example, found in the final pages of the book:

`Our sense of purpose will be satisfied by the thrill of our own evolution, by the elation of receiving intuitions and then watching closely as our destinies unfold. Each of us will first observe the other's energy field, exposing any manipulations'."

In an essay entitled "The Latest Fashion in Irrationality, (*The Atlantic Monthly*, July 1996), Wendy Kaminer says that The Celestine Prophecy, like other books of its ilk, is characterized by

`a general disdain for reason, enshrined by the therapeutic culture…the truth lives in what you feel, not what you know in your head.' She goes on to say that these books promise a `spiritually enlightened culture ot peace and harmony. [where] critical thinking will be unnecessary. Truth will be self evident and people will simply intuit answers to problems like pollution.' This poses a serious challenge for a religious tradition like ours which believes that our ability to reason is what makes us most like God. It should also cause some anxiety to our students and faculty, who have just set themselves to nine months of rigorous intellectual exercise. If the same results can be achieved through intuition, why bother?

Eventually, the Prophecy says, we will `live in all the most beautiful places on earth, among five hundred year old trees and carefully tended gardens…Guided by their intuitions, everyone will know precisely what to do and when to do it, and this will fit harmoniously with the actions of others.' The appeal of these books rests in their promise that the transformation of the world will take place mysteriously as we move to ever higher levels of vibrations, apparently with no greater effort on our part than an openness to new fields of energy. We become passive recipients of a kindly universe that exists only to please us. As ministers, scholars and citizens, especially when voter apathy and cynicism about politics are at record levels, we should be concerned that such political and spiritual passivity opens the way for demagoguery, anarchy, or worse.

The final problem is that these books have a definite tendency toward gnosticism. Gnosticism was present at the birth of Christianity and has resurfaced every so often in movements that are defined by a secret and arcane knowledge that usually defies rational analysis. It is a knowledge that is possessed by only a few of the enlightened. This is clearly at odds with the Christian Gospel which is to be preached to the ends of the earth and which is available to all who have ears to listen.

Gnosticism, because it is essentially private and mysterious, leaves no objective truth to bind us together and no common purpose to moderate our differences. In the Gnostic world we learn `truth' not by study and intelligent discourse with others, but by private encounters with angels, lost documents, or death itself.

Celestine Prophecy is not dangerous, but it is an inadequate response to the spiritual and political needs of our culture."

Michael Galligan examines the question of evil's historical answers from the faith and admits that none are particularly satisfying or without problem. In looking to a direction for future investigation of the question, the best he can do is, first, throw in the towel, and say that,

> "...one may have to give more attention to the implications of the faith *that* the reality of the goodness attested to by the Gospel shall prevail than to the explanation of *how* this goodness came to be enmeshed in fatal combat with opposing forces and how it shall extricate itself."
>
> ["God and Evil", New York: Paulist Press, 1976, p. 69]

His second suggestion is one that lies behind much of the Church's response to the question of evil, an approach which is not heard as much today in the Catholic Church but certainly is heard in its cruder form in other denominations. It is that,

> "...when dealing with the problem of God and evil I do not think that we can evade the intuition that something has gone seriously wrong in the universe which cannot be expressed by mere ratiocination no matter how sophisticated. It may very well be that the idea of structures or forces of being with vaster ontological range and effect than man himself enjoys could offer us the vehicle for the symbolic expression of this persistent sense of the `surd' nature of evil." ["God and Evil" p. 73]

He is careful to say that we should not personify these forces, but others are not so cautious and we end up with demons and devils behind every negative occurrence. Evangelical faith-healers like John Wimber are often strong in belief in demonic power and possession, first because they believe it, second because this lets god off the hook of being the cause of evils, and third because this makes prayer for healing more attractive, if not to say better business.

> [See John Wimber with Kevin Springer, "Power Healing", San Francisco: Harper & Row, Publishers, 1987, Chapter 6, "Healing the Demonized". His whole book is a well-reasoned explanation for faith healing and presents a good understanding of the thinking of those involved in this ministry.
> A favorite name for the adversary is "Lucifer", a fallen angel, said to have rebelled against god and was cast into Hades or Sheol. This Lucifer is then identified with "the ancient serpent" and the "Devil". Interestingly, the name "Lucifer" comes from Isaiah 14:12f but there it refers to the king of Babylon, not to the Devil.
>> [See Ronald F. Youngblood, "Fallen Star: The Evolution of Lucifer", *Bible Review*, Vol. xiv, No. 6, December 1998, p. 22]
>
> Wimber's book was written in 1975, which is a fair ways back, but the picture has not changed dramatically since.]

The novel "Sula" explains why the eccentric title character is tolerated, even though feared and different. The people knew that god

> "...was not the God of three faces they sang about. They knew quite well that He had four, and that the fourth explained Sula. They had lived with various forms of evil all their days, and it wasn't that they believed God would take care of them. It was rather that they knew God had a brother and that brother hadn't spared God's son, so why should he spare them?"
> [Toni Morrison, "Sula", New York: Plume, 1973, p. 118]

Of course if our knowledge today leaves little room for one intervening god in the universe there is even less for two.

There are other believers who admit that there is evil in the world and try to face up to the problem of how god could allow it to be part of god's creation, or, being a part, why god does not intervene. This is the central question of Christianity and of this book. Virgil Elizondo states the question, "Have you ever been face-to-face with such misery and human suffering that you instinctively ask; `my God, how can you allow this?' Some ask, `if there is a God, how can He be so cruel?'" And Elizondo immediately gives us part of his answer, "Yet the real question should be, why has man done so little?"

["The Human Quest", p. 125]

This is not an attempt to shift the blame, though some try to do so by claiming that god created everything good but humankind has corrupted it. Elizondo takes another approach in claiming that suffering must be part of creation since it is itself so creative. Elizondo points out,

"Bishop Fulton Sheen, once said that the great tragedy was not that there was so much suffering in the world, but that so much suffering was wasted because it had no meaning...To me, suffering is incomprehensible and absurd without the light of god's redeeming and recreating love. Without the message of the cross-resurrection, suffering and death are absurd."

["The Human Quest" p. 126]

Many would say that *with* the cross-resurrection theme suffering and death are still absurd. After all, one can serve god and one's neighbor in many ways that do not require suffering and death, or even, very often, more than minor inconvenience. Why is it so necessary to Sheen and Elizondo and the many generations of theologians that they represent that there be suffering and misfortune? If these were not part of our world, just what great good would we be missing? There are even apologists for pain and suffering as a good which god knowingly placed into god's creation as something necessary. Elizondo admits that it seems to be a contradiction that god loves and then asks us to suffer,

"...but there is no contradiction. To love is to suffer. Love forces us to be open. Because of selfishness, we are prone to turn in to ourselves rather than open to others. This opening of oneself and going beyond oneself to the other is a painful process."

["The Human Quest" p. 126]

He claims even more, saying that,

> "...through suffering that is accepted, material deficiencies and limitations take a secondary role. Consciousness is liberated, allowing us to come to an ever more perfect union with the supreme consciousness, so that the individual reflects what the universe will be at the end of time. The evolving of consciousness is an evolving towards spiritualization, not in the sense of spirit as opposed to matter, but spirit as the creative life of love." ["The Human Quest" p. 134]

In the world in which we live this is unquestionably true for believers in god or some cosmic consciousness or the "force" or whatever, but that is not the question. The enigma is why that is the kind of world which we inhabit and which presumably god intended, for with this argument we are way beyond any claim that god never really wanted suffering to be a part of god's universe. We are now considering the claim that suffering is necessary for growth and creativity in the universe itself. Are there not other processes that bring about growth? How about trust and love? Admitting my "sinfulness" or error on the way to growth is only painful because of our screwed-up sense of self-defense and aggression against perceived attack, even an attack against our good name and social standing. Could we not have evolved with an innate understanding that correcting errors in ourselves is a good and enjoyable thing? What's more, are we saying that there is not growth in "heaven"?

Some, indeed, do hold that there is no growth in heaven since heaven by definition means that we are "perfected". An argument against this is that humans seem to need a challenge, but then we have no idea if this is true of a "perfect" human. Indeed, apologists for the traditional heaven have often pointed out that god has no need to change yet is happy and fulfilled and so may we be. However, this is not the dynamic god which many theologians and Scripture scholars have discovered in recent decades, nor the god of these pages.

Presumably, heaven could contain unlimited opportunity for the realization of the potential of the self without the limitations of temptation to the wrong and without pain and suffering. Elizondo's whole argument seems like an attempt to explain what is there by saying "it must be so", "this is the best of all possible worlds". If we lived in a universe (or a heaven) without evil and suffering would we feel compelled to invent them? I doubt it.

Or maybe, it is argued by some, god could have done otherwise, but is using the presence of evil in the world to teach us something. Augustine thought that god could even use the suffering of children to soften the hearts of the parents or increase their faith.

> [St. Augustine, "De Libero Arbitrio", in "St. Augustine, The Problem of Free Choice" Trans. by Dom Mark Pontifex, Westminster: The Newman Press, 1955, p. 210]

The Catechism thinks so and claims that,

> "With infinite power God could always create something better...With infinite wisdom...(God) willed to create a world `in a state of journeying' toward its ultimate perfection." ["Catechism of the Catholic Church" #310

It also states,

> "But in the most mysterious way God the Father has revealed His almighty power in the voluntary humiliation and resurrection of His Son, by which He conquered evil. Christ crucified is thus `the power of God and the wisdom of God. For the foolishness of God is wiser than men, and the weakness of God is stronger than men.' (1 Cor. 1:24-5) It is in Christ's resurrection and exaltation that the Father has shown forth `the immeasurable greatness of His power in us who believe.' (Eph 1:19-220"
>
> ["Catechism of the Catholic Church" #272]

This means, of course, that there is something about the presence of evil in the world that is good and it was intended by god and is used by god to make us better people and to show forth god's power over it. This seems like throwing everyone into a shark-infested pool with ourselves as the only one with shark repellant and then saying that we did it to show our beneficence. We want them to learn how to deal with sharks. Sorry about those who were eaten during the lesson! The Catechism must finally reach the conclusion that it doesn't know what to say about the presence of evil, even though it tries to absolve god from blame. One just has to take in the whole picture and believe that somehow it all makes sense, it claims, and concludes that,

"...No quick answer will suffice. Only Christian faith as a whole constitutes the answer to this question: the goodness of creation, the drama of sin, and the patient love of God...*There is not a single aspect of the Christian message that is not in part an answer to the question of evil.*" ["Catechism of the Catholic Church" #309

Well! How good can god get than to give us the answer to the problem which god created? Of course we will have to wait until after death before we can understand the answer, but think how much we will have grown by then. That is unless we are among the infants who die or those deliberately kept illiterate and brutish by a political or economic regime, or the many others who are the victims of the course being taught by god for the benefit of those who can afford to attend the class or are lucky enough to avoid the hazards of life. For the rest, the Catechism says,

"We firmly believe that God is master of the world and of its history. But the ways of His providence are often unknown to us. Only at the end, when our partial knowledge ceases, when we see God `face to face' will we fully know the ways by which - even through the dramas of evil and sin - God has guided his creation to that definitive sabbath rest for which He created heaven and earth." ["Catechism of the Catholic Church" #314

Thank goodness that, the Catechism continues,

> "God is in no way, directly or indirectly, the cause of moral evil. He permits it, however, because He respects the freedom of His creatures and, mysteriously, knows how to derive good from it." ["Catechism of the Catholic Church" #311]

419. "Is There A God" p. 98

420. "The Faith of a Physicist", p. 83

421. The discussion has moved from horizon to horizon over the centuries to solve this question, but the church seems to have settled on human freedom as the key to an answer. This was expounded by Augustine and many theologians today continue in his tradition,

> "...to uphold the notion that evil was introduced into the world by some primordial deviation in which the first man or group of men exercised freedom in violation of God's will. In so interpreting the origin of evil, they conform to the basic Augustinian program of protecting God from responsibility for evil by demonstrating man's responsibility for its initiation."
> [Michael Galligan, "God and Evil", p. 24]

The Christian apologist C.S. Lewis points out that even though god is omnipotent, even supreme power does not extend to having the ability to do the impossible.

> "His Omnipotence means power to do all that is intrinsically possible, not to do the intrinsically impossible. There is no limit to His power. If you choose to say `God can give a creature free will and at the same time withhold free will from it,' you have not succeeded in saying *anything* about God: meaningless combinations of words do not suddenly acquire meaning simply because we prefix to them the two other words `God can.'"
>
> [C.S. Lewis, "The Problem of Pain", New York: Macmillan Publishing Co., Inc. 1962, p. 28]

Author...

American General Life Insurance Company, Houston, TX
New York, NY

The above listed life insurance company as selected on page one
under any policy that it may issue. No other company is responsi
insurance company which was selected on page one.

Authorization to Obtain and Disclose Information and Declara

I give my consent to all of the entities listed below to give to the (
affiliated service company), and affiliated insurers all information
confinements for physical and mental conditions; use of drugs or
information could include items such as: personal finances; habi
Vehicles; court records; or foreign travel, etc. I give my consent
practitioner; any hospital, clinic or other health care facility; any i
support organization; my employer; or the Medical Information Bur

I understand the information obtained will be used by the Compan
existing policy. Any information gathered during the evaluation
organizations performing business or legal services in connection
or entity required to receive such information by law or as I may fu

I, as well as any person authorized to act on my behalf, may, upon
revoked at any time by sending a written request to the Company,

This consent will be valid for 24 months from the date of this applic
AGLC or affiliated insurers to obtain an investigative consumer re
receive, upon written request, a copy of such report. ☐ Check if

I have read the above statements or they have been read to me.
that this application: (1) will consist of Part A, Part B, and if applic
that any misrepresentation contained in this application and relied
(1) it is within its contestable period; and (2) such misrepresentat
Limited Temporary Life Insurance Agreement (LTLIA), I understan
new policy issued by the Company, unless or until: the policy has b

Lewis goes on to try to honestly face all the questions, including why animals suffer, if they do. His answers are, of course, speculative, including examining whether animals might not be compensated for being devoured by being raised to immortality. Still, his most important argument is surely that if humans are to be free then there must be the possibility of the wrong choices and in the end this is the cause of evil in the world. It is a somewhat satisfactory answer when dealing with misery and evil caused by human agents, though there is always the objection that god could whip off a miracle and change the hearts of the evildoers, or better yet blast them from the face of the earth. We used to be able to take some comfort that god was making headway in changing the world from the barbaric place it used to be in the days of Attila the Hun and his ilk, but then came the British Raj, the displacement of aboriginal peoples in the Americas and Australia, and certainly the Holocaust and other ethnic cleansings. These occurrences test the faith of the strongest.

In the movie *Sophie's Choice*, the title character says, "I knew that Christ had turned his face away from me. Only a Jesus who didn't care for me could kill the people that I loved and leave me alive with my shame." God seems to be losing ground in the battle against evil, but the teaching of the church is that this is all due to human free will and its capacity for evil and cannot be laid at the doorstep of god. A popular book, "Conversations with God", questions god about evil and god answers that,

> "...the world exists the way it exists because *you have chosen for it to*...I will do nothing for you that you will not do for your Self. *That* is the law and the prophets."
>
> [Neale Donald Walsch, "Conversations with God: Book 1", New York: G. P. Putnam's Sons, 1995, p. 50.]

This has always been part one of the answer, "you bring it on yourself and I respect your freedom to do so." Another modern catechism takes the same position and insists that we must own up to our own responsibility for evil and not blame god. After all, as John Wijngaards says,

> "...perhaps we might question God's wisdom in granting human beings autonomy and freedom-but asserting that God could make us truly free in such a way that we would always be paragons of virtue is plain poppycock."
>
> "John Wijngaards, "How to Make Sense of God" Kansas City: Sheed & Ward, 1995, p. 243]

422. "Is There A God" p. 107

423. Some have claimed that there is really no such thing as evil, but as Ninian Smart, in a collection of philosophical essays points out,

> "Nor can we evade the issue by saying that perhaps after all evil does not really exist, that it is an illusion. Equally futile is the attempt to say that evil is something merely negative...This is to ignore the fact that when I feel toothache [it really hurts]...We are quite right to speak of actual feelings and states of affairs."
>
> ["Philosophers and Religious Truth" p. 154]

Some Thomists have claimed that evil is merely the absence of good and an absence is not something real, and god cannot be held responsible for "creating" something that does not exist. Tell it to the dentist. However, Smart, who is trying to be honest in his approach to the question and not minimize evil as something non-existent, still reaches the conclusion that, "we therefore see that creativity itself demands the possibility of moral evil...[It] must be accepted as part of the price."

["Philosophers and Religious Truth" p. 159]

We are right back to god at least allowing evil because of all the growth it brings about, or at least makes possible. Old habits die hard, and besides, moral evil is not the problem here. On the other hand, the same author again comes back to reality and admits that,

> ".it is...human beings, of which we are speaking. Is there not a contradiction in saying that, in the interest of the growth of morality and creativity, it may be necessary for individuals to be destroyed? For surely the highest point in morality is to recognize the claims and the worth of the individual."
>
> ["Philosophers and Religious Truth" p. 163]

Nevertheless, he cannot find an answer other than the usual, that god does not cause but allows evil in what may be the best possible cosmos for creativity and allying godself with sufferers in Christ.

Karl Rahner can do no better than the usual answer to the question by claiming that death and suffering are natural consequences of living in this world and god has stepped in to conquer them.

> "All suffering and all sickness, which seem to the profane eye a decay of life, become for the believer a summons and an acceptance and therefore a spiritual conquest of suffering, sickness and death. Thus suffering grows akin to love and accordingly embraces everything in God's world-except sin, the negation of his love. But suffering is no sin and dying is no sin, but a natural law of earthly

life, the universal and inescapable evil that becomes sanctified and transfigured for Christians just as it was for Christ: `He who believes in me, though he die, yet shall he live" (Jn. 11, 25)."

[Rahner, "Do You Believe In God" p. 94]

God is bringing good out of evil, but it is an evil for which so far we must still say god is responsible.

Pope John Paul II wrestles honestly with the problem and forthrightly phrases it with all the vehemence of those who accuse god of being unjust in creating a world such as ours. He answers that god is indeed impotent in the face of evil, having made godself impotent, but through the cross and a seeming defeat brings about a victory that every human can achieve. He tells of Pilate's asking of Jesus-God if he is really responsible, really omnipotent, really a king. But then the Pope's answer:

> "What is truth?" (Jn 18:38), and here ended the judicial proceeding, that tragic proceeding in which man accused God before the tribunal of his own history, and in which the sentence handed down did not conform to the truth. Pilate says: `I find no guilt in him' (Jn 18:38), and a second later he orders: `Take him yourselves and crucify him!' (Jn 19:6). In this way he washes his hands of the issue and returns the responsibility to the violent crowd. Therefore, the condemnation of God by man is not based on the truth, but on arrogance, on an underhanded conspiracy. Isn't this the truth about the history of humanity, the truth about our century? In our time the same condemnation has been repeated in many courts of oppressive totalitarian regimes. And isn't it also being repeated in the parliaments of democracies where, for example, laws are regularly passed condemning to death a person not yet born?"

["Crossing the Threshold of Faith" p. 65]

The blame is placed on humanity rather than on god, god being declared innocent by the verdict of Pilate; on the creature rather than the creator. The creator is excused because god has put godself, in the suffering of Jesus, under the same sentence of liability to evil as god's creatures. And those who are responsible do not disappoint. They crucify the just one, but there is no mention that it is the just one who must bear the ultimate responsibility for the injustice in the first place by the creative act. Humans must accept their share of the blame, but declaring god excused by one statement of an ancient Roman who is clueless about what is occurring before him is not a good verdict.

M. Scott Peck, a famous author on religious questions, agrees with the Pope that ultimately it is humanity that must take the responsibility and not try to place it on god. God had to give us the freedom to do evil if we were to have free will.

> "You can't have evil unless you have choice. When God allowed us free will, He inevitably allowed the entrance of evil into the world. It so happens that...Genesis 1 also has something to say about evolution as well as good and evil. It tells how God first created the firmament, and then the land and the waters, and then the plants and the animals. It is the same sequence as that suggested by geology and paleontology. As far as scientists can determine, that is the sequence of evolution, even though we can't say it all happened in seven days. A whole new meaning to Genesis 1 occurred to me when I remembered that God first created light and He looked and it was good. So He created the land and saw that it was good. And so He separated the land and the water and saw that was good too, so He went on to create the plants and the animals. And when He saw that they too were good He then created human beings. Thus, I think the impulse to do good has something to do with what creativity is all about. Similarly, the impulse to do evil is destructive rather than creative. The choice between good and evil, creativity and destruction, is our own. And ultimately, we must take that responsibility and accept its consequences."
>
> ["Further Along The Road Less Travelled", p. 109]

In the aftermath of the terrorism of September 11, 2001, "Our Sunday Visitor" sent out a help for Catholics in the Church's response to the questions raised, especially, "where is God when evil strikes?" The response is based on the free will argument.

> "God is present always and everywhere. Not commandeering our free wills, but present with strength, hope, and love. God is with the victims at the moment of tragedy. God is with the survivors. God is with their rescuers."

The other part of the response is also familiar,

> "While acts of great evil such as terrorism can seem overwhelming, as Catholics we know that death is not the end, and that God awaits all those who perish."
>
> ["What the Church teaches when evil strikes", *Our Sunday Visitor*, Huntington: 2001]

It is with some reluctance that I point out that this is precisely what the terrorists may have believed and we would say that they are mistaken.

Peck's is certainly a nice try to be scientifically relevant but I suspect evolutionists would not agree with the parallel. The real question, however, is if you cannot have evil without choice does it follow that you cannot have choice without evil? If god could create a universe without evil but with free will and opts to create what we in fact experience, maybe the reason is that god enjoys a little destruction now and then, too, in addition to all the goodness. "God knows" there is plenty of destruction in the universe, from annihilating matter and antimatter to colliding galaxies to one animal feeding on another. I still wonder if anyone would think that this argument accepting the possibility of evil in return for free will would be necessary if suddenly the universe cared about us and protected us and all our tendencies were to do good rather than evil. Would there really be less creativity?

Oscar Lukefahr, the author of a popular catechism, claims that, "if believers have to answer the question, 'If God exists, why is there evil in the universe?' then nonbelievers must answer a far more difficult question, 'If there is no God, why is there good in the universe?" ["We Believe" p. 6] There is no denying that there is goodness and creativity in the universe, it is in fact a very inventive place. Nevertheless, we have seen that it can all be explained without the intervention of a creative designer.

There is however a far more serious objection to his question. Who is to say that there *is* any good in the universe? In fact, it looks for all the world as if there is nothing that has any moral value at all, either good or evil, except from our subjective viewpoint. If there is nothing objectively good, then by Lukefahr's own admission there need not be a god making goodness possible. Think of something good that could happen to you. To how many people is that relevant? In a thousand years will it make any difference that it happened to you or that it did not? Think of something good you could do. When the universe is either crushed out of existence in a Big Crunch or when the nearest quark is 10^{14} light years from the next and any interaction is impossible for all practical purposes, will it make any difference that you did that good thing so many eons ago? None at all. From the viewpoint of the universe there is not good or evil, neither moral nor physical, except only from our own subjective perspective, and though we would like to think that our judgment matters, the people right next door to us may care very little, much less does the rest of the universe care.

Scientists, even agnostics and atheists, have weighed in on the question of evil. We can use the word "evil", of course, since it means something to us. As Robert Wright says,

> "The concept of `evil,' though less metaphysically primitive than say, `demons,' doesn't fit easily into a modern scientific worldview. Still, people seem to find it useful, and the reason is that it is metaphorically apt. There is indeed a force devoted to enticing us into various pleasures that are (or once were) in our genetic interests but do not bring long-term happiness to us and may bring great suffering to others. You could call that force the ghost of natural selection. More concretely, you could call it our genes (*some* of our genes, at least). If it will help to actually use the word *evil*, there's no reason not to."
>
> "Natural selection designed two things for narrow self-interest - cold reason and warm moral impulses - and somehow, when combined, they take on a life of their own."
>
> ["The Moral Animal" p.373]

But not a real life, though the effect is real within us. One must keep in mind that one is speaking metaphorically, as we might when using the term "soul" or "spiritual" and not referring to something which is independently real.

George Greenstein speaks for many when he responds to the philosophical and theological discussion on the presence of evil by saying that,

> "Evil has always been a conundrum for religious thinkers. How an omniscient and a benevolent God could allow it to exist is a problem no religion has put to rest. But evil is not a problem within the concept of a universe fitted for life. Nothing in that concept should be taken as proving that we should not suffer. There is no conflict with the overall fitness of the universe if an individual happens to lead a life of unrelieved misery and pain. It would not even be a conflict if a thermonuclear war were to break forth, eradicating every last vestige of life upon the Earth. The religious view, I suspect, would tend to celebrate life's appearance upon the cosmic scene. It would uphold the fitness of the universe as an example of God's goodness. But was it really such a good thing that the universe conspired to bring forth life? Why be thankful for the fact? We are accustomed to believing that life is good and nonlife bad, but that is only because of our built-in will to survive.

And this, in turn, exists within us for evolutionary reasons: Any species lacking that will would have been doomed to early extinction. But if we were capable of backing away from our biologically ingrained urge to live, we might well conclude that the appearance of life within the cosmos was no cause for rejoicing. The dog writhing in the gutter, its back broken by a passing car, knows what it is to be alive. So too with the aged elk of the far north woods, slowly dying in the bitter cold of winter. The asphalt upon which the dog lies knows no pain. The snow upon which the elk has collapsed knows not the cold. But living beings do. Are you alive? Then you can feel pain. Are you conscious? Then you can feel more pain. The lower animals, at least, appear not to be subject to mental anguish. Given the bare necessities of survival, they seem content. But we humans suffer in more sophisticated ways. Perhaps we even suffer more than the dumb animals. No one among us has led a life free from grief, loneliness, and despair. The more that cosmic evolution has progressed, the more suffering has come forth in the world. There may be no cause for thankfulness in the fact that against all odds the cosmos succeeded in bringing forth life. It may be only the truth."

[George Greenstein, "The Symbiotic Universe", William Morrow and Company, Inc., 1988, pp. 194-195]

I have quoted Greenstein at some length to give the flavor of the response. Religious people, and many who are not religious, will not like that response since it gives no reason for the presence of "evil" and suffering, or of good, and leaves no room for hope. It merely concludes that it is simply the truth that we are here and under these conditions. No judgment is made on the universe for it, too, merely exists. This is a very bleak picture and many would like reality to not be so stark and unpromising and so we search for more hopeful answers, contending that, "God is in charge after all, and therefore it will all come out well in the end." "He never gives you more than you can bear." "God has made all creation subject to futility so that the salvation of god will be more fully appreciated." "The presence of free will and choice requires that there be evil (not just the possibility but the actuality, evidently) in the universe." "In fact, the evil is really a good thing since it helps us be creative and good in combating it." "Evil is just an illusion and someday it will fade away." Or so it is said.

A very ingenious explanation, to be sure, but unhappy as we may be with Greenstein's assessment, has anyone ever been happy with the answers from religion? Has any pastor ever given Susan the orthodox Christian reasoning and found her satisfied? Resigned, perhaps, but certainly not happy. "Submission to the will of god," has brought about much deep spirituality, it is true, but the questions inevitably remained. To say that god does these things we call evil, or "allows" these things, which amounts to the same, and that the explanation is a mystery, can be assented to but that is all. Of course to say that god *does not* and *cannot* do these things, which as is evident by now is the conclusion I am pursuing, is scarcely any more satisfying. To say that god cannot be held responsible since god is powerless to control these evil events is also to say that god cannot cause the blessings, either. Steven Weinberg points out that, "If there is a God that has special plans for humans, then He has taken very great pains to hide His concern for us. To me it would seem impolite if not impious to bother such a God with our prayers." ["Dreams of a Final Theory", p. 251]

We are not going to rejoice over this situation any more than over the explanations previously given which make god's purposes good but unknown. In the last chapter we must ask ourselves why, if god is so impotent, we should take any notice at all. Here I am merely saying that if the faith is going to say anything about the relationship of god and our universe vis-a-vis suffering and evil, and it must, it cannot go on saying what it has always said.

Richard Dawkins tells of a bus accident in Britain that killed a number of children. He cites the explanation given by a priest who is obviously aware of the challenge that science has brought to faith in responding to such occurrences. The priest says, "The simple answer is that we do not know why there should be a God who lets these awful things happen. But the horror of the crash, to a Christian, confirms the fact that we live in a world of real values: positive and negative. If the universe was just electrons, there would be no problem of evil or suffering." Dawkins is not buying this. The "values" are neither positive nor negative and there is certainly no need of God's permission before anything happens. He continues,

> "On the contrary, if the universe were just electrons and selfish genes, meaningless tragedies like the crashing of this bus are exactly what we should expect, along with equally meaningless good fortune. Such a universe would be neither evil nor good in intention. It would manifest no intentions of any kind. In a universe of blind physical forces and genetic replication, some people are going to get hurt, other people are going to get lucky. and

> you won't find any rhyme or reason in it, nor any justice. The universe we observe has precisely the properties we should expect if there is, at bottom, no design, no purpose, no evil and no good, nothing but blind, pitiless indifference."
> ["River Out of Eden" p. 131f]

One cannot dismiss god's accountability for the presence of evil as a believing parent did who wrote to a magazine about a series of miscarriages, "God gave us our children and he took them from us. We know we shall meet them in heaven one day. I'm not angry or looking for answers. I trust god is protecting our children." ["*Ligourian*", February 1995 p. 16]
A noble sentiment, "the Lord giveth and the Lord taketh away," but one that leads many other parents to indeed curse god for taking their child prematurely. The faith has never had a better apologist than this parent who voices the connection the faith has always made. God brought about the event, whether good or evil, and if it is evil god is against it and will make it both understandable and right someday.

John R. Aurelio in "Mosquitoes in Paradise", subtitles his book, "A New Look at Genesis, Jesus, and the Meaning of Life." He begins promisingly to look at the question of theodicy and certainly asks the right questions. In presenting the usual answers he says,

> "Without impugning the motives of their authors or demeaning them for their efforts, let us rather examine the inadequacies of these solutions and refrain from using them in the future." [New York: Crossroads, 1985, p. 7]

He then rejects the standard answers: "God does not will evil, He permits it;" "God lets evil happen for some good reason known to Him;" "Sickness and pain are the work of the devil;" and even, "There but for the grace of God go I." Aurelio is a hospital chaplain so he certainly represents what pastors tell their people. Yet he builds up to his response by, first, asking "Must suffering always be perceived as bad?" [p. 50] Then he presents the usual explanation that evil is the result of our sin, not god's action. He reaches the conclusion that since god has given us reason to hope that sickness and death will not be the final chapter in our lives, what does it matter that we have some suffering now? However, even he is not satisfied with his own answer. He tells of the death of a child whose parents respond in much the same way as those mentioned in the text, "God's will be done." [p. 112] However, this response is called, "a lot of bull" by another parent and Aurelio can only respond with his last paragraph,

> "How can you say in a few words what even the faltering words of this lengthy manuscript cannot say adequately? How can you say that the deceiver has done his job well; that we value this life more than the next; that it is the Spirit that God breathed into us that makes us great and not this paltry body; that faith is a gift freely given but severely tested only in fire; that death has been swallowed up in victory; that eye has not seen nor ear heard nor can the mind of man imagine all that God has prepared for those who love him? `Oh God. Do not put me to the test!' Lord, I believe. Help my unbelief!" [p. 113]

I could not have phrased it any better nor made the standard answer more unsatisfactory.

Hugh Calkins, also a hospital chaplain, has a simple answer to this most complex question,

> "The answer lies clearly before us. Suffering lets us share in the redeeming work of Jesus." ["Living with Illness", Chicago: Thomas More Press, 1973, p. 12]

He means just what he says. This is what he would have his suffering clients believe. I understand why he does so; this is the way he has been taught. And I understand why he may have success with this explanation. No one wants to think that their life doesn't count, so finding meaning in suffering is a positive approach to their questions. However, it still makes god liable for choosing to teach this person or their caregivers something through a very negative (not to say evil) means.

While we are on the subject, another popular explanation of "what the pastor will tell the people" is found in Timothy O'Connell's "What a Modern Catholic Believes About Suffering and Evil", Chicago: The Thomas More Press, 1972. He begins his treatment of the question by saying that he will not "prove" anything and that the question is a "mystery" and, "because it is a mystery, evil does not have an answer or a solution toward which we will move in this book." [p. 20] And sure enough, though he brings in the free will argument as well as the god-knows-what-is-good-for-us argument, and the god-is-involved-in-the-struggle-along-with-us argument, the best he can do is that Christ has conquered evil and so the best we can do is be a part of the struggle as well.

I for one am not mollified.

Dawkins' explanation that it is just a matter of the luck of the draw and these parents happened to draw the short stick, too bad but that's the way it is, is not going to be anywhere near as consoling. But what *does* the faith say to the parent who blames god or to Susan who is asking the same question? And what do we say to Dawkins if he loses a child? This, of course, will be the topic of the last chapter.

424. "Demon-Haunted World", p. 12

425. "Demon-Haunted World" p. 197

426. Carl Sagan, "Billions & Billions", p. 214

427. "Billions & Billions" p. 215

428. "The Demon-Haunted World" p. 330

A reviewer of Richard Dawkin's book "Unweaving the Rainbow: Science, Delusion, and the Appetite for Wonder", New York: Houghton Mifflin, 1998 says the books is,

> "Dawkin's eloquent reply to this depressing charge [that any meaning we may happen to see in life is illusory]. Biologists may be unsentimental about their subject, he says, but that doesn't make them unfeeling. And there may be no grand design in life, but there is plenty to be awestruck about. `To accuse science of robbing life of the warmth that makes it worth living is so preposterously mistaken, so diametrically opposite to my own feelings and those of most working scientists, I am almost driven to the despair of which I am wrongly suspected'."

The reviewer concludes that, "With Carl Sagan gone, the English-speaking world needs a new poet laureate of science. Richard Dawkins now bids for the title."

With the reviewer, my impression remains that the sciences can offer plenty to be amazed about. That is until we ask, "but does it have any meaning beyond the universe?"
It also strikes me that very few have access to the pursuit of scientific knowledge. To what can they look for the sense of purpose and awe?

See Wade Roush, "Barcodes in the Stars", *Technology Review,* January/February 1999, p. 87.

See Ian Barbour, "Religion in an Age of Science", p. 5

429. Stephen Jay Gould, "Rocks of Ages", New York: The Ballantine Publishing Group, 1999

430. Stephen Jay Gould, "Nonoverlapping Magisteria", *Natural History*, Vol. 106, No. 2, March 1997, p. 62

Kenneth Miller, "Finding Darwin's God", p. 170f takes a dim view of Gould's praise of the Pope.

For Matt Ridley, "Genome", p. 24, the Pope's pronouncement is a subject of ridicule, jesting that the Church is reconciled to evolution by the position that God injected a soul into humanity. Speaking of the human chromosome #2 being a fusion of two chromosomes in apes, he says,

> "Perhaps the ontological leap came at the moment when two ape chromosomes were fused, and the genes for the soul lie near the middle of chromosome 2."

431. Stephen Jay Gould, "The right can't figure out Darwin", *St. Louis Post Dispatch*, Tuesday, June 9, 1998, copyright The New York Times.

432. "Nonoverlapping Magisteria", loc. cit.

433. "Three Scientists and Their Gods" p. 190

434. Edward O. Wilson, "Consilience" New York: Alfred A. Knopf, 1998, p. 264

435. "Consilience", p. 265

436. See Kenneth Miller, "Finding Darwin's God", p. 180f for a theistic scientist's view of Wilson.

437. Stuart Kauffman, "At Home In the Universe", New York: Oxford University Press, 1995, pp. 4-5

438. Daniel Dennett, however, feels that Kauffman does leave too much room for those who would insist on a place for god or at least an intelligent designer. See "Darwin's Dangerous Idea", p. 227 where he says,

> "Kauffman wants to stress that the biological world is much more a world of Newtonian discoveries (such as Turing's) than Shakespearean creations, and he has certainly found some excellent

demonstrations to back up his claim. But I fear that his attack on the metaphor of the tinker feeds the yearning of those who don't appreciate Darwin's dangerous idea; it gives them a false hope that they are seeing not the forced hand of the tinker but the divine hand of God in the workings of nature.

Kauffman himself has called what he is doing the quest for "the physics of biology" (Lewin 1992, p. 43), and that is not really in conflict with what I am calling it: meta-engineering. It is the investigation of the most general constraints on the processes that can lead to the creation and reproduction of designed things. However, when he declares this a quest for "laws," he feeds the antiengineering prejudice (or you might call it "physics envy") that distorts so much philosophical thinking about biology."

439. "At Home In the Universe", p. 304

440. Francis Bacon, "New Atlantis", New York: Walter J. Black, 1942

441. "At Home in the Universe", p. 302

442. "At Home in the Universe", p. 304

443. Lee Smolin, "The Life of the Cosmos", New York: Oxford University Press, 1997, p. 299
Smolin speculates that gravity is the source of the energy for organization, not only in solar systems but in ecological systems as well.

444. Ibid. p. 300

445. David F. Noble, "The Religion of Technology" New York: Alfred A. Knopf, 1997, p. 208

446. See "The Conscious Universe" p. 10f

447. "Darwin's Dangerous Idea" p. 18

448. "Darwin's Dangerous Idea" p. 520
Lee Smolin in "The Life of the Cosmos" reaches a different approach to Nietzche. See p. 299f

449. Rev. Dr. Steven C. Kuhl, "Darwin's Dangerous Idea...And St. Paul's", "*Creation and Evolution: Proceedings of the ITEST Workshop October, 1997*", St. Louis: ITEST Faith/Science Press, 1998. According to Kuhl,

> "According to Dennett, if evolution is true, then `Meaning' cannot be established by appealing to a `Meaning' already given (or guaranteed) by God from the beginning. For no such Meaning (Giver) exists). Rather, Humankind *alone* is responsible for creating its world of Meaning."

I agree with Kuhl that some design in creation is not a basis for finding meaning (Dennett) nor for building a theology of God the Creator, but, with Paul, the Mercy of God,

> "which is the basis for a truly meaningful (justified) existence in the world, does not exist `naturally,' but was established historically in the life, death and resurrection of Jesus Christ..., is presently being made available to all humanity through the `Word' and is appropriated by humanity personally `through faith'. This can provide a basis for meaning."

Not for Dennett, of course. Humanity's meaning can only be found in humanity. Paul sees this as impossible, Dennett says that he sees it as possible but not guaranteed. Kuhl seems to agree with my assessment of Dennett and the others reviewed that this is not much comfort to the individual human and maybe not much comfort to humanity collectively

> "From Paul's theological perspective, however, Dennett's humanistic atheism which leaves humankind accountable *only* to itself, is a rather dubious, empty solipsistic kind of accountability." (p. 86)

450. "The Last Three Minutes" p. 154
In his book, "The Fifth Miracle", Davies cites Jacques Monod's conclusion that,

> "the ancient covenant is in pieces: man at last knows that he is alone in the unfeeling immensity of the universe, out of which he has emerged only by chance. Neither his destiny nor his duty have been written down."

However Davies continues that,

> "if it transpires that life emerged more or less on cue as part of the deep lawfulness of the cosmos - if it is scripted into the great cosmic drama in a basic manner - it hints at a universe with a purpose." [p. 27]

He rejects strict reductionism which he had presumed in "The Cosmic Blueprint" and now believes that,

> "It seems to me very unlikely that all that is necessary is for the right chemical reaction or the right molecule to turn up. Real progress with the mystery of biogenesis will be made, I believe, not through exotic chemistry, but from something conceptually new." [p. 260]

Indeed,

> "For three hundred years, science has based itself on reductionism and materialism, leading inevitably to atheism and a belief in the meaninglessness of physical existence." [p. 264]

Have Theists found a convert? Is Davies going to conclude with a designer god? Don't believe it! His final paragraph speaks of a view alternative to Monod, a view,

> "undeniably romantic but perhaps true nevertheless, the vision of a self-organizing and self-complexifying universe governed by ingenious laws that encourage mater to evolve towards life and consciousness. A universe in which the emergence of thinking beings is a fundamental and integral part of the overall scheme of things. A universe in which we are not alone."

If he were going to put god in, that would be the place and there is no mention of god. Just where Davies would go from here is a mystery to me, but I suspect that he is arguing more for a view closer to Kaufmann's than to Aquinas.

451. Cited in Dennis Overbye, "Lonely Hearts of the Cosmos", New York: HarperCollins Publishers, 1991 p. 225

452. "Wrinkles in Time" p. 296

453. "Wrinkles in Time" p. 296

454. "Wrinkles in Time" p. 297

455. Richard Dawkins, "Unweaving the Rainbow", op. cit.

456. Review by Melvin Konner in *Scientific American,* Vol. 280, No. 3, March 1999, p. 107

457. "The Astonishing Hypothesis" p. 260

458. "Dreams of a Final Theory" p. 255

459. George Orwell, "1984" New York: New American Library, 1949, p. 228f

460. "Dreams of a Final Theory" p. 261

461. Richard P. Feynman, "The Meaning of It All", Reading: Helix Books: 1998, p. 46

462. Ursula Goodenough, "The Sacred Depths of Nature", New York: Oxford University Press, 1998, p. xvii.

463. Ibid. p. 11

464. Ibid. p. 13.

465. Ibid. p. 169

In a review of Professor Goodenough's book, Barbara Smuts acceptingly points out that,

> "For [Goodenough], experience qualifies as religious if it entails emotions like awe, wonder, gratitude or joy, regardless of whether or not the person associates such emotions with traditional religious creeds, deities or supernatural phenomena." ["*Scientific American*", Vol. 280, No. 5, May 1999, p. 100]

466. Ibid. p. 171

467. Professor Goodenough, "*Washington: Washington University Magazine and Alumni News*", Vol. 69, No. 3, Fall 1999, p. 17 continues her reflections by admitting that,

> "When my awe at how life works gives way to self-pity because it doesn't work the way I would like, I call on assent - the age-old religious response to self-pity, as in `Why, Lord? Why This? Why ME?' and then, `Thy Will Be Done.' As a religious naturalist I say `What is, is" with the same bowing of the head, the same bending of the knee. Which then allows me to say `Blessed Be to What Is' with thanksgiving. To give assent is to understand, incorporate, and then let go. With the letting go comes that deep sigh we call relief,

and relief allows the joy-of-being-alive-at-all to come tumbling forth again."

Letting go and assenting to the inevitable and unavoidable is somewhat different than letting go to a loving god.

468. "Skeptics and True Believers", p. 213

469. "Skeptics and True Believers", p. 214

470. "Skeptics and True Believers", p. 234

471. Ibid. p. 246

472. On the difference between pantheistic and panentheistic, see Eugene Mallove below.

C.S. Lewis has an opinion (naturally), about this kind of god. In "Miracles: A Preliminary Study", New York: The Macmillan Company, 1947, p. 30f he points out that it will not serve to,

> "...believe in a certain kind of god-a cosmic consciousness to which [the universe] gave rise: what we might call an *Emergent* God...But I am afraid it will not do. It is, of course, possible to suppose that when all the atoms of the universe got into a certain relation...they would give rise to a universal consciousness. And it might have thoughts. And it might cause those thoughts to pass through our minds...This cosmic mind would be, just as much as our own minds, the product of mindless Nature. We have not escaped from the difficulty [of needing a prime mover sort of being], we have only put it a stage further back."

Obviously, Lewis makes some assumptions with which I cannot agree. His is a very lucid presentation of the usual argument, however, he must conclude that,

> "it is never possible to prove empirically that a given,...event was or was not an answer to prayer...the skeptic can always point to its natural causes and say, 'Because of these it would have happened anyway,'...The efficacy of prayer, therefore, cannot be either asserted or denied without an exercise of the will - the will choosing or rejecting faith in the light of the whole philosophy. Experimental evidence there can be none on either side."

473. A.N. Wilson, "God's Funeral", New York: W.W. Norton & Co., 1999, p. 335

474. "God's Funeral", p. 337

475. "God's Funeral", p. 336

476. "God's Funeral", p. 354

477. "God's Funeral", p. 354

478. BRIAN SWIMME is described as "Director of the Center for the Story of the Universe" of the University of Oregon. In his "The Universe is a Green Dragon". op. cit. he certainly waxes eloquent about the beauty of the creative power of the universe bursting forth in human beings. He is a little short on the question of evil in this universe, and very short on a personal god. He explains that humans have a great concentration of Being. "Being folds itself into concentrated fullness, then erupts in an explosion of joy. The artist sends forth her works; the parent lavishes care upon his children." One may think he is speaking of the traditional creative god, but he is not. When asked by his Socratic questioner youth if this thinking is new he answers,

> "Not really. The understanding of Being within the cosmic creation story, yes. But Being's innate urgency to unfold has been appreciated in different forms. Classical theologians spoke of Supreme Being's ontological desire to pour forth goodness, to share and ignite being spontaneously. The explained the human desire to share life and being as participating in Supreme Being, in Divine Reality." (p. 145)

I, like Swimme, deny that we are the result of god's gift of Being in the fact of our physical existence. However, I must ask of him as I have of the others cited, "Is that all? As grand as you make it sound to be a part of the exuberance of the creative activity of the universe, it is still the universe and by your own admission the universe has no intrinsic worth except to itself and to us within it. The universe has no goal or purpose, it just is. Is that really all there is? Left to its own devices, the answer must be a discouraging `yes', but have we been left to our own devices?"

Dr. Swimme has another book, "The Hidden Heart of the Cosmos", op. cit. It was published by the Maryknoll Order's Orbis Books for use in their Catholic missionary activity so they must feel that there is something here that speaks of god, though they add a disclaimer on the copyright page that the material is not the usual orthodox approach. After reviewing the standard cosmological picture he gets metaphorical, speaking of the "all-nourishing abyss" which has few anthropomorphic allusions, though "nourishing" provides some. He also repeats Einstein's use of the title, "the Old One" several times which certainly does hint at a personal being above the universe and he capitalizes the "Great Power" that was expressing itself through Einstein but these powers are `gravity,' or `the second law of thermodynamics,' or `the strong nuclear interaction'." (p. 109)

Once again I have no problem with Swimme's prose and poetry about our place in the universe and the universe's place through us. But the Maryknoll's are kidding themselves if they think they can take one more step from his position to that of a personal loving god they have preached all over the world.

JOHN R. SEARLE

In his book, "Mind, Language, and Society", New York: Basic Books, 1998, Searle brings up the concept of god just long enough to say,

> "I believe that something much more radical than a decline in religious faith has taken place. For us...the world has become demystified. Or rather, to put the point more precisely, we no longer take the mysteries we see in the world as expressions of supernatural meaning...The result of this demystification is that we have gone beyond atheism to a point where the issue no longer matters in the way it did to earlier generations." [p. 34]

He speaks of Bertram Russell's answer to the question of what he would say to god should he find such a being to exist. "Russell answered without a moment's hesitation. `Well, I would go up to Him and I would say, `You didn't give us enough evidence'.'" [p. 37]

This certainly qualifies as agnosticism and his book makes no further mention of the divine.

ROBERT WRIGHT

We have seen a number of citations of Wright's thought as expressed in "Three Scientists and Their God" and in "The Moral Animal". In his latest book, "Nonzero", New York: Pantheon Books, 2000, he specifically asks if the evidence, "that biological and cultural evolution have some hallmarks of design" [p. 319] is any reason to believe in, "the values that people associate with God? In one sense, the answer has to be no." As close as he comes is in saying,

> "...that biological evolution has an arrow...and that this arrow points toward meaning isn't, of course, proof of the existence of God. But it's more suggestive of divinity than an alternative world would be..." [p. 323].

Then, "...cultural evolution's movement toward [a] moral threshold isn't proof of a benign

> universal architect...But...this cultural development is closer to being evidence of divinity than its opposite would be." [p. 327]

Still, these hints are not nearly enough for him.

> "I gave up so long ago on an omnipotent and benign deity that I'll take a few wisps of good karma and hope they signify something larger." [p. 332]

He really doesn't believe it.

Wright makes one remark in passing that I resonate with. At one point he says,

> "Personally, I prefer a Zoroastrian scenario. Or, perhaps, a scenario in which a good God, though not confronting an active, satanic force, is in some other sense of limited power. Maybe in creating the universe, He (She, It) faced metaphysically imposed design constraints." [p. 319-20]

We have seen others who suggest that god cannot, for example, create free will without allowing for the possibility of evil and suffering. This is not seen as a limitation on god, just a metaphysical impossibility. I certainly do not agree with this but, jumping the gun a little, I certainly see a place for a god with other than the usually ascribed powers.

479. The results of an interesting survey of scientists about their beliefs was published by Edward J. Larson and Larry Witham, "Scientists and Religion in America", *Scientific American*, Vol. 281, No. 3, September 1999, p. 88, repeating a survey done in 1914 and in 1933. Then as now about 40% of scientists reported that they believed in, "a God in intellectual and affective communication with man...to whom one may pray in expectation of receiving an answer" and in "personal immortality", the two survey questions. Among members of the National Academy of Sciences the figure

for disbelief was much higher. For many it was the question of evil in the world that led them to their positions.

Also personally interesting to me is the report's mention that astronomer Jocelyn Bell Burnell,

> "...finds a place for both science and religion in her life. `I don't think God created the world in any physical sense...But that's not to say there isn't a God.' For her, God just *is* - a private, interior experience tells her so. She said that some people want to put God in that little nanosecond gap after the big bang, but her faith did not need God in any gap."

This would certainly be my position as well.

The article cites a number of other scientists and theologians and is worth looking into.

480. Evelyn Waugh's "The Loved One", Boston: Little, Brown & Company, p. 85, pokes fierce satire at the commercialization of death, and hence the ability (read honesty) to take death for granted. Not everyone is able to be as brutally candid about death as the narrator of his story who writes a poem in honor of an acquaintance who has died by his own hand. I cite it here because I like it, but also because without faith it is the most honest thing that can be said of the dead.

> "They told me, Francis Hinsley, they told me you were hung
> With red protruding eye-balls and black protruding tongue
> I wept as I remembered how often you and I
> Had laughed about Los Angeles and now 'tis here you'll lie;
> Here pickled in formaldehyde and painted like a whore,
> Shrimp-pink incorruptible, not lost nor gone before."

481. George Johnson, "Fire in the Mind", New York: Alfred A. Knopf, 1995 does a masterful job of examining the findings of science today and raising questions that perhaps only a faith understanding can approach. He gives few answers to his own questions, but is worth reading for both the science and the limits and questions it faces and raises.

482. David Noble in "The Religion of Technology" says that there has not been a breach but that religion has been co-opted by science to justify its vision of a new, transcendent world and its faith in the power, not of a divine being, but of technology and science. The high priests of this religion

merely replace the high priests of the old. He may well have a point, but for those who are still the priests of the old, science often looks more like the enemy.

483. Leo R. Ward, C.S.C., "God and World Order" St. Louis: B. Herder Book Co., 1961. See chapter 12f

484. John Courtney Murray, S.J. "The Problem of God" New Haven: Yale University Press, 1964 p. 98

485. On Einstein and his thoughts on god and religion, see Max Jammer, "Einstein and Religion" Princeton: Princeton University Press, 1999.

See also "The Quickening Universe", p. 227f

> "Science has many `prophets,' but none whose vision of nature seems more congenial to a quickening universe than Albert Einstein. He stood on the shoulders of Isaac Newton and in a sense may have seen the body of God-beyond dead, absolute time and space to a vibrant, living universe of spacetime tortured and twisted by matter. Because Einstein, the humble and intensely private seeker of truth, was so reluctant to broadcast his views, few know him as the God-intoxicated man that he was. His writings and spoken words, sprinkled along the path of his seventy-six years, reveal an intensely religious person. While he lived, paradoxically some reviled him as an atheist, but they apparently didn't understand. In some ways Einstein has suffered the same fate of his seventeenth-century intellectual forebear, philosopher Benedict Spinoza, a similarly God-obsessed and misunderstood maverick. Einstein's God was not the personal God of western religions, nor did his theology match religions of the Orient. He spoke and wrote of having a `cosmic religion,' beliefs that he claimed were difficult to describe to anyone who is entirely without them. Central to his religiosity was, in his words, `a rapturous amazement at the harmony of natural law, which reveals an intelligence of such superiority that, compared with it, all the systematic thinking and acting of human beings is an utterly insignificant reflection.' He did not believe in a personal God, writing in his 1931 essay *The World as See It*, `I cannot conceive of a God who rewards and punishes its creatures, or has a will of the kind we experience in ourselves. Neither can I nor would I want to conceive of an individual that

survives his physical death.' Mystery, but not untutored mysticism, was key to Einstein's religious sentiment. In words impossible to paraphrase he wrote, `The most beautiful experience we can have is the mysterious. It is the fundamental emotion which stands at the cradle of true art and true science. Whoever does not know it and can no longer wonder, no longer marvel, is as good as dead, and his eyes are dimmed. It was the experience of mystery-even if mixed with fear that engendered religion. A knowledge of something we cannot penetrate, our perceptions of the profoundest reason and the most radiant beauty, which only in their most primitive forms are accessible to our minds-it is this knowledge and this emotion that constitute true religiosity; in this sense and in this sense alone, I am a deeply religious man.'"

486. "Is God Dead" in "*Concilium*", Vol. 16 p. 57

487. "Religion, Science, and the Search for Wisdom: Proceedings of a Conference on Religion and Science September 1986", Ed. by David M. Byers, Washington: United States Catholic Conference, Inc., 1987, p. 7

488. In my treatment here, the page references are to the Proceedings of a Conference on Science and Religion as published by the National Conference of Catholic Bishops. Since this is especially relevant to this discussion, being a serious effort by the Church to react to the scientific scene, I am reacting to it at greater length. To follow the discussion, it would help to have the text at hand.

In the opening address, Bishop James A. Hickey says all the usual things for an enlightened bishop of the 20th century, but continues to insist that the faith knowledge of god, "a God whom we know more intimately as our knowledge grows of the world he created and sustains." (p. 11) is indeed knowledge of a creating-sustaining god. My opinion should be clear by now that this is an assumption which we need not make in order to know something of god.

In the first presentation, Rev. Ernan McMullin of Notre Dame University tackled the big question of, "what is God's relationship with the world?" (p. 16) He says, and I certainly agree,
"the primary focus of the...Hebrew Bible...is on salvation history...and not on cosmology, on Yahweh's role as cosmic Creator." (p. 17) This god is one

who, “holds all things in existence”, Israel came to believe, but more importantly is also, “…a Being to whom his creatures can confidently look for redemption.” (p. 21) Rev. McMullin contrasts the Greek approach to what is real, restricting it to the natural, by and large. With Christianity, Augustine linked the ideas of creation, Incarnation and Trinity. Beyond the Old Testament concept, poor as it was, this Trinitarian god is more than a mere arranger of pre-existing materials, and contrary to the Greeks, god is not part of nature as a Prime Mover, but rather a transcendent being. For this god, there are no chance events (p. 26) Everything was created in one act and merely moves from potentiality to actuality according to the plan and Mind of god.

By the 13th century, the thought of Aristotle was introduced to Christianity and especially with Aquinas was made to agree with the Biblical picture of a Creator. The Prime Mover becomes a First Cause. Following on Greek philosophy, the claim was made that one could demonstrate the existence of god from an examination of god’s works in creation. This made the proofs dependent on features that “science alone can never explain”. Acquinas’ proofs are therefore at risk from philosophical arguments as well as whenever science could explain the phenomenon without recourse to divine intervention or causality.

McMullin gives a brief treatment of the period following and then treats the anthropic principle idea. The needed fine tuning of the universe appeals to theists leaving room for a Designer god, though one not involved in the details of evolution of this universe. McMullin rightly points out that this is still a “gap” argument, explaining by an appeal to god what science can not as yet explain. (p. 39) And this is pretty much where he leaves it. His conclusion is that, “we may, then, lack an argument for God’s existence that would convince a science-minded generation.” (p. 41) He also says, “I challenge the idea that the existence of God can be proved by arguing from some observable fact in the natural order which can *only* be explained by invoking god as the beginning of a sequence.” (p. 43) Ultimately he seems to conclude, to the chagrin of other theists present, that Faith and Science are talking about the same universe, but faith had better listen to science and adjust itself accordingly. Precisely my conclusion.

The physicist Dr. Freeman Dyson does his best to allow for some place for god in the universe but succeeds only in finding a god *of* the universe. His god is an extension of the universe and bears little resemblance to the god of the Bible. His "metascience" is an attempt to allow room for theology but it resembles nothing so much as unfounded speculation. Since that is what much of theology has been, some may welcome it. Unfortunately, the response to Dr. Dyson by Rev. Paul M. Quay, S.J. is on this order.

Edward Wilson begins his address by saying that, "I believe that traditional religious belief and scientific knowledge depict the universe in radically different ways. At bedrock, they are incompatible and mutually exclusive." (p. 82) His next words are hardly more conciliatory when he denies any natural law morality and finds the basis for morality written in our genes. He then asks, "Given the combination of mutation and natural selection, the biological equivalent of watches [referring to Paley] can be created without a watchmaker. But, did blind natural selection also lead to the human mind, including moral behavior and spirituality?" (p. 85) His answer, backed by examples is "yes" and I certainly concur. Still, he points out, genetic evolution alone is not enough to provide us with the material with which to make good ethical decisions, and religion as a cultural phenomenon, has a unique role but only as a human construct.

In response, Fr. Thomas M. King, SJ points out that Wilson leaves some room for religion in the present dispensation but that as science answers more questions, it may well be phased out. He claims that science, at least a materialist metaphysics such as Wilson allows, cannot give values, but does not answer where religion gets the basis for its values. He proposes Divine revelation or the concept of natural law as embodying the mind of the Creator as this basis, I presume. Finally, he wants to leave room for wonder and therefore the "God of the gaps" but gives no basis for his desire or the necessity of fulfilling it.

The panel responses to Wilson and King included a bishop who held out for the "possibility" that god has something to do with ethical guidance. I would argue that god does indeed give such guidance but that we must be ever careful in deciding when this has occurred and when our cultural biases lead us in certain directions. Wilson responded that the bishop's examples are faulty.

A theologian also, like King, wants to have more solid values and rejects that "survival" favored by Wilson is the ultimate value. In response, Wilson does not back down from the position that science may yet explain the origin of values and spirituality, though he does point out that if it cannot, then we will be forced, "into a much more determined effort to look for divine or external influence. This search would be forced on us as a scientific enterprise, not a theological one." (p. 98) I don't know if anyone really believes that Wilson and others would seriously suspect that this is going to be the outcome or allow any thought of such a thing to cross their minds unless they were at a meeting called by Catholic bishops. I certainly don't believe it. Wilson does say something that I resonate with, "How does one determine that [a divine plan] is out there other than a leap of faith?" (p. 98) Incidentally, one of his arguments at least would find less favor among fellow scientists. In response to a question about altruism in caring for elderly members he says that it is a value to a species to keep the line intact. Dawkin's selfish gene cares not a whit for the species, but does care about the individual's line.

In a Eucharistic homily during the conference, Bishop Edmund Szoka spoke of the mysterious wisdom revealed in the Incarnation. No quarrel there. Unfortunately, he repeats the assumption that the universe reflects its creator and the further assumption that there is something beyond human wisdom to which we must be open, both put forward as givens needing no explanation. Lord knows what the agnostic and atheist members thought of this.

A psychobiologist, Dr. Roger Sperry, speaks of a new appreciation for the "mind" as an emergent quality of the brain. There must not be a dichotomy, he insists, and as long as there is not then, "with reductionist fallacies now corrected in the new microdeterminism, there seems no logical reason why scientific belief cannot be fused with religious belief." (p. 114) Just what religious beliefs he has in mind was not clear to me. He speaks of values and "oughts" but just how god would figure in this is not mentioned. Religion as a system for upholding moral values is not the religion of most believers. I doubt if he has in mind the existence of a soul or the need for a divine creation of mind, since he insists that the new approach he espouses, "...denies that these mental-spiritual phenomena can exist separately in an unembodied form apart from the functioning brain." (p. 114) If his "spiritual" phenomena are an emergent quality of the brain, then the divine is not involved, as I argue in the text.

Dr. Sperry also calls for a collaboration of science and religion in the matter of global ethics. Again there is no mention of god as the basis for religion's contribution to the discussion of what "ought" to be. There is no appeal to a creator, for example, as a basis for an environmental ethic. Indeed, he says that, "...humanity's creator becomes the vast creative force system in `evolving nature,' which includes human nature.) (p. 117) I certainly agree that Christianity's reverence for creation and its creator is helpful in setting a tone for humanity's use or abuse of the environment, but its reason for taking such a position is not proved by its appeal. However, I also believe that god is at work among humanity (and any denizens of other worlds who can respond to god), building on what nature has provided us. This is closer to traditional religious faith, but removed from what Dr. Sperry calls religion.

In response, Fr. Joseph Bracken, SJ seems to have the same reaction I did in claiming that Dr. Sperry's position would lead to pantheism if incorporated into a religious position, and the pantheistic god is not the immanent/transcendent god of Christianity. The discussion brought up ideas of a Cosmic Mind and god's presence in the universe, but again the theists present had a problem with restricting this presence to the universe. What about transcendence? As one member, Fr. McMullin, put it, in Sperry's ideas, "God, insofar as he can appear in a natural system at all, is simply the upper state of that system. This is an entirely immanentist or pantheist view...The notion of God that you are proposing is that of a higher level within nature itself." (p. 132) Sperry not only does not answer this objection, he seems to not understand it, nor does Wilson. This is understandable for non-theistic scientists. The theists seem more bothered that their point is not being heard rather than that these human beings are not able to hear it. In my opinion, as long as we try to carve out a place for the traditional god in the scientific view we are not going to be understood.

The next speaker, Dr. Jerome J. Lejeune, attempts to reconcile science and religion in the manner of the creationists, (though Ian Barbour generously says Lejeune is not in this camp) but finding correlations between Scripture and science that satisfy no one. Few modern scientists, or biblicists for that matter, really believe that the authors of Genesis knew of the luminous era following the Big Bang so that light preceded the formation of the sun and stars. Nor do many believe that god was revealing this fact by the way the events of Genesis are numbered. That is precisely what Lejeune attempts to do, to preserve the "science" of Genesis and other stories with modern science, thus doing violence to today's understanding of Scripture and to the findings of cosmologists. In the response to Lejeune's presentation, Rev. Benedict Ashley, OP shows the same previously-held assumptions that "what we have always believed" of god and of the universe must be correct in the main, though science forces some minimal corrections.

This did not satisfy Fr. McMullin even, much less Edward Wilson who says pointedly, "I must say that I recognize very little of modern biology in what Dr. Lejeune is saying. I find the model that he is presenting...singularly unconvincing and without a shred of evidence." (p. 160) Lejeune himself is finally reduced to saying that even if his arguments are true, it tell us nothing of god. Only his faith tells him of god, which is my point exactly.

In the last presentation, Dr. Ian Barbour labors mightily to find some way for the two disciplines to speak to one another but in the main fails. He admits, finally, that, "We should not expect science and its view of nature to contribute too much to theology." (p. 182) but calls for an effort to do what Acquinas once did as he, "brought Scripture into relationship with the best philosophical and scientific thought of his day." On the other hand, "I would not look to science for the establishment of fundamental ethical norms such as social justice, individual freedom, or even respect for nature." and he leaves this to religion. Just how it can provide a basis for decisions among people who have no respect for its views is problematic.

To me it is telling that in the following discussion there is no input from the scientists present. Perhaps by that time they had despaired of finding any connection with which they could agree. This would not surprise me since I have met a number of scientists who would love to have some reason to allow room for faith but have found little. It is obviously my opinion that we can give them none except Faith and the personal and communal experience of god working with us.

I can find no official continuation of the effort analyzed here to draw scientists and theologians into discussion. It continues to be my hope, therefore, that the work to which this is a commentary may give new ground for such activity.

489. "Quantum Questions", p. 163

490. "The Conscious Universe", p. 188

491. "The Conscious Universe", p. 59

492. There is much similarity to Diarmund O'Murchu, discussed below.

493. "Molecules of Emotion", p. 305

494. "Molecules of Emotion", p. 307

495. "Molecules of Emotion", p. 310

496. Tipler explains that a human consciousness arising from the brain, now reproduced in a computer, would *be* that human being. I certainly agree. The famous scientist popularizer/science fiction author Isaac Asimov has the reverse situation in his book, "Foundation and Earth", Garden City: Doubleday, 1986 when the question is asked about how would you feel making love to a robot which was completely indistinguishable from a human being. The character's answer is that, "It seems to me...that a robot that can in no way be distinguished from a human being *is* a human being. If you were such a robot, you would be nothing but a human being to me. This scenario will reappear in the final chapter on resurrection.

497. "The Physics of Immortality" p. 219

498. Timothy Ferris in "The Whole Shebang" p. 350 notes that, "Tipler called his theory FAP, the final anthropic principle - that `laws of physics allow life to exist *forever*.' The science writer Martin Gardner memorably, if unkindly, dubbed it `CRAP-the Completely Ridiculous Anthropic Principle.'"

Chet Raymo in "Skeptics and True Believers" is perhaps even more blunt, saying,

> "Is Tipler's theory the end of the clash between science and religion? Hardly. Is it the most dramatic scientific proof of all time? Give me a break. This is the kind of pseudo-physical mysticism that gives physics a bad name." p. 114

Tipler's ideas do not resonate with theologians, either, a typical comment being, "One wonders if the author is serious."

> [Zachary Hayes, O.F.M., "God and Theology in an Age of Scientific Culture", *New Theology Review*, Vol. 8, No. 3, August 1995, p. 5f. Hayes' article does a good job of examining the relationship of science and theology from a faith viewpoint.]

499 Ernest J. Sternglass, "Before the Big Bang", New York: Four Walls Eight Windows, 1997.p.16, then p. 262

500. "The Cosmic Blueprint" pp 140-141

501. "At Home In Universe" p. 33

502. "A History of Modern European Philosophy", p. 815

503. "History of Modern European Philosophy" p. 832

504. "A History of Modern European Philosophy" p. 837

505. "The Cosmic Blueprint" pp. 141-2

506. Mark Schoof, O.P. "A Survey of Catholic Theology 1800-1970", Paramus: Paulist Newman Press, 1970, p. 116

507. Michael H. Murray, "The Thought of Teilhard de Chardin" New York: The Seabury Press, 1966, p. 28-29

508. Andre Ligneul, "Teilhard and Personalism" New York: Paulist Press Deus Books, 1968, p. 8
See Ernan McMullin's treatment of Chardin and his historical and philosophical mileau in "Religion, Science, and the Search for Wisdom", p. 36

509. For his thoughts on "hominization" and the development of a "noosphere" in addition to the so called biosphere, see Pierre Teilhard De Chardin, "The Future of Man", New York: Harper & Row, Publishers, 1964, which followed on his publication of "The Phenomenon of Man."

510. Karl Popper and John Eccles, "The Self and Its Brain", Berlin: Springer International, 1977.

See also Karl Popper, "The Open Universe"

For an extended treatment of Popper, see Bryan Magee, "Confessions of a Philosopher", New York: Random House,1997, chapter 11.

In my local Catholic newspaper, *The St. Louis Review*, one columnist, Antoinette Bosco, wrote on August 29, 1997 an article titled, "Scientific study reinforces religion". The authorities she cites are John Eccles and John Polkinghorne plus some unnamed "eminent physicists" who "have acknowledged that their discoveries seem to point to the conclusion that we and all of creation are not here by accident. All has been designed by a supernatural architect." I would not expect complete answers in such a column in such a forum, but this means that the Catholic laity are still being told that they can believe in a creator-god without fear of contradiction by any really "eminent scientists." I fear the reality is otherwise, but I fear even more the reaction of these faithful when they find out that the majority of "eminent scientists" would not accept any argument from design.

511. Julian Huxley, "Essays of a Humanist", New York: Harper and Row, 1964;
"Man Stands Alone", New York: Harper & Brothers, 1941;
"Man in the Modern World", New York: Mentor Books, 1962

512. "Evolutionary Humanism and the Faith" in "Concilium", Vol. 16, p. 56f

513. Joseph Ratzinger "Introduction to Christianity", Trans. by J. R. Foster, New York: Herder and Herder, 1970, p. 43

514. "A Survey of Catholic Theology 1800-1970" p. 208

515. M.B. Martin, S.J., "Christianity and Its Cultural Bondage", St. Louis: Robert Bleile & Assoc. 1972, p. v-vi

516. "Before the Beginning" p. 98

517. "Before the Beginning" p. 99

518. "God and the New Biology" p. 117

519. "God and the New Biology" p. 111
Ian Barbour, "Religion in an Age of Science", p. 27 puts Peacocke in context.

520. "God and the New Biology" p. 107

521. "The Anthropic Cosmological Principle" p. 183
The two author/scientists commenting on Peacocke's work suggest that his argument is, "so general that it would be consistent with any scientific result, and so, although interesting, it is completely useless." Remember that the strong anthropic principle which they espouse has also been found not helpful.

522. Peacocke also finds a place for the traditional assumption of the possibility of miracles, but tempered by science, especially the discovery of the role of chaos. He would like to see the two principles of the chaos present in the world and the control exercised by god combined and says,

> "In human life we must accept, for the stability of our own mental health and of our faith, that reality has a dimension of chance interwoven with a dimension of causality-and that through such interweaving we came to be here and new forms of existence can arise. This acceptance of chance as part of the mode of God's creativity is more consistent with the fundamental creativity of reality than the belief- stemming from a Newtonian, mechanistic, determinist view of the universe with a wholly transcendent God as the great Lawgiver - that God intervenes in the natural nexus for the good or ill of individuals and societies. We must learn to accept these conditions of creation and of creativity in the world..."
>
> ["God and the New Biology" p. 99-100]

Richard Swinburne, another theologian-philosopher we have seen, concludes that

> "I am, however, inclined to think that we do have enough historical evidence of events occurring contrary to natural laws of a kind which God would have reason to bring out to show that probably some of them (we do not know which) are genuine miracles...We are rational to believe, while allowing the possibility that evidence might turn up later to show that we are mistaken."
>
> ["Is There A God?", pp. 120-121]

523. "Toward a Theology of Nature" p. 33
Ian Barbour, "Religion in an Age of Science", p. 18 has Pannenberg in context with other theologians.

524. "Toward a Theology of Nature" p. 35

525. "Toward a Theology of Nature" p.37

The same assumption underlies the position of Howard J. Van Till, a noted creationist, who says that,

> "..when we ask about the origin of the entire universe, however, we are using the term in a different and far more profound way [than speaking of the origin of a feature within the universe such as the sun]. In that case we are asking not merely about the formation of its structures; we are asking the ultimate question of cosmogenesis: What is the source, or cause, for the existence of the whole cosmos? We are not asking *how* things were formed, but *why* there exists something rather than nothing...The question of origin (in this ontic sense) lies outside the domain of natural science."
>
> [Howard J. Van Till "The Scientific Investigation of Cosmic History" in "Portraits of Creation" p. 122]

526. "Toward a Theology of Nature" p. 48

527. In Tipler's "Physics of Immortality", Page 13 he refers to Pannenberg.

> "Wolfhart Pannenberg has suggested [in "The Spirit of Life" *In Faith and Reality*, 1977, and in 'Theological Questions to Scientists" *Zygon*, 1981] that there may exist a previously undiscovered universal physical field (analogous to Teilhard's "radial energy") which can be regarded as the source of all life, and which can be identified with the Holy Spirit. There are no undiscovered `energy' fields of significance to biology; conservation of energy and the size of the energy levels in biology preclude it. However, I shall argue in Chapter VI that the universal wave function (provided it satisfies an `Omega Point' Boundary Condition) is a universal field with the essential features of Pannenberg's proposed new `energy' field."
>
> Page 183ff "Let us suppose that the Omega Point Boundary Condition gives a unique wave function. Then it would mean that the laws of physics and every entity that exists physically would be generated by the Omega Point and its living properties. For these properties determine the universal wave function, and the wave function determines everything else. In any interpretation and with any boundary condition, the universal wave function is the unique field which gives being to all other fields-the electroweak fields, the gluon fields, the quark fields, the lepton fields, indeed all the usual physical fields. With the Omega Point Boundary Condition, this

all-determining field becomes ultimately personal. So we have an all-pervasive physical field which gives being to all being, which gives life to all living things, and which itself is generated by the ultimate life which it defines.

In several papers the German theologian Wolfhart Pannenberg has suggested that there exists an undiscovered physical field, an all-pervading physical field which can be regarded as a transcendent source of life. I claim the universal wave function with the Omega Point Boundary Condition is a good candidate for such a field. The Omega Point Boundary Condition explicitly requires that the wave function force the physical universe to give rise to life, and it requires that this life persist into the Omega Point. With the Omega Point Boundary Condition, the universal wave function thus brings life into existence and sustains it in existence. Furthermore, the universal wave function is not restricted to living things, but it is everywhere. It has the power of self-transcendence as Pannenberg defines it: `...self-transcendence is to be regarded at the same time as an activity of the organism and as an effect of a power that continuously raises the organism beyond its limitations and thereby grants it its life' [1977 citation]. An excellent description of the relationship between an organism in the universe and the universal wave function with the Omega Point Boundary Condition. Pannenberg has also pointed out that `In biblical traditions...the life-giving power is seen as an agent that influences the organism from the outside' [1981 citation]. In the biblical traditions, this life-giving power is the Holy Spirit. I am thus in effect proposing that we identify the universal wave function constrained by the Omega Point Boundary Condition with the Holy Spirit. I claim this identification is reasonable, for, as discussed above, a wave function is the all-pervasive physical field which creates and guides all the directly observed physical fields, and the wave function is made explicitly Personal by the Omega Point Boundary Condition. Thus the universal wave function constrained by the Omega Point Boundary Condition is an omnipresent invisible field, guiding and creating all being, and ultimately Personal - these are the traditional defining properties of the Holy Spirit. One could also identify the universal wave function constrained by the Omega Point Boundary Condition with what Teilhard called `radial energy.' For, as Pannenberg points out, `In Teilhard's perspective there is only one spirit permeating and activating all the material processes and urging them beyond themselves in a process of progressive spiritualization of converging unification towards a centre of perfect unity which in providing the end of the evolutionary process proves to be its true dynamic origin.' [1977 citation]. This is exactly what the universal wave function with the Omega Point Boundary Condition does. According to the Omega Point Boundary Condition, the structure of the phase paths (more precisely, their ultimate future) gives probability weights-guidance, so to speak, not rigid control-to all paths. The ultimate future guides all presents into itself. But this guidance is *not* determinism."

Pannenberg may get as much of a hearing from other physicists as Tipler on immortality. Not much.

528. "The Faith of a Physicist", p. 76

529. John Polkinghorne, *The Modern Interaction of Science and Theology* in "The Great Ideas Today", Encyclopaedia Britannica, Inc., 1995, p. 34

530. "The Faith of a Physicist", p. 77

531. John Polkinghorne, "Quarks, Chaos & Christianity", New York: Crossroad Publishing Co., 1996, p. 99

532. "Quarks, Chaos, and Christianity", p. 94-95

533. "Religion in an Age of Science", p. 270

534. ibid.

535. "The Quickening Universe" p. 239

Chet Raymo in "Skeptics and True Believers", p. 131f examines panentheism.

Matthew Fox, "Original Blessing", p. 88f recommends it.

Frans Jozef van Beeck, "God Encountered", p. 38f situates it in the usual Catholic approach to god as Creator.

536. "Science & Religion" p. 147

537. "Science & Religion" p. 149

538. “Science & Religion” p. 151

539. “Science & Religion” p. 154

540. “Science & Religion” p.156

541. “Science & Religion” p. 160 Kenosis is the self-emptying love that Paul speaks of in Phillipians, for example.

542. “Science & Religion” p. 154

543. Timothy McDermott, “Design vs. Chance: Eavesdropping on Aquinas and Darwin”, *Theology Digest*, Vol. 43, No. 4, Winter 1996, p. 317

544. “Soul” p. 175-6

545. Diarmuid O’Murchu, “Quantum Theology”, New York: The Crossroads Publishing Company, 1997

546. “Quantum Theology”, p. 49

547. “Quantum Theology”, p. 107

548. “The Whole Shebang”, p. 310

549. STEVEN C. KUHL
One example, cited above under the critique of Daniel Dennett’s position, is Steven C. Kuhl, op. cit. He points out that,

> “...not only Darwinism but the very `fact of evolution’ *does* challenge traditional theism in legitimate and profound ways...That, however, is not, in my judgment, a problem, but an opportunity: an opportunity to reintroduce Paul’s theism into the citadel of theology which, for the most part, has been thoroughly dominated by traditional theism.”

Kuhl points out that in the authentic biblical and Christian theology,

> “...humanity’s unique standing is *within* the creation, *as part of* the creation, not outside or apart from the creation, which is the same point that evolutionists like Dennett make...” (p. 93) Theologically, this puts humanity, “uniquely equipped...to serve as the creation’s

representative...before God, opening up an axiological in the world...Nowhere in all of Scripture are the nature and implications of this representative role of humankind *within* the creation more thoroughly worked out than by Paul himself in Romans 8." (p. 94)

Kuhl shows this very well.

On the subject of god as creator/designer of the universe, Kuhl points out that though Paul would probably not deny this role for god, it is not Paul's focus. That world, the pre-Fall world, can give no meaning to humanity's existence.

> "For Paul, Meaning, in any deep sense is not grounded in creation, origins, or `first things,' but in Christ, in eschatology, `last things.' Meaning is a matter of a new creation." (p. 96)

In a third point, Kuhl takes up teleology. In traditional theology, the universe, or at least humanity, is going somewhere because it was designed from the beginning to do so. This, of course, can be held no longer since,

> "...there is no evidence of an overarching blueprint guiding the direction of life. Life proceeds in response to natural changes that are not only unpredictable but blind, void of any beatific vision. The most compelling evidence of this is the fact of extinction..." (p. 97)

Not a problem for Paul, according to Kuhl.

> "For Paul, the predominant `telos' of the world is not that of the completion of an edifice, but of judgment. The telos of the world is that of a trial, a dynamic process through which `the whole world may be held accountable to God. (Rom 3:19)"

I found this rewarding reading and would like to hear more from this theologian. It certainly departs from traditional thinking and language which, of course, I insist we must do if the Faith is to have anything to say. At the same time, it is based on Paul's writings in the New Testament, and you don't get more "traditional" than that.

C. S. LEWIS

C. S. Lewis is somewhat dated now but was certainly aware of the challenge that science was presenting to the faith in the 30's and 40's and tried to respond in his "Miracles". Don't look here for new answers but rather for the old assumptions.

PATRICK GLYNN

Patrick Glynn, "God: The Evidence", op. cit. is a former atheist turned believer and it is the anthropic principle that turned the tide for him. He is a source for the many factors that must be "just so" for the universe to exist as it does. He is rather big on near-death experiences and the evidence for god from faith's effect on human lives and health. He even brings in Pascal's wager. His conversion to faith from atheistic scientist is based more on philosophical and sociological reasoning than any familiarity with scripture or the Church. His starting point, the anthropic principle, is not a very strong basis for faith for for me, though it may be for him.

GERHARD STAGUHN

Gerhard Staguhn, "God's Laughter", New York: HarperCollins, 1992 was one of the first works I read on the relation of science and religion. It is an excellent overview of many of the areas I have covered in the text, beginning with a short history of cosmology such as I learned it in the seminary, a very interesting chapter on Nicholas of Cusa, relativity, quantum mechanics, and even the Tao. He also has a chapter on Planck, Heisenberg and others and their thoughts on science and religion. In his conclusion he points out that faith is not knowing in the sense of scientific, or even logical, knowledge and points out that science cannot satisfy our metaphysical longing, while rightly asking, "but does this longing have to be satisfied at all?" [p. 236] He answers honestly that it does not. He flirts with quantum mechanics as showing some direction to an answer when he says,

> "…thus in quantum mechanics we now witness a kind of recovery of the spiritual in physical phenomena after materialists had long hoped that everything, even the spiritual, could be explained by purely mechanical actions between matter…however…matter can no longer be explained materialistically. Even matter receives its order by something spiritual that can only be described mathematically." [p. 241]

Obviously, "this does not leave room for an omnipotent personal God or other supernatural beings." [p. 245] and his answer to the quest is the last line of his work where he cites a Jewish proverb saying, "Man thinks, and God laughs." [p. 251]

That is just about the same answer that Job gave to the problem of evil. As I said, Staguhn's book is an excellent overview of the problems, but is not long on answers.

WALTER BRUEGGEMANN

Walter Brueggemann, "Texts Under Negotiation", New York: Fortress Press 1994. This evangelical theologian looks for a response to "postmodernism" but unfortunately for our purposes, though he is a lyrical apologist, just states that the evangelical,

> "...community operates with a powerful, poignant memory, a memory that affirms that our past has originated through and been kept for us by a faithful, sovereign God who calls into being things that do not exist." (p.29)

This presumes that this is a memory of an actual event rather than a memory of an inherited assumption. It also presumes that it is a memory of a cosmological event rather than a faith-community creating history, to which I would certainly admit. Since he does not want to deal with, "...exasperating questions of creation and science," he would rather begin with a faith which sees what science seeks to explain as, "an inscrutable mystery of providence." (p. 29)

For Brueggemann the origin of each person is part of this inscrutable mystery and he goes so far as to say that, "I did not `evolve,' but was loved and named by one even beyond mother and father." (p. 32) In trying to understand the origin of the world, one should "not seek to accommodate scientific learning" and though he says he does, "not denigrate the gains of science in understanding the world," he would rather cite psalms about God's sovereignty over it.

When he gets to the origin of community we are on more common ground but then we come to eschatology and he has,

> "...the very creation that decays has ordained within it - because it is God's creation and not `nature' - an urging toward God's newness that is promised and sure." (p. 45)

This may all be reassuring for the beleaguered, "small group of ecumenical Christians, cast in a sea of fearful, religious reactionism" (p. 36) in Anderson, South Carolina that he is worried about but to my mind he is trying to build their faith on fragile ground and has nothing at all to say to those who are reacting against such facile faith. His conclusions about what kind of life the church must live in response to modern challenges rings true, but for the wrong reasons and based on untenable, internally proven assumptions.

KAREN ARMSTRONG

Karen Armstrong, "A History of God", New York: Alfred A. Knopf, 1993, gives a historical treatment of the development of ideas and approaches to god in many cultures. Among many points, she shows that,

> "for centuries monotheists in each of the God-religions had insisted that God was not merely another being. He did not exist like the other phenomena we experience. In the West, however, Christian theologians had got into the habit of talking about God as though he really *were* one of the things that existed. They had seized upon the new science to prove the objective reality of God as though he could be tested and analyzed like anything else. Diderot, Holbach and Laplace had turned this attempt on its head and come to the same conclusion as the more extreme mystics: there was nothing out there. It was not long before other scientists and philosophers triumphantly declared that God was dead." (p. 345)

On the future of God, she has little hope for a resurrection of anything like the traditional faiths (whatever they are) and says that,

> "the aimlessness, alienation, anomie and violence that characterize so much of modern life seem to indicate that now that we are not deliberately crating a faith in `God' or anything else - it matters little what - many people are falling into despair." (p. 398)

Nevertheless,

> "human beings cannot endure emptiness and desolation; they will fill the vacuum by creating a new focus of meaning. The idols of fundamentalism are not good substitutes for God; if we are to create a vibrant new faith for the twenty-first century, we should, perhaps, ponder the history of God for some lessons and warnings." (p. 399).

For me, we do not need so much to create a replacement for god as we need to recognize a god revealing godself, without burdening this god with our own assumptions. We will inevitably "create" a new god and new assumptions but we can *try* to avoid this.

MATTHEW FOX

Matthew Fox thinks he has common ground for science and religion, especially in "The Coming of the Cosmic Christ" and "Original Blessings" Santa Fe: Bear & Company, 1983, but his presumption of god the creator is blatant.

KENNETH MILLER

Kenneth R. Miller, "Finding Darwin's God", is a thoroughgoing evolutionist whose book responds to Wilson, Dawkins, Dennett, Barbour and many others I have addressed here. His own position, however, seeks to preserve god as a creator with a lot of patience, using evolution as a tool.

> "The vastness of the universe itself gives a hint athat this was exactly God's approach. If a Creator were to fashion a world in which the constants of matter and energy made the evolution of life *possible*, then by forming millions of galaxies and billions of stars with planets, he would have made its appearance *certain*." (p. 275)

He is also big on the free will argument and that God is constrained to "allow" evil and mayhem because that is the way the world evolved and we are very much a part of this world. Ironically, and maybe consciously so, Miller says that he believes in Darwin's God as reflected in the last line of "On the Origin of Species" which has the Creator breathing life into the universe and allowing it to evolve. "There is grandeur in this view of life" says Darwin. Stephen J. Gould, whom Miller cites at length but not in agreement has long written articles for the magazine *Natural History* entitled, "This View of Life." Gould would probably disagree that this is what Darwin meant.

DENIS EDWARDS

Edwards has a fine little book entitled, "The God of Evolution", New York: Paulist Press, 1999 which sums up much of the modern theologians reaction to science from a very welcoming position. His central position is that the,

> "universe can be understood as unfolding `within' the trinitarian relations of mutual love. Creation takes place and flourishes within the divine life." [p. 30]

He is no pantheist, however and sees the universe as the work of god, not some kind of emanation. Its nature is its own and not divine, though he comes close.

My quarrel with him begins, not surprisingly, when he tries his hand at theodicy, the question of evil in a universe created by a good god. The answer for Edwards is,

> "God may not be free to overrule natural process. A God who creates through physical process may well be committed to the integrity of the process. If this is the case, then God is not free or able to simply abolish all suffering. God, in creating, accepts the limits of physical processes and of human freedom." [p. 44]

This is all to the good if there were any evidence of god creating at all, and this brings us to the most questionable part. Since divine action is "physically undetectable", then there can be no "purpose" at work behind the scenes. His way out of this dilemma is to redefine purpose.

> "But it is important to recall the theological principle that tells us that when we use words like *achieving purposes* and *action* of God, we are using analogies taken from the human experience of purposeful action. Does not the doctrine of divine transcendence suggest that God might achieve purposes in a way that radically transcends all human notions of achieving purpose?" [p. 53, italics his]

We are then firmly in the realm of faith in an area not open to scientific inquiry. I can hardly say that he is wrong, but then he cannot claim to have anything to say to the non-believers. Of course neither can I except, "come join us and possibly see for yourself." However, I do not have to claim, against the evidence, that god has anything to do with creation. I suppose it comes down to preferring mine over his since it does not require recourse to "analogy" or beg any questions.

Edwards has another little gem, "Jesus and the Cosmos" which we will examine in the next chapter when we investigate the identity of Jesus.

JERRY KORSMEYER

Jerry D. Korsmeyer, "Evolution & Eden", op. cit., maintains the position that god is a creator and, "order requires an orderer," [p. 105] yet holds that god's power is limited.

> "Realizing that God's power is only persuasive is mentally wrenching for some Christians who derive a great sense of comfort from the idea that God's providential power means that God can jump in and save them from any possible situation, if only God wills to do so…Rather, the realization that God's power is persuasive suggests that action to overcome natural evil must come from our action as God's partners, responding to God's grace-filled urging. God can't persuade a cancer cell from establishing itself and growing in a critical site in a human body, because such response is not a possibility for such a cell. God can persuade a doctor to take an action to kill or remove such a cell, resulting in preservation of human life." [p. 107]

I recommend his book, at least in the area of understanding a limited god.

RUSSELL STANNARD

Also recommended is Russell Stannard's "The God Experiment", op. cit. Stannard is a physicist and gives an honest investigation about whether religious ideas can be maintained. He succeeds in making room for god and god's activity in the world but certainly leaves it up to faith as to whether or not god is really present. However, his answer to the question of evil which I believe to be central, is the same as Job's. He puts the best face on it, saying that,

> "...a God who was fully comprehensible to the mind of man would be a product of the mind of man. The fact that we do *not* fully understand God strikes me as the mark of his independence - his independent reality. He is not the sort of God anyone would have been inclined to dream up out of their own imagination. He is too problematic for that." [p. 101]

His other conclusions are in the same vein. It is not only that this proves nothing about god, for I would not expect him to even try for proof, but I believe that the god he describes, especially on the matter of free will and suffering, is precisely the kind of god we would dream up. This god fits the evidence of our experience and absolves god from responsibility yet leaves room for god to whip off a miracle for us now and then. I hope that I have demonstrated that in fact it does not relieve god of guilt. Still, I recommend the book for its honesty in the search if not for its hidden assumptions.

FR. GEORGE COYNE, S.J., the director of the Vatican Observatory, is reported to have told the U.S. bishops' Conference on Science and Human Values that there is only a conflict between science and religion when we yield to the, "temptation to equate God with ultimate explanation rather than with infinite love." I have not followed up on this, so both the reader and I may find some support in *Origins*, January 9, 1997.

FRANS JOZEF VAN BEECK

Frans Jozef van Beeck, "God Encountered" op. cit. gives a scholarly review of the history of the belief in god as creator and does not reach any conclusion that god is a creator or is not. He does point out the problems with the facile assumption that god is a creator. Near the end [p. 163] he says things like, "if God does create...", and, "if and whenever God does self-manifest..." leading me to think he has his doubts. Indeed, I think he leans toward the conclusion that I will discuss in the next chapter and would say that god can only be known through faith and not through demonstration from science or philosophy.

550. In chapter 2 we were invited to chose from various positions we believe. Those who chose the first answers in the various sets were considered to have a more traditional faith while those who chose the latter answers were considered to either have a more scientifically literate faith or no faith at all.

For these latter the intervening chapters probably reinforced your beliefs. For those, like Susan, who had not been exposed to ideas which pose a challenge to this traditional faith, these chapters may have been unsettling. Now that you have read them, let's take a look at the choices I proposed and see if you have changed.

1a. There is a cause for everything that happens.
1b. Nothing can come from nothing so there must be a cause.
1c. There is a cause for most things that happen.
1d. There is a cause for almost everything that happens.

I hope I have demonstrated that while in the world we know there is a chain of causality, certainly in the quantum world this is not necessarily true and it is not true that the existence of the universe or the course of evolution has some ultimate cause outside of itself. I pick c or d.

2a. There is a logical explanation for everything.
2b. There must be a logical explanation for everything.
2c. There is a logical explanation for most things, but some just happen.
2d. There is not a logical explanation for everything.

As in the first choices, there is a logical explanation for most things that happen, which is the whole point of science and philosophy as well. However, there is no logical explanation why, for example, there is a god, nor a universe. I would pick c or d.

3a. There is a first cause which itself has no cause.
3b. There is a first cause, which must keep on working.
3c. There is a first cause, which need not continue to operate.
3d. There is no first cause.

This is a rather philosophical idea and the reader may not have understood its implications in chapter 2. By now I hope that the idea that there must be some first cause other than chance (by my definition of chance) cannot be maintained. I pick d.

4a. If God does not keep things in existence they don't exist.
4b. If God does not move things, they do not change.
4c. Things exist until something destroys them.
4d. Things move without direction from God.

Another philosophical position which has infiltrated itself into popular religious apologetics that without god there would be nothing. Also untenable. I am with c or d.

5a. Something which "must exist" is required so other things exist.
5b. Nothing would exist unless this "something" makes it exist.
5c. Things can just "happen", there is no need for a Creator.
5d. Things did just happen, there is no Creator.

In all honesty we should by now all be in position c. Things did just happen, but this does not disprove that there is a creator. However, there is no need for a creator.

6a. There is a reason why things exist.
6b. There is a reason why things exist, but it may be known only to God.
6c. There is no reason why things exist.
6d. There need not be a reason why things exist.

Here again we should all be on position d. We cannot say that there is not some reason in the mind of god why things exist, but there certainly can be existence which has nothing to do with god.

7a. There is a plan for human beings, including one for me.
7b. There is a plan for the universe, but not necessarily one for me.
7c. There is no grand plan for humanity.
7d. There is no plan for the universe.

The faith position certainly hopes that answer a or b is true and that god is working with us and on us to bring us to something better than we might reach merely through the workings of evolution. Since we are part of the universe, perhaps this divine plan becomes a plan for the universe as well. However, it is not a plan which was in place at the inception of the universe or of humanity. God is working out of god's hip pocket on this.

8a. There is someone in control, working out this plan.
8b. There is someone in control, but it is a loose control
8c. There is no one in control, things just happen.
8d. There is nothing in control except physical or mathematical laws.

Manipulation and control of the universe seems to be outside of god's capability, so answers c. and d. are possible, though scientists would probably prefer answer d. Quantum physics throws some confusion into this over c. however, and tips the weight for me to c.

9a. The one in control often works miracles.
9b. The one in control sometimes works miracles.
9c. There is no one in control and there are no miracles, but things occur which cannot be explained.
9d. There is no one in control and everything can be explained.

Answer d. is going to have to be my choice.

10a. Prayer works.
10b. Prayer works, but God expects us to do our part.
10c. Prayer has no effect
10d. Prayer has no effect and it is harmful to believe it does.

If we are speaking of prayer "working" by influencing god to have an effect on the physical universe, then c. or d. are correct. I would lean to d. since I have too often seen the negative effects of people believing that if they pray correctly or enough or use the right formula, or just "have faith" they can expect a miracle. When one does not occur the effects can be very destructive. But let's wait until the final chapter before we decide if prayer "works".

11a. The one in control always does good.

11b. The one in control can do good but often stays inactive.
11c. There is no one in control, so they cannot change things.
11d. There is no in control except physical laws.

If we are speaking of control as control over the physical universe then d. would be the preferred answer. There may be other ways to influence and therefore have some effect, which we will see in the last chapter.

12a. God has told us what is good and evil.
12b. God has told us what is helpful and harmful.
12c. There is good and evil but based on physical reality.
12d. There is no good or evil.

At this point in our investigation we would have to go with c. or d. and they are really the same answer. Whether we can use god as a source of morality will also be investigated shortly.

SO WHERE ARE YOU NOW?
If you, like myself have some real problems with the traditional explanations of god's working in the world because of the evidence I have included in this work, where does that leave us? Let's get into the last chapter and see.

551. The god of the Judeao-Christian tradition is in many ways not so different than the god of other religions. A god who shows love and mercy for the people is not unknown in other faiths. Still, with the revelation of the god of the Hebrews there is a new dimension brought into sight. The message of the Exodus and Sinai is a message of covenant love for the chosen people.

> "The Sinai narrative tells the story of Israel going to meet her God, but it is not just the story of the first generation of Israelites. It is the story of Israel in every generation. It was the story of the current generation who, even as they listened to the narrative, felt themselves standing at the foot of Sinai ready to listen to the voice of God in the liturgical celebration." ["The God of Exodus", p. 21]
>
> "Beneath its bilateral aspect, the covenant was...the result of a purely divine initiative."

[Thierry Maertens O.S.B. "Bible Themes", Vol. 2, Notre Dame: Fides/Claretian, 1964, p. 141]

God revealed love for the people and promised to be with them and bring them to the promised land or situation. This love for the people invited a response, not only of love for the god, but of love for the other members of the covenant people. This is soon overshadowed by another event, the giving of the ten commandments, and the whole moral code of the Judeo-Christian tradition. The truly amazing thing about the Sinai event, or the process that is described in mythical shorthand by the Sinai story, is that god has chosen this people to be god's own; the covenant that god makes with them is unprecedented and unique. In typical human fashion, they soon downplay the relational and concentrate on what it means to their daily behavior. They are far from contemplatives meditating on the mystery of god's love for them. They are practical people who want to know, "what do I have to do about this?"

[There is not a simple answer to the question, "What Really Happened at Sinai?" as an article by that name by Baruch J. Schwartz, *Bible Review*, Vol. XIII, No. 5, October 1997, p. 20 makes clear in an account of the four renderings of the giving of the law]

550 Jean Guitton reports that Pope Paul VI was very concerned with having an honest dialogue with the world, including the world of science and that in a conversation with him about the nature of a "dialogue", Paul VI said, among other things, that what was needed is an equality among the participants. In response to the question of how that could be achieved he responded,

"Quite simply; certainly not by situation, information, authority, age, talent, not even by genius, but by an equal love of truth. It is that common love of truth which is the only reason for the existence of the sincere dialogue of which I am speaking, and which has little in common with worldly dialogues, which only seek a display of one's wit and, if one is able, that of others."

["The Pope Speaks: Dialogues of Paul VI with Jean Guitton", New York: Meredith Press, 1968 p. 167.]

Certainly Paul tried to dialogue with other religious traditions. How successful he was in achieving a dialogue with science is debated.

The *St. Louis Review,* of July 13, 2001 reported that now Cardinal Avery Dulles called for "respectful internal dialogue, but not the kind that tolerates dissent from Church teachings" at a lecture at Georgetown University. Dulles decries the kind of "liberal model" which ends up with both conflicting opinions intact but the combatants able to live in peace. So far, so good, but the tone is definitely that the Church is in possession of the truth and therefore any dissenting opinion must be wrong and should not be tolerated. "Catholic liberals sometimes promote [the liberal model] in order to persuade ecclesiastical authorities to tolerate dissenting opinions or doctrinal indifference." This seems to me a far cry from the kind of dialogue Paul VI called for and begs the question of whether or not the dissenting views (such as mine) might be more correct than the accepted views.

It is not only religion that is challenged to be honest in the face of modern science. See Paul R. Gross and Norman Levitt, "Higher Superstition: The Academic Left and Its Quarrels with Science", Baltimore: The Johns Hopkins University Press, 1994. It is interesting that this work does not mention God, religion or theology except cargo cults, creationism, Pius IX's condemnation of scientism and modernism, and a quote from an attack by the Church on Galileo. Evidently there are no Catholics on the academic left.

See "Skeptics and True Believers", p. 171

553. "God and the New Biology" p.97. Citing D. J. Bartholomew.
Daniel Dennett in "Darwin's Dangerous Idea" p. 180 points out that life is,

> "Nothing but a cascade of algorithmic processes feeding on chance? And if so, who designed that cascade? Nobody. It is itself the product of a blind algorithmic process."

This kind of chance is not what is being called for here. Dennett also mentions that,

> differ on whether the question [why is there something rather than nothing] makes any intelligible demand at all. If it does, the answer `because god exists' is probably as good an answer as any, but look at its competition: `why not?'"

554. Chaim Potok, "In the Beginning", New York: Fawcett Crest, 1975, p. 415

555. "Catechism of the Catholic Church", p.23

556. "Catechism of the Catholic Church", #66

557. "Theology for the Third Millennium", p. 187

558. Luke 13:4

559. This is a whole field of theological endeavor that I will not go into much here. Process metaphysics as developed by Alfred North Whitehead and furthered by many since, including Teilhard de Chardin, is a study in its own right and I do not mean to dismiss it so abruptly. However, North's own starting point is with a God who has a relationship with the physical universe.

> "Viewed as primordial, he is the unlimited conceptual realization of the absolute wealth of potentiality. In this aspect, he is not *before* all creation, but *with* all creation."

["Process Theology", Ed. by Ewert H. Cousins, New York: Newman Press, 1971]

Whitehead's God is not the conscious, designing god of the Deists, but more like a prime mover who is the underlying potentiality that is actualized in the universe. For his followers however, especially Christians, god becomes a being in process of becoming, like the universe, but still maintains the role of a conscious person such as is found in Christian scripture and tradition. This is not the god of whom I speak here who rather *forges* a relationship with the universe which exists on its own, whether or not god pays attention to it.

> [Ian Barbour, "Religion in an Age of Science", p. 28 has process theology in context. His entire chapter 8 is about "Process Thought", then see p. 260f
> See van Beeck, "God Encountered", p. 157]

560. Marcel Sarot, "Evil, tragedy, and feminist theology: new impulses for theodicy", p. 29f points out that,

> "Many feminist theologians see the problem of theodicy as a symptom of the failure of white male theology; i.e., they reject the notions of God's *omnis*cience, *omni*potence, and *perfect* goodness. Thus the conflict between these and the presence of evil in the world never occurs."

His own attempt to use feminist insights to excuse god from responsibility by use of the term “tragedy” is not very convincing. He likes the thought because,

> “the term `tragic’ suggests that it makes no sense to ask who is responsible. Something that is tragic has not been staged by evil forces or a demonic God; it was unavoidable.”

Perhaps the feminist theologians he cites explain this better, but changing the word used does not seem a suitable solution.

561. Elizabeth Kubler-Ross, “On Death and Dying” New York: Macmillan Publishing Company, 1969, speaks of hope in the context of terminal illness without once mentioning hope of life after death. The hope she is concerned with is hope of a remission or cure.

However, in a subsequent work, “Questions and Answers on Death and Dying” New York: Macmillan Publishing Co., Inc. 1974, p. 160f, she speaks of the hopes of the patient as hopes of dying with dignity, of having left a good effect behind them, and finally, of being received by God. She also reports that people of real faith in life after death,

> “...truly religious people with a deep abiding relationship with God have found it much easier to face death with equanimity. We do not often see them because they aren’t troubled, so they don’t need our help.” [p. 138f]

Robert Nowell, “What a modern Catholic believes about Death”, Chicago: The Thomas More Press, 1972, holds that, because death has been overcome by Jesus Christ,

> “we can begin to glean some idea of what should be the Christian attitude to death—that of one self and that of others. At its root is an ability to laugh at death because through our faith we know that death has been cheated of its prey. In this there is room for a certain measure of black comedy, of what the Germans call *Galgenhumor* or gallows humor, but without that being twisted into the savage desperation that rests on the belief that death is in fact annihilation, does represent the crumbling into dust of all one’s hopes and effort. The Christian laughter at death is an affirmation of life, not an expression of stoic despair.” [p. 77]

This is just as true in my version of hope as it is in the traditional view.

This is not to say that all this laughing at death is not the product of the mind’s wishful thinking. After all, the same untroubled attitude is true of those who really believe in reincarnation. On this subject, see the chapter entitled, "The Death That Ends Death in Hinduism and Buddhism" by J. Bruce Long, Ph.D. in Kubler-Ross's "Death: The Final Stage of Growth", Englewood Cliffs: Prentice-Hall, Inc., 1975.

A joyful approach to death is also found in Buddhism as described in chapter 14 of "The Monk and the Philosopher", op. cit.

Again, it is faith that makes this attitude possible, not any kind of scientific proof.
See "We Are But A Moment's Sunlight", Ed. by Charles Adler and Sheila Morrissey Adler, New York: Pocket Books, 1976 for a collection of literary works dealing with death and dying, many of which illustrate some aspect of human reaction as described by Kubler-Ross.

562. James Baldwin in "Nobody Knows My Name" New York: Dell Publishing Co., Inc., 1961, p. 100, was speaking of putting aside racial prejudice when he said,

> "Any real change implies the breakup of the world as one has always known it, the loss of all that gave one an identity, the end of safety. And at such a moment, unable to see and not daring to imagine what the future will now bring forth, one clings to what one knew, or thought one knew; to what one possessed or dreamed that one possessed. Yet it is only when a man is able, without bitterness or self-pity, to surrender a dream he has long cherished or a privilege he has long possessed that he is set free—he has set himself free—for higher dreams, for greater privileges."

The sentiment is true of many areas of human thought, including this reconsideration of the object of our hope.

563. Charles Rich, "Reflections from an Inner Eye", Huntington: Our Sunday Visitor, Inc. 1977, p. 72f

564. "Companion to the Cosmos" p. 336

565. David Z. Albert, "Quantum Mechanics and Experience", Cambridge: Harvard University Press, 1992, p. 38

566. "Alice in Quantumland". See chapter 4 and following for a good discussion of the Copenhagen Interpretation.

567. There are cave-dwelling bacteria which,

> "...derive all their energy from inorganic chemical reactions—metabolize the hydrogen sulfide in the water and the oxygen in the cave's air to produce sulfuric acid..."
> [Louise D. Hose, "Cave of the Sulphur Eaters", *Natural History*, Vol. 108, No. 3, April 1999, p. 54]

Peter Tyson, "Neptune's Furnace", *Natural History*, Vol. 108, No. 5, June 1999, p. 42f,

> "The idea that life on Earth may have begun in an environment resembling today's hydrothermal vents and then migrated upward to populate the world is one of several current theories of how life got its start. The first evidence of life, in the form of organic carbon associated with phosphate minerals in rocks from Greenland, dates to about 3.8 billion years ago."

Kevin Krajick, "To Hell and Back", *Discover*, Vol. 20, No. 7, July 1999, p. 76, reports on microbes in South African gold mines subsisting on iron, manganese and sulfur

See chapter 7 of Paul Davies, "The Fifth Miracle" on the evolution of life underground rather than as a surface phenomenon.

568. Herman Melville, "Moby Dick" New York: Airmont Publishing Co. Inc., 1964, p. 369

569. There is a book by Rosemary Haughton titled, "The Passionate God", New York: Paulist Press, 1981. As the author says in the introduction, her

> "thesis is that we can begin to make some sense of the way God loves people if we look very carefully at the way people love people, and in particular at the way of love we can refer to as 'passionate' because that kind of love tells us things about how love operates which we could not otherwise know...if we say 'passion' we evoke something in motion—strong, wanting, needy, concentrated towards a very deep encounter."

This is the kind of "fascination" that I have in mind as well.

Haughton makes some of the assumptions I object to and says that her work is about,

> "the entire, mysterious and infinitely complex system of inter-relationships which is creation, and the Creator in creation."

However, this creator/created business is minor in her treatise and I recommend it to anyone interested in the topics of Incarnation, the eucharist and the Church, among many others.

[van Beeck, "God Encountered", p. 156 puts Haughton in context]

Neale Donald Walsch, "Conversations with God, Book 3", Charlottesville: Hampton Roads Publishing Company, Inc., 1998, p. 49 says that,

> "...the soul yearns to [be and express Who You Really Are]; yearns to know itself and its own experience. This yearning to know is life seeking to be. This is God, choosing to express. The God of your histories is not the God who really is. That is the point. Your soul is the tool through which I express and experience Myself."

This is similar to my position that god learns something of existence by exposure to the human experience. However, his god is presumed (assumed) to be the Creator/Designer and he has god saying,

> "My purpose in creating you, My blessed creatures, was so that I might have an experience of Myself as the Creator of My Own Experience." (p. 50)

This seems to me to be a little selfish on god's part. I prefer to be discovered than invented so that god can have a new experience.

Walsch's idea of god reminded me of Tolstoy's who toward the end of life told his daughter,

> "God is that infinite whole of which man is conscious of being a finite part. *Man is his manifestation in matter, space and time.*" (italics mine).

Henri Troyat, "Tolstoy", Garden City: Doubleday & Company, Inc. 1967, p. 680

Dennis Edwards, "The God of Evolution" refers to Richard of St. Victor's theology which,

> "suggests that it is friendship which is at the heart of things." Edwards says, "I find this a fruitful way to approach an understanding of the God of evolution." He also refers to Sallie McFague who, "offers the image of God as Friend of the universe, and she sees human beings as creatures called into friendship with the Friend of the universe."
>
> [He refers us to Sally McFague, "Models of God: Theology for an Ecological, Nuclear Age", Philadelphia: Fortress Press, 1987]

Edwards also favorably cites Elizabeth Johnson, "She Who Is", New York: Crossroad, 1997 who,

"...portrays the divine economy in terms of trinitarian friendship: the Spirit befriends us, making us friends of God; Jesus-Sophia is the incarnation of divine friendship inviting us to table and calling us to be not servants but friends; and `the creative love of Mother Wisdom reaches throughout the universe and all of it embedded individual lives with a friendship brimming with desire for the well-being of the whole of her creation.' Johnson states that, `the love of friendship is the very essence of God.'" [p. 23. See van Beeck, p. 157.]

[Walter Kasper, "The God of Jesus Christ", and Catherine LaCugna, "God For Us", New York: HarperSanFrancisco, 1991 are both cited as proposing relationality as central to understanding god and the universe.

Pamela Kirk, "Women and God in the Church: Critique and Construction", *New Theology Review*, Vol. 8, No. 3, August 1995, p. 19 examines the feminine theologians mentioned, as does Linda A. Moody, "Constructive Theological Understandings of God", p. 29f of the same issue.]

As an aside, the ancient myth, *Enuma Elish*, preserves the story of the Babylonian Noah, called Utnapishtim (spellings differ) who upon landing on dry land after the deluge makes a sacrifice to the gods. According to the legend, the gods flock to the scent of the sacrifice like flies to carrion. I don't know if the ancients were trying to say that their gods were just rapacious or if perhaps they sensed that their gods (or god) were attracted to humans and their activities.

In any case, I like the ideas expressed by those cited above, though all these authors presume god as creator, but I prefer the more intense "fascination".

570. The Princeton Engineering Anomalies Research laboratory is trying to find out if human brains can influence machines. There is evidence that this is possible. Mice have been trained to operate machinery through the action of probes implanted in their brains. Whether the same can be done without probes but just by the natural effect of brain activity remains to be seen. If it turns out that there is something to it this would certainly be on the order of an "emergent quality", something that happens to be present though it did not evolve for this purpose. There have been no machines to influence until recently.

Rupert Sheldrake is a maverick scientist who talks of "morphic fields" through which humans may communicate, and indeed through which animals may communicate with their masters. He is not well received by many other scientists, however.

See Brad Lemley, "Heresy", *Discover*, Vol. 21, No. 8, August 2000, p. 60.

See an interview by John David Ebert, "The Nature Channel", "*tne Reader*, No. 103, January-February 2001, p. 67

571. Most people seem to process the same information in the same manner. I presume that what I call red as the result of my cells picking up certain frequencies of light is the same as what you call red. There are, however, those with Synesthesia who may perceive "red" when exposed to the letter "T". There is much room for chance in the evolution of the senses.

See Susan Hornik, "For Some, Pain is Orange", *Smithsonian*, Vol. 31, No. 13, February 2001, p. 48

572. Bernard Lonergan is quite difficult to read and understand, I have been told, and I have not read all his work. From reviews and commentaries, including Terry J. Tekippe, "What Is Lonergan Up to in *Insight*?", Collegeville: The Liturgical Press, 1996, I gather that his work is epistemological about how we understand and how we can understand. I believe that what he is speaking of in the term "insight" is similar to what I propose as a form of "intuition". Readers knowledgeable in Lonergan may want to elaborate on this for me.

However, there is a chapter in Tekippe's book which disturbs me somewhat. In speaking of the existence of god which he explains that Lonergan addresses in chapter 19 of "Insight", Tekippe says, "…there are only two logical possibilities:

> Either God exists, or
> God does not exist.
> Second,
> If God exists, then the universe is intelligible;
> If God does not exist, then the universe is unintelligible."

Granted that he does not purport to prove the existence of god from this argument but rather seems to invite the readers' "insight" into the question to lead them to a conclusion; and granted that I do not know how Lonergan himself handles this question; still, I would like to point out a glaring

assumption on their part. According to my conclusions on the relationship of god and the universe, the second part of the second postulate is not necessarily true, and in fact I do not believe that it is true. If god is not the creator or the prime being responsible for the existence of the contingent universe, then god may not exist and yet the universe is intelligible. God can and, I believe, does bring meaning and purpose into the universe and into human lives, but did not put any such into it in the first place. In this case, the first part of the second postulate is also faulty since god may exist and yet the universe can BE absurd and perhaps unintelligible at root.

573. Denis Edwards, "Human Experience of God", New York: Paulist Press, 1983, p. 13.

My first reaction on reading him was that it would be very hard to separate the action of the unconscious human mind from this pre-conceptual experience. However, he agrees with me, I believe, in holding to this "pre-conceptual experience" as non-intellectual and will not allow an intellectual proof of the existence of god. Even in visions and hearing the voice of god he cites John of the Cross and Teresa of Avila, among others, in cautioning against putting too much credibility in such experiences since they can indeed come from sources other than god.

Teresa and John have become popular once again and there are a number of modern books about them and their methods of prayer. John of the Cross can be very impenetrable, partially because of the overlay of 16th century Spanish spirituality which can be very dark. However, John's insistence on meditation and opening oneself to what Edwards calls a pre-conceptual experience, even though this can lead through some "Dark Nights" is essentially what I am saying.

> [See Sister Eileen Lyddon, "Door Through Darkness: John of the Cross and mysticism in everyday life", Hyde Park: New City Press, 1995 for an understandable explanation of John's works.]

I was reminded of a study of religious experiences that I read many years ago and looked it up again. It is Antoine Vergote's "The Religious Man", Trans. by Sister Marie-Bernard Said, O.S.B., Dayton: Pflaum Press, 1969. It is the report on religious experience based on psychological interviews and, though dated, was a reinforcement of Edwards' ideas on the experience of god.

574. John Main, "The Heart of Creation: The Meditative Way", New York: Crossroads, 1988. In the introduction Laurence Freeman OSB says,

> "Pure prayer means the transcendence of all thoughts and images. Not looking at God, but into God. It means seeing God with the imageless vision of faith, which is the power (and gift) which realizes our union with Christ at prayer in us. We see God through his eyes when we stop trying to see him through our own inadequate vision. We know God with the mind of Christ when, by renouncing the great hum, an gift of self-consciousness, our mind becomes one with his." [p. xiii]

This all seems so complicated until we read books such as this on meditation and realize that the only thing required is to take the time to sit quietly and repeat a mantra over and over. This certainly seems to me to be the opening to pre-conscious experience of Oraison or to the kind of intuition mentioned in the text.

Main is true to the tradition he works from in saying that, "we were created for light not shadow,...we were created to become light, to *be* light." [p. 65] However, this is not central to his teaching on meditation.

575. Andrew Newberg, M.D., Eugene D'Aquili, M.D. and Vince Rause, "Why God Won't Go Away: Brain Science and the Biology of Belief", New York: Ballantine Books, 2001. On page p. 168 we read,

> "...if the unitary states that the brain makes possibles are, in fact, glimpses of an actual higher reality, then religions are reflections not only of neurological unity, but of a deeper absolute reality."

And on page 174-5,

> "The neurobiological roots of spiritual transcendence show that Absolute Unitary Being is a plausible, even probable possibility. Of all the surprises our theory has to offer - that myths are driven by biological compulsion, that rituals are intuitively shaped to trigger unitary states, that mystics are, after all, not necessarily crazy, and that all religions are branches of the same spiritual tree - the fact that this ultimate unitary state can be rationally supported intrigues us the most. The realness of Absolute Unitary Being is not conclusive proof that a higher God exists, but it makes a strong case that there is more to human existence than sheer material existence. *Our minds are drawn by the intuition of this deeper reality*...As long as our brains are arranged the way they are, as long as our minds are capable of sensing this deeper reality, spirituality will continue to

> shape the human experience, and God, however we define that majestic, mysterious concept, will not go away." [Italics mine.]

I imagine that this report will meet with opposition from some scientists to whom the very mention of the possibility of god is odious. They will surely take the position that our religious experiences are merely evolutionarily developed survival mechanisms, controlled by genes or memes or both. At the same time, some believers will probably make the point that the brain was wired by god to receive these messages of the divine presence, an article of faith that is hardly proven. Still, I personally am happy to receive what I consider support of the possibility of an emergent quality of the brain, not created by god, yet able to sense in some diffuse manner a reality outside not only itself but outside the material universe.

Dr. Keith Crutcher in a paper for ITEST titled, *Is There a God-Spot in the Brain* gives a balanced view, recognizing that the current state of research cannot prove anything one way or another. He leans to a theistic interpretation, and presumes that there is purpose to the universe as created by god, but he keeps his beliefs in balance. I imagine the paper and others on the subject will be published by ITEST in St. Louis in the near future.

576. Dr. Freeman J. Dyson in "Religion, Science, and the Search for Wisdom" Ed. by David M. Byers, Bishops' Committee on Human Values, NCCB, says he has a problem with what *materialism* means. As I point out in the idea of an essence, as well as the explanation of transubstantiation and Incarnation, *what* a thing is depends on us, not on its constituents. There is not the essence of something, but merely what that something is for us. As Dr. Dyson points out, the same is true of matter. "Speaking as a physicist, I judge matter to be an imprecise and rather old-fashioned concept" (p. 51) He then speculates that,

> "If God exists and is accessible to us, then his mind and ours may, likewise, differ from each other only in degree and not in kind. We stand, in a manner of speaking, midway between the unpredictability of matter and the unpredictability of God. Our minds may receive inputs equally from matter and from God."

577. Josef Fuchs, "How can Christians believe?", *Theology Digest,* Nol. 44, No. 1, Spring 1997, p. 23f. Fuchs is certainly an orthodox theologian, but still insists that,

"The God question itself is not without problems. Many people might require strict arguments for the existence of God in order to believe. But in what sense is that still faith?...Despite everything, all doubt is not completely eradicated, nor is the temptation to refuse faith—a risk that always remains. There is no compelling proof. However, there is a well-founded and supportive faith, and thus also there is a meaningful and conscious stand in the church of open trust in this faith."

578. Numbers 14:8; Deuteronomy 10:15; Psalm 103:13f; Psalm 104:34. There are many more passages where the people are described as taking delight in Yahweh, but it works the other way as well.

579. For example, Scholastic theologians attribute a quality analogous to "beauty" to god since there is beauty in the universe. There is not, of course. There is "beauty" in our minds as the result of an evolutionary development, but then a pile of manure may be beautiful to a dung beetle. That this tells us anything about god is questionable. However, that we can see beauty in things, or especially people, that evolution has not prepared us to see is an extension of our ability that god may well have taught us. To say "God is beauty" because there is beauty in the universe should rather be expressed by saying that our concept of beauty has been expanded by god, if we believe that, and therefore "God is beauty".

580. Some have tried to prove the existence of god by appealing to the Anthropic Principle and making god the being who observes the quantum state of each part of the universe and thereby collapses the probability wave. Suffice it to say that the arguments which deny this necessity to humans also deny it to god.

581. See Piers Vitebsky, "What is a Shaman?" *Natural History*, Vol 106, No. 2, March 1997, p. 34. He notes that,

"Shamanic religion may date from the time of the earliest known Paleolithic drawings, which were made some 30,000 years ago by our hunting ancestors."

Also, "In shamanic thinking, every element of the world around us, whether human, animal, tree, or rock, is imbued with spirits. Spirits are conscious, often anthropomorphic, and can also be interpreted as representing the essences that underlie surface appearances."

See Piers Vitebsky, "Dialogues with the Dead", Marjorie Mandelstam Balzer, "Soviet Superpowers" and Laurel Kendall, "The Shaman's Apprentice" and other articles in the same issue.

Theodore C. Levin and Michael E. Edgerton in "The Throat Singers of Tuva", *Scientific American*, Vol. 281, No. 3, September 1999, p. 80 note that,

> "According to Tuvan animism, the spirituality of mountains and rivers is manifested not only through their physical shape and location but also through the sounds they produce or can be made to produce by human agency. The echo off a cliff, for example, may be imbued with spiritual significance. Animals, too, are said to express spiritual power sonically."

In animism there is no question that these forces are real and "personal".

Animism is still found in Indonesia in "syncretic harmony with devout Catholicism…Each May, after the priest has blessed the fleet and prayers have been offered to Kotekema, the spirit of the sperm whale, the hunting season begins."

> [Fred Bruemmeer, "Sea Hunters of Lamalera", *Natural History*, Vol. 110, No. 8, October 2001, p. 56.]

The Quechua of Peru still tie a condor to the back of a bull, for their own reasons. In the capture, a horse is used as bait and, "during their mission, the hunters carry out ceremonies directed to the holy mountains and to Mother Earth so that a condor will be attracted to the carcass." [Andre Affentranger, "The Condor and the Bull", *Natural History*, Vol. 110, No. 6, July-August 2001, p. 50]

582. Mary Roach, "Ancient Altered States", *Discover*, Vol. 19, No. 6, June 1998, p. 52 writes of a theory of the archeologist Dave Whitley that petroglyphs made by ancient Native Americans are actually connected with shamanistic visions. Several stages of the behavior of a mind altered through drugs or trances will produce visions which may be recorded on the rocks. Visions in the brain are no longer seen by the subject as visions but,

> "to them it was a parallel reality; they had entered the realm of the upernatural…Subjects stop using similes to describe their experiences and assert that the images are indeed what they appear to be."

My contention is that god may be able to make use of these normal processes in some manner as described in the text.

583. We have considered the problem of evil under a number of headings and concluded that god is not responsible for the evil in the world. Certainly not the physical evil since god has no control over physical events. It is conceivable that god could be liable in regard to evils brought about through human actions inspired by god or good human actions which could have been influenced but were not, seemingly the only effect god can have on the universe. Some blame the German Christian church, in all of its denominations, for the evil of the holocaust, not only because some of its members participated in it, but also because the Church itself did not publicly condemn the actions and policies of the government. Indeed, it sometimes supported the policies.

> [see Daniel Jonah Goldhagen, "Hitler's Willing Executioners", op. cit.]

The Vatican has also been attacked for its failure to speak out about the arrests and deportations of Italian Jews. Even if you only have the opportunity to influence the actions of others and do not do so, you accept some degree of liability for them. By the same token, if god can effectively influence people to do some good or refrain from some evil, and does not do so, then god is responsible for the evil in the world. But it is precisely our faith that god always urges us to the good, and is constantly at work in so urging. The witness of the Bible is to god's covenanting with the people and the most important sign of that covenant is the moral direction that is given to the people. Again and again the prophets point out that one cannot claim a relationship with god without good relationships with all the people god cares about, which is precisely everyone. Once again this god may seem rather impotent. As has been mentioned, some say that there does not seem to be much progress over the thousands of years of recorded interaction of god with the people of the Covenant, or over the millions of years preceding that. On the other hand, a case can be made, though not a scientifically verifiable case, that humanity is much farther advanced thanks to god's interventions than we would have been without. Israel of the Bible is the most ready example. Again and again we can make comparisons between Israel and its neighbors and find progress in its moral fiber. Israel may not have been the most intellectual, the most efficient, the most cultured of peoples, but as a people they often are the most moral, at least on paper. Even though individuals, and sometimes the whole people, do not live up to their own ideals of the treatment of aliens and slaves, orphans and widows, they at least had the ideals. And where did they come from? The Bible claims that they did not come from the innate goodness of the people but from god. They acknowledge that "my father was a wandering Aramean" and that it was god who took them to a higher plane, not themselves. Given the human proclivity to take the selfish route (proposed by selfish genes?), their assertion makes sense.

584. The god who shows love and mercy for the people is not unknown in other faiths. However, with the revelation of the god of the Hebrews there is a new dimension brought into sight. The message of the Exodus and Sinai is a message of covenant love for the chosen people.

> "The Sinai narrative tells the story of Israel going to meet her God, but it is not just the story of the first generation of Israelites. It is the story of Israel in every generation. It was the story of the current generation who, even as they listened to the narrative, felt themselves standing at the foot of Sinai ready to listen to the voice of God in the liturgical celebration."
>
> ["The God of Exodus", p. 21]

"Beneath its bilateral aspect, the covenant was...the result of a purely divine initiative."

[Thierry Maertens O.S.B. "Bible Themes", op. cit., p.141]

God revealed love for the people and promised to be with them and bring them to the promised land or situation. This love for the people invited a response, not only of love for the god, but of love for the other members of the covenant people. But this is soon overshadowed by another event, the giving of the ten commandments, and the whole moral code of the Judeo-Christian tradition. The truly amazing thing about the Sinai event, or the process that is described in mythical shorthand by the Sinai story, is that god has chosen this people to be god's own; the covenant that god makes with them is unprecedented and unique.

> [There is not a simple answer to the question, "What Really Happened at Sinai?" as an article by that name by Baruch J. Schwartz, *Bible Review*, Vol. XIII, No. 5, October 1997, p. 20 makes clear in an account of the four renderings of the giving of the law.]

In typical human fashion, they soon downplay the relational and concentrate on what it means to their daily behavior. They are far from contemplatives meditating on the mystery of god's love for them. They are practical people who want to know, "what do I have to do about this?"

585. Meeting a god, or a god's utterances, in trees is a well-known phenomenon in ancient times. See. Shawn O'Bryhim, "An Oracular Scene from the Pozo Moro Funerary Monument", *Near Eastern Archaeology,* Vol. 64, No. 1-2, 2001, p. 67.

586. There is an ongoing debate about the date of the Exodus and the historicity of the events as described by the Scripture. See for example, Abraham Malamat, "Let My People Go and Go and Go and Go", *Biblical Archaeology Review*, Vol. 24, No. 1 p. 62f

587. Dennis J. Murphy, M.S.C., "His Servants the Prophets" Collegeville: The Liturgical Press, 1965, p. 9

588. Carrol Stuhlmueller, C.P., "Thirsting for the Lord" Ed. by Sr. M. Romanus Penrose, O.S.B. New York: Alba House, 1977, p. 225

> "The `classical prophets,'...were persons who were so fully and consistently a member of their community and so perceptive and articulate that as a result they bring the internal challenge of the community's conscience—its divinely inspired hopes and ideals—to bear upon the external form of the community's life style and work."

589. There is evidence that the religious reforms of Hezekiah, which had the support of the prophet Isaiah among others, was really only a part of his plan to centralize the resources of the country in preparation for a revolt against Assyria. Those who seemingly spoke for god at that time and later evidently were deluded, willingly or not. However, it seems that the faith of the community was not so fooled, since there never was the centralization of cult in Jerusalem that was desired.

See Oded Borowski, "Hezekiah's Reforms and the Revolt against Assyria", *Biblical Archaeologist*, Vol. 58, #3, September 1995, p. 148f

590. On the development of the synagogue see Steven Fine, "This Holy Place: On the Sanctity of the Synagogue During the Greco-Roman Period", Notre Dame: University of Notre Dame Press, 1997

591. Louis Monden, S.J., "Faith: Can Man Still Believe?", Trans. by Joseph Donceel, S.J., New York: Sheed & Ward, 1970, p. 215

A popular commentary on Luke's Gospel, Richard Rohr, "The Good News According to Luke", New York: The Crossroads Publishing Company, 1997, points out right away that,

> "What is written in Luke's Gospel describes a believing community's reflection on the mystery of the risen Lord at the time Luke was writing, reflecting their questions, problems, and needs. I am not denying that this sacrred text is the word of the Lord, but it is more accurate to say the Bible contains the messages of God as they are received, understood, prayed upon, and used to lead a community of believers." [p. 14]

592. See Geza Vermes, "Jesus the Jew", New York: Fortress Press, 1990

See A.N. Wilson, "Jesus: A Life" New York: Fawcett Columbine, 1992, Chapter 1, "Jesus the Jew", p.3f

See John Dominic Crossan & Jonathan L. Reed, "Excavating Jesus", New York: HarperSanFrancisco, 2001

593. Maya Angelou, "Wouldn't Take Nothing for My Journey Now" New York: Random House, 1993, p. 73

594. The giving of a moral code was one impetus for the development of the idea of the writings of the people as the writing of the word of god. Not the only one, certainly, and perhaps not the main reason for identifying the writings of the historians and prophets as the word of god. And that is precisely what the words of the Bible are considered.

> "...God has spoken, using words. God opens Himself out and reveals Himself to us as one person to another, using the personal, or interpersonal, means of communicating," the written word.
> [Luis Alonso Schokel, S.J. "The Inspired Word" New York: Herder and Herder, 1965, p. 41]

This is not to say that god sat down and wrote the words. The process of inspiration of the authors of the various parts of the Bible is extremely complex and has induced the writing of many studies on it. The word Inspiration is not used in Scripture in the sense of god inspiring the writing of revelation, nor is it used in the early Church.

> "For the appearance of Inspiration in an official document of the teaching Church we must wait till 1441 a.d. In the '*Decretum pro Jacobitis*', the Council of Florence says: 'One same God is author of Old Testament and New, that is of the Law and the Prophets and the Gospel, since the Holy Men of each Testament spoke under the

inspiration of the same Holy Spirit'—*eodem sancto Spiritu inspirante*...But of what Inspiration is, there is no explanation of this official sort by the teaching Church until the nineteenth century."
[F.J. Sheed, "God and the Human Condition", op. cit., p. 90]

Some Christians define inspiration as dictation by god, leaving out the human author by and large. Catholic Christians were in this number until recently until it was realized that, "God spoke through the human authors of the Bible, inspiring them to write using their own talents, abilities and styles. God did not just dictate messages to them as a ventriloquist uses a dummy."
[Citing just one of many Catholic commentaries on the Scripture which take this approach. Oscar Lukefahr, "A Catholic Guide to the Bible", Ligouri: Ligouri Publications, 1992, p. 19. See p. 30ff for elaboration on the Catholic idea of inspiration.]

It is certain that some if not all authors of words that found a place in the Bible could scarcely know that the words and meanings were anything but theirs. The scribes in the royal palaces who wrote the histories of the kings, putting their own slant on the facts to make their man look good, were part of the process and their works would be edited and commented upon and used in bits and pieces by later authors that were perhaps more aware that they were writing of things of god rather than human affairs. Were all involved in the process inspired? No one can say for sure.
[In "Divino Afflante Spiritu", and encyclical from Pius XII in 1943, biblical scholars were urged to notice that the inspired author,

> "...so uses his faculties and powers, that from the book composed by him all may easily infer `the special character of each one and, as it were, his personal traits.' [cites Benedict XV from 1920] Let the interpreter then, with all care and without neglecting any light derived from recent research, endeavor to determine the peculiar character and circumstances of the sacred writer, the age in which he lived, the sources written or oral to which he had recourse and the forms of expression he employed. Thus can he the better understand who was the inspired author and what he wishes to express by his writings."

The encyclical also urges scholars to take note of the style and mode of the writing and what it meant to the author. It does use the words "inspired" and "inspiration" and seems to presume that there was one inspired author for each of the books. It does not take up this question directly. From the tone, however, one can conjecture that if Pius had been asked who in the process of the production of the final work was inspired, he would have been open to saying that perhaps more than one. This is certainly the position of scholars immediately prior to the Second Vatican Council. John J. Castelot, S.S., "Meet the Bible!", Baltimore: Helicon Press, 1960, in a book used in many seminaries defends the position that, "*A man is inspired to the extent that he contributes to the composition of an inspired book.*" p. 26 (italics his)

Pierre Benoit, O.P. points out that,

"Vatican Council II…in the Schema for the Constitution: *De Divina Revelatione*, discussing and confirming the assertions of previous documents which had come to be cited. The Council reaffirmed that the inspiration of the sacred books means that God is their author, that the two testaments are unified, that there is a sovereign action of the divine influence upon all the faculties of the human interpreter, that the inspiration of Scripture is integral and that an infallible teaching flows from it, and that it is necessary to examine closely the human language that God has seen fit to be used in his service." ["Inspiration and Revelation", in "The Human Reality of Sacred Scripture", "Concilium" Volume 10, New York: Paulist Press, 1965, p. 8f]

Of other biblical authors, especially in the New Testament, some claim that,

"…they conceived themselves to be…men to whom the revelation-event was entrusted in a special way and who recognized, formed and proclaimed the true claim made by the tradition of revelation. Because they had met the risen or exalted Christ and by the power of the Holy Spirit manifesting him in his truth, the authors of these writings knew themselves to be authentic and authoritative witnesses of this event, who wrote down authentic, authoritative and 'inspired' testimony to it."

[Heinrich Schlier, "The Relevance of the New Testament", New York: Herder and Herder, 1968 p. 27]

No matter the process of writing, it was the acceptance by the Jewish people and later of the Christian people of a canon of scripture that put the final approval on identification of the writings as the word of god. They had to be! Who is going to follow the commandments or the beatitudes on the word of another human being? In the popular mythology one can find plenty of people in California who will listen to any guru who chances down the block, but most people are not going to listen to anyone short of god. And again this all follows from the presumptions that there is a god who has involved godself in the workings of the universe and that this god has a special interest in us.

This interest and revelation continues after the period of biblical formation, as is certainly the Catholic understanding, in the Tradition of the Church, with a capital "T".

> "The Bible is the Word of God insofar as it has been committed to writing under the inspiration of the Holy Spirit, and Tradition is the Spirit-directed reflection upon that written Word, after the fact…Consequently, when we say that we must judge everything according to the Gospel, we do not mean only `the Book' but also the living expression of that Gospel in the history and experience of the Church."
>
> [Richard P. McBrien, "Who is a Catholic", Denville: Dimension Books, 1971, p. 97. See the following pages for a discussion of the Catholic position on Tradition after Vatican II.
>
> See also, Oscar Lukefahr, "A Catholic Guide to the Bible", Ligouri: Ligouri, 1992, p. 33

Pierre Benoit, shows that,

> "…scriptural inspiration ceases to be the charism of one isolated individual working in a vacuum and taking down on paper `truths' wafted to his ears. It is the last moment of a long thrust of the Spirit which, after having forged a divine-human exploit which reaches its peak in the coming of the Song and, after having announced in diverse manners the message of the Father up to the time of the last appeals of his Son (Heb. 1:1), places all this in sacred writings which are to reach men of all times and all places. The Scriptures cease to be a collection of more or less timeless truths like those which so many religions offer; instead it becomes the book of a people, a people ageless and holy; it becomes a `family record', a chronicle always fresh that records God's actions and the challenges he presents to his children. The Bible is truly the book of a chosen People, the Church." [Benoit, "Inspiration and Revelation", p. 17]

Benoit is one of the authors of the preface to "The Dynamism of Biblical Tradition", in "Concilium", Vol. 20, New York: Paulist Press, 1967, which has a number of articles on the role of Tradition in the formation of the Scriptures. A little dated, but still helpful.

595. Vatican II, "Dei Verbum" #9. Cited in the "Catechism of Catholic Church", #81

596. Paul Wilkes, "The Good Enough Catholic", New York: Ballantine Books, 1996, p.xxi

597. "Catechism of the Catholic Church", #84

598. See Karl Rahner, "Nature and Grace", New York: Sheed and Ward, 1964, chapter 4, p. 83
for the understanding of this Conciliar insight by a contemporary theologian.

599. See Dava Sobel, "Galileo's Daughter", New York: Walker and Company, 1999.
See the article by the same name in *Astronomy*, Vol. 28, No. 4, April 2000, p. 46

600. See Johannes Jorgensen, "Saint Francis of Assisi" Trans. by T. O'Conor Sloane, Garden City: Image Books, 1955

601. A whole line of reformers, their impact on society, and some indications of how they were formed by the ideas of their culture can be found in Chapter 2, *From individual to institutional reform*, in John P. Dolan, "History of the Reformation", New York: Desclee Company, 1965, p. 56f

602. "The Screwtape Letters" p. 158

603. William Damon, "The Moral Development of Children", *Scientific American*, Vol. 281, No. 2, August 1999, p. 73f.
See also his book, "The Moral Child: Nurturing Children's Natural Moral Growth", New York: Free Press, 1990.

604. It is true that in the scripture we can find reflection on god's motivation for giving a moral code of conduct residing in god's

communication of self to the people. Since god is a holy god, a loving and caring god, god invites the people to be the same, for their own welfare. Unfortunately, not being holy, loving and caring, but human, which might include these traits but also many more, some negative, the people do not respond very well to god's prompting. They do, however, react well to threat. So we find the assumption that god's motivation is the same as a human king, including sanctions for disobedience. God has given a command, not an invitation, and there will be hell to pay if god is not obeyed. "God is a jealous god" and does not take kindly to being ignored.

> [Exodus 20:5; Exodus 34:14; Deuteronomy 4:24; Deuteronomy 5:9; Deuteronomy 6:15; Joshua 24:19; Nahum 1:2]

Even if the command is not obviously loving it is to be followed without question. Saul is deposed from his kingship because he does not follow the ban on conquered peoples and spares the king of a foreign nation contrary to god's decree that every person should be destroyed and all their property, including the cattle and goods. Later generations will find real difficulty in whether god could ever give such an order, either denying that god would do so or pointing out that the lord of life and death has the right to choose the manner of each one's demise, and in fact makes these determinations daily. So their numbers were all up at the same time. So what? However, even though god's saving action in the Exodus from Egypt was accomplished through some rather nasty things done to the Egyptians and ending with the destruction of their army in the waters of the Red Sea, god is not revealed primarily as a god of destruction and threat so much as a god of deliverance and caring. And knowing the superiority of these traits over the negatives, god wants them for god's people as well. The commandments, and all the elucidation they will receive through the following centuries, are an invitation to the moral life that is in fact the happiest and best kind of life. Unfortunately, they are perceived more as a law code, testing god's people in their response to god's dominion. In correction of this misinterpretation, god sends the prophets who, while continuing to threaten since they too are products of their culture, will interweave in their message more or less consciously the revelation that god originally intended, namely that it is living in the covenant relationship with god and neighbor that is real life.

> "Through the prophets, God forms his people in the hope of salvation, in the expectation of a new and everlasting Covenant intended for all, to be written on their hearts."
>
> ["Catechism of the Catholic Church", # 64]

"In the course of its history, Israel was able to discover that God had only one reason to reveal himself to them, a single motive for choosing them from among all peoples as his special possession: his sheer gratuitous love. And thanks to the prophets, Israel understood that it was again out of love that God never stopped saving them..."
["Catechism of the Catholic Church", #218]

605. Heather Pringle, "New Women of the Ice Age", *Discover*, Vol. 19, No. 4, April 1998, p. 62 presents a newly-emerging picture of women as far more active and aggressive than they have been portrayed in the past.

606. W.H.Van de Pol, "The End of Conventional Christianity", Trans. by Theodore Zuydwijk, S.J., New York: Newman Press, 1968, p. 133

I was very happy to see that Bishop Oscar Lipscomb of Mobile, writing for the "*Extension*" magazine, Vol. 94, No. 10, January 2000 when asked about people being more educated today and the effect on the faith said,

"We will never get away from a faith context, and this is where the challenge comes: We can never really give proof in the strict sense because basically it comes from God's revelation of Himself. You never can really prove God exists, but you can have enough evidence of His presence in your life to appropriate it and make it work."

607. W. H. van de Pol, "The End of Conventional Christianity", p. 295f

For a contemporary explanation of the development of faith in bodily resurrection, see John Dominic Crossan, "Jesus A Revolutionary Biography", op. cit., p.160f

608. Donald P. Merrifield, S.J., "Two Modes of Thought: Science and Religion", St. Louis: "*The ITEST Bulletin*", Vol. 30, No. 2, Spring, 1999, p. 2, speaks of receiving,

"a criticism of [Stephen Jay] Gould's approach [of the 'Nonoverlapping Magisteria' found in an article in *Natural History* magazine and in his book *Leonardo's Mountain of Clams and the Diet of Worms*] by a Catholic theologian who asserts theology involves both 'facts,' presumably empirical, and 'theories' which explain them. I, not a theologian, wonder if it is 'facts' or 'faith' which the theologian seeks to explain, following the Anselmian

> definition of theology as `*fides quaerens intellectum*' or `faith seeking understanding."

Merrifield suggests that,

> "it might be possible to speak of `religious realities,' based, however, in faith, of which Gould does not speak."

I certainly agree that many of the teachings of the Church are based on faith and are not empirically provable and that we should recognize them as such.

609. Romans 8:24-5?

610. "Catechism of the Catholic Church", #153

611. John 13:34; John 15:12; Romans 12:10; 1 Thessalonians 3:12; 1 Thessalonians 4:9; 1 Peter 1:22; 1 John 3:11; 1 John 3:23; 1 John 4:7; 1 John 4:11; 2 John 5

612. See Paul's discussion of power and weakness in 1 Corinthians 1:25f, also in 2 Corinthians 12:9f.

613. "The Mind's Eye", p. 66

614. It is interesting that in a magazine for popular explanation of the faith called, *The Catholic Answer*, Vol. 11, No. 6, January/February 1998, Jonathan Neiss, in an article entitled "Science and Belief", tries first to say that there must be something spiritual about us humans because even though the chemicals of our bodies change, "I" remain. He also dredges up the old "ex nihilo" argument in claiming that since electricity cannot think, therefore,

> "...we cannot merely be some kind of brain activity, a kind of electrical activity, and yet exhibit a free will that electricity cannot exhibit. The reason that we can will the course of electrical forces...is that we are a consciousness (at a causal level) deeper than the physical forces of the world."

However, his main argument against "science's" seeming denial of religion is the effect that faith has in the life of the person. He seems to know on some level that his main argument is faulty, and to know on some level where the true argument is. He is not alone today.

My diocesan paper reports that a study by Mark Regnerus of the Center for Social Research at Calvin College in Grand Rapids, Michigan titled

"Making the Grade: The Influence of Religion Upon the Academic Performance of Youth in Disadvantaged Communities", and another by Bradford Wilcox titled, "Good Dads: Religion, Civic Engagement, and Paternal Involvement in Low-Income Communities", find that those influenced by membership in religious organizations (read "Church") are better family men and better students than those who are not. This, of course, could be the result of purely cultural influence and not of any action by god. But, then, it could also be god. In my experience it is both.

> [Mark Paterson, "Studies suggest depth of church influence in society, *St. Louis Review*, January 11, 2002]

615. Thomas Hart, "What Does God Want?", *Catholic Digest*, Vol. 62, No. 9, July 1998, p. 96f speaks of,

> "...two theologies regarding God's will. Some believe God plans a detailed map for each of us. God has already willed where we should live, whom we should marry, what our career should be, what accidents will befall us, and when we should die...Others believe God's will consists of broad strokes only, the particulars left to us...The broad strokes are the basic values by which God has us orient ourselves in life...Out of the broad-strokes theology I draw on a simple formula for making these seemingly impossible decisions. The place to look for God's leadings is inside ourselves, because what God wants for us is the unfolding of our true self."

I am with him up to this point. Unfortunately, he adds,

> God who made us; the seeds of our destiny are planted within us. When we have found what we most deeply want, we have found God's will for us."

If there was not a prior assumption that God has created our DNA and placed these tendencies in our genes, or possibly that our brain or soul has been inoculated since our creation, I believe this author would reach the same conclusion about finding god working within and about rejecting the "all planned out theology".

616. Number 21:5f

617. Which can be carried to some ridiculous extremes. In "To Kill a Mockingbird", New York: Warner Books, Inc. 1960, the author, Harper Lee tells of the "footwashing Baptists", a particularly rigid group who,

"...believe anything that's pleasure is a sin. Did you know some of 'em came out of the woods one Saturday and passed by this place and told me me and my flowers were going to hell?
Your flowers, too?
Yes ma'am. They'd burn right with me. They thought I spent too much time in God's outdoors and not enough time inside the house reading the Bible." [p. 49]

This approach is not uncommon among the "Calvinists", though it is often not as pronounced. Catholic theology takes the opposite view that anything that is a human pleasure is good unless it is abused, though many Catholics have been influenced by the opposite view.

618. See *Time*, September 15, 1997. The civilization of Nubia is recognized as being very developed three thousand years before that of Egypt. This was not recognized until recent years and one reason was,

> "...racial prejudice which turned many in the field away from cultures emanating from deeper in Africa. Prominent Egyptologists...thought they were excavating the remains of an offshoot of Egyptian culture. They didn't believe black Africa was capable of producing high civilization."

A similar outlook is reported by Weber Ndoro, "Great Zimbabwe", *Scientific American*, November 1997, p. 94f. He reports that an early explorer, Karl Mauch in 1871,

> "...concluded very quickly that Great Zimbabwe...was most certainly not the handiwork of Africans. The stonework was too sophisticated, the culture too advanced. It looked to Mauch to be the result of Phoenician or Israelite settlers."

619. 2 Corinthians 13:11 speaks of the god <u>of</u> love. God also loves, asks love, challenges us to love, gives the spirit of love. The love of god dwells in us.

Finally in 1 John 4:8, and 16, the Church reaches the conclusion that god <u>is</u> love.

> [See Marie-Louise Gubler, "Whoever...hates his brother is still in darkness (1 Jn 2:9)", *Theology Today*, Vol. 46, No. 3, Fall 1999, p. 203f for an understanding of how this love is found primarily in the community of the Church.]

I have used the word "love" many times, including in chapter 3 on the physiology of love. We use the word in many contexts, from "I love chocolate" to "I love my children" and yet it means something completely different in various contexts. If I say "I love you" in the context of a sexual encounter in which I have no emotional attachment, this is quite different from saying it to the one I married, yet we use the same phrase, "making love" for both activities. We also speak of "tough love" and "smothering love", the English are liable to call anyone "love" and then of course there is "love" in tennis. We humans seem a little confused about the meaning of the term. Most uses, however, seem to have something to do with an emotion, and hence a hormonal response. When the hormones cease, usually love does as well. We are also aware however, of a different kind of love which may arise from hormones but which seems to transcend them. This is probably the kind of love which god hopes to help us improve, without neglecting the others. God is happy for us that we love ice cream and sex, but there is much more to it as numerous saintly authors attest.

620. John Updike, "The Centaur", Greenwich: Fawcett Publications, 1962, p. 220

621. Matthew 16:24; Mark 8:35; Luke 9:24; John 12:25

Marc Oraison, "Morality for our Time", op. cit. as far back as 1968 was calling for a redefinition of the basis of morality and concludes that,

> "According to the convergence of the natural findings of science and reason on the one hand and the transcendent facts of revelation on the other, man attains his happiness by responding with his whole being to the call of God, which according to St. John is love. It is by striving in every existential situation to recognize and love his neighbor in a manner that promotes the other and himself in an authentic and intersubjective relationship that he responds to this call of God." [p. 76]
>
> "Morality consists therefore in studying the concrete exigencies of charity (of which the moral law ought to be one expression), namely our relationships with our neighbor and with God." [p. 77]

622. Matthew 5:48

623. For the scientific view of the continuing argument on this question, check Carl Sagan, "Billions and Billions", chapter 15

624. On the other hand, the pro-life proponents are opposed to any form of abortion, "from the moment of conception." If a two-celled embryo is split

in many species, including humans, two individuals often result. Were there two individuals before the split? How did the egg know that it was going to split and so to begin as two instead of one? For those who speak of the implanting of a soul, this would seem to say that souls are not normally implanted until there are at least two cells. The "morning after" pill would then seem to be morally permissible.

625. Daniel C. Maguire, "Sacred Choices: The Right to Contraception and Abortion in Ten World Religions", New York: Fortress Press, 2001 shows how the pro-choice position finds justification in all the world's religions, including Catholicism.

626. There is a character from Joseph Conrad's "Victory", Garden City: Doubleday Anchor Books, 1915, p. 49 whom we will meet later, but there is another personality, or rather a non-personality, in his story, a Mrs. Schomberg who is a seeming complete non-entity. Yet it is she who thwarts her domineering husband and helps the hero escape with the girl. Another person involved in the affair returns some evidence that would be incriminating of Mrs. Schomberg if she were perceived as capable of the crime, but she appears to have no personal identity so she is not even suspected. However,

> "As to Mrs. Schomberg, she sat there like a joss. Davidson was lost in admiration. He believed, now, that the woman had been putting it on for years. She never even winked...The insight he had obtained almost frightened him; he couldn't get over his wonder at knowing more of the real Mrs. Schomberg than anybody in the islands, including Schomberg himself. She was a miracle of dissimulation."

We are all "miracles of dissimulation" except to god who sees, often when we do not even see ourselves, as we are. At least god sees the goodness.

627. Barbara Ehrenreich, in "How `Natural' Is Rape?", *Time*, Vol. 155, No. 4, January 31, 2000, p. 88, reviews a book titled, "The Natural History of Rape" in which the authors claim that rape is just another mechanism evolved to have offspring. She makes a good case for rejecting this position from evolutionary grounds. Though she must allow that there could be merit in the position, it certainly deserves to be called reprehensible.

My position is certainly that even if such behavior is somehow "natural" it is surely one of those inherited traits that god would counsel us to put aside.

628. Some reviewers of this book have wondered about this sentence. I am not trying to approve of prostitution in general. It is often brought about by unjust economic systems which force women (and men) into the only means to gain income available to them, or by other societal, familial or psychological conditions that should be remedied. Rather, I am thinking of actual conditions in the real world when men or women who have little chance of attracting a mate find buying time with another the only way to have anything resembling intimacy. An ugly, elderly, impoverished man, for example, may happily pay for some kind of human contact, even if they have to lie to themselves about it to make it meaningful. Prostitutes will say that it is not rare to have the client request conversation instead of sex. (That costs extra of course since it takes more time.)

Prostitution is certainly a symptom of societal ills, but the "best little whore-house in Texas" had its saving graces, too.

629. Abel Jeanniere, "The Anthropology of Sex", New York: Harper & Row, Publishers, 1967, p. 129

630. Leonard Foley, O.F.M., "Believing in Jesus: A Popular Overview of the Catholic Faith", Cincinnati: St. Anthony Messenger Press, 1981, p. 170

631 In a collection of the teachings of Catholic Bishops on Racism entitled, "Love Thy Neighbor As Thyself", Washington: United States Catholic Conference, Inc. 2001 there is a wonderful collection of homilies, letters and teachings. However, many base their position on god's creative action. See the "Statement of the Catholic Bishops of Louisiana," p. 17, the homily at St. Christina Church by Cardinal Francis George, and many others. At the same time, other reasons for treating others with dignity are mentioned, sometimes without the reference to the creation. These include the teaching of Jesus to love your neighbor as well as Paul's invitation to love all those who have been baptized in Christ. These latter are going to be more potent reasons for excluding racism than the "Image of God" argument, at least for Christians, since they are based on the example of Jesus and therefore the example of god.

632. This thought is often called "annihilationism" or "conditionalism" and is opposed by "traditionalism" which holds for an eternal punishment of an existing and damned being. For an excellent report on the arguments of

both sides in the Evangelical tradition, which is also occurring in the Roman Catholic and other traditions.

See Robert A. Peterson, "Undying Worm, Unquenchable Fire", *Christianity Today*, Vol. 44, No. 12, October 23, 2000, p. 30. The article also refers to the relevant scripture passages.

633. I have often felt that Edmond Rostand's "Cyrano de Bergerac", New York: Bantam Books, 1923, while an interesting, even fascinating character, would be rather hard to live with. Indeed, after a famous "No thank you" speech upholding his right to be his own person and not the toady of another (p. 75) he admits that, "It is my pleasure to displease. I love hatred." And indeed he brings nothing but misery to everyone he contacts. Maybe god would resurrect such a fellow, but I would think twice.

634. This is well put in, "Readings in Faith and Science", St. Louis: ITEST Faith/Science Press, 1997, p. 114,

> "For about a century, we have abandoned the idea that heaven is a place; we smile at the image of sitting on a cloud strumming a harp. We have faith in God that He has something much better in mind for heaven. Now it's necessary to take an even bigger leap of faith and concede that heaven is not a time either...By repeating the word `afterlife' so often, we have remained dependent upon time much too long. Any image of God that is confined by time is too limiting; similarly, any image of our ultimate relationship with God that is dependent upon time is doomed to fail. We have to start imagining heaven as a totally different kind of existence...Heaven is not a time-dependent entity, any more than it is a specific place. Time isn't frozen; time isn't continuous, boring and repetitive; time simply isn't an issue at all. It's not one of the dimensions, not part of the heavenly existence. God has something else in mind, and remember, God is a lot smarter than we are. God gets along just fine when outside the realm of time and space."

635. Will, the lead character in Ray Bradbury's "Something Wicked This Way Comes", New York: Bantam Books, 1962 asks his father if there is anything that does not make him sad. The answer is,

> "Death makes everything else sad. But Death itself only scares. If there wasn't Death, all the other things wouldn't get tainted."

636. "Catechism of the Catholic Church", #365

637. “Catechism of the Catholic Church”, #366

638 There is nothing wrong with using the term “soul” of course, though we must recognize that it is a poetic way or a shorthand manner of referring to the whole complex of emergent qualities of the human brain and thought processes (including hormonal, etc) which characterize human life and potential. We do not have to insist there be a real entity of “soul” to use the figure of speech.

639. James Wood, “The Broken Estate”, New York: Random House, 1999

640. “The challenge of the millennium” Editorial in *Our Sunday Visitor*, Vol. 88, No. 36, p. 2

641. Ray Kurzweil theorizes that, “in the second half of the 21st century we'll routinely be able to scan a person's brain and `reinstantiate' that person in a computer.” *Technology Review*, Vol. 103, No. 1, January/February 2000, p. 80

642. Hans Kûng, “Eternal Life”, Trans. by Edward Quinn, Garden City: Doubleday & Company, Inc., 1984, p. 113

643. “Eternal Life”, p. 227

644. Sebastian Moore, “No Exit”, New York: Newman Press, 1968, p. 91

See also Bruce Chilton, “The Son of Man—Who Was He?”, “*Bible Review*”, Vol. XII, No. 4, August 1996, p. 34

See also “Excavating Jesus”, p. 124 which makes the same point in that John the Baptist's death marked the end of his movement, but Jesus' did not.

645. “Wouldn't Take Nothing for My Journey Now”, p. 75

646. See “Catechism of the Catholic Church”, articles 2626 and following

647. “Catechism of the Catholic Church”, #2633

648. Jean Laplace, S.J. “Prayer According to the Scriptures”, Brightons: Alpine Press. p. 4

649. “The Power of Myth” p. 209

650. “Story of a Soul, The Autobiography of St. Therese of Lisieux”, Trans. by John Clarke, O.C.D., Washington: ICS Publications, 1976, p. 242

Tertullian, one of the very early “Fathers of the Church” would certainly not deny that god can manipulate events, yet though he says that, “it will gain for us all that we ask of God,” and believes that in response to prayer god was willing to, “rescue from fire and beasts and hunger,” still he says, “no longer does prayer bring an angel of comfort to the heart of a fiery furnace, or close up the mouths of lions, or transport to the hungry food from the fields.” Now it, “gives the armor of patience to those who suffer...strengthens the power of grace so that faith may know what it is gaining from the Lord, and understand what it is suffering for the name of God.”

> [“On Prayer”, cited in “Touching the Risen Christ: Wisdom from the Fathers”, Ed. by Patricia Mitchell, Ijamsville: The Word Among Us Press, 1999, p. 33]

John Chrysostom echoes these sentiments when he describes prayer as, “the longing for God, love too deep for words, a gift not given by man but by God’s grace...When the Lord gives this kind of prayer to a man, he gives him riches that cannot be taken away, heavenly food that satisfies the spirit.” [Ibid. “From a homily”, p. 36]

651. “God’s Encounter With Man” p. 85

652. “God’s Encounter With Man” p. 115
A similar idea is noted from The Swiss theologian Otto Karrer, who is cited as saying,

> “Christian prayer...does not seek to make God inclined toward and ready to hear human beings (God is ready and so inclined), but the opposite: human beings are to be inclined toward and ready for God.”
>
> [See Herbert Frohnhofen, “The Holy Spirit: source, goal, and fruit of our prayer”, “*Theology Digest*”, Vol. 46, No. 1, Spring 1999, p. 22]

653. Elie Wiesel, "Night", records that after experiencing the abandonment of a father by
a son, "...in spite of myself, a prayer rose in my heart, to that God in whom I no longer believed. My God. Lord of the Universe, give me strength never to do what Rabbi Eliahou's son has done." [p. 87]

654. "Sula", p. 107

655. "Every Eye Beholds You", p. xiv-xv

656. "Pilgrim at Tinker's Creek" p. 269

657. Annie Dillard, "For the Time Being", New York: Alfred A. Knopf, 1999, p. 167

658. Pope John Paul II, "Celebrate 2000", Selected by Paul Thigpen, Ann Arbor: Servant Publications, 1996, p. 52f

659. Carmen Bernos De Gasztold, "Prayers from the Ark", Trans. by Rumer Godden, New York: Penguin Books, 1976

660. Michel Quoist, "Prayers" Trans. by Agnes M. Forsyth & Anne Marie de Commaille, New York: Sheed and Ward, 1963

661. Rick Hamlin, "Finding God on the A Train", New York: HarperSanFrancisco, 1997

662. "Finding God on the A Train", p. 145

663. Macrina Wiederkehr, "A Tree Full of Angels", San Francisco: Harper & Row, Publisher, 1988. A favorite example of her method is that of contemplating,

> "a community of empty trees sways outside my window, dancing in the early morning light...delighting me with golden streams of warmth...How many parties have you hosted for the squirrels, watching their restless little bodies leaping through your limbs."

And though she credits them with, "obeying an inner energy divine," the thought they inspire is not for the majesty of the creator, but that,

> "I want to be, like you, a welcome home for every guest. I yearn, like you, to embrace all the colors of my life."

The assumption that god created these trees is not essential to the thought. Her fascination with the universe of which she is consciously a part enables her to allow god to inspire deeper values within her.

664. "An African Prayer Book" Selected by Desmond Tutu. New York: Doubleday, 1995. p. 79

665. Cathleen Medwick, "Teresa of Avila: The Progress of a Soul", New York: Alfred A. Knopp, 1999. An excellent modern biography.

666. "The Autobiography of St. Teresa of Avila", Trans. by Kieran Kavanaugh, O.C.D. and Otilio Rodriguez, O.C,D., New York: Book of the Month Club, 1987, p. 343

667. Ibid. p. 449

668. Mother Teresa, "No Greater Love", Ed. by Becky Benenate & Joseph Durepos, Novato: New World Library, 1997, p. 7f

669. Miriam Pollard O.C.S.O., "The Laughter of God", Wilmington: Michael Glazier, 1986, p. 17 & p. 22

Other examples of god tinkering with our lives are plentiful. I ran across what is described as the text on a bronze plaque in a rehabilitation center in New York which says,

> "I asked God for the strength to achieve success; he made me weak so that I would humbly learn to obey.
> I asked for health to do so many important things; he made me infirm so that I could do better things.
> I asked for wealth so that I could be happy; he gave me poverty so that I could be wise.
> I asked for power so that people would value me; he gave me weakness so that I could know my need of God.
> ...I received nothing that I asked for, but I got all I had hoped for.
> Almost in spite of myself my fumbled prayers were answered. I am the most richly rewarded of all."

I suppose this may be consoling for a weak, sick, poor, powerless person such as we maybe are, or may be sometime. But it really does a number on god.

670. Louis Evely, "That Man Is You", Westminster: The Newman Press, 1965, p.17

671. Carol Schuck-Scheiber, "Healing", *Company*, Vol. 14, No. 4, p 12

672. "Victory", p. 11

673. ibid. p. 13

674. There has always been an understanding that sin cannot be defined only by reference to the breaking of a law. It came to the fore around the time of the Second Vatican Council in the 1960's. One example is Robert P. O'Neil and Michael A. Donovan, "Sexuality & Moral Responsibility", Washington: Corpus Books, 1968, who say,

> "We propose a redefinition [from nominalistic casuistry, with law and jurisprudence as the framework for...conclusions] of the concept of serious sin in terms of orientation rather than of act." [p. 35]

and then proposes,

> "...the following working definition: Subjectively grave (mortal) sin is a fundamental orientation of mind, consciously and deliberately maintained in opposition to what one simultaneously perceives clearly to be the true purpose and aim of one's life." [p. 47]

His work is an early application of psychology and anthropology to morality.

675. "Islam", ed by John Alden Williams, New York: George Braziller, 1962, p.61f

676. In the orthodox view, it is imperative that god give us some special guidance, for without it we are little better than the beasts. Human behavior can often be described as "beastly", or a person is said to be an "animal", but at the same time this is considered an aberration. We are not like the rest of the animals (we have a soul) and our behavior can be guided. The Catechism states that,

> "By his reason, he is capable of understanding the order of things established by the creator. By free will, he is capable of directing himself toward his true good. He finds his perfection `in seeking and loving what is true and good.'"
> ["Catechism of the Catholic Church", #1704]

We were created in order to respond to god and to respond in freedom. The animals respond by being what they are, but they must be so, they have no choice, and their response is not very satisfying. We are created to respond to the covenant and to do so freely. Unfortunately, since we are not programmed like animals, we are therefore free to choose not to respond or to choose the evil and the false. So god communicates through the written word as well as the spoken word of the prophets (which is often written down as well). The prophetic message, itself flawed by its human heralds, is not heard any better than the original covenant. If it is observed at all it is by the "little people", the *anawim* who are normally those who best respond to this god.

[See Leopold Sabourin, S.J., "The Psalms", New York: Alba House, 1970, p. 95]

Let them love their neighbors and trust in their god. The princes of the land will still trust in their horses [Psalm 20:7] and the wealthy will still trust in their riches. [Psalm 49:6; Mark 10:24; 1 Timothy 6:17] It is the model of the powerful and the rich that will influence future generations, not that of the "anawim".

[The word *anawim* has a complex history. See Sr. M. Rose Eileen, C.S.C., *The Spirit of the `Anawim*, "Contemporary New Testament Studies", Collegeville, The Liturgical Press, 1965, p. 68f
See the comment on its use in the Psalms in "The Jerome Biblical Commentary" Ed. by Raymond Brown, et al., Englewood Cliffs: Prentice-Hall, 1968, p. 573
See brief discussions in Sigmund Mowinckel, "The Psalms in Israel's Worship, Trans. by D.R. Ap-Thomas, New York: Abingdon Press, 1962, p. 12, 30 etc.]

The covenant theme, so drastically altered by the captivity in Babylon, will give way to the themes of Messianism and legalism and the prophetic corrective will give way to Wisdom and apocalypticism as the literary genre in which god reveals godself.

What is god to do to renew the covenant with god's people? God has sent his servants the prophets, now he will send the son. [Matthew 21:33f; Mark 12:1f; Luke 20:9f]

In Jesus, god becomes human.

[A contemporary retelling of the Gospel story interwoven with the current understanding of its meaning can be found in Jean Vanier, "Jesus, the Gift of Love", New York: Crossroad, 1994.
About Jesus he says,

> "Again and again Jesus says that everything he says and does is from the Father: he himself proceeds and comes forth from the Father; he and the Father work continually one with the other; he is sent by the Father; he is for the Father; he is the beloved Son; he is in the Father and the Father is in him; he is one with the Father." (p. 73)

With our understanding of the Church teaching on the Incarnation we understand this as Trinitarian, but the same could be said of any good Jew, especially a prophet. However, the author makes it clear on page 101 for example, taking his cue from the Gospel of John, that Jesus is the "I am", the physical presence of the invisible God, though he recognizes that "this revelation became clear to John only after the resurrection of Jesus and after he had been filled with the Holy Spirit" (p. 103). Most of the work however is about a very human, weeping, teaching, human Jesus.]

How can this Jesus be the physical presence of god? Following Aristotle, through Aquinas, and therefore explaining the ground of being through "nature", "form" and "matter", it was explained that Jesus had two natures, one divine and one human and that the human nature was composed of soul or form and body or matter and the accompanying "accidents" that made it sensible.

> "Aquinas departs from Aristotle in an important point related to this discussion. For Aristotle, "the form or essence of a material thing does not include matter. Only the concrete entity does. And since it is this concrete entity or substance that is defined by the natural philosopher, an act of addition is required to grasp form and matter together. St. Thomas, however, does not require such an operation, for according to him the essence of a material thing at once includes both form and matter."
> [Armand Maurer, "St. Thomas Aquinas, The Division and methods of the Sciences", Toronto: Pontifical Institute of Mediaeval Studies, 1986, p. xxx]

For Aristotle's treatment of substance, matter, essence and form, see his "Metaphysics", Book VII in any translation, such as "Aristotle: On Man in the Universe", Ed. by Louise Ropes Loomis, New York: Walter J. Black, 1943, p. 24]

Through a series of responses to heresies this was elaborated even before Aquinas to bring out that the human was truly divine and the divine fully expressed itself through the human. And what was the role of this Jesus? Not to deliver a new law code; not to fulfill the grosser forms of messianism; not to deliver a new wisdom as an esoteric knowledge reserved to the privileged; and certainly not to bring about the end time; though he would be made by others who followed to do all of these things. No, it was to renew the Covenant relationship of god and god's people and between individuals among those people. This was not done by dying, as some forms of Christianity would later believe, but by showing that living by the Covenant is real living or living to the utmost. By the resurrection from the death that his life style, not to mention the evil in the world in which we all live, had brought him, he is not rewarded for obeying god, as some would have it. If this were so, then the commandments would indeed be a test, arbitrary demands made to see how we would react. This would mean what we felt all along, namely that some things the commandments prohibit would really be good for us if we were only allowed to do them, like live for ourselves alone, not be concerned about the misfortunes of others, live for today, and the like; it would mean that these are true and god only prohibits them to see if we will obey. If the resurrection were a reward for unquestioning obedience, then the commandments need not be the wisest way to live if only they carried no arbitrary sanctions. But Jesus' resurrection was not the reward of a life spent in obedience, it was the continuation of a life lived in true wisdom, a life truly lived well. Contrary to the popular understanding that it was the death alone which bought our salvation, Jesus' death is not an end to be hoped for so as to prove his loyalty, it is an evil to be endured and conquered, and escaped if one can do so without selling out one's covenant relationships with god and neighbor. And selling out Jesus could not do. To be true to the covenant, Jesus would have to put himself in harm's way and he did. He was obedient to god's invitation that he live well, not god's will that he die. And for Jesus and his disciples, to live well is to love and care. Jesus then renews the Covenant, or in another view initiates "a new covenant in his blood" by believing that it is by living in relationship with god and neighbor that he himself truly lives, being willing to die by the same principles, and being proved correct by the resurrection.

[Gerard Rosse, "The Cry of Jesus on the Cross", Trans. by Stephen Wentworth Arndt, New York: Paulist Press, 1987, p. 125 points out that in Philippians 2:6-8 Paul adds the words, "death on a cross!" to a hymn extolling Jesus' acceptance of death. Rosse says that, "in exhorting the community to the life of unity Paul does not present so much as a model the obedience lived by Jesus in general but places the accent on his obedience to the point of death *on the cross*." I do not believe that Paul would deny that the death was an evil nor that the life of obedience should not be a model. Paul rarely mentions the historic Jesus but is concerned with the Risen Christ.]

Dermot Lane expresses the relationship of his death to his life in this way,

> "The impending prospect of death arising out of the mounting opposition now appears as the only way in which the truth of his words and deeds will emerge. The only avenue left for the realization of God's Kingdom on earth is in and through the awesome reality of death itself. To this extent death for Jesus appears to be God-willed, taking on a divine necessity *vis-à-vis* the realization of his mission. That death could assume such a positive significance is something that could have emerged from Jesus's direct or indirect acquaintance with the servant songs of Isaiah."
>
> ["The Reality of Jesus", p. 42.]

Lane goes on to say that,

> "An awareness of death as the instrument for the realization of God's purpose and Kingdom would have taken place during the final stages of the earthly life of Jesus. This type of thinking seems to loom large during the Last Supper when Jesus says: 'Truly, I tell you, I shall not drink again of the fruit of the vine until that day when I drink it anew in the Kingdom of God' (Mk 14:25). Further, Jesus seems to associate the Last Supper with his death as bound up with the realization of God's plan. As such the Last Supper at this stage is best understood as a prophetic act taking place within the apocalyptic setting of the rest of the life of Jesus."

It is not just Jesus' death, then, that is redeeming, it is the example of his life and teachings as well, for he lived by what he taught. Just how much of what Jesus is said to have taught in the New Testament are his actual words, the *ipsissima verba*, is open to question with a widely held view that it is "not much". The "Jesus Seminar", a group of biblical scholars who have tried to determine what is contained in the Gospels that Jesus himself might have said found only one instance in Mark and six in Matthew that they judged to be very likely a quotation of Jesus' words or close to them.

[See "The Five Gospels" New York: Macmillan Publishing Company, 1993. There are many other instances that may well be from Jesus, but there is some doubt about them.

Few dispute that it is Jesus who is behind the beliefs of the Christians who would follow. Paul and others would elaborate on the meaning of his life and message, draw conclusions and implications that Jesus himself may not have realized about himself, but they did not invent him. Jesus himself evidently believed, as every Jew was invited to believe, that he had been incorporated into the covenant people of god; that god was a holy god and god's people should be a holy people; and as a Pharisee that god is a god of life who would save him from the death of a meaningless life on earth and at the end of his life from physical death through resurrection. Little or nothing of what he said was new to those familiar with covenant language and little or nothing of what he did, miracles included, was not expected of one who lived a covenant faith and morality. Still, the unfaltering manner in which he lived and said these things was impressive to those who experienced him. [Matthew 7:29; Mark 1:22] And of course his trust in the covenant was most sorely tested by the threat of death unless he desisted, something he would not do, though the unbelief of the people and especially his own disciples must have run a close second as a severe test to him. Even the unbelief of his own family could not deter him since he loved these people and the most loving thing he could do for them was to give them an example of the covenant life.

[Mark3:21 records that, "When his relatives heard of this they set out to seize him, for they said, 'He is out of his mind'." This group possibly included his own mother, for her arrival is also noted in the same context.]

So thoroughly did he believe and exemplify life as a child of the covenant god, which every person was invited to live, that gradually the Church would recognize that he was the son of god in a unique manner. He *was* god incarnate, in the flesh. To live in covenant relationship with Jesus was to live in covenant with god. To watch Jesus' reaction to the trials and temptations of life was to watch god be true to god's own nature.

[See "The Passionate God", p. 7f]

677. "The Good Enough Catholic" p. 4

678. See the "Catechism of the Catholic Church", #464f

679. Anthony J. Saldarini, "Understanding Matthew's Vitriol" in *Bible Review*, Vol. XIII, No 2, April, 1997, p. 32.

> "Matthew does not present Jesus as God in the manner of later Christian Trinitarian theology. Instead, he draws on designations and titles from Jewish scriptures...Thus Matthew tells us that Jesus was sent to Israel by God as anointed leader ("messiah," as the term is rendered in Psalm 2:2), the son of Man (as in Daniel 7) and son of God (as in 2 Samuel 7:14 and Psalm 2:7). Matthew accords Jesus a very high status, but he does not say Jesus is God, Rather, Matthew writes as a member of the late-first-century Jewish community, using the words, symbols and thought patterns of his age in an attempt to establish Jesus' way of life as the authentic way of living the Jewish tradition."

Another approach found in James K. Hoffmeier, "Son of God: From Pharaoh to Israel's Kings To Jesus", *Bible Review*, Vol. XIII, No. 3, June 1997, p. 45, understands Jesus as the "son of God", for example in the baptism narratives (Mtt. 3; Mark 1; Luke 3),

> "as a type of royal anointing ceremony like that undergone by Hebrew kings. The Gospels' use of Psalm 2:7, a royal coronation psalm, certainly supports this hypothesis. Early Hebrew kings were anointed by a prophet, the spirit of God came upon the individual (see 1 Samuel 10:1,10; 16:13) and, as Psalm 2 declares, the king is called the son of God. In other words, new Testament sonship Christology may be rooted in Hebrew concepts of kingship rather than in Hellenistic religious ideas imported from the Greek world."

It is these Hellenistic religious ideas which permeate the patristic writings and certainly those of Aquinas and the Scholastics.

680. See the "Anchor Bible", Vol. 36, "*To the Hebrews*", Translation, Comments and Conclusion by George Wesley Buchanan, Garden City: Doubleday & Co., Inc., 1972

681. This is in fact the title of E. Schillebeeckx's book, "Christ the Sacrament of the Encounter With God", New York: Sheed and Ward, 1963

682. Much of what is believed about the Sacraments by many Catholics arises from the theological position that these rites operate, *ex opere operato*, which is to say they bring about an effect just by being done.

> [See the excellent discussion of this concept in E. Schillebeeckx, O.P., "Christ the Sacrament of the Encounter with God".
> See also Godfrey Dickmann, O.S.B., *Two Approaches to Understanding the Sacraments,* "Readings in Sacramental Theology", Edited by C. Stephen Sullivan, F.S.C., Englewood Cliffs: Prentice-Hall, 1964, p. 1f.]

The origin of this concept was for a good purpose, to assure the faithful that the Sacrament achieved its purpose independent of the minister. The condition of the minister, how well the ritual is performed, whether or not the minister even believes in the efficacy of his or her own actions, is all independent of the effect of the divine presence in the Sacrament.

> [Juan Luis Segundo, S.J., "The Sacraments Today", Vol. 4, New York: Maryknoll, 1974, p. 59 is a good post-Vatican II treatment of the *ex opere* concept.]

This is certainly a worthy motive, but it has had some unfortunate consequences. The first is that it has often fostered an understanding of the workings of the Sacraments as almost magical. Indeed, the words of Consecration of the Bread and Wine of the Mass, *hoc est enim corpus meum,* in Latin, were corrupted into the magical incantation, "hocus pocus", obligatory before pulling the rabbit from the hat. Baptism is believed to remove "original sin" by magically washing a black mark (or a white mark for African American Catholics) from the soul. Confession (even if now called Reconciliation and much ignored) removes sin magically, expecially if it is done in a little dark box, whether or not you really intend to cease and desist the negative action.

> [Graham Greene, "The Heart of the Matter", Harmondsworth: Penguin Books, 1948, p. 221 has the confessor remind the penitent adulterous hero that he must have a real purpose of amendment, since, "nobody can begin to forgive the uncontrite. It's better to sin seventy times and repent each time than sin once and never repent." Greene is a good apologist for the faith and the absolution is not given since the man has not the intention of stopping the affair. Ironically he is the most devout and religious figure in the book.]

The Sacament of Anointing is meant to mediate Jesus' healing to the whole person, but priests have been requested to anoint specific areas (breast cancer presents interesting scenes) and DIY anointing with Lourde's water or St. Anne's oil are tried to do the same magic. Confirmation also uses oil and is sometimes thought to transform the person into a Spirit-filled disciples with little effort on their part. For most families, however, Confirmation is not so much magic as a rite of coming-of-age. The Ordination of the clergy also involves oil, but this is not part of the Sacrament, the true sign being the imposition of the hands of the Bishop. Still, young priests are sometimes described as especially holy (if clueless) since, "the oil hasn't dried on his hands." In Marriage, it is really the relationship of the couple that is the sign and symbol, but most identify the ring in this role. Evidently magic does not work without a lucky charm.

The other unfortunate effect of the *ex opere operato* explanation of the Sacraments is that, while the importance of the human minister manipulating the physical symbols is minimized, the importance of god's doing so is maximized, and my conclusion is that god cannot have an effect on such physical objects, at least the innanimate ones. For a look at the beginnings of the more modern understanding of the Sacraments, see the following and note how the emphasis began to shift from strictly god's activity to god's activity in the Church. Or see any of the more recent that they inspired.

Bernard Cooke, "Christian Sacraments and Christian Personality", New York: Holt, Rinehart & Winston, 1965.

Joseph Powers, "Eucharistic Theology", New York: Herder & Herder, 1967

Edward Schillebeeckx, "Christ the Sacrament of the Encounter with God".

Monika Hellwig, "The Meaning of the Sacraments", Cincinnati: Pflaum/Standard, 1972.

Karl Rahner, "Meditations on the Sacraments", New York: The Seabury Press, 1977]

In the following treatment of a more up-to-date explanation of the Sacraments, which could well be given by someone who continues to presume that god can manipulate physical entities of the universe. Note that such power is not necessary in order that the Sacraments communicate what god wants to impart.

Baptism is the first sacrament experienced in life in the Church. If the person is an infant at the time, the experience may be for the parents and family and only vicariously experienced by those baptized in infancy by participating at a later date in the same ritual for a brother or sister or another relative or in watching it for adults at the Easter Vigil. So we may want to go back to the baptism class for parents of new-borns. When pushed to explain the meaning of the sacrament, parents most commonly present the following picture. Adam and Eve were the first humans, though how this squares with the picture presented by evolution is somewhat of a mystery to them, best left unexplored. They lived in a state of "original justice", though the parents in the baptism class do not present it in these words. ["Catechism of the Catholic Church", #374 and #375]

Adam and Eve sinned by eating the fruit from the forbidden tree, commonly thought of as an apple, though why it must be this particular fruit is unknown. It is vaguely understood that the original sin is one of disobedience. God discovers their sin, ejects them from the Garden of Eden, and imposes punishments on them, especially the divine disfavor, the necessity of working for a living, their liability to death, and revocation of their free pass to heaven. Somehow the effects of their sin are transmitted to their offspring and thence to the generations to follow, right down to this baby.

> [For a good summary of the position of the Church on this topic in 1969, which is certainly the faith taught to most of today's adult Catholics when they were children, including the new insights which were just beginning to be taught at the time, see Herbert Haag, "Is Original Sin in Scripture?" Trans. by Dorothy Thompson, New York: Sheed and Ward, 1969
>
> In his introduction to Haag's book, Bruce Vawter points out that, "only with considerable foot dragging has [the Church] now begun to come to terms with the Darwinian revolution that lies in the background of this book." Judging by the present understanding of this idea of original sin among the faithful and among their teachers, the foot dragging continues.
>
> E.L.Doctorow in "City of God", blames Augustine for the concept of Original Sin, calling it, "a nifty little act of deconstruction—passing it on to the children, like HIV."]

When asked how this inter-generational influence is transmitted most profess ignorance, or more commonly admit that the question has never come up, but some will explain it by the conveying of a black mark on the soul. That there is such a thing as a soul is taken for granted and that it can be marked is also not questioned. The color of this evil is open to question, however, African-American Catholics preferring to think of it as a white mark on a previously beautiful black soul and Caucasians reversing the image. However, the more knowledgeable among them may explain it differently. They will probably begin with the assumption of a state of original justice, but then add a more subtle explanation of how such sin might be transmitted to descendants. No such concept as "soul" is required, much less a marking of it. We can see today the effect of the sins of the parents on the lives and values of the offspring, whether considered individually or culturally. If one's parents have no work ethic, insisting on making an honest living, but rather live by preying on others, the likelihood that their children will live by these same values is great. If the father's vocabulary is liberally sprinkled with obscenities, it may take a special act of the will for the son to refrain. If serial polygamy is accepted and practiced in the society, it is very difficult for a young couple to understand that the marriage commitment could be for life. We pass on these cultural values, these black marks, even more readily than we pass on the good. Consequently, anyone born into that culture is already affected by the "sin" just as surely as if a mark had been left on the soul. There is nothing magical about this explanation. It is in accord with the findings of modern psychology and group dynamics. Obviously a first parent was not affected by the mores of their ancestors or their culture, so some other culprit must be introduced, most commonly the serpent who entices them to do evil. Once they do, you can dispense with the serpent since we will pass on our tendencies and values to our descendants and spread them around our society. This more advanced understanding of original sin, the soul and the effect of sin on the new generation may or may not be new to the parents in our baptism class, but they are usually relieved when they understand that there may be a more normal explanation for the phenomenon that the old "black mark" was meant to convey. They are usually a little more reluctant to put aside the concepts of the soul and the special creation of human beings, however. Most are also relieved that god does not seem so intent on condemning to eternal separation from god most of the people born into the world, least of all a newborn child. God seems much more intent on counteracting the presence of this "original sin" than condemning one for it. However, many a priest, having explained to the parents, and sometimes to irate grandparents, that the child should not be baptized because of the parental lack of involvement with the Church, has been accused of putting the child in mortal peril lest they die with original sin on their soul. Even after hearing the explanation of original sin and the manner in which Baptism has an effect on it, and agreeing that it makes no sense to baptize their child because they are not going to be a part of the parish community, they opt for the magic effects of the sacrament. Lesson taught but not taken.

The symbolic effect of the water made perfect sense when it was seen as washing a mark of whatever color from the soul. With this removed, however, a new symbolism is necessary. Certainly there is the meaning of washing, but not from a sin incurred by some ancestor, before whom there was no sin. Rather it is a sign of incorporation into a holy people, the entire people washed from the stain of alienation from Gospel values.

> "Baptism dedicates and consecrates the believer to Jesus Christ and, more than this, it establishes him in a living union with Christ. But baptism is also and in some sense primarily—a rite that links onto the people of God those who respond with faith to the divine call. Christians realized this from the very beginning and it is an idea to which St. Paul has given definitive expression."
> [Jean Giblet, "Baptism-The Sacrament of incorporation into the Church, according to St. Paul", in "Baptism in the New Testament", Trans. by David Askew, Baltimore: Helicon, 1964, p. 161]

There was always the question of how an infant could be guilty of personal sin and the answer was that original sin was not personal but communal.

> [The preacher in John Steinbeck's "The Grapes of Wrath", New York: Bantam Books, 1939, p. 88, accepts the story of Adam's sin but sees it as a trait to which all humans are liable. He explains that, "I got to thinkin' how we was holy when we was one thing, an' mankin' was holy when it was one thing. An' it on'y got unholy when one mis'able fella got the bit in his teeth an' run off his own way, kickin' and draggin' and fightin'. Fella like that bust the holiness. But when they're all workin' together, not one fella for another fella, but one fella kinda harnessed to the whole shebang—that's right, that's holy."]

This is certainly the case in this symbolism. The adult may be washed free from personal sin and its consequences, but this is done by the acceptance of the community's values. The more meaningful symbolism of the water is in the dying to a life without the presence of Christ and to the Johannine "world", either the life the child would have lived without baptism or the world that the adult actually did inhabit for a time. Immersing the person in the water until they drown to this style of life and then bringing them out in the rush of the waters of birth is the most important symbol of the reality of what god is doing in their lives. The clothing with a new garment, anointing with the oil of consecration and presentation of the flame of faith are all symbolic, not of a reinstatement into a life of original justice that never historically existed, but into the here-and-now life of the Church community. The effect of baptism is then seen as having an impact on the identity of the person, a new birth of a new person, not by a change in the essence of the personality by quasi-physical change in the soul/substance, but by a re-ordering of relationships, to god and to the community.

Confirmation also affects the identity of the person by re-ordering the relationship to the community. "Catechesis for Confirmation should strive to awaken a sense of belonging to the Church of Jesus Christ, the universal Church as well as the parish community."

["Catechism of the Catholic Church", #1309]

The sign of Confirmation is usually seen as the anointing with oil of the head of the individual, a sign that has a long history. In days of less than lax personal hygiene it was the custom to perfume the important visitor to ones quarters so that he would smell the perfume rather than the family and the animals lodged on the first floor of the family dwelling. Only the important visitors, of course, (hence the limitation imposed by the masculine pronoun in the previous sentence, an important female visitor being hard to imagine). Perfume was expensive. You could normally tell if a person was important by whether they were anointed with perfume upon entering the house. Egyptian murals often portray party-goers with a cone on their heads, not a hat but a sweet smelling gelatin pomade which would melt, covering the hair with perfume.

[Lionel Casson, "Ancient Egypt", Time-Life "Great Ages of Man" series, Time, Inc. 1965, p.111.

See also a wall painting reproduced in John Ruffle, "The Egyptians", Ithica: Cornell University Press, 1977, p. 185]

The Babylonians perfumed their beards, a custom the Jews may well have adopted and the use of oil as a health measure to restore oil to sun-ravaged skin was very ancient.

["Harper's Encyclopedia of Bible Life", Madeleine S. & J. Lane Miller, Harper & Row, New York 1978, p.86]

In a reversal, it became customary not just to anoint the person in recognition of their importance but also to confer importance by the anointing with perfume. All the kings of Judah were anointed and indeed it was one of the essential rites of coronation. It confers the divine grace and presence and therefore, the political office.

> [Roland de Vaux, O.P., "Ancient Israel", New York: McGraw-Hill Book Co., 1961, p. 102f]
>
> "Anointing brings the spirit of Yahweh upon the person and impels him to some extraordinary deed."
>
> [John McKenzie, S.J., "Dictionary of the Bible", Milwaukee: The Bruce Publishing Co, 1965, p.35]

In Confirmation the symbolism of recognition of the importance of the person anointed is rather clear, though in eighth graders their importance is usually more *in potentia* rather than in actuality, except perhaps to the family. The meaning of the divine force impelling the person to a renewal of the rebirth of baptism into the new identity of the mature Christian is what is today being taught. In practice it is the time of preparation and all the interest and encouragement expended by the parish towards the candidates, culminating in the presence of the bishop, a charismatic figure if only because of the miter and crozier, that is the effective sign. The anointing with oil is little understood unless extraordinary effort is made to make it understandable. In fact the now omitted slap on the cheek is what the parents of today's young Catholics best remember about their own Confirmation beyond the presence of the bishop. A better sign of the recognition of the individual's importance to the Church and the urging of the Spirit for the person to put themselves at the service of the gospel would seem to be the acceptance by the adult community *after* the reception of the sacrament and the insistence by the Church that the person *now* put their talents at the disposal of the community. Unfortunately this incorporation into the parish's activities is too often done before the sacrament through "community service projects" which, again too often, sometimes degenerate into setting up tables for the parish fish fry. After the confirmation, it is often neglected. Sometimes the preparation programs rely on the operation of the sacramental signs *ex opere operato* to too large a degree. If one just goes through the motions of Christian service, but without thought and introspection, and writes the right letters to the parish and bishop and then goes through the ritual with minimal understanding, the Holy Spirit will do something (magical) in them.

The same is sometimes true of the sacrament of Reconciliation. While it began as a means of re-incorporation of the person into the community of those who are living by gospel values following a public abandonment of those values, today the emphasis is sometimes on offenses against god rather than against god's people or rejection of god's inspirations. Yet the sacrament can be a powerful means of growth. Bernard Haring's treatment, "Shalom: Peace" at the time of Vatican Council II brought to focus a renewed understanding of the sacrament which theologians have developed ever since. His insights did not, however, prevent a general lessening in the frequency of the celebration of the sacrament, which may or may not turn out to be a good response. Possibly it is just a necessary response due to the negative understandings and practices observed in the previous centuries."

> [Bernard Haring, C.Ss.R., "Shalom: Peace", New York: Farrar, Straus and Giroux, 1967]

The Anointing of the Sick is primarily a sign for the person afflicted by physical or mental illness or by the effects of aging, all of which bring our own mortality and fragility to our minds, of the empowering and healing presence of god to them in their situation. The oil of the anointing is meant to remind the person, and others who participate in the celebration of the sacrament, of medicine from god. Since most of our medicine today comes in the form of gel caps or injections, it is not so apparent to the senses that the oil used by the priest is medicine, but in the days of castor oil, cod liver oil and even "snake bite oil", the connection would have been more apparent. The rubbing on the body of the medicine may remind us today of Ben Gay for sore muscles, but in the days of mustard plasters and medicinal mud packs, the symbol of medicine was more easily grasped. Today these signs can be appreciated with a little reminding and education, but the most apparent sign is the gathering of the Church around the sick bed or the personal attention of the priest during a Mass of Anointing or Healing celebrated in the church, accompanied with special hymns and prayers which tell the person that the other members of the parish are with them, and thus can remind them that god is with them as well.

> "There is a very real sense in which the sacrament of the anointing of the sick has a highly accentuated community dimension. In illness, particularly one as we near the end of our lives, we should never have to stand alone." [Bishop Donald Wuerl, "Anointing the Sick-Christ's Healing Hand" "Columbia, Vol. LXXVIII, No. 2, February 1998, p. 10]

The aspect of physical cure is largely downplayed in favor of spiritual healing and the healing of isolation from the community.

The Eucharist, of course, is the sacrament most familiar to Christians, especially Catholics, for they engage in it frequently. The central doctrine involved is the transformation of the bread and wine into the body and blood of Christ, which is believed to be an ontological change and not merely a symbol.

> [In an ontological change, remember, it is not the appearance of the object that changes, necessarily, but what the thing really "is".]

In the Sacraments of Initiation there is considered to be an ontological change, but this time in the recipient of the sacrament. The baptized person is considered to have received a "character" or a "mark" which changes who they are. "It is no longer Paul that lives, but Christ lives within me." [Galatians 2:20] It is a new person that emerges from the baptismal waters, profoundly different from the one that went in. Confirmation does not change the person in so profound a way, but specifies the change already made at Baptism but now to be lived in a new way as an adult member of the Church. In the Eucharist, the person is not so much changed but strengthened in their identity, but the bread and wine are considered to have undergone a change in their identity, their essence, the "what-ness". This has not always been the focus of theologizing on the Eucharist. In the early church the central meaning was on the reality of the unity of the Church itself. Cyprian says that,

> "The sacraments of the Lord demonstrate sufficiently the firmness of Christian unity. For when the Lord calls the bread made from many mingled grains his Body, he means that all the Christian people he bears in his heart must be united." [Epis 76, Patres Latinae" 3, 1142].

The emphasis was on the effects of the sacrament and the meaning of its symbol, not on the means by which this effect was produced. As time went on the emphasis came to be placed on the physical reality present being "really" the body and blood of Christ. This was proper but soon turned improper as it was so misinterpreted as to bring about stories of bleeding hosts. In reaction to this the Scholastics spoke in terms of "substance" and "accidents", a terminology used by the Council of Trent as a "convenient and proper" way of explaining what the Church understood about Christ's presence. In today's theology it is recognized that there is much leeway to speak of "transfinalization" and "transignification", in addition to "transubstantiation" as all being terms which can speak to the manner and possibility of Christ's "real presence". Piet Schoonenberg concludes that the encyclical "*Mysterium Fidei*" which treated the topic of the manner of Christ's presence in the Eucharist, "only denounces as inadequate that symbolism and that change of meaning which exclude a real presence and therefore a real, substantial change of bread and wine. A "symbolism" or "transignification" explanation that maintain this realism are not condemned..."

[Piet Schoonenberg, S.J., "TRANSUBSTANTIATION: How Far Is This Doctrine Historically Determined?" in "Concilium", Vol. 24, New York: Paulist Press, 1966, p. 91. See the whole article for a consideration of the history of the doctrine of transubstantiation.]

"Vatican Council II's *Constitution on the Sacred Liturgy*...rather than repeating Trent's declaration that Christ is `really present' by the `transubstantiation' which takes place by the power of the words of consecration spoken by the priest...stresses the fact that it is the presence of Christ *throughout the entire liturgical action* which gives the liturgy its value as an act of worship and sanctification." (Italics mine.)

[Joseph M. Powers, S.J. "Faith and the Eucharist" in "Bread From Heaven", Ed. by Paul J. Bernier, S.S.S., New York: Paulist Press, 1977, p. 115f]

The "Catechism of the Catholic Church" seems to want to pull back from this advance when it repeats the use of the term "substance" and concludes that, "This change the holy Catholic Church has fittingly and properly called transubstantiation." [# 1376]

Recent polls show that many Catholics explain the Eucharist as a "symbol" only, harkening back to the early days of the Church, but when reminded that the official teaching of the Church maintains that there is more to it than this, most readily agree, whether they can explain it or not. Those whose business it is to explain it have pulled back from making the explanation of the "real presence" in the bread and wine the focus. LaVerdiere says, "...it becomes obvious that any contemporary malaise concerning the real presence of Christ in the Eucharist may stem from our divorcing Christ's presence as nourishment from his presence as participant."

[Eugene A. LaVerdiere, S.S.S., *The Presence of Christ in the Eucharist* in "Bread From Heaven", p. 100]

Unfortunately a misguided attempt to answer this lack of understanding has recently grown stronger in some reactionary circles which emphasizes the presence of Christ in the bread or host as an object for adoration apart from the liturgy of the Eucharist and often apart from the community.

[Weigand, "Feeding His Lambs", p. 52. In complaining about the general lack of understanding of the real presence cites as one cause, the "...idea of the Mass as only a banquet in which one shares by receiving the Body of Christ in order to manifest, above all else, fraternal communion." The bishop does not seem to agree with St. Basil cited above, or with John or Paul for that matter.

See Bishop Donald Wuerl, "The Eucharist: The Real Presence" in "*Columbia*" Vol. LXXVII, No. 11, November 1997, p.12 for a combination of insistence on the real presence and the practice of adoration of the reserved bread.]

The practice stems from the twelfth century and is considered suspect by many, though both conservatives and traditionalists favor it.

[Father Benedict J. Groeschel, C.F.R. and James Monti, "In the Presence of Our Lord" Huntington: Our Sunday Visitor Publishing Division, 1997, p. 119]

All of the sacraments are usually explained first by reference to the rituals and symbols that are seen and sensed, rather than as an act of god become visible. This is certainly true of the life-style sacraments of Matrimony and Ordination to the clergy. Ask about Marriage and the symbol of the ring is most often mentioned, followed by the exchange of vows. Ask about Ordination and the anointing of the hands or perhaps the laying on of hands by the bishop are the symbols mentioned. However, the most powerful signs of god's activity are the lives of the people themselves. God wants to communicate god's love, fidelity, giving of self, sharing of life, the depth of the desired relationship with the individual and with the Church. How can we understand these things except by seeing them in a manner which is apparent to human sensibilities? We can see them in the life style of the married couples that god sends into our lives.

[I am presuming that we are dealing with a monogamous relationship, however these same things can be said about the relationships of couples in polygamous or polyandrous families if there is a covenant relationship between the individuals.

See John L. McKenzie, "Did I Say That?", Chicago: The Thomas More Press, 1973 p. 131 for a discussion of the non-necessity of insisting on monogamy from the "natural law".]

Only then can we make the leap of faith that says that just as these two love one another, have committed themselves to one another, will always forgive as well as cherish each other, and always make the other and their union the top priority in their lives, so has god done with us, god's people.

[This was not always understood in the Church but has gone through a long process before reaching the recognition of marriage as a sacrament, and certainly before reaching the ideas presented here. For an exhaustive study of the development of the understanding of marriage as a sacrament see, E. Schillebeeckx, O.P., "Marriage, Human Reality and Saving Mystery", Trans. by N.D. Smith, New York: Sheed and Ward, 1965.]

How can we understand the concept of the covenant to which we have been invited unless we can see what covenant means as the couple lives it out? We do not first say to ourselves, "I see what marriage is all about because I understand what our relationship with god is all about." No, we understand what god has offered after we understand what a good human relationship can be. This is best understood if we think about the effect the marriage has on the children. Those whose parents show them what love is can understand when they are told in religion class that god loves them. Those who have never seen genuine human love cannot grasp the concept. Indeed, some cannot relate to god as "Father" since they never had a relationship, or what they had was a bad one, with their own father. So the reality that takes on a divine dimension is not the ring or the speaking of the vows but the relationship and the life-style. By extension, friendships which are not sacramental marriages can also be used by god as a means of communication of god's own intentions towards god's people. There certainly are parent-older child relationships lasting for many years, high-school pals who remain friends for decades, common-law marriages or just plain friendships that make the concepts of love and caring operational in the understanding of others who are exposed to them. The chauffeur who drove Miss Daisy and the police partners in "Lethal Weapon" all exhibit to some degree the covenant love god offers to us. These unions are not recognized as sacramental, but then the Church never meant to limit god's love through the concept of sacrament, just to make it more noticed.

Even the concept of the Trinity becomes more understandable the more the marriage partners truly live the sacramental union. Any number of theologians have maintained that the Trinity is reflected in creation and though a three-fold process of knowing, willing and loving is seemingly the preferred example, many have see the relation of husband and wife and the two becoming one as certainly a reflection of the process producing the divine trinity.

[See for example Robert A. Brungs, S.J. "You See Lights Breaking Upon Us", St. Louis: by the author, 1989, p. 113f. His argument is typical of many theologians and is stated thus:

"It can simply be asked why we were created in two sexes. There is a very simple answer to that simple question: that alone images the Blessed Trinity. If we accept the notion that the creation somehow images the Creator, and if we accept the fact that those forms of life we like to call "higher' are composed of two sexes, then we are bound to say that our sexual differentiation is connected to our imaging of the Trinity. This is easy enough to say. It is even true. But the meaning and the understanding (and the overwhelming implications) are more difficult to grasp. It's easy enough to say that is the way things are; and so they are. It's more difficult to understand the meaning of the simple facts as they relate to our salvation and glorification. Take, for instance, a significant question, perhaps a significant objection: how can we show that sexual differentiation into two sexes is a truly Trinitarian image, rather than simply a triadic image? That is a legitimate and significant question. Indeed, to maintain the kind of position being developed here, it is essential to show that the union built into the relationship between the sexes is indeed Trinitarian and not simply triadic. Let us begin out of the earlier discussion about the aloneness of the good. Let's state that there is a strict and necessary connection between a Triune God and a covenantal creation. The good is, as we noted earlier from C.S. Lewis, becoming more and more set off not only from evil but also from other good. It is becoming more and more isolated from other goods. But, as St. Thomas and many others tell us, *bonum est diffusivum sui,* i.e., good spreads itself out to the other(s). The only way for the perfect Good to so express itself, it would seem, is in a community of Persons within the unicity of Being. We know the words are proper, even though we do not understand the mystery they represent. But we do know that the infinite Good is Father, Son and Spirit. We know that God is diffusivum sui in three infinite Persons while remaining one God. Even though we do not understand the Three-Persons-in-One-God, we do know and accept the fact. God as goodness is also diffusivum sui on the created, finite level. God has flung out from himself, if we may so speak, myriads of partial reflections of his goodness, beauty and love. Every being reflects a part or a piece, so to speak, of God. But if each is alone and remains alone, there can be no imaging of union-in-multiplicity. If all there is in the created universe is multiplicity, there can be no imaging of the Trinity. If the union sought and achieved is the union of sameness, of a loss of self in another, there can be no Trinitarian imaging either, because multiplicity will have been lost. In order to image the Trinity we need two beings to remain each themselves while entering into a union that is not either the one or the other, but a union that is in fact both. It is in such a unity that we find a Trinitarian rather than a triadic unity. In classical theology the Holy Spirit is said to be that Love which binds the Father to the Son and the Son to the Father. So, in the world of the finite we must look for the finite expression of the spirit in the real uniting of two who each remain themselves. Creation is good—a point on which we have God's own word. It is not ambivalent as the classical pagans or the contemporary Gnostics would have it. It is not evil as the Manicheans would maintain nor irrelevant as the purveyors of the so-called modern mind would assume. We can truly appreciate the beauty and splendor, the goodness and the glory of creation only by accepting, welcoming, embracing and fully affirming its Trinitarian character. This is done primarily through the affirmation of the validity of the sacramental symbol of sexuality, as Pope John Paul 11 has been at pains to point out. The original covenant of God with man (and with creation through us)—the sacrament of creation, as John Paul II calls it—is a fleshly covenant. That it is `of the ground' can be seen in both accounts in Genesis. It is `in the flesh' that we find our individuality; it is also and especially `in the flesh' where we live and worship God. It is, however, not only a fleshly covenant; it is more highly specified: it is from the beginning a marital covenant."

All we have said of Marriage is true of the Sacrament of Ordination. It is the life style of the priest, deacon or bishop that is to make understandable some aspect of god's relationship with the Church. The word "love" may not immediately spring to mind from being exposed to these persons, but "service of the other", availability, openness to all, especially to the marginalized, the poor and the troubled, certainly should. God's love is "like" the love we see in Marriage, but it is also "like" the love we see in ordained ministry. Pope John Paul II asked for appreciation of this role when he said to priests,

"The priestly personality must be for others a clear and plain sign and indication...The people from among whom we have been chosen and for whom we have been appointed want above all to see in us such a sign and indication, and to this they have a right."

> [John Paul II, "To All the Priests of the Church On the occasion of Holy Thursday, 1979", in "Set Apart for Service", Boston: St. Paul Editions, 1982, p. 91]

Certainly the Religious Brothers and Sisters give the same sign of god's presence and the meaning of covenant that those ordained do, but so too do the "mothers of the Church", the widows, retired, the single who dedicate themselves to the service of others, either in pastoral ministry or through their careers in "the world". Even within the role of the ordained clergy there are different expressions of the same basic theme. The deacon's witness is primarily of service to the poor and troubled, though they also have a liturgical and teaching role. The priest, though most of their time is spent with individuals, and often it is the poor and troubled, add a dimension of service through leadership and teaching and certainly prayer and worship which is associated with them more than with the deacon. The bishop is usually a more distant figure experienced in special liturgies, but when they are functioning properly bishops show the entire local Church's care for the poor, the welcoming of all, the teaching office in the Church, and especially the unity of each person in the Church with every other member, as well as the roles of leadership and fidelity to the values of the Gospel. The Pope is merely another bishop, but one who shows the role of bishop on an even wider scale and with more intensity.

> ["Merely another bishop" is not meant to be a slur on the role of Pope. Popes like to call themselves "an equal among equals" with other bishops and patriarchs. To many of the faithful, however, Popes are beloved but remote figures. After his 1999 visit to St. Louis, "Time" magazine reported (February 8, 1999) a lay woman's admission that, "the Pope is rarely mentioned in our household and rarely mentioned as part of our Catholicism." She feels free to disagree on birth control and the ordination of women to the priesthood. Yet by the end, the article must point out that the universality of the Church, "of which this man is still the vibrant center, overcomes her."]

Popes in the 19th and 20th centuries have certainly appropriated to themselves powers which were formerly exercised by local bishops or national or regional assemblies of bishops and it remains to be seen if this is a trend that will last in the face of reforms initiated by the Second Vatican Council. In spite of a papal tendency to limit the scope of these reforms, it is possible that the cat is out of the bag once more and the role of the bishops, indeed the role of every individual baptized Catholic, will be more appreciated and allowed than the forces of centralization might like.

None of these roles, of the married in loving and caring and sharing, or the clergy in serving, teaching, guiding and praying, are restricted to those who are the sacraments of them. Every Christian is expected to do all of these things. It is just that their particular life style may or may not show them as well as another's. Certainly a celibate priest may not be the most understandable example of the depth of love possible in human relationships, although his love may be quite deep. It is just not as likely to be as recognizable as that of husband and wife.

Thus, even though there is still an aura of "magic" about the sacraments, especially for those who rarely experience them, they are really understood by most Catholics as moments of contact with god, even if they cannot explain how this happens. And many modern Catholics do indeed have a rather sophisticated understanding of what a sacrament is, at least when they are prompted. Many theologians and pastors are involved in just such prompting, and have been for some time, so that,

> "to envisage the Christian sacraments...in their profound significance on the one hand as an actual motion of the Christ of glory drawing us by sacramental signs into the mystery of the redemption eternalized in his holy humanity, and on the other as a sharing by faith in this mystery, is to re-assess them in the eyes of the faithful, to save them from the routine and automation which too often empty them of their salvific vigor."
>
> [Clement Dillenschneider, C.Ss.R., "The Dynamic Power of Our Sacraments", Trans. by Sr. M. Renelle, S.S.N.D., St. Louis: B Herder Book Co., 1966, p. 23]

683. I have an ally in Victor S. Johnston, "Why We Feel" Cambridge: Perseus Books, 1999, who in the introduction says,

> "We believe that we live in a world that is full of sounds and colors and smells and tastes because this is what we experience every day of our lives and there appears to be no reason for thinking otherwise. In [the book] I propose that we must abandon this common-sense view of reality and eventually accept the fact that our conscious experiences depend on the nature of our evolved neural processes and not on the nature of the events in the world that

activate these processes. That is, although the external environment is teeming with electromagnetic radiation and air pressure waves, without consciousness it is both totally black and utterly silent. Conscious experiences, such as our sensations and feelings, are nothing more than evolved illusions generated within biological brains." [p. vii]

Obviously his purpose is to demonstrate this and not support my argument about transignification, but his work shows that we make our own "reality", which is all the reality we know or need.

684. I found some confirmation of this line in, of all places, *The Catholic Answer*, a somewhat conservative publication, Vol. 12, No. 4, September/October 1998, p. 32f in an article by Jack Vogel, "Science and the Eucharist". At one point he says,

"...even the most agnostic physicists would have to agree that, at bottom, the faith required [for the Real Presence] is a belief that one very mysterious thing [quantum reality] has been changed into Another of equal mystery. Further, they would have to concede that in neither case would the human senses give the slightest indication of the underlying reality."

He does not reach my conclusions, but his point is well taken.

685. Nathan D. Mitchell, "Real People, Real Presence", *GIA Quarterly*, Vol. 11, No. 3, Spring 2000, p. 8

Maurice Nedoncelle in "God's Encounter with Man", p. 140f draws out the implications for the liturgy of this identification of the Church as the Body of Christ.

686. 1 Corinthians 11:29

687. Luise Schottroff, "Sexuality in John's Gospel", *Theology Digest*, Vol. 45, No. 2, Summer 1998, p. 103f points out that,

"Jesus' relationship to God as his heavenly Father is represented as a consummate relationship of mutual love. The Father loves Jesus (Jn 5:20; 3:35, etc. *philein* and *agapan* are not delimited from each other but interchangeable), and Jesus loves the Father (14:31). Jesus is called the only begotten (or only generated, 1:18), but the emphasis is on the love relationship, not on God's begetting; the same obtains for the Father-Son metaphor."

It is not my purpose to examine the ontology of the Trinitarian relations, but the god we are discovering is perfectly capable of engendering a loving relationship far beyond relationships as we know them. "Incarnation" may well be a metaphor for this relationship.

688. Denis Edwards, "Jesus and the Cosmos", New York: Paulist Press, 1991, p. 66

689. Denis Edwards' work, "Jesus and the Cosmos", does a fine job of explaining Karl Rahner's much more extensive investigation of the question of the Incarnation. It makes it clear that Jesus is a product of the universe but one that is surpassingly open to the presence of god.
"Yet the universe could not reach this goal of itself. It occurs through God's grace." [p. 67]
Humanity opens itself to god most fully in Jesus and god's grace or presence can find its most full communication in Jesus. Yet,

> "It happens through the very same reality that is offered to us all, God's self-communication in grace. God's giving of God's self to all of us, and the incarnation in Jesus Christ, can only be understood together, and they both occur through the same reality, grace." [p. 69]

I have no problem with this at all and applaud when Edwards writes,

> "God's self-offering (grace) and human acceptance of this offer occur in such an absolute way that we can say that this is not only something God accomplishes, but it *is* God. While we remain receivers of God's offer, we can say that Jesus *is* God's offer. Jesus not only communicates God's pledge to us, Jesus *is* the pledge of God to us." [p. 69, italics his]

However, for the reasons stated I must adhere to my own explanation of how this is possible. Edwards approvingly quotes Rahner in saying, "God lays hold of matter when the Logos becomes flesh." It is true that both Edwards and Rahner approach both creation and incarnation in a manner that I believe would accept my approach and take the positions they do to remain orthodox, though their conclusions are quite distant from the Thomism of most bishops. Edwards writes, in italics,

> "The ancient faith of Chalcedon can be expressed anew, then, from within a cosmic and evolutionary context, in the following fashion: In God's self-bestowal in Jesus of Nazareth, first, *God accepts the cosmos definitively and irrevocably,* and, second, *the cosmos*

accepts God definitively and irrevocably, and, third, these two acceptances are manifested in our history *as constituting a real unity in the one Jesus of Nazareth.*" [p. 75]

It is quite possible that if they were not quite so intent on preserving the ancient faith and were a little freer to allow that god cannot manipulate matter and need not be a creator in anything like the usual sense, they would be correct. After all, when Rahner says that, "God lays hold of matter" he need not mean any more than god finally finding a wide open door through Jesus and throwing godself into it. And when Edwards sees humanity's seeking god and god's seeking humanity finding a common ground in Jesus he need not explain it as Aquinas would. In fact he pointedly does not.

690. As we have seen elsewhere, it is generally believed that god can be known through the works of creation. The universe is an image of the nature of god, though only through analogy. It is realized that the knowledge of god gained through this process is very tentative and open to much misinterpretation. God is much more clearly seen when intervening in the workings of creation, often by showing mercy or compassion by averting some natural catastrophe. God is even better seen when inserting godself in the workings of human history, particularly in the history of the chosen people.

> "The Israelites understood all of reality as history, that is, as historical actions of their God."
>
> [Wolfhart Pannenberg, "Toward A Theology of Nature", op. cit., p. 82]
>
> "...The contingent events of Israelite history were experienced as the way in which Yahweh relates to his promises."
>
> ["Toward A Theology of Nature" p. 85]

For Israel, the events of their history are never seen as the result of the forces of sociology but are all considered a part of god's plan for them as a people. In fact,

> "It was not in the wonders or in the order of nature that Israel first came to know God, but in his saving actions in history on Israel's behalf. It was because God could exercise such obvious mastery over the events and course of history as he displayed in his saving actions for Israel that in time Israel came to see that he exercised a similar - but wider - mastery over the whole of nature."

[Robert Butterworth, S.J., "Theology of Creation", op. cit., p. 26]

They did not start from a philosophical, much less scientific, knowledge of god's mastery of nature, they started from a perceived experience of god's use of nature for their welfare that made them define god as nature's master. Their prophets were not so much those who foretold the future as those who pointed out as god's work what the people had experienced, even if what they had "experienced" was only vicariously (or if it had never happened at all).

> "The prophets were always there to interpret the events to the people, for it was only through their words that reality became truly real."
>
> [Virgil Elizondo, "The Human Quest", p. 48]

Christians, too, see the record of god's revelation through the history of Israel as relevant to their own faith.

For Christians, the definitive revelation of god was not to be found in the witness of creation nor even in the saving events directed toward Israel, but in the person of Jesus the Christ. "In many and various ways God spoke of old to our fathers by the prophets, but in these last days he has spoken to us by a Son." [Hebrews 1:1-2]

> "Christ is the fullness of God's revelation. Christ represents - in fact, personally and actually is - God's revealing Word to men. Whatever God is believed to have revealed about himself and his saving plan in man's regard is ultimately revealed in his Christ."
>
> ["Theology of Creation" p. 23]

The Son, the Second Person of the Trinity, takes on our humanity, becomes "incarnate," but remains god. So to encounter Christ is to encounter god. Whatever there is to know about god can be known in Christ, just as meeting a person in the flesh is much more revealing of the person than reading their biography. Of course, I may not understand much of what I see standing before me and may not know the person at all without a serious dialogue, a true "encounter" with the person. The same is true of Christ. We have spent two millennia trying to understand what or who we have experienced in meeting Christ, but he is completely present to us even if we are not completely open to him. In the orthodox view, if Christ is present to us then god is present.

691. It is not problematic to Msgr. M. Francis Mannion who writes a column in *Our Sunday Visitor*. When asked whether it is arrogant to say that Christ is lord of another planet responded,

> "Fundamental to the biblical and Christian revelation is the affirmation that all that exists, everything in space and time, all that God has created, is subject to Christ. Christ is Lord of the whole universe." (April 22, 2001)

This lordship, and relevance to inhabitants of another world, is predicated on god as creator, which is just assumed.

692. For an expansion of this observation and the effect it has on many people's idea of god, see Richard R. Gaillardetz, "Recovering the Sacred Mystery: how to Connect Liturgy and Life", *Modern Liturgy*, Vol. 24, No. 7. September 1997, p. 10f.

693. The "Catechism of the Catholic Church" states,

> "In order to articulate the dogma of the Trinity, the Church had to develop its own terminology with the help of certain notions of philosophical origin: `substance,' `person' or `hypostasis,' `relation,' and so on. In doing this, she did not submit the faith to human wisdom, but gave a new and unprecedented meaning to these terms, which from then on would be used to signify an effable mystery, `infinitely beyond all that we can humanly understand'."

694. There are many explanations of the Pauline theology of "in Christ" or εν χριςτω which can be found in many sources.

The Rev. Dr. Michael Hoy in a paper entitled, "A Theology of the Body: Body, Genes and Culture, Who's Holding the Leash?" in "*Christianity and the Human Body: Proceedings of the ITEST Workshop, October, 2000*", St. Louis: ITEST FAITH/SCIENCE PRESS, 2001, does an interesting comparison of the views of Dawkins, Dennett, Wilson, et. al., and Holmes Rolston, "Genes, Genesis and God", Cambridge: Cambridge University Press, 1999, representing the faith. Dr. Hoy leans in favor of the scientists. However, he finds space for the views of faith by our being united with the body of the resurrected Christ. Dr. Hoy seems to hold the orthodox view that Jesus is god enfleshed, saying,

> "..what is new is that Jesus the Christ himself became "leashed" with our nature…a new being, a new body, God generating Godself in the flesh." [p. 13]

However, his explanation of god's working through Jesus the Christ works even without the view of god's operation on the physical. We are "saved" by our identification with the body of Christ.

Dr. Hoy does not like Rolston's idea that god works by inserting "information" into the universe, an idea that on the surface it seems I would have to agree with against Hoy. However, in his view,

> "The joy of faith is that under this new ownership, this new leashing, there is `freedom' from the only trajectory available to us in our natural determinism. There is no *Geist* of our own that can adequately capture it. It is ours only by faith, through the working of the Holy Spirit (*der Heilige Geist*), in the promising gospel of Christ's new leashing."

This can, I believe be explained as being available to us through the working of god influencing humanity and does not require Jesus to be both Divine and Human in any other way than I have laid out.

Other participants in this workshop explain the concepts of "creation", "soul", "original sin", "salvation" and like concepts without recourse to the Thomistic philosophy and with an eye to explanations consistent with modern science.

695 . Somewhere I ran across the explanation that the problem with the Arians came down to a Greek "ι" iota. The Arians wanted Jesus to be described as "*theios*" or God-like and as "*homoiousios*", of like substance. The orthodox proclaimed Jesus to be "*theos*" and "*homoousios*" or "God" and "the same substance." My thanks to whoever that "somewhere" is. We see the result of the 325 a.d. Council decision whenever we see the symbol of a fish on the back of a vehicle, and they are quite common. "Fish" in Greek is "*ichthus*". The iota is the first letter of iesus, Jesus. The chi (ch) is the beginning of Christ. The theta (th) comes from "*theos*", the upsilon (u) from "`*uios*", Son, and the sigma (s) from "*soter*", Savior. I suppose I can still use the fish symbol even though I lean to "*theios*" of godlike, since they both begin with theta.

696. Raymond E. Brown, S.S., "Jesus God and Man", New York: Macmillan Publishing Co., Inc., 1967, p. 103-4

697. Michael Figura, "The Holy Spirit and the Church", *Theology Digest*, Vol. 46, No. 1, Spring 1999, p. 17

698. "The Good Enough Catholic" p. 42f

699. "The Silence of St. Thomas", p. 38f
The same incident is referred to, though slightly differently, in "The Mind's Sky", p. 89

700. Karl Rahner, "Theological Investigation" Volume IV, Trans. by Kevin Smyth, Baltimore: Helicon Press, p. 87

701. Jerome Murphy-O'Conner, "What Really Happened at Gethsemane?", *Bible Review*, Vol. XIV, No. 2, April 1998, p. 28f, points out two sources for the story of the agony in the garden. In one,

> "...disciples projected onto Jesus the emotions that they imagined they would experience if they suddenly realized their death was imminent...The intensely human Jesus revealed by [this] Source A- a leader on the verge of a nervous breakdown- proved to be more than some other Christians could accept. In consequence, they wrote a different version of what happened in Gethsemane, which...lacks the explicit statement that Jesus `became filled with terrified surprise and distressed from shock' (Mark 14:33b)
> ...In Source B Jesus is sufficiently composed to make scriptural allusions."
> [p. 36-7]

702. "Catechism of the Catholic Church", #253

703. "Catechism of the Catholic Church", #254

704. St. Augustine, "The Trinity" Trans. by Edmund Hill, O.P., Brooklyn: New City Press, 1991, p. 282. But see all of Book IX.

705. See Elizabeth A. Johnson, "She Who Is", op. cit., Part III, pp. 124-190
For another feminist approach to the Trinity, see Ivone Gebara, "Longing for Running Water", Minneapolis: Fortress press, 1999, chapter 4.

706. "She Who Is" p. 179

707. Catherine Mowry LaCugna, "God For Us", p. 2

708. "God for Us", p. 168-9

709. The traditional understanding goes like this. It is his lifestyle that Jesus believed would allow him to truly live the covenant life on earth and to continue to live it after the resurrection. Where, then, does Jesus continue to live? One can take a Ptolemaic view in which the sun, stars and planets revolve around the earth, and speak of Jesus enthroned at the right hand of the Father in heaven, located vaguely "up" above the celestial spheres in which the heavenly bodies are embedded but which we no longer believe are present. Or one can take the view of the Scriptures that Jesus lives precisely where god wishes to live, in covenant relationship with god's beloved. And this means in the Church. Many images are used to explain that the Church is not just a human assembly but something more. The image of the building erected with Jesus as the cornerstone says something of the importance of each member of the Church. The image of the vine and the branches adds the complementary vision of each life being lived in common with the life of Jesus. The image of the People of God, popular since the Second Vatican Council, though certainly found in the Scripture from early times as the *qahal Yahweh*, the Εκκληςι∝ from which we develop "ecclesiastical", develops further the dignity of each member along with the relationship established among all the members.

> [For a good contemporaneous reflection on the Council's reassessment of the identity of the people of the Church in its Schema 13, "The Church in the Modern World", see "Eyes On the Modern World", Ed. by John G. Deedy, Jr., New York: P.J. Kennedy & Sons, 1965.

Still, it is the image of the Body that most clearly answers our question of the locale of the Risen Christ. If Jesus was the historical incarnation of the divine life, the Church is the locus of that physical presence now. It is in the life of the Church as a living body of human beings and in the life of each member of that Church that Jesus continues to live in covenant relationship with god and with other beings. Prior to the Church, prior to Easter/Pentecost and all that those code words involve, god could be found somewhat removed from god's people in the work of creation, whatever that might involve; in god's influence on the physical world; in god's influence on the cultural and moral life of the people; and in the words and deeds of the law and the prophets. God makes this presence even more accessible in the physical presence of the man, Jesus, so that to come face to face with

Jesus of Nazareth was to come face to face with god, something that was considered either impossible or lethal, with few exceptions.

> [In Exodus 32:30, Moses is amazed that he has seen god face to face and lived. The same is mentioned in Ex. 33:11 and Deut. 5:4. In Exodus 34:33 we are told that Moses veiled his face when speaking with the people because he had spoken with the Lord face to face and not died, but his own face then presumably became a threat to the people. Gideon in Judges 6:22 is worried because he saw merely an angel of the Lord face to face. Deuteronomy 34:10 notes that no other prophet like Moses has ever seen god face to face. 1 Corinthians 13:12 promises that we will see "face to face" but only in heaven.]

Even more thought provoking is the position that Mary is truly the θεοοκος, the *theotokos*, the mother-bearer of god, and so to suckle Jesus was in a real way to suckle god.

But Jesus, the historical presence in a human being, is no longer accessible as he was to the apostles. Where in our age can you experience such an intense physical presence of god? Or has the death of Jesus precluded this possibility? The answer is in the Church, which though a corporate entity in a legal and linguistic sense, is also a corporate entity in a physical sense. As in no other group of human beings there is a unity among the members with god as its basis, the very thing that god invited us to have at Sinai. "The Body of Christ takes up space on earth," was the way that Dietrich Bonhoeffer expressed it. "A truth, a doctrine, or a religion need no space for themselves. They are disembodied entities."

> [Dietrich Bonhoeffer, "The Cost of Discipleship: New York: Macmillan Publishing Co., Inc. 1959, p. 277]

The physical presence of god is in the Body of Christ, the living Church.

In this view, how is this physical presence manifested, and how can it be experienced? The life of the Church is still a human life, analogous to the life of a corporation or a fraternal society. It is prey to all the human foibles that afflict the individual human person. But it is precisely in working through these that god can influence human reality. Some have always been scandalized by the presence of human limitation and even human evils in the body of the church. The parable of the sower and the seed [Mark 4:3; Luke 8:5] was later allegorized to address this very problem. [Matthew 13:18f; Mark 4:14; Luke 8:11]

Why are there weeds in the field (of the Church)? Because the harvest has not yet occurred. For others this very weakness of the Church has been its strength, following on the lesson of Paul who prayed for relief from his weakness only to realize that his weakness made it even clearer to others the source of his strengths. [2 Corinthians 12:7f]

In our own day we are sometimes treated to stunning views of the shortsightedness of bishops and Popes, not to mention that of the people right around us, and including ourselves. We are so attached to our own biases that god has a hard time moving us on to another and better understanding. Yet move the body god does since it is god's own body, god's own personality, as well as ours. For example, in 1767 even a Catholic bishop could and did own an African slave. In 1967, only a White deacon in the church, myself and a classmate, could be found at the side of a new bishop being installed in his See, and that deacon was known as "temporary" since he would soon be ordained a presbyter. In 1980, a married African-American professional man who was a "permanent" deacon, could be found at the side of a newly installed bishop and be the cause of no particular notice. The body has been moved and in the direction of greater covenant relationship. The weakness of the Church is the foil for its strength. It is behind the scene of human limitation that the presence of god is experienced and it is in living with this body, indeed as a member of it, that god is seen face to face. And in the Church's better moments this presence is even more intensely expressed and experienced.

710. Werner Bulst, S.J., "Revelation", Trans. by Bruce Vawter, C.M., New York: Sheed and Ward, 1965, says,

> "In Catholic Scholastic theology, both fundamental and dogmatic, supernatural revelation is defined with highly consistent unanimity as *locutio Dei (attestans)*, that is, 'God speaking out of the treasury of his own understanding, communicating to men truths which otherwise would be attainable by them only with difficulty or not at all.' This is, moreover, an 'attesting' speech: it does not produce understanding of the matter communicated by commands faith in it."

Our use of the word "reveals" or "revelation" certainly includes this notion of something known from god which we may not have known otherwise, such as the Trinity, or with great difficulty, the depth possible to human love.

711. See the "Declaration on the Question of the Admission of Women to the Ministerial Priesthood" issued by the Sacred Congregation for the Doctrine of the Faith issued prior to the Pope's teaching.

712. Congregation for the Doctrine of the Faith, "Declaration on the Question of the Admission of Women to the Ministerial Priesthood" (Inter Signores), "*Origins*", February 3, 1977

The Reuters News Service carried a story on March 11, 1998 which stated that,
"Two senior Vatican cardinals reaffirmed...that women cannot become deacons of the Roman Catholic Church. `Christ was a man. It seems to me that this is the fundamental and theological reason,' Cardinal Pio Laghi said."

713. Congregation for the Doctrine of the Faith, "Reply to the *Dubium* concerning the teaching contained in the Apostolic Letter *Ordinatio Sacerdotalis*", #12 of the briefing questions.

714. Galatians 3:28

715. "She Who Is" p. 155

716. *Time* magazine, November 30, 1998, p. 8, cites a "Vatican functionary" as saying that some of the women who wish to be priests and even women who merely function in priest-less parishes, "...are power-hungry witches...They have no concern for the church and for souls." This person, at least, is not very fascinated with an emerging ministry.

717. See Letha Scanzoni and Nancy Hardesty, "All We're Meant to Be", Waco: Word Books, Publisher, 1974 for an early example. There has been much work done since.

718. John 1:1

719. But then scripture scholars and theologians have largely abandoned this anyway. Anthony Salidarini, "Counting Time", *Bible Review*, Vol. XIV, No. 4, August 1998 p. 22 is an example when he says that,

"the nature of God's intervention and rule (kingdom) varies in the Bible because the biblical authors had to imagine the future on the basis of God's past action."

They could only speak of the new in terms of the old. His conclusion is,

"when we finish counting, imagining, evaluating and longing for a better future, we find ourselves in the present...The Bible urges us to hope: We live in the present, a present open to the promises and potential of the future, a future given to us by God."

720 I may be wrong on the numbers. *Time* had an article titled "Apocalypse Now", by Nancy Gibbs, on the phenomenon of the "Left Behind" series of books which presumes, along with a reported 59% of Americans that the Book of Revelation makes predictions about the future, that these will come true, and that they are immanent.

[Vol. 160, No. 1, July 1, 2002, p. 40.]

721. See "Sacramentum Mundi", Vol. 2

722. See *Bible Review* or other popular journals for examples such as the words for weights and measures, etc which are archaic and would not have been in use at the times the stories were written or edited, but which were in use in the times spoken of.

William G. Dever in "Save Us From Postmodern Malarkey", *Biblical Archeology Review*, Vol. 26, No. 2, March/April 2000, p. 28 has a rather antagonistic tone about the so called "minimalists", but he gives a number of examples of correspondence of the Scripture with archeological finds.

In the same issue, see Philip Davies, "What Separates a Minimalist from a Maximalist? Not Much", and other articles, plus their endnotes to get the gist of the debate.

On the other hand, without reference to the debate, Elizabeth Bloch-Smith and Beth Alpert Nakhai, "A Landscape Comes to Life: The Iron Age I", *Near Eastern Archeology,* Vol. 62, No. 2, June 1999, p. 62, concludes that from the archeological evidence, "...were it not for the Bible, no late thirteenth-early twelfth century Israelite invasion would be suspected. For the most part, language, dress, pottery, architecture, and cultic features demonstrate continuity from the late Bronze Age into the Iron Age."

723. Rolf Rendtorff, "What We Miss by Taking the Bible Apart", *Bible Review*, Vol XIV, No. 1, February 1998, p. 42f, makes the point that,

> "the Bible was the sacred scripture of Israel. `Israel' in this context, refers to a community of faith. Therefore, to read the Hebrew Bible as `Scripture' means, first of all, to read it as a religious document that served a religious community. From this viewpoint, the main question is no longer `How did this text emerge and develop?' but `What is the message of the text in its final form?' Only in this form did it serve as sacred scripture for a religious community." [p. 44.]

He gives and example from Genesis. Chapters 1 and 2 are obviously from different sources, but a final editor has put them together. It is this final understanding that is Scripture. The author carefully adds, "I should stress that this does not involve a departure from the historical-critical method, only a change in the focus."

724. Victor Hurowitz, "From Storm God to Abstract Being", in *Bible Review*, Vol. XIV, No. 5, October 1998, p. 40f

725. Carolyn R. Higginbotham, "The Egyptianizing of Canaan" in *Biblical Archaeology Review*, Vol. 24, No. 3, p. 42.
See also John Ruffle, "The Egyptians", p. 151.

726. James A. Sanders, "Spinning the Bible" *Bible Review*, Vol. XIV, No. 3, June 1998, p. 23. Sanders points out that even the manner in which the canons are structured not only differs between the Protestant canon and the Jewish Tanach but that there is a reason. For example,

> "In the tripartite Jewish canon, the Prophets play the crucial role of explaining the uses of adversity in the hands of One God...The function of the Prophets in the Christian canon is not so much to explain God's uses of adversity as to point to Jesus Christ."

727. See 1 Samuel 9 where Saul wants Samuel to tell him where the donkeys are. Note also vs. 11 which calls attention to the development of prophets from seers.

728. "Islam" or any other commentaries on Islam

729. See "Buddhism" ed. by Richard A. Gard, New York: George Braziller, 1962, p.44f or other commentaries, including those on the sutras of the Tibetan Kangyur.

730. See "Hinduism" ed. by Louis Renou, New York: George Braziller, 1962 or other commentaries.

731. "Black Elk Speaks: Being the Life Story of a Holy Man of the Oglala Sioux", as told through John G. Neihardt, Lincoln: University of Nebraska Press, 1961

732. see "Afro-American Folktales"

733. Fr. John C. Ganly M.M., "Kaonde Proverbs", private printing.

734. See Katharine J. Dell, "Wisdom Literature makes A Comeback", *Bible Review*, Vol. XIII, No. 4, p. 26f

> "The Book of job contains few proverbs, Instead, its deep concern with what happens to good and bad people relates the book to the Wisdom quest. Job is a good man, who has lived according to the maxims of wisdom advocated in Proverbs. Then, unexpectedly, calamity comes. He loses his family, his wealth, his status and his health. How is he to understand his new predicament and still retain his faith in God? The Book of job suggests some possible answers. One, given in the prologue, is that this is a test set up by God and Satan to see if job will remain faithful despite calamity. The friends who come to comfort Job-Eliphaz, Bildad and Zophar-suggest that if his life is now a mess, he must have done something wrong or been less wise or righteous than he thought he was (Job 4:7-8, 8:3-4, 11:4-6 etc.). Job's wife tells him to curse God and die (Job 2:9); he might as well relinquish his faith and his life. But Job is determined to understand and will not give up. At the book's climax, God appears to Job. He tells Job that God's ways are greater than those of human beings; God created the heavens and the earth, and he knows the answers to the "big" questions about the universe. Who is Job that he should know all the answers? There are reasons for suffering that human beings cannot understand; Job has to be satisfied with that."

735. Luke 13:4-5a

About the Author

A Catholic priest and pastor for 35 years, the author has delved into the question of God and evil through the practical questions, especially “why is God doing this to me?”

Combining faith and theology with an eclectic knowledge of modern science led him to unorthodox but effective answers to this age-old question, tested in the trenches of real life situations.

Printed in the United States
1529200001B/105